Near Infrared Spectroscopy:

Proceedings of the 10th International Conference

Edited by
R.K. Cho and A.M.C. Davies

NIR *Publications*

Near Infrared Spectroscopy:

Proceedings of the 10th International Conference

Edited by
R.K. Cho and
A.M.C. Davies

Published by:
NIR Publications, 6 Charlton Mill, Charlton, Chichester, West Sussex PO18 0HY, UK.
Tel: +44-(0)1243-811334, Fax: +44(0)1243-811711, E-mail: info@nirpublications.com,
Web: www.nirpublications.com.

ISBN: 0 9528666 3 3

British Library Cataloguing-in-Publication Data
A catalogue record for this book is available form the British Library

Printed in the UK by The Cromwell Press, Trowbridge, Wiltshire.

Preface

After a gap of twelve years, the International NIR conference returned to the far East. This time the venue was Kyonjgu, Korea. This allowed the Koreans to be the most numerous delegates while the Japanese provided the largest number of oral and poster presentations. However, unlike our Conference in Tsukuba, when the number of delegates from distant countries was quite small, this conference was truly international as is reflected in the pages of these Proceedings.

There are two important differences between this and previous volumes in this series. You will not find a subject index but you will find a CD in which every paper is presented in Acrobat™ format and every word is available for searching. I am confident that you will find this a much more useful tool than the usual index.

The other change has been a procedural one. I have had a team of sub-editors, who you will find listed overleaf who have been very helpful to me in each reading and, where required, correcting manuscripts. I would like to record my gratitude to them. I am also extremely grateful to Gill Stockford who has been responsible for all the logistics of the production of the Proceedings in addition to her normal task of creating the printed page from the authors' submissions. Ian Michael has, as usual, been in overall charge of the production and printing.

Tony Davies
Norwich Near Infrared Consultancy
75 Intwood Road
Cringleford
Norwich NR4 6AA

Sub-editors

The following gave invaluable help in the production of this volume:

Graeme Batten
Charles Sturt University, Australia

Ian Cowe
Foss Tecator, Sweden

Gerard Downey
The National Food Centre, Ireland

Tom Fearn
University College London, UK

Sumio Kawano
National Food Research Institute, Japan

Fred McClure
North Carolina State University, USA

Karl Norris
Beltsville, USA

Yukihiro Ozaki
Kwansei-Gakuin University, Japan

Contents

Theory and Principles

Instrumentation and Sample Presentation

Artificial Neural Networks

Interpretation of NIR Statistics

Food

Precision Agriculture

Environment

Polymers and Chemicals

Textiles and Petrochemistry

Biomedical

Pharmacy and Cosmetics

Emerging Aspects

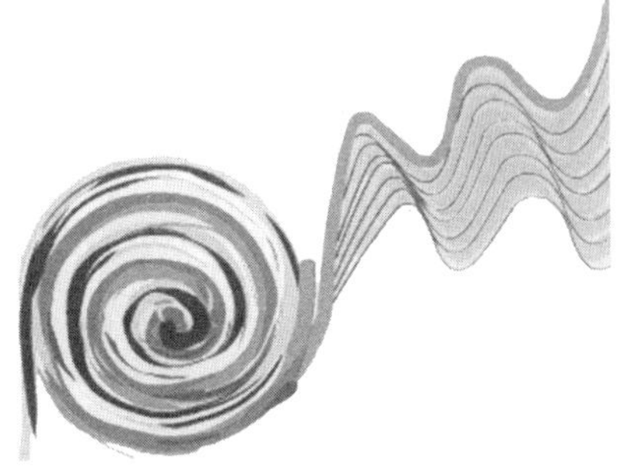

Theory and Principles

Theory and principles of near infrared spectroscopy

Franklin E. Barton, II

USDA-Agricultural Research Service, Richard B. Russell Agricultural Research Center, PO Box 5677, Athens, GA 30604-567, USA

History

The history of near infrared (NIR) begins in 1800 with Herschel.[1–3] His experiments to find a way to filter heat from a telescope demonstrated that there was light radiation beyond what we know as the visible spectrum. This discovery was largely ignored for the better part of a century and even longer before modern instruments were used to acquire spectra. In the mid-1950s Wilbur Kaye, with Beckman Instruments, published two papers which put NIR spectroscopy on a firm footing.[4,5]

These manuscripts described the instrument[4] and the theoretical basis for the spectra in di- and tri-atomic molecules from group theory and selection rules.[5] The NIR was not viewed as containing relevant structural information. The separation of NIR from mid infrared (MIR) was further enhanced when commercial instruments appeared combining the visible (vis) and NIR from Perkin-Elmer and Cary. The rationale was to get the most from sources and detectors rather than of building an instrument that could be used for the whole vibrational spectral region. The emergence of NIR into the analytical world as an accepted technique began with the work of Karl Norris of the US Department of Agriculture, Agricultural Research Service in the early 1960s.[6,7] This was followed by the USDA/ARS National NIR Research Project, which developed, into a worldwide network of collaborating laboratories. This research project's efforts culminated in the publication of *Handbook #643* and two AOAC International Official Methods.[8–10] From this point NIR spectroscopy flourished and expanded well beyond the agricultural realm into pharmaceuticals, industrial, process control, food processing, remote imaging spectroscopy and other diverse applications. The programme of this, the 10th ICNIRS, shows the breadth of the technology in the papers and posters to be presented.

Vibrational spectroscopy

There are many good text books on vibrational spectroscopy and the list in the literature cited reflects both my years in the area and my book shelf.[11–16] A soon to be published compendium by Chalmers and Griffiths and a new Edition of Williams and Norris are the most up-to-date references available[17,18] on vibrational spectroscopy and NIR, respectively. For anyone just beginning to use spectroscopy, any general text will suffice to give you the pertinent information to understand the technology. For NIR, some of the best sources of information and specifics about applications can be found in the Proceedings of these conferences. Several of them are available from NIR Publications.

There are two Laws which govern the basics of vibrational spectroscopy, "Hooke's Law" and the "Franck–Condon Principle". Hooke's Law states that the frequency of a vibration (<) is equal to the reciprocal of 2 *pi* times the speed of light times the square root of the force constant (5×10^5 dynes cm^{-1}) times the sum of the two masses divided by there product for a simple two body harmonic oscillator.

$$\tilde{v}(\text{in cm}^{-1}) = \frac{1}{2\pi c}\sqrt{\frac{k(m_1 + m_2}{m_1 m_2}} \tag{1}$$

Hooke's Law can be used to calculate the fundamental vibrations for diatomic molecules in the MIR, but the NIR is comprised of combination bands and overtones. If everything were simple, NIR bands could be calculated from combinations of the MIR fundamentals and 2,3, 4, etc. times the fundamental frequency to produces the overtones. However things are not simple and so we need to understand how the Franck–Condon Principle, given below, applies to NIR spectroscopy.

Franck–Condon Principle, a classical interpretation

"*When a molecule vibrates, the probability of finding a given atom at a certain point is inversely proportional to its velocity when it is at that point. Therefore the atoms in a vibrating molecule spend most of their time in configurations in which the kinetic energy is low—that is, the configurations in which the potential energy is nearly identical with the total energy, or at the intersection of the vibrational energy level with the potential energy surface of the molecule. Thus the photon is most likely to be absorbed when the nuclei are stationary or are moving slowly. Furthermore, the excitation resulting from the absorption of the photon cannot be transferred immediately to the nuclei. The nuclei will therefore tend to continue moving slowly immediately after the absorption process. Thus in the excited state the nuclear configuration also tends to be close to the intersection of the vibrational energy level with the excited potential energy surface. Therefore transitions tend to take place between vibrational levels in which the nuclear configurations are the same in both states, and they tend to occur when the nuclear kinetic energies are small*".[19]

Thus, these small variations give rise to anharmonicity, which causes the combination and overtone bands to appear at imprecise multiples of the fundamentals. The more vigorous the vibration the greater the anharmonicity. One of the best examples is the water molecule. The principle fundamental vibrations of water are shown in Figure 1. The combination band of water, which occurs at 5180 cm^{-1} (1930 nm), is thought to arise from the asymmetrical stretch at 5180 cm^{-1} (2660 nm) and twice a rocking motion vibration at 660 cm^{-1} (15150 nm), i.e., 5180 cm^{-1} + 660 cm^{-1} + 660 cm^{-1} = 5076 cm^{-1} or (1970) nm. This is 10–40 nm longer in wavelength than we are accustomed to seeing for the water band, the difference is anharmonicity. The same types of calculation can be made for the C–H stretch, N–H stretch, C=O and C–O–H which, along with O–H stretch, comprise the bulk of NIR active vibrational bands.

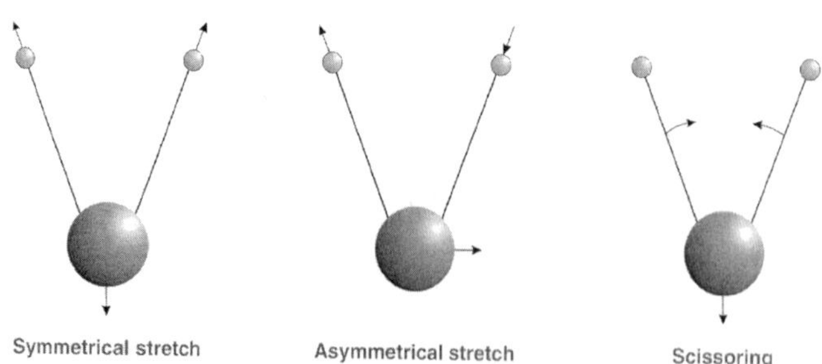

Figure 1. Fundamental vibration modes of water.[20]

Instrumentation

In my early days of NIR spectroscopy, instruments were not computer controlled and in some ways simpler but harder to use than the modern spectrometers we use today. The first NIR instrument I used was a Beckman DU, which did not even have a recorder to plot the spectrum! For looking at the MIR the Perkin-Elmer Infracord was the first general purpose IR spectrometer. The Cary 14 was the first recording NIR I used. This instrument was the one Karl Norris used in his research combined with a Hewlett–Packard calculator to do the statistics, later replaced by a mini-computer.

In a recent article, Tony Davies detailed the history of the many instrument companies that have produced near infrared instrumentation.[21] Throughout the 90's the main research instrument was a dispersive monochromator and still is. Only the names have changed. Like all industries, the instrumentation industry for NIR spectroscopy underwent growth, acquisitions and mergers. The figure in Davies' 1999 article shows who became whom and approximately when the changes took place and the table in the article lists the prominent manufacturers of instrumentation both in the near and mid infrared. As the 90s came to an end, the NIR community was still heavily into dispersive instruments and the mid infrared was totally into FT instruments. There are reasons for this difference of perspectives. For many years the agricultural community used NIR to measure compositional analysis of agricultural commodities and were concerned with taking the spectrum of a series of samples to the same exacting specifications. Those in the MIR community were interested in taking the best spectrum of the analyte. These two positions were somewhat mutually exclusive and remain so today. The NIR user is still "model" and "statistically" oriented and the infrared user is still concerned with functional groups resolution and spectral interpretation. In the United States our laboratory has been the only NIR group primarily concerned with interpretation. FT-Raman is quickly bridging this separation. Its "NIR" ease of use and sampling with the information content of mid-infrared satisfies both communities. One additional aspect has also remained constant. In 1990 we were asked to say where we wanted to see technology head, both in those areas we would research and those requiring input from instrument and software makers.[22] In an article for the "Spectroscopy Across the Spectrum" meeting, a system of interconnected instrumentation was described (Figure 2). This goal has largely been accomplished due in great measure to local area networks and PC's. Currently in our system we can accomplish this task from any PC to any of our instruments, across spectrometer types, manufacturer or software applications. The modern laboratory of today is pretty much as we predicted in 1990. The "LAN" has replaced the term "HOST" but the rest is as shown in Figure 2. The individual instruments all have their own data station (DS) and multiple experiments can be called up from remote locations. It is important that

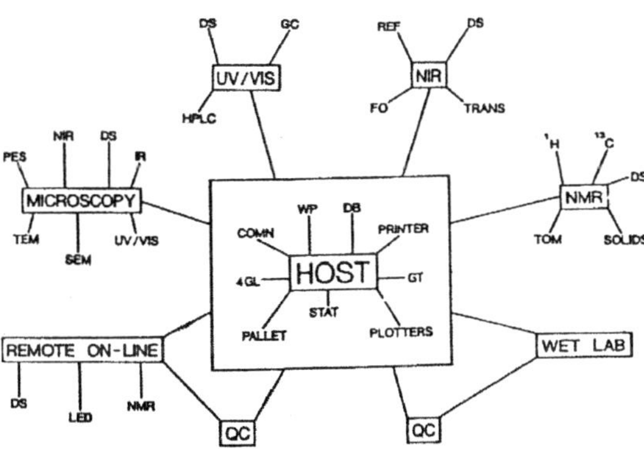

Figure 2. The 21st Century laboratory as predicted in 1990.

data can be moved from one station to another. Today we can even monitor and control experiments from home. The system we currently have in place consists of microscopy (all the way from UV to MIR and NMR), NIR, FT-IR, FT-Raman, UV/VIS, chromatography, mass spectrometry and the reference laboratory data stations. All the peripheral devices found around the "HOST" in Figure 2 can be found somewhere on the "LAN". Some of the "networks" that exist today for NIR instrumentation can contain many NIR instruments. The Grain Inspection Packers Stockyard Administration (GIPSA) Federal Grain Inspection Service (FGIS) maintains over 125 instruments in their network which monitors grain quality all across the United States.[23] While networks come in all sizes and levels of diversity, we have managed to make them functional to serve agriculture. Many of the networks of instruments are in use by regulatory agencies by numerous countries and the results are not published in the literature. However, some of the findings of one such networks in Europe have been published.[24–26] These authors discuss the network in Belgium and how calibrations are transferred from one instrument to another. In this case, the software used and the procedures followed are those of Shenk and Westerhaus.[27]

When the 1990's began, dispersive instruments were the principle tools used for NIR research and for spectroscopy in agriculture. Major manufacturers included Foss NIRSystems, Bran+Luebbe, Foss Grainspec and Tecator which eventually became part of Foss. The instruments produced by these companies were the workhorse instruments for agriculture and the instruments that formed numerous forage and grain networks in the US, Europe, Australia and Asia. The instrument has remained unchanged, but the sampling accessories have been greatly augmented. These sampling devices, new to NIR, would be "at home" in visible and traditional infrared laboratories. This is particularly true for fibre optic probes and liquid sample devices. In the 80s our only choice was a ground sample in a small cup in a spinning device. The improvement in signal-to-noise (S/N) allowed non-spinning devices with unground samples to yield the same spectral results. Fibre-optics permitted sampling of bulk containers and in line processes. Small instruments were also to be found from companies like Ocean Optics. Bran+Luebbe introduced an FT-NIR and an accoustic optical tunable filter (AOTF) instrument that gave them perhaps the broadest base of NIR technologies. They were joined by others, Buchi, Buhler, (now one company) and Bomem for FT instruments while Brimrose and Rosemount introduced AOTF instruments The diode array spectrometer was introduced about midway through the 1990s. The Perten DA 7000 and the Zeiss Corona spectrometers are two instruments that can function in the processing arena with "no moving parts". The Zeiss Corona has recently been mounted in a combine by a major harvesting equipment manufacturer for in-field use. The ATR System of Analytical Spectral Devices also brought another mid-IR technique to NIR.

In the mid 1990's the NIR equipment caught up with Mid-IR in another way. Perkin-Elmer introduced the first solely NIR FT-NIR microscope. Micro-spectroscopy usually thrives on "Information content" which is often lacking in NIR spectra. However, the combination with an FT instrument could provide a unique system for certain purposes. The field of FT instruments has expanded during the last five years. The major manufacturers all have an instrument at this time. The Perkin-Elmer instrument utilises a methane gas internal reference and software, which can create "virtual instruments", easing calibration transfer problems. The Bruker instrument has a sample wheel which permits multiple samples (40 for small cells, others as designed) and since it is possible to use both the internal integrating sphere or the external InGaAs detectors, one can theoretically sequentially obtain both the reflectance and transmittance spectra on the same sample. However, set-up limitations in the software for the transmission system require modifications before this technique can be used. The new Nicolet Antaris instrument offers a modular approach to experiments by replacing sampling modules. As such, these new instruments promise to enhance what we have traditionally done with NIR and open new vistas for research and industrial applications. Along with all these new instruments comes new chemometric software for calibration and model development.

Chemometrics

When I began in NIR spectroscopy, multiple linear regression was the technique of choice to develop models for prediction as well as for kinetic studies. My earliest ventures required "punch cards" and the IBM 1620 computer, a system that took up a large room but today can be replaced by a $50.00 calculator. In 1996 there were two excellent reviews of chemometrics published.[28,29] The authors of these reviews examined some 25,000 citations before selecting the 1400 they used. The Brown, *et al.* review, containing over 1200 citations, describes the success of the First International Chemometrics Internet Conference in 1994. For the first time the full papers were available for scientists to read and comment upon over the course of four months. The review covered software, tutorials, books and journal articles on vibrational and magnetic resonance spectroscopy. The second review, by Workman *et al.*, was the first part of a three part series which covered the general techniques of chemometrics but also quantitative analysis and advanced chemometric techniques.[30,31] Both reviews cited the tremendous advantages and use of partial least squares (PLS) as the best or preferred "whole spectrum" modelling technique. Beyond these similarities, the two reviews then went into different directions. Brown *et al.* described many chemometric applications and gave a thorough listing of those who were publishing in the areas. Workman *et al.* went into more detail on the particular application of the techniques to specific problems with fewer references (about 200). Taken together these reviews are a body of knowledge that anyone should find invaluable.

In many ways chemometrics is similar to the spectroscopy it serves. It is as much an art form as a science. This has been obvious to me each time I am asked the question "which software package is best?" The answer must always depends on the perspective of the authors of the package as to what they emphasised as important and what is included in the software package. The greatest change in chemometrics was initiated in the late 1980s when Galactic Industries Corporation developed the LabCalc software package and the GRAMS software packages that have succeeded it over the years.[32] These packages are a bridge for all the different formats of data collection in all the popular spectrometers and chromatographs. Files can be converted to a standard format and handled by the GRAMS package or exported to JCAMP or ASCII and moved to other packages if desired. The other big change was moving all chemometric software packages to the Windows operating system. This too made the moving of data from one package to another easier.

Conclusion

The theory and principles of NIR are not comprised of just the physics and electronics. The theory and principles must include the chemometrics and the reference analysis as part of the total technique. Our use of NIR spectroscopy is expanding into many new applications. I believe, in the near future, we will see sensors utilising multiple regions of the electromagnetic spectrum with embedded computational capabilities, which will greatly enhance our ability to classify, identify, measure and provide assurance of quality. There is a pyramid of growth for any instrumental technology, i.e., after a period the technology is replaced. This will not happen for NIR spectroscopy since its growth has always been dependent on a synergism of technologies, which together provide us with powerful new analytical tools.

References

1. W. Herschel, *Phil. Trans. Roy. Soc.* (London) 284 (1800).
2. W. Herschel, *Phil. Trans. Roy. Soc.* (London) 293 (1800).
3. W. Herschel, *Phil. Trans. Roy. Soc.* (London) 538 (1800).
4. W. Kaye, *Spectrochim. Acta* **6,** 257 (1954).
5. W. Kaye, *Spectrochim. Acta* **7,** 181 (1955).
6. K.H. Norris and W.L. Butler, *IRE Trans. Biomed. Electron.* **8,** 153 (1961).

7. K.H. Norris, *Trans. ASAE* **7,** 240 (1964).
8. G.C. Marten, J.S. Shenk, and F.E. Barton, II (Eds) *USDA Handbook* #643 (1989).
9. F.E. Barton, II and W.R.J. Windham, *Assoc. Off. Anal. Chem.* **71,** 1162 (1988).
10. W.R. Windham and F.E. Barton, II, *J. Assoc. Off. Anal. Chem.* **74,** 324 (1991).
11. Sadtler, *The Atlas of Near Infrared Spectra*, Sadtler Research Laboratories, Philadelphia, PA, USA (1981).
12. P. Williams and K Norris, *Near Infrared Technology in the Agricultural and Food Industries*, American Association of Cereal Chemists, St. Paul, MN, USA (1987).
13. J.W. Cooper, *Spectroscopic Techniques for Organic Chemists.* John Wiley and Sons, New York, NY, USA (1976).
14. J.W. Robinson, *Practical Handbook of Spectroscopy. CRC Press Inc., Bocca Raton, FL, USA (1991).*
15. G.W. King, *Spectroscopy and Molecular Structure.* Holt, Rhinehart and Winston Inc., New York, USA (1965).
16. T. Bellamy, *The Infra-red Spectra of Complex Molecules.* Chapman and Hall, London, UK (1975).
17. J.M. Chalmers and P.R. Griffiths (Eds), *Handbook on Vibrational Spectroscopy.* John Wiley and Sons Ltd, Chichester, UK (2001).
18. P. Williams and K. Norris, *Near Infrared Technology in the Agricultural and Food Industries*, 2nd Edition. American Association of Cereal Chemists, St. Paul, MN, USA (2001).
19. W. Kauzmann, *Quantum Chemistry.* Academic Press, New York, New York, USA, p. 667 (1957).
20. R.M. Silverstein and G.C. Bassler, *Spectrometric Identification of Organic Compounds*, 2nd Edition. John Wiley and Sons Inc., New York, USA, p. 66 (1967).
21. A.M.C. Davies, *NIRnews* **10(6),** 14 (1999).
22. F.E. Barton, II and D.S. Himmelsbach, In "*Analytical Applications of Spectroscopy II*," A.M.C. Davies and C.S. Creaser (Eds), Monograph. Royal Chemical Society, Thomas Graham House, Cambridge, UK (1991).
23. D.B. Funk, Personal communication, 6th NIRT Workshop, Kanas City, MO, USA, May 15–17 (2000).
24. E. Bouveresse, D.L. Massart and P. Dardenne, *Analytica Chimica Acta* **297,** 405 (1994).
25. E. Bouveresse, D.L. Massart and P. Dardenne, *Anal. Chem.* **67,** 1381 (1995).
26. P. Dardenne, J. Andrieu, Y. Barriere, R. Biston, C. Demarquilly, N. Femenias, M. Lila, P. Maupetit, F. Riviere and T. Ronsin, *Annales De Zootechnie* **42,** 251 (1993).
27. J.S. Shenk, S.L. Fales and M.O. Westerhaus, *Crop Sci.* **33,** 578 (1993).
28. J.J. Workman, P.R. Mobley, B.R. Kowalski, *Appl. Spectroscopy Rev.* **31(1/2),** 73 (1996).
29. S.D. Brown, T.S. Sum, F. Despagne and B.K.Lavine, *Anal. Chem.* **68,** 21R (1996).
30. R. Bro, J.J. Workman, P.R. Mobley and B.R. Kowalski, *Appl. Spectrosc. Rev.* **32,** 237 (1997).
31. P.R. Mobley, B.R. Kowalski, J.J. Workman and R. Bro, *Appl. Spectrosc. Rev.* **31,** 347 (1996).
32. Galactic Industries Corporation, 348 Main Street, Salem New Hampshire, 03079, USA (1991).

Sources of non-linearity in near infrared spectra of scattering samples

Donald J. Dahm
Department of Chemistry and Physics, Rowan University, Glassboro, New Jersey, USA

Introduction

The goal of our work is to convert spectra, which are readily obtained, to data that is a combination of linear functions of the various factors that affect the spectra.[1] Ideally, we would convert the spectra to the *absorbing power* of the material in the sample. The absorbing power of a material is the hypothetical fraction of incident light that would be absorbed if every absorption site, in some defined amount (usually given by a thickness) of material, were illuminated with the full incident intensity. The hypothetical fraction may be greater than 1.

In a mixture of absorbers, the absorbing power of the mixture is proportional to the absorbing power of each component times its concentration: that is, it follows Beer's law. In real samples, the incident light is not equal at each absorption site but, for a homogeneous material, the absorption power may be obtained from the Bouguer–Lambert law, which says that the intensity traveling through a material will fall off in accordance with the exponential function: exp($-kd$), where k is the absorbing power and d is the distance travelled by the light (Figure 1). Taking a *logarithm* of this exponential yields a straight line, the slope of which is proportional to the absorbing power and called a *linear absorption coefficient*. The relationship is strictly true only for light travelling *within* a continuous material.

Examples and discussion of sources of non-linearity

Experimentally, we do not make a measurement within a material. When we make absorbance measurements on a real sample (especially a highly-scattering sample), a spectrum collected is not the

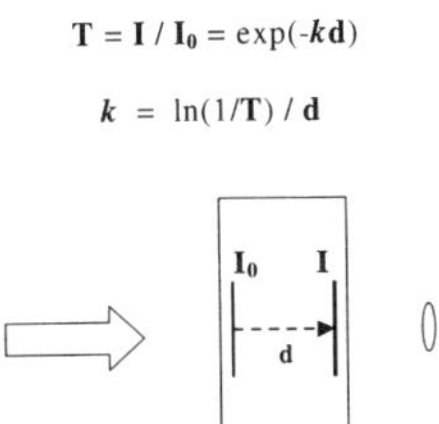

Figure 1. The transmitted intensity decreases as an exponential function of distance within a homogenous particle. The absorbance (negative logarithm of the fraction of light transmitted) is a straight line. The slope of the line is proportional to *k*, which is called the absorbing power of the material making up the particle.

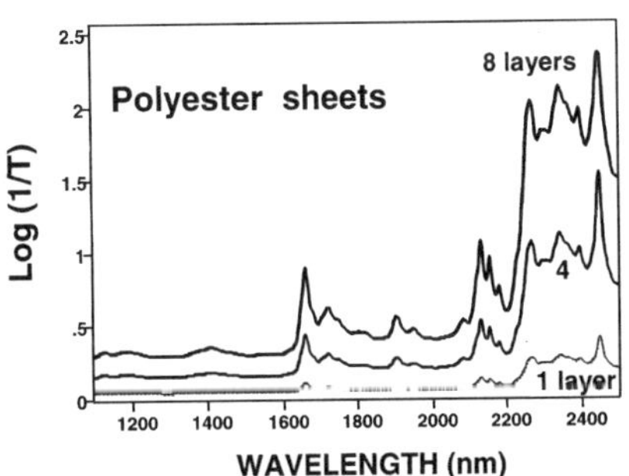

Figure 2. For samples consisting of plane parallel sheets, the absorbance [log(1/*T*)] in transmission is approximately proportional to the number of sheets in the sample.

absorbing power of the material and, generally, is not even linear with the absorbing power. [By the term "is linear with", we mean "can be related by an *offset* and a *multiplier*".] Below we will show how non-linearity can arise. [We mention in passing that the term *absorption coefficient* is also commonly used to mean the *log(1/T)* spectrum obtained from a material as it is presented in a particular transmission experiment. Usually, in a series of measurements the presentation is kept the same so that this difference between the experimental coefficient and absorption power may be corrected for by a calibration.]

Our first illustration of non-linearity involves increasing the number of layers in a transmission measurement. The spectra in Figure 2 show the log (I_0/I) [also referred to as log ($1/T$) or *absorbance*] spectra of 1, 4 and 8 identical polyester sheets, each with a rough, light diffusing surface. At first glance, the spectra appear to be approximately proportional to the sample thickness, given by the number of layers. However, as can be seen in Figure 3, where each spectrum is presented full scale, the shapes of the spectra are different. This is an indication of non-linearity.

If we had increased the thickness of the sample by increasing the thickness of a single layer, we would not have observed this. But by introducing the additional surfaces in the sample, we give the light the opportunity to reach the detector after it has bounced back and forth between the layers. This increases the effective path length within the sample. If there is very low absorption, the extent of this bouncing is determined by the amount of reflection or scatter that occurs at each surface. However, if there is very high absorption, there will be little increase in path length, because most of the light is consumed in the first pass through the layer. This leads us to our first generalisation:

Non-linearity will occur whenever the effective path length of light within a sample is a function of absorption.

Our second generalisation is:

Non-linearity will occur whenever light that has not been subjected to the absorption process reaches the detector.

Two very commonly encountered sources of such light are:

- in transmission: stray light; and
- in remission (reflectance): front surface reflection. (The reflection from the first surface encountered by the incident radiation has not been subject to absorption by the sample)

In Figure 4, we illustrate the effect of 10% of the incident light that has not been subjected to the absorption process striking the detector. Notice that the absorbance is suppressed, but the zero point is not moved (because there is no absorbance to suppress). The effect is to reduce the absorbance at the high absorption levels far more than at lower levels, introducing a non-linearity.

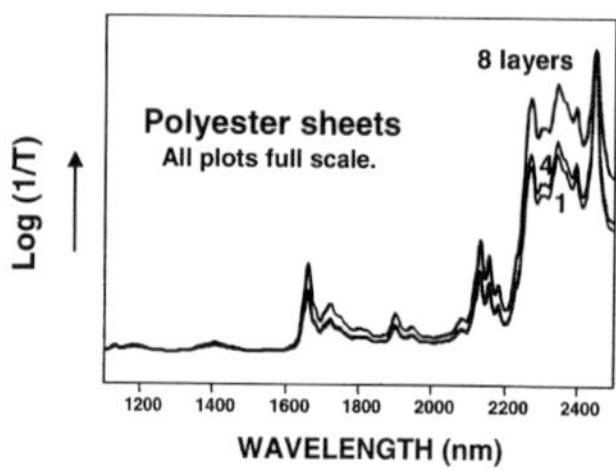

Figure 3. Displaying full scale the absorbance of each sample in Figure 2 shows that the presence of surfaces within the sample causes non-linearity in absorbance among the spectra, resulting in absorbance curves with different shapes for differing numbers of sheets.

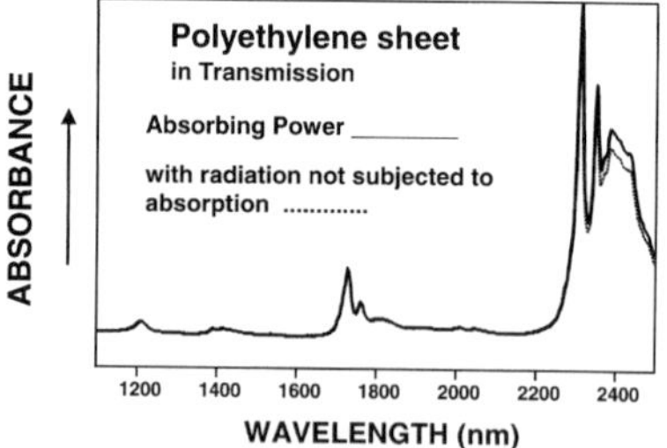

Figure 4. Illustration of non-linearity that would be exhibited in a transmission spectra of a single sheet, if 10% of the incident radiation reached the detector without being subjected to the absorption process.

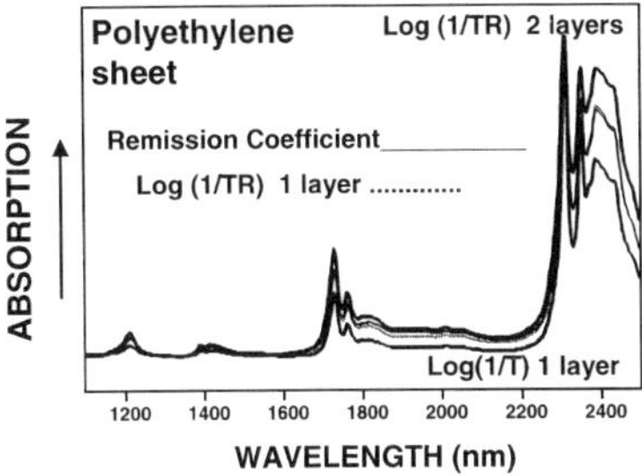

Figure 5. The reflection from the front surface observed in remission causes a change in shape when going from a transmission spectrum (bottom curve) to a transflectance spectrum (middle curve, dotted) of a single sheet. The absorption coefficient observed in remission (middle curve, solid), to a good approximation is proportional to the transflectance spectrum of a single sheet, but not to the transflectance spectrum of samples with additional sheets (top curve).

The effects of both these generalisations may be observed in Figure 5. Here we see that the absorbance spectrum in transmission [log(1/*T*)] of a single layer (shown as the bottom curve) has a different shape than the absorbance spectrum obtained in transflection [log(1/*TR*)] of the same sample (shown as one of the two middle curves). The number of layers for a sample observed in transflection is effectively twice that in transmission. Further, there is surface reflection observed in transflection. Thus the two curves are non-linear. The log(1/*TR*) spectrum of two layers (shown as the top curve) has yet a different shape, as further non-linearity is introduced by the increased number of layers.

In transmission, the absorbance spectrum of a single plane parallel layer yields an absorption coefficient that is a good approximation to the absorbing power. In transflection, the absorbance spectrum of a single plane parallel layer yields a spectrum that, to a good approximation, is proportional to the effective absorption coefficient of the sample as it would be observed in remission. This coefficient [2,3] (obtained as described in Reference 2) is also displayed as one of the middle curves in Figure 5. The effective linear absorption coefficient obtained in remission is not a linear function either of the absorbing power or the absorbance spectrum obtained in transmission. In transflection, as in transmission, there is further non-linearity introduced by increasing the number of layers in the sample.

The extent of non-linearity due to surface reflectance may be calculated using the Stokes' formulas[4] for absorption, transmission and remission by a single plane parallel layer. These results are compared in Figure 6 to the coefficient observed in remission.[3] [The coefficients obtained in transmission and remission have been adjusted to the same scale using the assumption of the Kubelka–Munk theory,[4] which is a simple factor of 2]. This illustrates that, for some simple cases, we can relate the absorbing power mathematically to the curves obtained in remission.

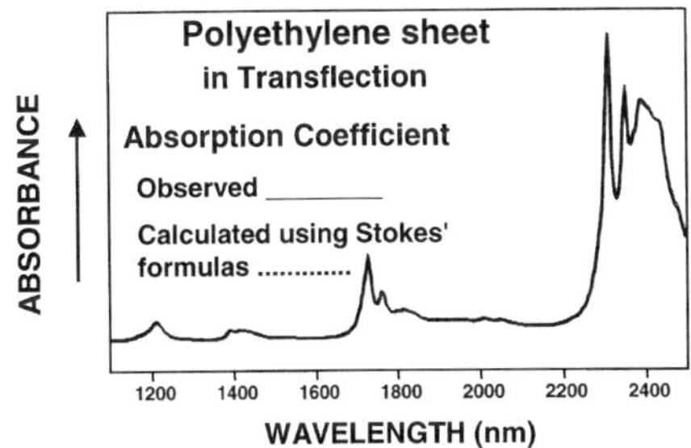

Figure 6. Using Stokes' formulas, developed in the 1860s, the spectrum observed from a single sheet in transflectance (solid) can be calculated (dotted) from the absorption coefficient observed in transmission.

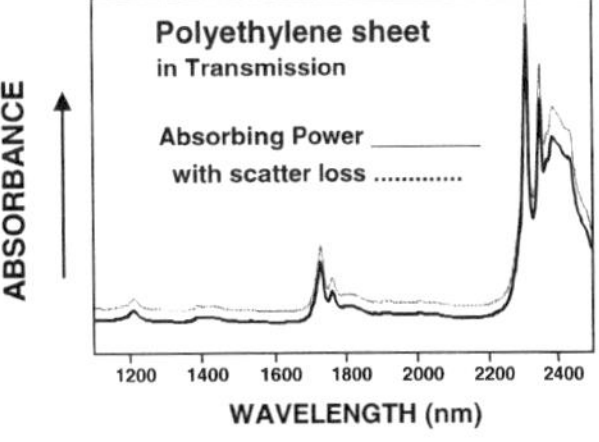

Figure 7. Illustration of the approximately linear change that would be exhibited in a transmission spectrum of a single sheet, if 10% of the incident radiation did not reach the detector (for example, due to scatter). Such a curve has essentially the same shape as that of the absorbing power of the material making up the sheet.

The above situation may be contrasted to a case that does not introduce significant non-linearity. In an absorbance measurement, any incident light that is not detected contributes to the measured absorbance. In Figure 7, we illustrate what would happen in an experiment if some of the incident radiation is lost through a mechanism other than the absorption process. The solid line represents the absorbing power of polyethylene. Displayed on the same scale is a spectrum (shown by the dotted line) that illustrates the effect of 10% of the incident light being scattered and missing the detector without having been absorbed. The absorbance curve is lifted up. If we displayed this second curve full scale, it would lie beneath the solid line, indicating that the shape remains essentially unchanged. This effect has introduced a change that yields a spectrum that is linear with the absorbing power.

Let us assume that this case is a transmission experiment and it is scatter that caused the intensity loss. The amount that the curve is lifted up is a measure of the amount of scatter loss. In this example it is 10%. It may be corrected for by:

subtracting an amount from the transmittance spectrum corresponding to 10% or 0.1, and

then dividing the remaining transmittance by 0.9 to put it on the scale of the incident intensity.

This is an example of a way to obtain the absorbing power of a material from a transmission measurement in the presence of scatter. Similar techniques have been shown to be useful in linearising a response in preparation for linear regression.[5]

There are, in fact, other mechanisms that introduce linear changes in the absorbance spectrum. An example is the change in void fraction in a sample. Of course, such a change in the presence of another effect that produces non-linearity, such as stray light, may produce two spectra that are non-linear with each other.

Since we have said above that:

(a) in transmission, the absorbance is a linear function of layer thickness; and

(b) changes in void fraction do not, in themselves, cause non-linearity in a transmission spectrum;

it is be tempting to conclude that a single layer of particles, in transmission, will yield a spectrum proportional to the absorbing power of the material making up the particles. We would like to now illustrate that particles do not behave as plane parallel layers and that this assumption is not correct.

In Figure 8 are two spectra of a very thin sample (essentially a mono-layer of particles, with spaces between them). The difference in the spectra is that one has the detector closer to the sample than the other. The loss due to scatter is greater for the one further away. In Figure 9, both of these are shown full scale. They have essentially identical shapes. This would seem to indicate that we have an absorbance curve, unaffected by scatter and, therefore, have a measure proportional to the absorbing power of the material. In Figure 10, the absorbance spectra of the particles is displayed full scale along with the

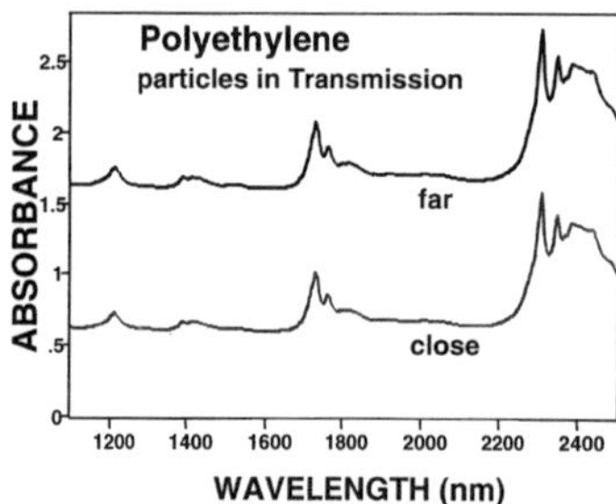

Figure 8. There is an observed increase in absorbance from a thin sample of particles when the distance from sample to detector is increased.

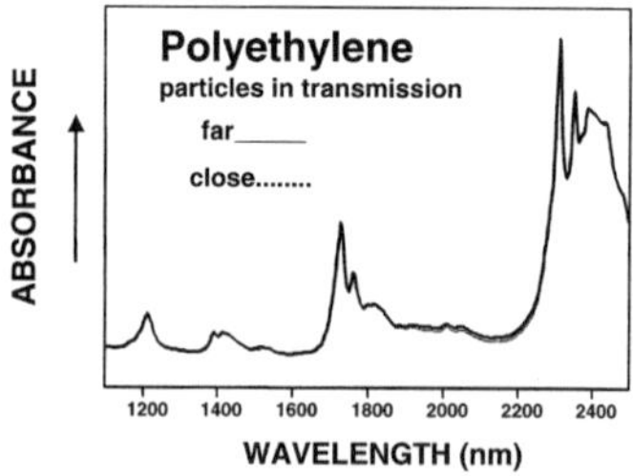

Figure 9. The shape of the spectra in Figure 8 are relatively unaffected by the change in distance from the sample to the detector and the curve has the shape, approximately, of the effective absorption coefficient for a particulate sample.

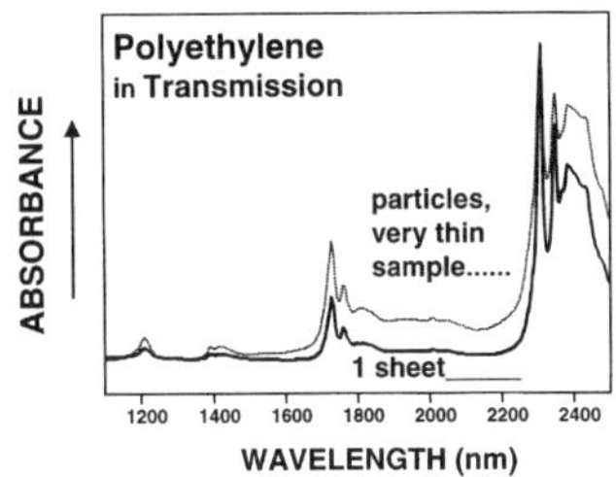

Figure 10. The effective absorption coefficient of a particulate sample is not the same as that of the material making up the particles, because a significant fraction of the light striking the surface of a particle reaches the detector (through the process of diffraction) without having been subjected to the absorption process.

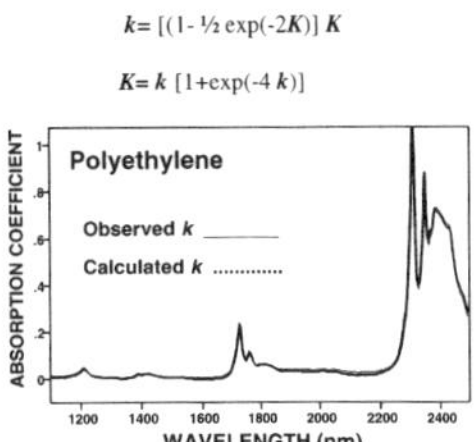

Figure 11. For a particular experimental arrangement, the absorption coefficient observed in transmission, *k*, can be related to the absorption coefficient observed in remission, *K*, by a function containing a single exponential term. The observed absorption coefficient has the same shape as the absorbance curve of a very thin sample.

transmission spectrum of a single sheet. Notice that they do not have the same shape. Why not? Particles so not reflect light, either externally or internally, in the same manner as plane parallel sheets. For example, small particles, unlike sheets, exhibit a significant amount of diffraction around the sphere.[6] Some of the light that strikes the surface of the particle continues toward a transmission detector without being subjected to the absorption process (analogous to the case of front surface reflection in remission). This is yet another source of non-linearity in a particulate sample.[7]

Correcting spectra for non-linearity

Now we would like discuss an approach to relate spectra obtained under one set of conditions to spectra obtained under another set. For example, if the effective linear absorption coefficient, *K*, obtained in remission is not the linear with the one obtained in transmission, *k*, can they be related by a simple function? In Figure 11 are shown the functions that relate such data as obtained on a Foss NIRSystems 6500. These are empirical functions, and were reported previously.[7]

So, if these functions can relate the absorption coefficients obtained in remission and transmission: Can they be used to more generally relate data obtained in remission to that obtained in transmission on the same sample? In Figure 12, we see data on an "infinitely thick" sample of polyethylene particles. The top and bottom curves are the absorbance spectra obtained in transmission and remission. The curve in the middle is that calculated by the formula at the top of the figure (which has the same form as the bottom one shown on the Figure 11). The simple one term function no longer works, and we need to add another term. The top curve is actually a bold spectrum observed in transmission along with a dashed curve that was calculated from the remis-

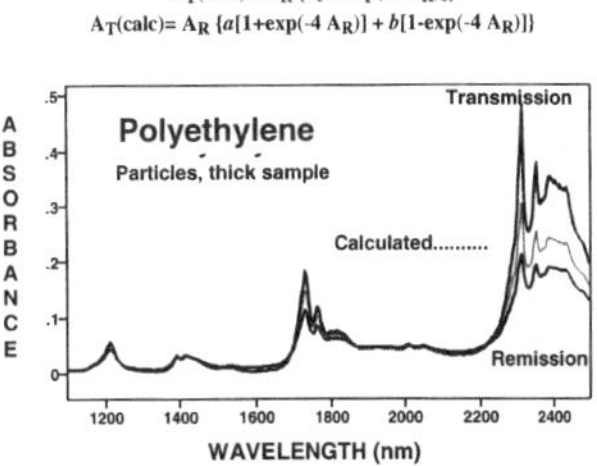

Figure 12. For the same experimental arrangement as in Figure 11, the absorbance observed in transmission (top curve, solid) can not be obtained from the absorbance observed in remission (bottom curve, solid) with a function containing a single exponential term (middle curve, dashed), but can be obtained with a function containing two exponential terms (top curve, dashed).

sion spectrum and the bottom formula in Figure 12. In general, we can relate curves obtained under various conditions by a two-term relationship, where:

- one term has full effect at very high absorbance values, and no effect at low absorbance values, and
- the other term has full effect at very low absorbance values, and no effect at high absorbance values.

We are now doing theoretical work to guide us in the selection of the form of these functions, but what is important is that:

If we adjust the *a* and *b* factors to fit the ends of the data, it seems possible to find a factor in the exponential (replacing the "4") that will fit the data in between.

Conclusion

So where does that leave us in the quest for linear spectral data?

We hope to be able to use the measurements at low absorbance levels to obtain effective linear absorption and remission coefficients. (It is not always possible to obtain such data at higher absorbance levels, when a single particle is "infinitely thick".) Then by a combination of theoretical relationships and empirical functions (obtained for a particular experimental arrangement), relate an absorbance curve collected either in remission or transmission to the absorbing power of the material.

We presume that if we had such linear data, principle component analysis would work better, and that results of calculations of "distance" between spectra would change significantly. The assumed advantages in linear regressions include: fewer PLS factors (comparable to the number of components) being required to adequately describe a data set, and eliminating the need to limit the spectral range to get a more linear response to amount of analyte.

Acknowledgment

The work described here was performed as part of a larger effort jointly with Kevin Dahm and Karl Norris. Experimental measurements were performed in the laboratories of Foss NIRSystems.

References

1. D.J. Dahm and K.D. Dahm, *J. Near Infrared Spectrosc.* **3,** 53 (1995).
2. D.J. Dahm and K.D. Dahm, *Appl. Spectrosc.* **53,** 647 (1999).
3. D.J. Dahm, K.D. Dahm and K.H. Norris, *J. Near Infrared Spectrosc.* **8,** 171 (2000).
4. D.J. Dahm, K.D. Dahm and K.H. Norris, *J. Near Infrared Spectrosc.* **10,** 1 (2001).
5. K.H. Norris, *NIR news* **12(3),** 6 (2001).
6. H.C. van de Hulst; *Light Scattering by Small Particles.* Dover Publications 200 (1981).
7. K.D. Dahm and D.J. Dahm; *Practical Limitations Imposed by Particle Size in Determination of Absorption Coefficients*. International Diffuse Reflectance Conference, Chambersburg, Pennsylvania, USA, August 14, (2000).

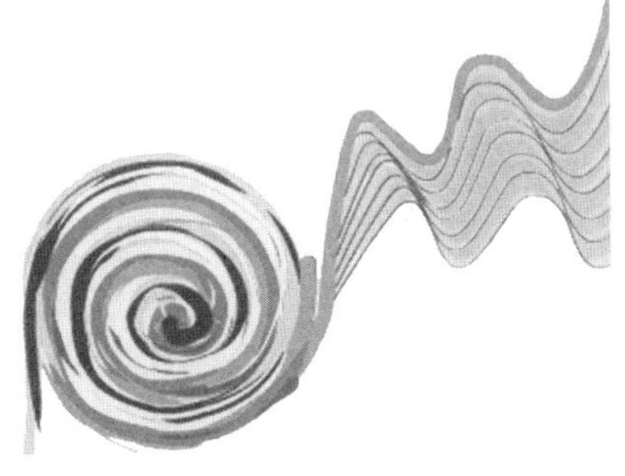

Instrumentation and Sample Presentation

Sample presentations for near infrared analysis of intact fruits, single grains, vegetable juice, milk and other agricultural products

Sumio Kawano

National Food Research Institute, 2-1-12 Kannondai, Tsukuba 305-8642, Japan

Indroduction

Sample presentation is very important for near infrared (NIR) spectroscopy. Figure 1 shows the four major modes of spectral data acquisition: (1) transmittance, (2) reflectance, (3) transflectance and (4) interactance. In the transmittance mode [Figure 1(a)], incident light illuminates one side of the sample and an unabsorbed portion of the incident light is collected by the detector on the opposite side. This mode is widely used for liquids and low density solids. Incident light, in the case of the reflectance mode [Figure 1(b)], illuminates the surface of a sample, some of the light enters the sample and is diffusely reflected to the detector. In this mode, the sample is usually opaque, like a powdered sample having more than 1cm depth. Transflectance [Figure 1(c)], originally developed by Technicon (now Bran+Luebbe, Germany) for the InfraAlyzer, was designed to study liquids in an instrument designed to measure reflectance. Hence, the liquid cuvette is "backed" with a diffuse reference. The incident light is transmitted and reflected through the sample. Interactance [Figure 1(d)] was developed by Karl

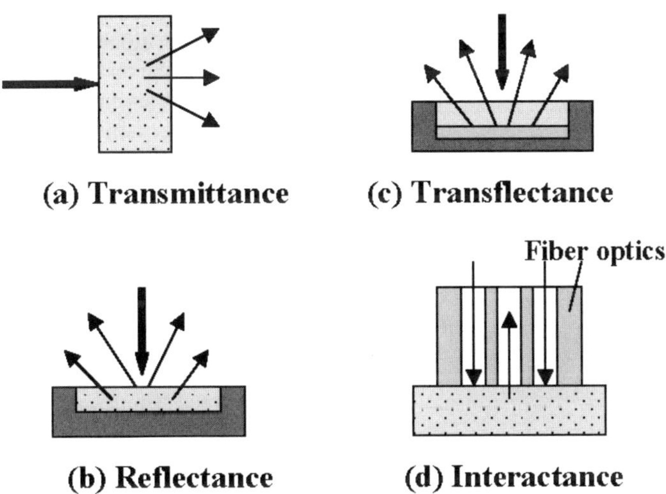

Figure 1. Sample presentation of transmittance, reflectance, transflectance and interactance.

Norris as a means for studying living plant and human tissue. Usually undertaken with concentric fibre optics, as shown in Figure 1(d), the sample-to-fibre optics is such that the incident light is forced into the sample to interact with the sample before making its way to the detector optics. Therefore, only the light transmitted through, or interacting with, the sample can be detected.

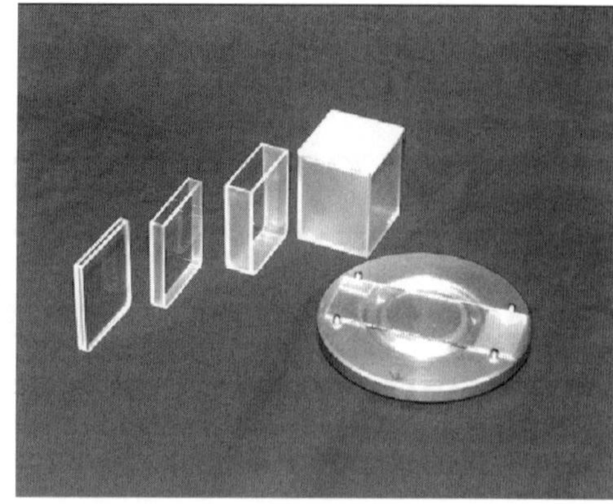

Figure 2. Sample cells for liquids; cuvettes and aluminium cell.

Sample cell for liquids

A typical cuvette made of quartz is shown in Figure 2. It is the popular mode for collecting spectra of liquids such as water, liquors and juices. The thickness (path length) of the cell depends on the wavelength region and samples used. In the case of water, 2–3 cm thickness is suitable for measurements at 960 nm, 2–3 mm thickness is suitable for measurements at 1450 nm and cells with a thickness less than 1mm can be used at 1930 nm. Sample thickness is very important when conducting spectral–transmittance measurements.

In a reflectance instrument, the aluminium cell specially designed for the InfraAlyzer 500 (right in Figure 2) is widely used. A small amount of the sample is taken and dropped onto the central portion. The cell covered with a glass plate may now be placed in the sample drawer of the instrument. The usual path length of the cell is 0.1 mm.

Sample cell for pastes

Spectra of pastes, such as dough, ground meat and fermented soybean paste (called Miso) can be acquired using the sample cell shown in Figure 3. Figure 3 illustrates the two types of sample cells for obtaining reflectance spectra (shown at the left in Figure 3, so-called "Open cup") and transmittance spectra (right, so-called "transport cell", or TC). The open cup, designed specially for the InfraAlyzer 500, may be positioned without a glass cover into the drawer of the instrument. The surface of the sample in the open cup should be smoothed. The TC was designed specially for the NIRS 6500. Samples are usually packed into polyethylene bags for measurements. Closing the cell forces the sample to a constant thickness of 1 cm. While the spectrum is continuously scanned, the sample is transported through the NIR beam.

Figure 3. Sample cells for pastes; open cups and high fat/high moisture transport cell.

Sample cell for a powdered sample

Powdered samples can be measured in a sample cup like that shown in Figure 4. The cell has a specially designed rubber pad to help produce a reproducible packing density. Density does affect scattering and scattering should be carefully

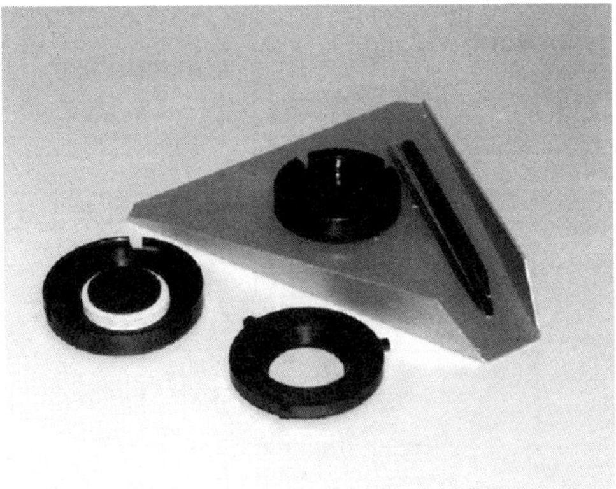

Figure 4. Sample cell for powdered sample.

Figure 5. Sample cells for whole grains.

controlled during any given NIR experiment. The cell in Figure 4 was designed for the InfraAlyzer 500. It has a circular quartz window, 3.5 cm in diameter, mounted in black plastic. The thickness of a powdered sample in this cell is approximately 1 cm. The reflectance cell in Figure 4 is shown with a cell holder and scoop.

Sample cell for whole grains

Whole grain samples can be scanned in the sample cells shown in Figure 5. The sample cell on the left was developed by NIRSystems for the NIRS6500 instrument. The sample cell has a loading funnel input that makes it easy to load. It is a "transport type" cell, mounted vertically with the NIR instrument which moves up and down slowly during NIR measurement to compensate for the heterogeneity of the samples. A single spectrum using this cell is usually the average of 30–50 scans. The circular cell, shown to the right in Figure 5, was designed for the InfraAlyzer 500. The cell, 85 mm in diameter and 12 mm in depth, is rotated quickly during acquisition of NIR spectra, another means of averaging heterogeneity.

Sample holder for fruits

Acquisition of NIR spectra of large samples such as fruit requires a specially designed "sample holder." Figure 6 is a typical of holder needed to make interactance measurements of fruits with fibre optics. Bifurcated fibre optics are required to isolate the incident and interactance energy. A cushion made of urethane foam is pasted on the end of the probe to hold the sample and exclude stray light. The holder requires only that the sample be positioned on the cushion. The sample holder in Figure 6 has been used to study the rela-

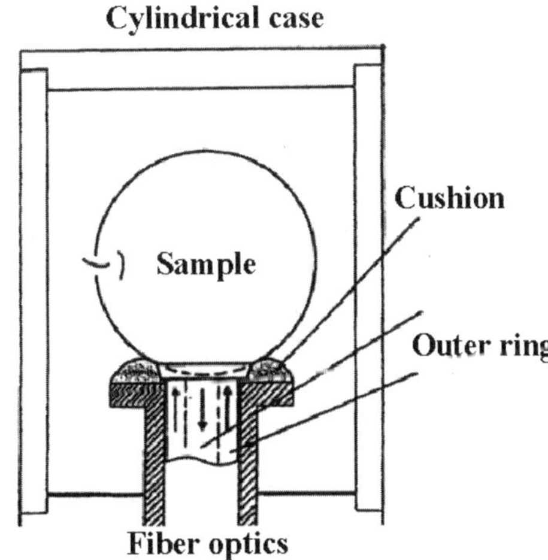

Figure 6. Sample placement for NIR interactance measurements using fibre optics.

tionship of NIR spectra to the Brix values of intact peaches.[1]

Useful interactance measurements are difficult to obtain from oranges because peel thickness inhibits interactance of NIR energy with orange meat. The transmittance mode is for more effective for this type of measurement. A schematic of a piece of apparatus for making measurements in the transmittance mode is given in Figure 7. The top of the sample is illuminated with monochromatic light through a fibre optic. The light passes through the sample and is measured with a silicon detector located below the sample. In transmittance spectroscopy, the spectra are affected by the variation in diameter of the samples measured. In order to reduce the sample size effect, $d^2\log(1/T)$ at each wavelength was divided by $d^2\log(1/T)$ at 844 nm, a wavelength having a high correlation only to fruit size. The corrected spectra, called "normalised 2nd derivative spectra", are nominally not affected by sample size. Multiple linear regression (MLR) based on normalised 2nd derivative spectra and Brix value, produced very good results (R = 0.989, SEC = 0.28°Brix, SEP = 0.32°Brix).[2]

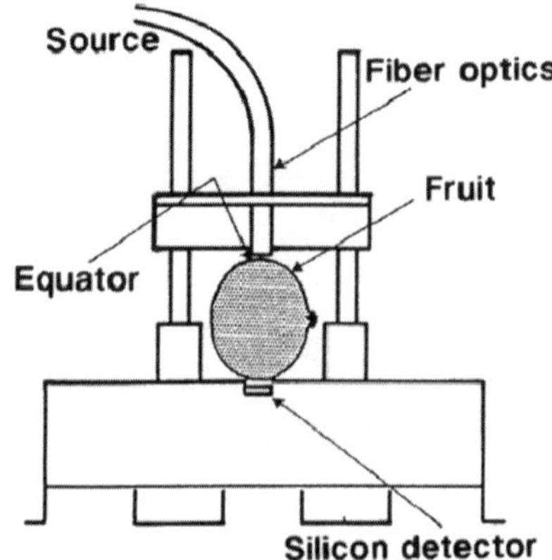

Figure 7. Sample placement for NIR transmittance measurement using fibre optics.

The other sample cells

Conventional liquid cuvettes are difficult to wash and dry. This laboratory has studied the use of test tubes for sample cells. Studies of milk, rumen juice and urine of milch cows have been conducted using test tubes.[3] Test tubes have been used to study spectral properties of spinach juice for determination of undesirable constituents. This laboratory has also worked to develop special sample holders for making measurements of single kernels of grain.

Conclusion

Recording of useful spectra requires many parameters to be monitored. However, selection of the most appropriate sample holder (cell) is absolutely essential.

Reference

1. S. Kawano, H. Watanabe and M. Iwamoto, *J. Japan. Soc. Hort. Sci.* **61,** 445 (1992).
2. S. Kawano, T. Fujiwara and M. Iwamoto, *J. Japan. Soc. Hort. Sci.* **62,** 465 (1993).
3. J.Y. Chen, C. Iyo and S. Kawano, *J. Near Infrared Spectrosc.* **7,** 256 (1999).

Choosing a near infrared instrument and a sample presentation option for plant and soil analysis

Graeme Batten,[a] Anthony Blakeney,[b] Susan Ciavarella,[c] David Lamb[a] and Sarah Spackman[a]

[a]*Farrer Centre, Charles Sturt University, LMB 588, Wagga Wagga, NSW 2678, Australia;*

[b]*Cereal Solutions, PO Box 201, North Ryde NSW 1670, Australia;*

[c]*Irrigation Research & Extension Committee, C/- Yanco Agricultural Institute, Yanco, NSW 2703, Australia*

Introduction

In Australia the yields of rice crops are amongst the highest in the world. A record average yield of 9.6 t ha^{-1} was achieved in the 2000–2001 season. Individual growers regularly achieve 12 t ha^{-1} (data provided by the Ricegrowers' Co-operative Limited, Leeton Australia). The average input of nitrogen fertilizer is 120 kg N ha^{-1} but when rice is grown after several rice crops as much as 240 kg N ha^{-1} may be applied.[1,2]

The cost of fertilizer together with the water used to grow rice and the declining price of rice are encouraging rice growers to adopt management practices which lead to higher yield ha^{-1}, higher yield kg N^{-1} applied, higher yield ML^{-1} water used to grow the crop and grain of marketable quality while exerting minimal impact on the environment.

Near infrared (NIR)-based analysis of rice crop shoot N status was developed as an aid to crop fertilizer management in the 1980's and provides a quantitative system to improve efficiency in the terms specified above.[3,4] We have reported on this application of NIR spectroscopy at previous NIR Conferences.[5,6]

The success of the shoot tissue testing service for rice has led to its adoption by over 40% of rice growers.[7] The benefits to the industry include an estimated 5% higher yield for those who have crops tested, reduced waste of fertilizer by those whose crops acquire adequate nitrogen (a saving to the industry and also protection of the environment from excess fertilization) and more marketable grain (slightly lower protein content). There are considerable flow-on benefits to farmers who do not use the testing services through a better understanding of crops requirements. The estimated benefits are some A$10 million per year to the industry at a cost of only about A$50,000 per year or a return of 100 : 1.

New developments

The present NIR-based plant shoot analysis services for rice and wheat depend on growers sampling their crops at defined physiological growth stages and collecting samples from up to nine small areas of crop. The total area sampled is < 1 m^2 from 20 to 30 ha of crop. This sampling intensity was chosen[8] after an examination of the variability in some crops. The current NIR testing procedure in-

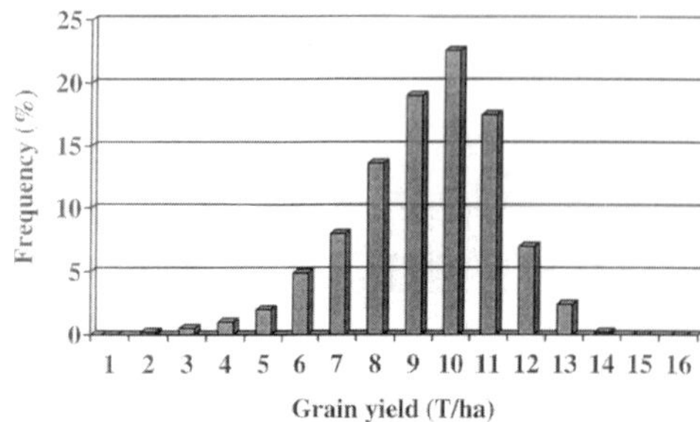

Figure 1. Yield range and distribution in a rice crop with an average yield of 10 t ha^{-1}. (Source—J. Lacy, personal communication).

cludes presenting the sample to the NIR instrument in a standard sample cup. The NIR instrument, therefore, only senses about 1 g of tissue taken from 20 to 30 ha of crop. Using larger cells, or the presentation system available on the Perten DA7000 instrument, would increase the amount of sample exposed to the instrument but would not reduce the error associated with sampling the crop in the field.

As we enter the 21st century, farmers are becoming familiar with yield maps which show the extent of crop variability. In a field of rice which yields an average of 10 t ha^{-1}, only 25% of the area will yield 10 t. The other 75% of the area will yield less or more than the average, possibly within the range of < 1 to 14 t ha^{-1} (Figure 1). Yield maps are a first step into the new 'precision farming' approach to crop management. To be of real benefit to farm managers, precision farming tools must provide information about variability within a field which is cost effective to obtain, is in absolute (not relative) units, indicates reasons for the variation; and can be collected in time to allow decisions about the current crop.

Crops are being assessed during vegetative phases of development using satellite or air-borne imaging systems with electromagnetic energy of visible and near infrared wavelengths.[9–12] To calibrate these images into actual data values, such as dry matter ha^{-1}, much field data recording work is necessary. In many situations the images from satellites cannot be obtained at the appropriate time or cost for timely crop management decisions to be made. Images of crops collected from an aircraft flying at an altitude between 1,000 and 3,000 m above the land surface may be more reliable and less expensive.

We have advanced the prediction of actual crop dry matter yields from a 4-camera airborne video system.[13] Each camera contains a 740 × 576 pixel array and is fitted with a 12 mm focal length lens. At 1524 m above the ground, this system achieves a resolution on 1 m (1 m × 1 m pixels) and an image area of 43.2 ha. Each camera acquires information in a preset spectral band governed by an interchangeable filter (25 nm band-pass). An on-board computer, fitted with a 4-channel frame-grabber board, captures and digitises the images from the cameras. In this study images were captured using filters with 440, 550, 650 and 770 nm wavelengths.

Images of the rice crops reported here were obtained between 11 am and 2 pm Australian Eastern Standard time at an altitude of 1400 m. Standard reference panels were imaged during each flight. Pre-processing of each image included shear correction, band-to-band registration and elimination of geometric and radiometric distortion. The normalised difference vegetation index or NDVI for each pixel was determined as follows:

NDVI = (red – infrared) / (red + infrared)

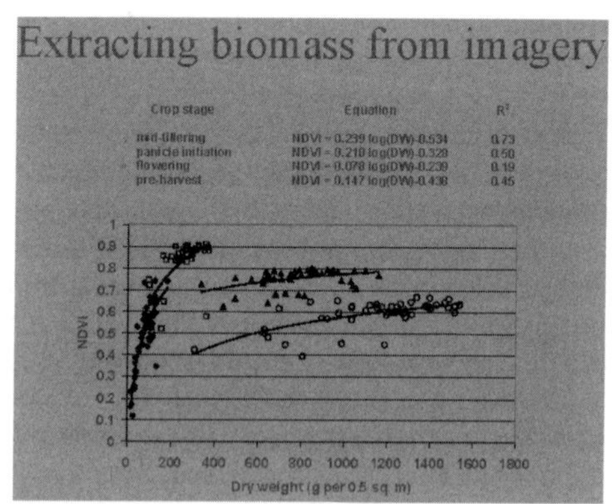

Figure 2. Correlations between normalised difference vegetation index (NDVI) and rice crop dry matter at four plant growth stages.

NDVI was found to be correlated with biomass at several crop stages (Figure 2). Final yield was estimated from mid-tillering NDVI with an error of < 1 t ha^{-1}. The correlations between NDVI obtained here and crop biomass at final grain yield are comparable with similar data reported by Ruan *et al.*[14] The stronger correlations between NDVI and dry matter during the earlier vegetative stages of crop development partly reflect the shoot to water ratio of the rice crop.

Future developments

At this time, the NDVI images are being used by rice growers to decide where to take samples for PI N tests and to indicate where nitrogen fertilizer rate should vary. In Australian rice crops deficiency of nitrogen is the dominant cause of low yields but sub-optimal yields due to inadequate phosphorus and zinc and now evident from recent reports by Batten *et al.*[2] and Dang *et al.* (2001, unpublished data). Extrapolation from NDVI data to biomass estimates may be valuable when calculating potential crop yields but further work is essential to predict the appropriate fertilizer type and amount to be applied.

A study by Brian Dunn (2000; personal communication) revealed that the final yield of a crop is explained largely by dry matter at the panicle initiation stage of development (R^2 = 0.74), while shoot nitrogen is a more informative predictor of final yield (R^2 = 0.80). Analysis of nitrogen, and also non-structural carbohydrate,[6] would provide a basis for more appropriate fertilizer recommendations. If crop composition data were made available, in conjunction with NDVI maps, crop managers could be in a better position to determine the appropriate fertilizers and the optimal rates of fertilizer, for small areas within crops.

NIR science and technology has much to offer in these new expectations of food producers who must continue to feed the current population of the world which will rapidly increase towards 8 billion over the next 25 years.

Conclusions

Techniques which detect the extent of variation within crops will enhance the average yields achieved by producers. Current sampling and subsequent analysis using NIR has been profitable for rice growers in Australia but large area surveys of crop dry matter variation, together with NIR analysis of crop nitrogen status, is suggested as essential to improving crop production on a larger number of rice farms. Yield-limiting factors (such as soil pH, salinity, nutrient deficiencies, water logging, disease, pests), must be understood before recommendations are made on fertilizer requirements.

Acknowledgements

We thank Brian Dunn for unpublished data, Jan Hubatka for assistance with the preparation of the paper and the RIRDC for financial support.

References

1. J. Lacy, W. Clampett, L. Lewin, R. Reinke, G. Batten, R. Williams, P. Beale, D. McCaffery, M. Lattimore, A. Schipp, R. Salvestro and J. Nagy, *2000 Ricecheck Recommendations*. NSW Agriculture, p. 20 (2000).
2. G.D. Batten, D. Reuter, M. Unkovich and C. Kirkby, National Land and Water Resources Audit Project, CD ver 1.1. CSIRO, Australia (May 2001).
3. G.D. Batten, A.B. Blakeney, M.G. Glennie-Holmes, R.J. Henry, A.C. McCaffery, P.E. Bacon and D.P. Heenan. *J. Sci. Fd Agric*. **54,** 191 (1991).

4. A.B. Blakeney, G.D. Batten and S. Ciavarella In *Temperate Rice - achievements and potential,* Ed by E. Humphreys, E.A. Murray, W.S. Clampett and L.G. Lewin. NSW Agriculture, Griffith, Australia, p. 477 (1994).
5. G.D. Batten and A.B. Blakeney, in *Proc. 2nd Int. NIRS Conference*, Ed by M. Iwamoto and S. Kawano. Korin Publishing Co. Tokyo. Japan, p.315 (1990).
6. A.B. Blakeney, G.D. Batten, S. Ciavarella and V.B. McGrath in *Leaping Ahead with Near Infrared Spectroscopy,* Ed by G.D. Batten, P.C. Flinn, L.A. Welsh and A.B. Blakeney. Near Infrared Spectroscopy Group; RACI, North Melbourne, Australia (1995).
7. G.D. Batten, A.B. Blakeney and S. Ciavarella, *IREC Farmers' Newsletter* **156,** 40 (2001).
8. G.D. Batten, A.B. Blakeney, P.E. Bacon, A. Williams and M.R. Glennie-Holmes, in *Rice Research 1988.* Yanco Agricultural Institute, NSW Agriculture and Fisheries, Yanco, Australia, p. 43 (1988).
9. D.W. Lamb, *Aust. J. Exp. Agric.* **40,** 725 (2000).
10. S. Spackman, D. Lamb and J. Louis. *Aspects App. Biol.* **60,** 99 (2000).
11. C. Yang, J.H. Everitt, J.M. Bradford and D.E. Escobar, *Trans. ASAE* **43,** 1927 (2000).
12. T.G. van Neil and T.R. McVicar, *Rice CRC Technical Report* P110501/01. Available at http://www/ricecrc.org.au (2001).
13. J. Louis, D.W. Lamb, G. McKenzie, G. Chapman, A. Edirisinge and I. McLeod, in *Proc Amer. Soc. Photogrammetry Remote Sensing 15*th *Biennial Workshop on Color Photography and Videography in Resource Assessment,* p. 326 (1995).
14. W.R. Raun, J.B. Solie, G.V. Johnson, M.L. Stone, E.V. Lukina, W.E. Thomason and J.S. Schepers, *Agron. J.* **93,** 131 (2001).

Standardisation of near infrared instruments, influence of the calibration methods and the size of the cloning set

Pierre Dardenne,[a*] Ian A. Cowe,[b] Paolo Berzaghi,[c,d] Peter C. Flinn,[e] Martin Lagerholm,[b] John S. Shenk[f] and Mark O. Westerhaus[f]

[a]*Centre de Recherches Agronomiques de Gembloux – CRAGx, 24 Chaussée de Namur, 5030 Gembloux, Belgium*

[b]*Foss Tecator AB, Box 70, SE-263 21 Höganäs, Sweden*

[c]*University of Padova, Agripolis, 350020 Legnaro Italy*

[d]*University of Wisconsin, 1925 Linden Dr., 53706 Madison, WI, USA*

[e]*Agriculture Victoria, Pastoral and Veterinary Institute, Private Bag 105, Hamilton, Victoria 3300, Australia*

[f]*Infrasoft International, 109 Sellers Lane, 16870 Port Matilda, PA, USA*

Introduction

A previous study[1] evaluated the performance of three calibration methods, modified partial least squares (MPLS), local PLS (LOCAL) and artificial neural networks (ANN) on the prediction of the chemical composition of forages, using a large near infrared (NIR) database. The study used forage samples (*n* = 25,977) from Australia, Europe (Belgium, Germany, Italy and Sweden) and North America (Canada and USA) with reference values for dry matter (DM), crude protein (CP) and neutral detergent fibre (NDF) content. The spectra of the samples were collected using ten different Foss NIRSystems instruments, only some of which had been standardised to one master instrument. The aim of the present study was to evaluate the behaviour of these different calibration methods when predicting the same samples measured on different instruments.

Material and methods

Twenty-two sealed samples of different kinds of forage were measured in duplicate on seven instruments (one master and six slaves). Table 1 reports the locations and the instrument modules used to take the spectra of the 22 samples. Table 2 lists the forage samples. The samples have been measured in duplicates on each instrument using the factory scanning parameters (16,32,16). Figure 1 represents the average spectra of the 22 samples measured on the master instrument.

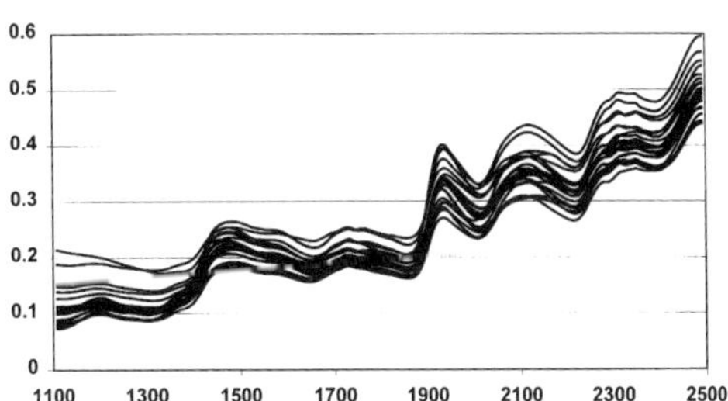

Figure 1. Log(1/*R*) spectra of the 22 sealed forage samples scanned on the master instrument

Table 1. Locations and NIRSystems instrument modules used to take the spectra of the 22 samples.

Abb.	Institution	Location	Instru.	Module
MA	Infrasoft International, LLC	Port Mathilda (PA)	6500	Spin/Drawer
AG	University of Wisconsin - Agronomy	Madison (WI)	6500	Spin/Auto
CW	Cal-west Seeds	West Salem (WI)	5000	Spin/Drawer
FG	Forage Genetics	West Salem (WI)	6500	Spin/Drawer
RR	Rock River Laboratoty	Watertown (WI)	5000	Spin/Drawer
US	US Dairy Forage Research Center USDA-ARS	Madison (WI)	6500	Spin/Auto
UW	University of Wisconsin - Marshfield	Marshfield (WI)	6500	Spin/Drawer

Table 2. List of the forage samples sealed in small ring cups.

1	Maize silage (Europe)	12	Lucerne hay (Australia)
2	Grass silage (Europe)	13	Cereal hay (Australia, species 1)
3	Lucerne hay (US)	14	Cereal hay (Australia, species 2)
4	Cereal hay (Australia)	15	Legume grass hay (Europe)
5	Legume grass hay (US)	16	Legume grass hay (Australia)
6	Fresh cut pasture (Australia)	17	Fresh cut lucerne (US)
7	Maize silage (Australia)	18	Fresh cut pasture (Europe)
8	Maize silage (US)	19	TMR (Europe)
9	Grass silage (Europe, species 1)	20	TMR (US)
10	Grass silage (Europe, species 2)	21	Native pastures (Australia, species 1)
11	Lucerne hay (Europe)	22	Native pastures (Australia, species 2)

Three sets of near infrared (NIR) spectra (1100 to 2498 nm) were created for each slave instrument. The first set consisted of the spectra in their **original form** (unstandardised); the second set was created using a **single sample standardisation** (Clone1) and the third using a **multiple (6) sample standardisation** (Clone6). WinISI software (Infrasoft International Inc., Port Matilda, PA, USA) was used to perform both types of standardisation.

Clone1 is just a photometric offset between a "master" instrument and the "slave" instrument. Clone1 procedure used one sample spectrally close to the centre of the population. A spectrum (sample No. 16) is selected from the 22 based on its smallest distance in the PCA space and the differences between each slave and the master is used to modify the other slave spectra.

The multiple sample standardisation[2,3] requires a selection of six samples covering the range of absorbances: samples Nos 3, 5, 9, 10, 19, 21 have been selected. Clone6 modifies both the *X*-axis through a quadratic wavelength adjustment and the *Y*-axis through a simple regression wavelength by wavelength.

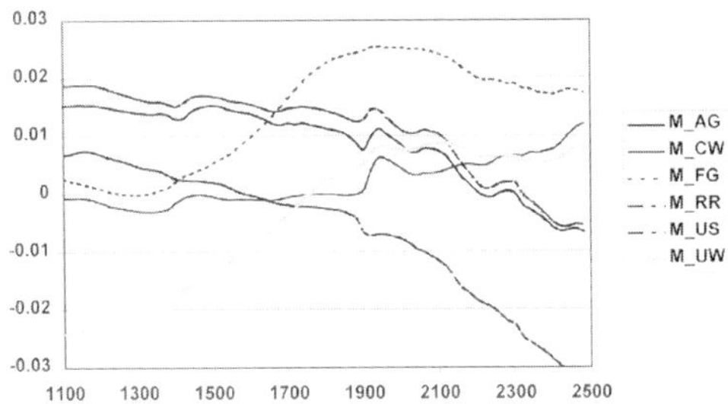

Figure 2. Spectra of the average differences between slaves and master [log(1/*R*)].

The remaining 15 samples were used to evaluate the performances of the different models. The predicted values for dry matter, protein and neutral detergent fibre from the master instrument were considered as "reference *Y* values" when computing the statistics *RMSEP*, *SEPC*, *R*, Bias, Slope, mean *GH* (global Mahalanobis distance) and mean *NH* (neighbourhood Mahalanobis distance) for the six slave instruments using the calibration models described in Berzaghi's paper.[1]

Table 3. *RMSC* between duplicates for each slave instrument and *RMSC* between instrument before and after standardisation.

	AG	CW	FG	RR	US	UW
Duplicates	59	77	131	107	105	250
Before STD	7038	3928	9153	7558	11910	12054
After Clone1	625	410	401	573	671	430
After Clone6	582	318	432	631	756	488

Table 4. RMS of RMSEP (master predicted as *Y*) across the six instruments based on the duplicates of 15 independent samples.

	DM			CP			NDF		
	UNSTD	Clone1	Clone6	UNSTD	Clone1	Clone6	UNSTD	Clone1	Clone6
PD-Local	0.88	0.32	0.28	1.66	0.42	0.43	2.99	0.93	0.53
GH	***4.23***	***1.97***	***2.00***	***3.50***	***1.92***	***2.08***	***3.23***	***1.75***	***1.75***
NH	***2.91***	***1.33***	***1.36***	***2.51***	***1.41***	***1.50***	***2.08***	***1.13***	***1.12***
MPLS	0.30	0.08	0.08	0.96	0.19	0.19	4.34	0.92	0.64
ISI-Local	0.70	0.26	0.18	1.29	0.32	0.22	3.44	0.88	0.69
GH	***3.85***	***1.94***	***2.00***	***2.08***	***1.23***	***1.25***	***2.02***	***1.25***	***1.37***
NH	***2.49***	***1.17***	***1.20***	***2.49***	***1.17***	***1.20***	***2.49***	***1.17***	***1.19***
ANN1	0.30	0.10	0.12	0.84	0.21	0.16	3.90	0.99	0.50
ANN2	0.51	0.12	0.12	0.86	0.21	0.28	4.14	1.02	0.65
RMS	2.34	1.10	1.13	2.00	1.00	1.05	3.28	1.14	1.03

Results

Before averaging, the *RMSC* (Root Mean Squares Corrected for the mean difference) between duplicate spectra have been calculated and the *RMSC*'s varied from 59 to 250 microlog indicating very repeatable scans and low noise values. After averaging duplicates, the *RMSC* were computed between the master and the salves. Figure 2 shows the average differences between master and slaves before

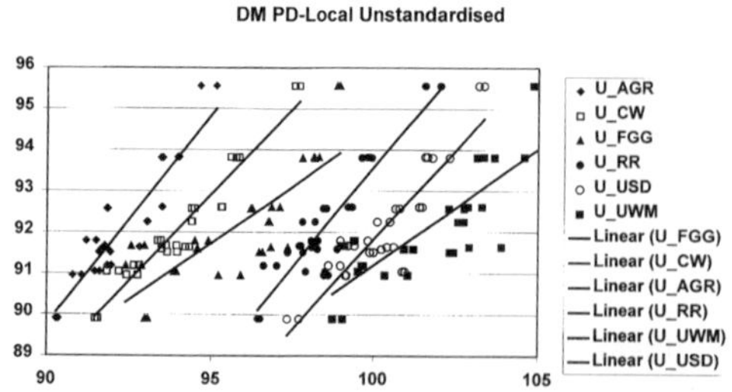

Figure 3(a). Scatter plot of the DM (dry matter) master vs unstandardised slave values for PD-LOCAL model.

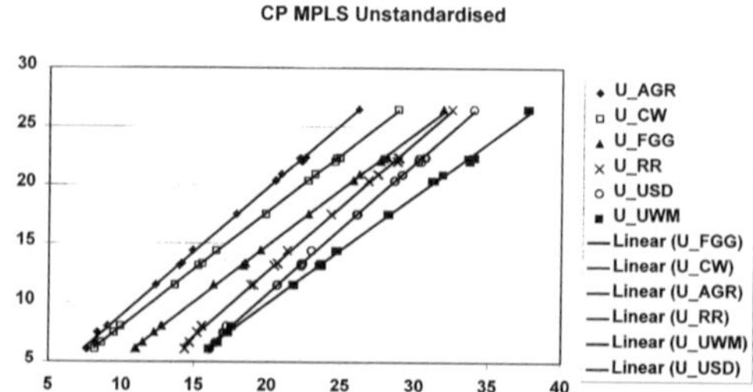

Figure 4(a). Scatter plot of the CP (Protein) master vs unstandardised slave values for MPLS model.

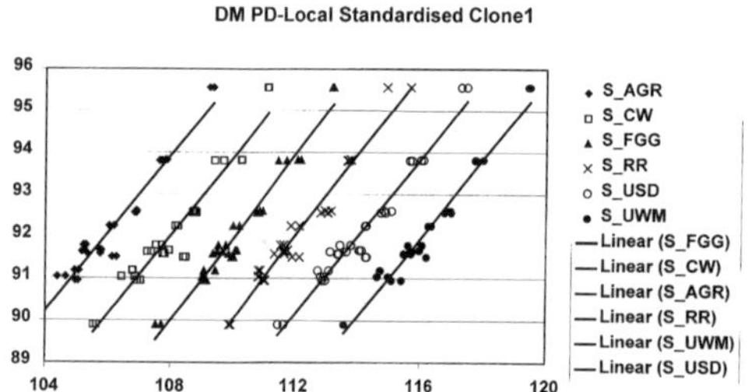

Figure 3(b). Scatter plot of the DM (dry matter) master *vs* Clone1 standardised slave values for PD-LOCAL model.

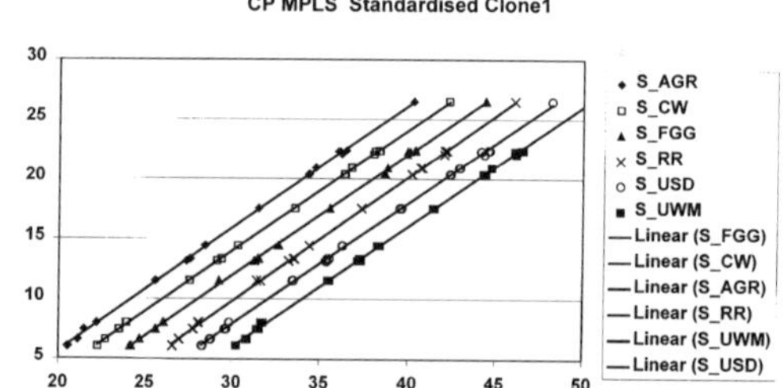

Figure 4(b). Scatter plot of the CP (Protein) master *vs* Clone1standardised slave values for MPLS model.

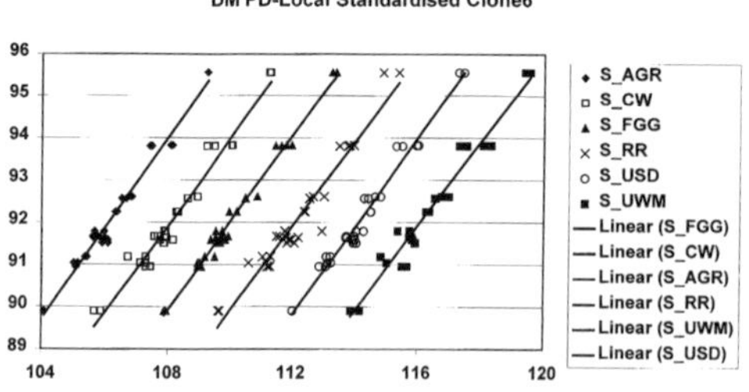

Figure 3(c). Scatter plot of the DM (dry matter) master *vs* Clone6 standardised slave values for PD-LOCAL model.

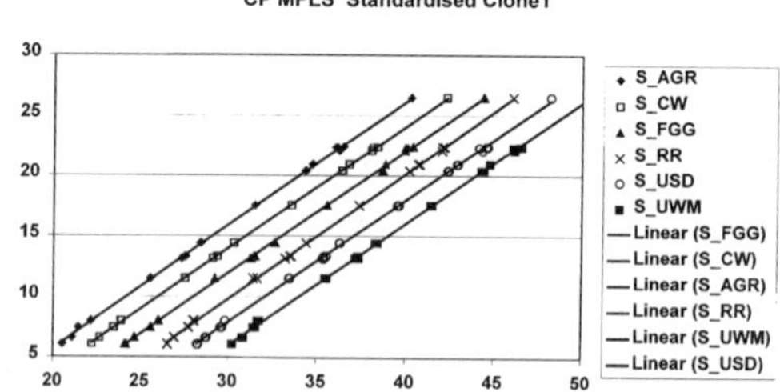

Figure 4(c). Scatter plot of the CP (Protein) master *vs* Clone6 standardised slave values for MPLS model.

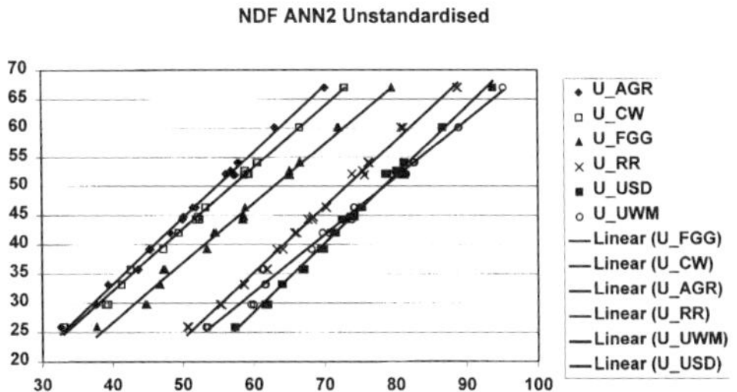

Figure 5(a). Scatter plot of the NDF master vs unstandardised slave values for ANN2 model.

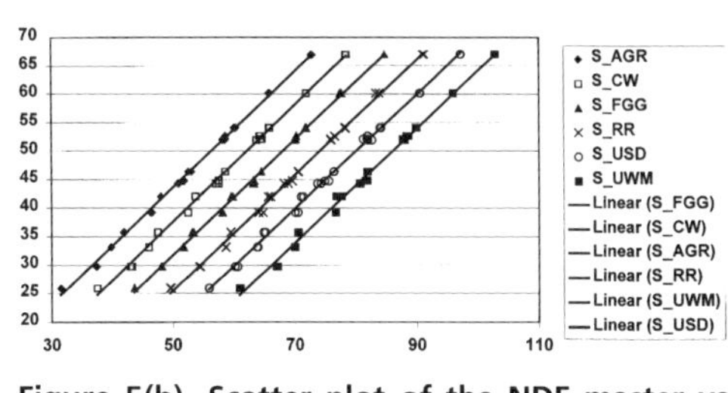

Figure 5(b). Scatter plot of the NDF master vs Clone1 standardised slave values for ANN2 model.

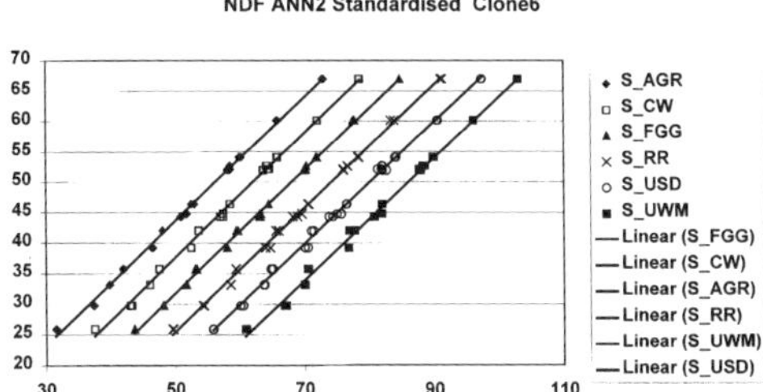

Figure 5(c). Scatter plot of the NDF master vs Clone6 standardised slave values for ANN2 model,

standardisation. The absence of a peak or the small peaks around 1930, indicate that the temperature effect has been minimised during the acquisition process. The *RMSC*s between the master and the slaves before and after standardisation are reported in Table 3. After cloning, *RMSC*s are highly reduced and lower than common *RMSC*s we can observe from a cup refilling effect.

Five prediction sets have been obtained from three calibration methods:[1] one set from PLS (ISI Modified PLS), two based on ISI-Local and two based on ANN (Foss-Tecator, SW). The design with five methods, three sets of spectra (unstandardised, Clone1 and Clone6), six instruments and three parameters leads to 270 comparisons. Table 4 reports only the *RMS* of *RMSEP* (master predicted values as *Y*) across the six instruments based on the duplicates of the 15 independent samples. Figures 3 to 5 illustrate the improvements in performance due to the standardisation. The predicted values have been shifted with constant values to be able to plot them. The *y* axis is always the predicted values from the master spectra for the corresponding models.

Conclusions

Calibration transfer without standardisation of the slave instruments gave unacceptable results. Significant biases and slopes were observed.

All calibration techniques gave satisfactory results after standardisation. The models used were based on very large data sets (> 10.000 samples) and they are considered as very robust. If the standardisation has a significant effect with these models, we can assume that the effect would be larger with calibrations obtained from smaller data sets.

Standardisation and even single standardisation corrected predictions for biases and slopes.

GH (global Mahalanobis distance) and *NH* (neighbourhood Mahalanobis distance) were reduced after standardisation and they were similar for all the instruments.

Clone6 gave better *RMSEP* than Clone1 for NDF. Otherwise for DM and CP Clone1 had similar results to Clone6.

References

1. P. Berzaghi, P.C. Flinn, P. Dardenne, M. Lagerholm, J.S. Shenk, M.O. Westerhaus and I.A. Cowe, in *Proceedings of the 11th International NIRS Conference.* Kyongju, Korea, 10–15th June, in press, (2001).
2. R. Biston and P. Dardenne, in *Proceedings of the 3th International NIRS Conference.* Brussels, Belgium, 24–29th June, pp. 655–662 (1990).
3. E. Bouveresse, D.L. Massart and P. Dardenne, *Anal. Chim Acta* **297,** 405 (1994).

The idea behind comparison analysis using restructured near infrared and constituent data (CARNAC)

Anthony M.C. Davies

Norwich Near Infrared Consultancy, 75 Intwood Road, Cringleford, Norwich, NR4 6AA, UK

Introduction

The idea for "Comparison Analysis using Restructured Near infrared And Constituent data (CARNAC)" was formulated in 1983. However it was only through the active participation of Professor Fred McClure that it could be demonstrated in 1986 at the IDRC at Chambersburg, USA and the FT Conference in Vienna.[1]

The idea of CARNAC is that prediction of quantitative information can be derived from databases containing near infrared (NIR) and analytical data rather than through some form of regression analysis derived from that database. The realisation of CARNAC involves the combination of several ideas: database compression, database modification and similarity analysis. Through these steps it is assumed that a few samples, which are very similar to an unknown sample, can be found in a modified database. Because the database has been modified to emphasise the spectral features of the analyte it is assumed that the analyte value for the unknown sample can be estimated from the analytical values of the selected samples.

Methods

Data compression

The original work on CARNAC[1] utilises compression by Fourier transformation.[2,3] In the 1980s it was assumed that data compression would be essential to obtain an analytical result in real time but because of the vast improvements in the speed of personal computers there is no absolute need for data compression in 2001. The use of principal components (PCs) for the data compression step was demonstrated at IDRC-2000.

Database modification

Unless the database is modified by some method that emphasises the analyte, selections of samples would be identical for the same unknown sample for different analytes and all selections would be dominated by the major analyte. The method used for most of our work used the coefficients generated by running an stepwise multiple linear regression (SMLR) program for the analyte of interest. The database and the unknown sample were multiplied by the regression coefficients as a means of emphasising the analyte contribution to the spectrum. Other methods for modification of the database that have been tested on CARNAC include multiplication of the database by the spectrum of the analyte and the selection of principal components that were correlated to the analyte.

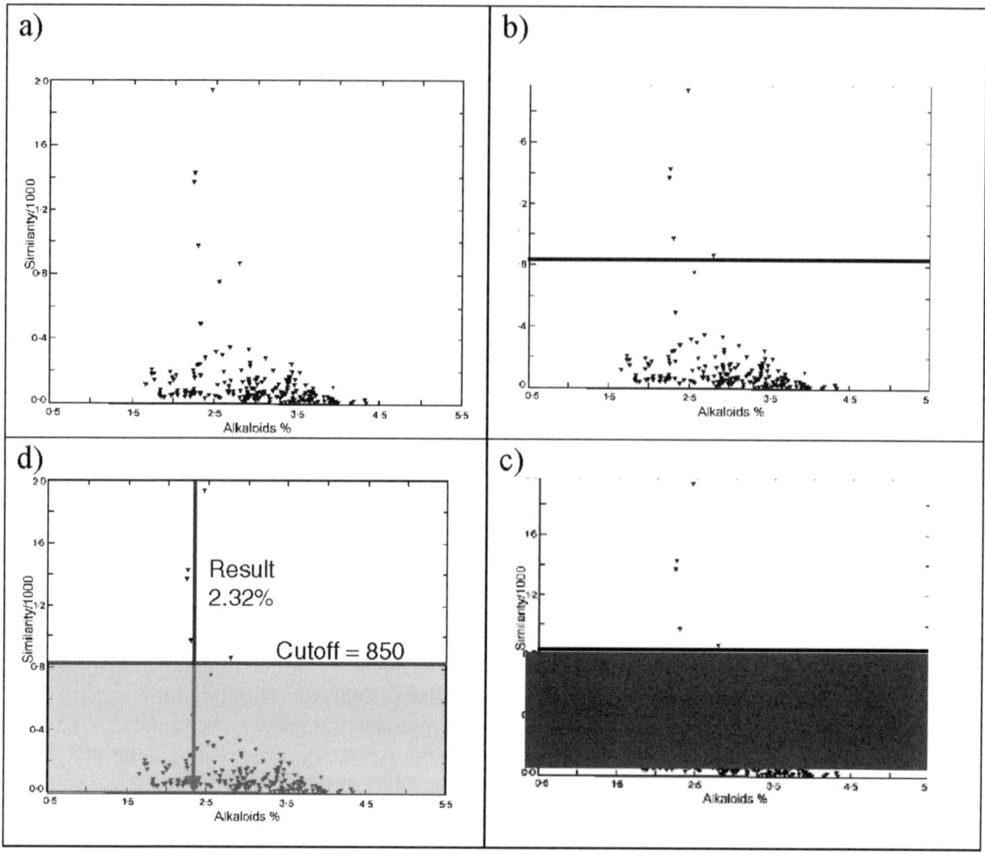

Figure 1. The main CARNAC process. (a) The SI values can be plotted against the value of the analyte for that member of the database. (b) Most samples are not similar to the unknown and they can be eliminated by drawing a horizontal line and (c) ignoring samples below it. (d) The result is calculated from a weighted average of the analyte values of the remaining samples.

Prediction of unknown samples

The modified unknown sample was compared to each member of the modified database by calculating a "similarity index"[1] (SI). The SI is calculated from the correlation coefficient (r) between the unknown sample and a member of the database The SI is calculated as:

$$SI = 1 / (1-r^2)$$

The few samples with very high similarity were then selected by using a minimum cut-off value and eliminating samples identified as outliers. The predicted analyte value for the unknown sample was calculated as the weighted average of the analyte values of the selected samples. This is shown diagrammatically in Figure 1.

The logic steps are shown in Figure 2 and 3.

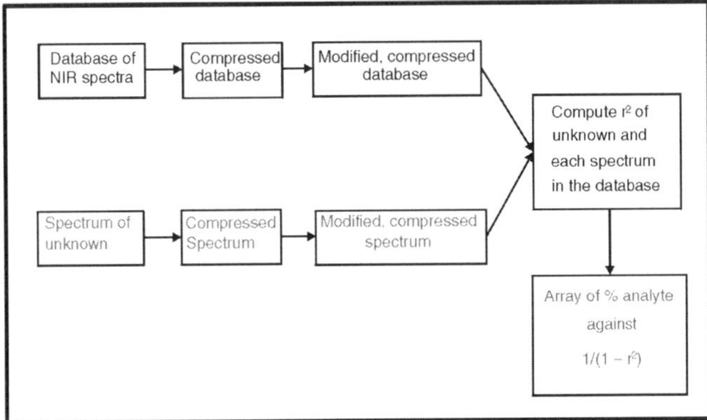

Figure 2. Logical steps for the first stage of CARNAC.

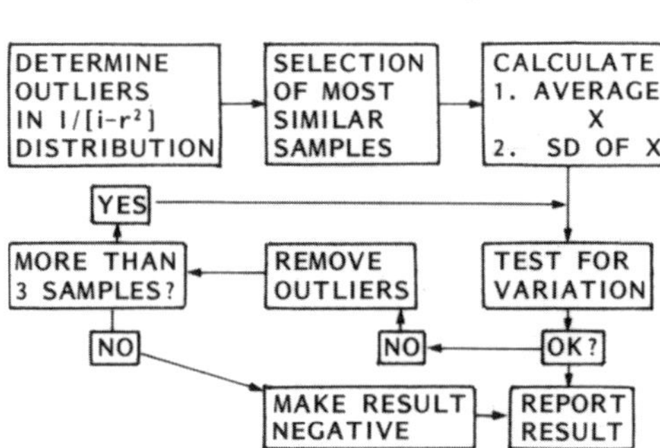

Figure 3. Logical steps for the second stage of CARNAC.

Results

Some results obtain by the CARNAC procedure are given in Figures 4 and 5.

Discussion

The most important of these results are those for caffeine in coffee. Caffeine is a minor component (1–2.5%); the method depends on the modification step to achieve these results.

When CARNAC was developed 15 years ago it was not really a practical proposition because of variation between instruments and the lack of interest in data compression by FT. My interest in CARNAC was rejuvenated by the success of a different, but related, technique developed by Shenk and Westerhaus called "LOCAL". In LOCAL a

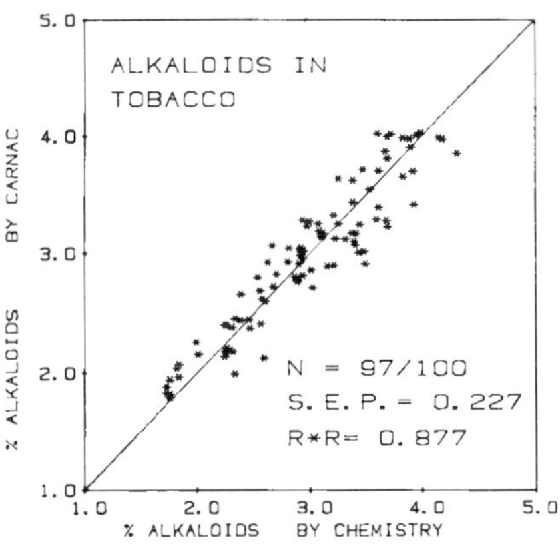

Figure 4. CARNAC results for alkaloids in tobacco.

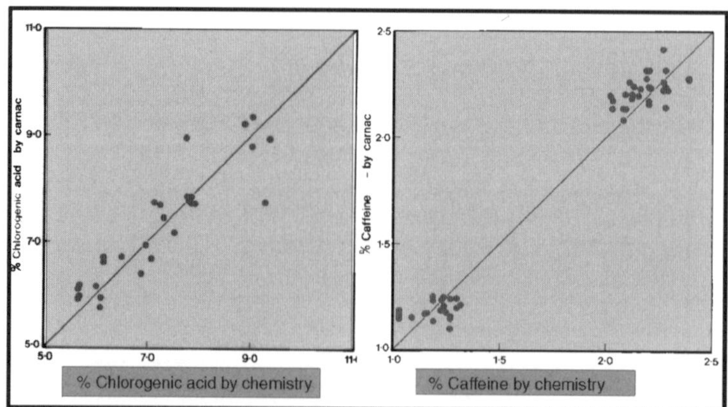

Figure 5. CARNAC results for chlorogenic acid and caffeine in coffee.

subset of samples which are similar to the spectrum of the unknown sample is selected from a compressed database and a PLS model is then constructed for these samples and used to predict the unknown sample. Thus, a new PLS calibration is determined for each unknown sample. In application, LOCAL has been successfully applied using very large databases (< 6000 in one example) and with several hundred samples being selected to form the PLS calibration set.[4–7]

Tom Fearn is collaborating with me to produce a modernisation of the CARNAC idea. A recent issue of *NIR news* contains a "Chemometric Space" article[8] by Tom, which discusses the background of these methods.

Acknowledgement

I would like to emphasise that the development of CARNAC would not have been possible without the enthusiastic collaboration of Professor Fred McClure.

References

1. A.M.C. Davies, H.V. Britcher, J.G. Franklin, S.M. Ring, A. Grant and W.F. McClure, *Mikrochim. Acta (Wien)* **I,** 61 (1988).
2. W.F. McClure, A. Hamid, F.G. Giesbrecht and W.W. Weeks, *Appl. Spectrosc.* **38,** 322 (1984).
3. W.F. McClure and A.M.C. Davies in *Analytical Applications of Spectroscopy*, Ed by C.S. Creaser and A.M.C. Davies. Royal Society of Chemistry, London, UK, p. 414 (1988).
4. J.S. Shenk and M.O. Westerhaus, *Crop Sci.* **31,** 469 (1991).
5. J.S. Shenk, P. Berzaghi and M.O. Westerhaus, *J. Near Infrared Spectrosc.* **5,** 223 (1997).
6. F.E. Barton, II, J.S. Shenk, M.O. Westerhaus and D.B. Funk, *J. Near Infrared Spectrosc.* **8,** 201 (2000).
7. P. Dardenne, G. Sinnaeve and V. Baeten, *J. Near Infrared Spectrosc.* **8,** 229 (2000).
8. T. Fearn, *NIR news* **12(3),** 10 (2001).

Near infrared spectroscopy—a tool for the evaluation of milling procedures

Szilveszter Gergely, Lidia Handzel, Andrea Zoltán and András Salgó

Department of Biochemistry and Food Technology, Budapest University of Technology and Economics, Müegyetem rkp. 3, Budapest, H-1111, Hungary

Introduction

Small-scale test methods that offer cheaper and faster results are increasing in the analysis of cereal quality. Recently, a number of different small-scale tests have been developed for the determination of physico-chemical, functional and rheological properties of wheat or wheat dough. These use miniaturised instruments with sophisticated sample preparation/handling methods and mechanical systems (RVA, 2 g mixograph, micro-Z-arm mixer, small-scale noodle maker, micro-baking method etc.). If test methods can be successfully scaled-down, then the sample size can be reduced significantly. These small-scale methods can be used either as basic research tools or in support of technology, and can also be essential in the early selection for quality traits in breeding programmes.[1] Micro methods can be very useful in the analysis of the effects of genetically modified (GM) materials or additives, as well as in the investigation of model systems. Recently, a micro-scale lab mill was developed for small-scale sample preparation providing flour and semolina samples from small amounts of grain (5–10 g) in a reproducible and reliable way.[2,3]

The aim of this study was to compare the milling action of a macro (QC-109) mill with the action of a micro-scale lab mill (FQC-2000) produced by Metefém Co. Ltd, Hungary.[4] The milling characteristics of the instruments were analysed both by near infrared (NIR) spectroscopy and by chemical and physical methods.

Materials and methods

Forty-four samples of a single variety of Hungarian hard, red, winter wheat were conditioned to 15% moisture content for 24 hours. The conditioned samples were milled, in parallel, through a macro lab mill (Metefém QC 109, 200 g sample) and a micro lab mill (Metefém FQC-2000, 10 g sample). The grist from the macro mill was separated by sieving into three fractions based on particle size (a: > 315 µm, b: 315–215 µm, c: < 215 µm), while the grist from the micro mill was separated into the following fractions (a: > 500 µm, b: 500–315 µm, c: 315–200 µm, d: < 200 µm). The mass distribution and ash content (modified ICC 104/1 method) of all fractions were determined. NIR spectra were obtained using an NIRSystems 6500 (NIRSystems Inc., Silver Spring, MD, USA), fitted with a sample reduction accessory. Reflectance spectra covering the 1100–2500 nm wavelength region were collected using NSAS 3.30 (NIRSystems Inc., Silver Spring, MD, USA) and were processed using the PQS32 1.18 (Metrika R&D Co., Hungary) software package.

Results and discussion

Mass distribution histograms, measured with macro and micro methods, showed very different characteristics. The macro method [Figure1(a)] produced a high yield (60–70%) of flour (fraction c) with 10–15% semolina (fraction b) and approximately 25–35% bran (fraction a). The micro mill [Fig-

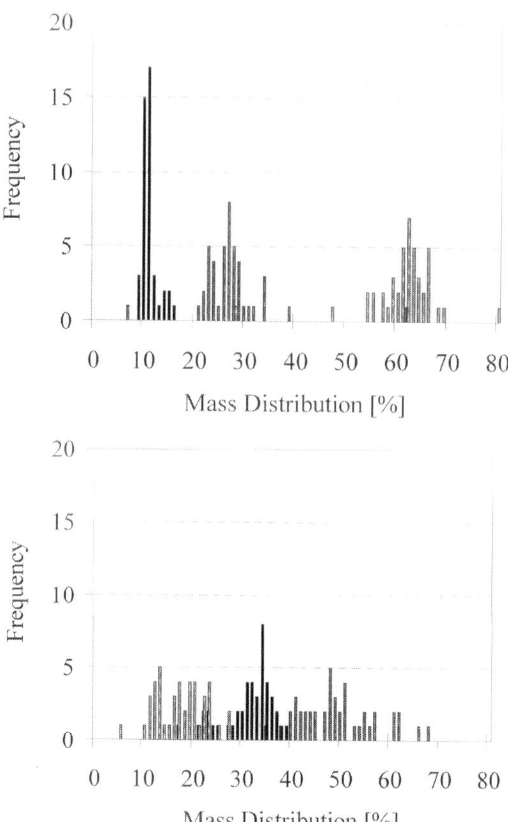

Figure 1. (a) Mass distribution of milled fractions using the macro mill and (b) mass distribution of milled fractions using the micro mill (dark grey = bran, black = semolina, light grey = flour).

ure 1(b)] showed a broader mass distribution in the fractions. The primary flour yield was only 15–25%, while the amount of semolina was between 30–40% and the amount of bran (which contains parts of bigger particles of semolina) was approximately 40–60%.

These results indicated differences in milling action. Seeds were crushed in the macro mill with a smaller milling angle (larger roll diameter) and the relative depth of grooves was bigger than with the micro mill. The shearing and crushing steps were more intensive. With the micro mill, the milling angle was relatively high, the milling surfaces were smaller and the whole milling process was relatively fast. As a consequence, the primary flour yield was lower. The efficiency of the milling process was tracked by the detection of the ash content of each fraction.

The distribution of ash content (Figure 2) indicated a high difference in ash for fractions produced by the macro and micro mills. The macro mill [Figure 2(a)] "separated" very clearly the bran fraction (ash content 3.2–5.5%) and produced a clear distinction in ash content between the semolina (0.8–1.5%) and flour (0.5–0.8%) fractions. The ash content of flour and semolina fractions produced by the micro mill [Figure 2(b)] were between 0.4–0.8% and their distributions were very sharp. The bran fraction showed a wide distribution in ash content but the absolute values (1.8–3.8%) are much lower compared with bran produced by the macro mill.

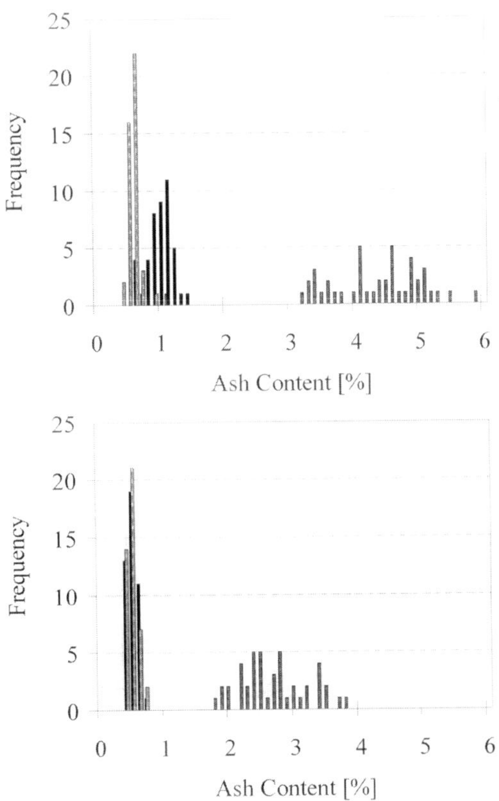

Figure 2. (a) Ash content distribution of milled fractions using the macro mill and (b) ash content distribution of milled fractions using the micro mill (dark grey = bran, black = semolina, light grey = flour).

These results clearly indicated that with the micro milling procedure the bran was not separated perfectly from the endosperm; as a result its ash content was significantly lower when compared with the macro method. The differences in the distribution of ash content were due to differences in milling action, indicating that the micro mill was significantly different in terms of efficiency of separation (rapid crushing, lower yield in flour) when compared with the macro procedure. In spite of this "crude" method, the micro mill produced good quality flour and semolina from a 5 g sample. The separation of semolina from the bran fraction can be improved and the efficiency of yield can also be increased.

The separation and chemical composition of fractions were detected by near infrared spectroscopy. NIR spectra were influenced by changes in chemical composition as well as the modification of particle size in different fractions.

Raw reflectance spectra of fractions obtained using the macro and micro mills were collected (data not shown). In the case of the micro mill, the spectra of semolina and bran fractions showed a higher variation, indicating the uncertain distribution (higher variance) of particle size.

In order to avoid the particle size effect, second derivative spectra were calculated (data not shown). In both cases (spectra from the macro and micro mills), two variable regions of the spectra (wavelengths between 1740–1770 nm and 2290–2340 nm) were observed. In the 1740–1770 nm region, the bran fractions showed characteristic twin absorption bands relating to the high lipid content (approximately 4%) of wheat bran. In the wavelength region between 2290–2340 nm, two compositional changes were observed. At 2290 nm, the starch content of fractions can be followed (high in flour and semolina, low in bran). At 2340 nm, the cellulose and hemicellulose components can be identified (bran fractions).

The two most variable spectral regions (around 1700 and 2300 nm) were used by the polar qualification system (PQS) software[5] for making discriminant models for the quality of fractions. This method calculates "quality" points of materials and their distributions. The results of PQS analysis of fractions for macro and micro milling procedure were shown in Figures 3(a) and 3(b). The macro mill method [Figure 3(a)] provided three significantly separated fractions, indicating the compositional differences between fractions. In the case of the micro mill [Figure 3(b)], only the bran fraction was clearly separated from the other three fractions. These results matched the observations obtained for mass and ash distribution of fractions.

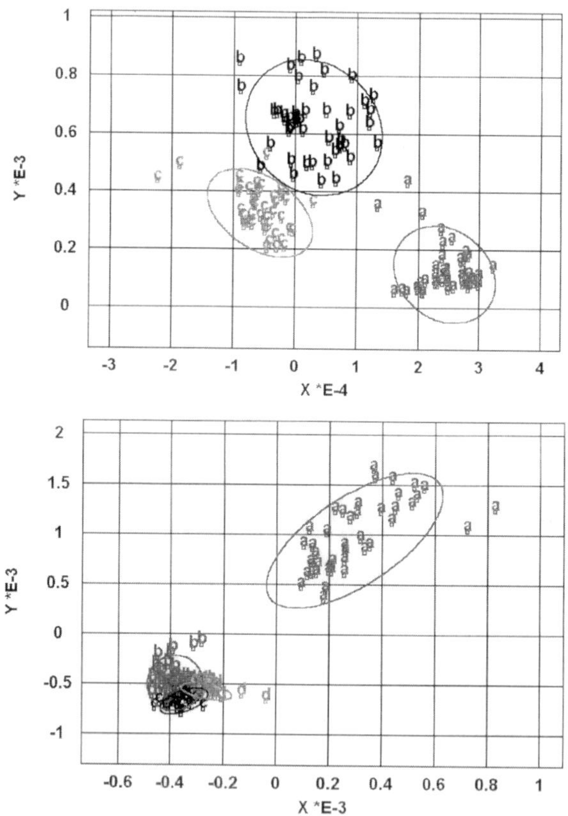

Figure 3. (a) Separation of fractions in "quality space" calculated using the PQS method (macro mill) and (b) separation of fractions in "quality space" calculated using the PQS method (micro mill) (dark grey = bran, black = semolina, light grey = flour).

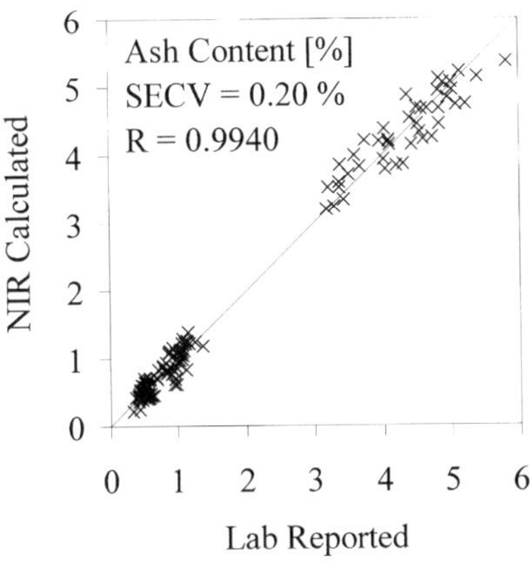

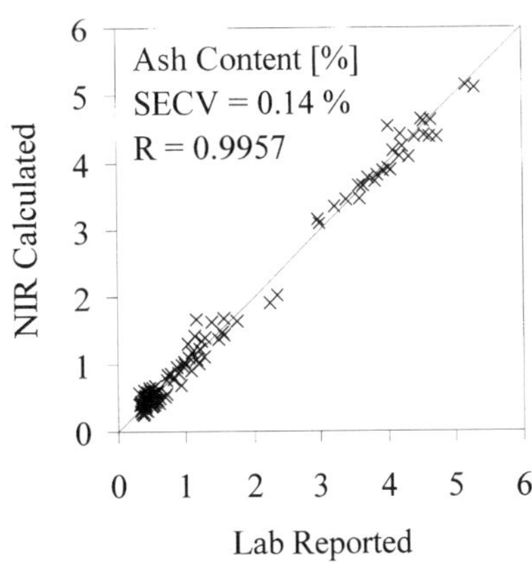

Figure 4. (a) Calibration equations developed for determination of ash content in fractions (macro mill) and (b) calibration equations developed for determination of ash content in fractions (micro mill).

Calibration equations were developed for the detection of ash content of macro and micro mill fractions from NIR spectra (Figure 4). The standard error of cross-validation (*SECV*) of equations developed for micro mill fractions showed a 40% better accuracy ($SECV = 0.13\%$, $R = 0.9957$) when compared with the model for macro mill fractions. Bran samples were separated clearly in both models.

Conclusions

The micro mill, with its small sample size (5–10 g), can replace lab grinders requiring several hundred grams of sample. Compared with the macro method, the flour and semolina yields are significantly lower but the quality is good. NIR is a tool with sufficient sensitivity in the detection and evaluation of milling action for both qualitative and quantitative measurements.

Acknowledgements

This research work was supported by the National Committee for Technological Development of Hungary (project number: 96-97-68-1354), and by the National Foundation of Science and Research Hungary (project numbers: T 031902 and A 143).

References

1. P.W. Gras and L. O'Brien, *Cereal Chem.* **69,** 254 (1992).
2. J. Varga, D. Fodor, J. Nánási, F. Békés, M. Southan, P. Gras, C. Rath, A. Salgó and S. Tömösközi, in *Proceedings of Gluten 2000 Workshop*, Ed by P. Shewry and A. Tathum. Bristol, UK in preparation (2000).
3. A. Salgó, J. Varga, S. Tömösközi, P. Gras, C. Rath, F. Békés, D. Fodor and J. Nánási, in *Proceedings of 11th Cereal and Bread Congress*. Surfers Paradise, Australia. in preparation (2000).
4. D. Fodor, J. Varga, S. Tömösközi, A. Salgó, J. Nánási and Gy. Veres, *Hungarian Patent Application* (1999).
5. K.J. Kaffka and L.S. Gyarmati, *J. Near Infrared Spectrosc.* **6,** A191 (1998).

Standardisation of near infrared spectra across miniature photodiode array-based spectrometers in the near infrared assessment of citrus soluble solids content

C.V. Greensill[a*] and K.B. Walsh[b]

[a]*Faculty of Engineering and Physical Systems, Central Queensland University, Rockhampton, Queensland 4702, Australia. E-mail: c.greensill@cqu.edu*

[b]*Plant Sciences Group, Primary Industries Research Centre, Central Queensland University, Rockhampton, Queensland 4702, Australia. E-mail: k.walsh@cqu.edu.au*

Introduction

Low-cost silicon photodiode array- (PDA) based near infrared (NIR) spectrometers have found application in the sorting of intact fruit by sugar content. However, while calibration transfer has been reported for relatively dry samples (< 10% water), little work has been published concerning PDA-based instruments using high water content samples (> 70%). PDA-based NIR spectrometers can vary in wavelength calibration and photodetector efficiency. Pixel related photo-detector output can be interpolated to yield a common wavelength scale across instruments. Correction of differences in photometric response between instruments is more difficult, an effect of differences in the signal-to-noise ratio between units associated with output at a given wavelength. Differences in illumination geometries associated with sample orientation relative to light source and detector also contribute to differences in the recorded absorbance spectra of a given sample from two instruments. To accommodate these differences, the absorbance spectra obtained on the slave instrument can be modified to appear as if originating from another instrument. The standard sample(s) used in such an exercise must be similar to the samples on which the predictions are to be used.[1]

A range of chemometric techniques have been applied to calibration transfer for NIR spectroscopy, although no calibration transfer methodology is recommended to suit all applications. We have previously briefly reviewed a number of these techniques, and applied them to the transfer of calibrations between Zeiss MMS1 PDA spectrometers used in the application of non-invasive assessment of SSC (soluble solids content) of intact melon fruit.[2] The techniques were assessed in terms of root mean squared error of prediction (*RMSEP*) (using Fearn's significance testing). Greensill and Walsh[2] concluded that a modified WT method performed significantly better than all other standardisation methods and on a par with model updating.

The following methodologies, involving collection of spectra from a set of 'standards' on both master and slave unit were applied in the previous study:[2] (1) slope and bias correction (SBC), (2) direct standardisation (DS),[3] (3) piecewise direct standardisation (PDS),[4] (4) double window PDS[5] (DWPDS), (5) orthogonal signal correction (OSC),[6,7] (6) wavelet transform-based standardisation technique (WT)[8] and (7) a photometric response correction and wavelength interpolative method and (8) a simple method involving wavelength selection. For cases where spectra of the same samples can not be collected on both mater and slave instruments, two methodologies were trialed by Greensill and Walsh:[2] (1) finite impulse response (FIR) and (2) model updating, using the Kennard and Stone[9] algorithm for selection of melon fruit spectra for model updating.

In the current study we trial the same techniques for the application of calibration transfer between PDA spectrometers used in the application of non-invasive assessment of SSC of intact mandarin fruit. In this application a *RMSEP* of < 1% SSC is required.

Experimental method

Standardisation

The performance of standardised calibrations, generated against SSC of mandarin (n = 100, 'Imperial' cultivar from Munduberra, Queensland) fruit tissue, was assessed. Spectra were collected using two MMS1 spectrometers with consecutive serial numbers from two production batches [designated 729, 730 (batch #1), and 845, 846 (batch #2)] giving four spectrometers in total. All samples were allowed to equilibrate to room temperature (27°C) overnight before spectral measurements were made. Wet chemistry was performed on the juice extracted from mandarin halves from each fruit using a commercial citrus juicer to extract juice and a Bellingham–Stanley RMF320 refractometer (~ 0.1% SSC accuracy) to determine associated SSC values. The mean and standard deviation of the SSC value was 9.80 and 0.45, respectively.

Single scans of 30 ms integration time were taken for each spectrum. A maximum count level > 10000 was maintained to minimise any variation in performance due to changing signal-to-noise ratio (SNR) of each system.[10] Spectral absorbance data (using a spectral window 730 to 930 nm) were pre-treated by mean centring. Partial least squares (PLS) multivariate linear regression calibrations were generated against mesocarp SSC using Matlab v5.3 (The Mathworks, Inc., USA) and PLS Toolbox, v. 2.0 (Eigenvector Research, Inc., USA). Calibration performance was recorded for the master instruments in terms of root mean square error of calibration (*RMSEC*), root mean square error of cross-validation [*RMSECV* using leave-one out (LOO) cross-validation segment selection] and standard deviation (STDev) of SSC. Calibration performance in terms of prediction on standardised slave spectra was recorder in terms of *RMSEP*.

The primary assessment for performance of calibrations was made on the significance of the variation in the *RMSEP* following the approach of Fearn[11] ($\alpha = 0.05$ and assuming bias negligible) (see also Snedecor and Cochran[12]). For each comparison of two calibrations, the R^2 of the correlation between residuals (predicted–actual SSC) and the 95% confidence limits on *RMSEP* are reported (Table 1). Since this assessment is always made in pairs, the standardisation technique achieving the best result in the respective data set was assessed against its two nearest neighbours (closest *RMSEP*s) (Table 1).

Algorithms to test each standardisation technique were implemented using Matlab v. 5.3 scripting (The Mathworks, Inc., USA) and the parameters relevant to each technique were incremented to achieve optimisation. Scripts assessing DS, PDS, DWPDS and FIR standardisation techniques used algorithms available in PLS_Toolbox software (Eigenvector Research, Inc., USA) for these standardisation assessments. A new OSC algorithm[7] was used for the OSC technique assessment. Assessment of the wavelet transform technique (WT) was based on a method proposed by Walczak,[8] but differed by the use of DS on the wavelet coefficients instead of directly univariately and linearly regressing one on the other. Wavelet coefficients from the first level decomposition were used in the DS association.

In all cases, except FIR which did not require this parameter, the number of samples used in the standardisation was varied between 3 and 25 to allow an optimum number to be determined. These were selected using the Kennard–Stone algorithm available in the PLS_Toolbox v. 2.0. Window size

Table 1. Significance testing of the results of a citrus population. A comparison of the technique with the lowset *RMSEP* against two nearest neighbours using Fearn's criteria to determine upper and lower significance limits of the *RMSEP* value (refer to Table 2).

Data Set	Method	RMSEP	R^2	Significant
729–730	WT	0.23		
	DS	0.22	0.73	N
	PDS	0.41	0.20	Y
	WT (50)	0.21		
	MU	0.22	0.59	N
729–845	WT	0.26		
	DS	0.2811	0.60	N
	PDS	0.48	0.28	Y
	WT (50)	0.21		
	MU	0.21	0.66	N
729–846	WT	0.28		
	DS	0.37	0.60	
	PDS	0.55	0.27	Y
	WT (50)	0.31		
	MU	0.30	0.75	N
730–845	WT	0.22		
	DS	0.27	0.61	Y
	PDS	0.60	0.1332	Y
	WT (50)	0.21		
	MU	0.24	0.36	N
730–846	WT	0.27		
	DS	0.43	0.36	Y
	PDS	0.41	0.35	Y
	WT (50)	0.33		
	MU	0.29	0.53	N
845–846	WT	0.30		
	DS	0.36	0.47	Y
	PDS	0.38	0.43	Y
	WT (50)	0.33		
	MU	0.26	0.73	Y

for PDS and DWPDS was varied between 3 and 21 (increments of 2). The window size for FIR was ranged from 3 to 41 in increments of 4. The number of OSC components was varied from 1 to 5.

Wavelength range varies slightly among instruments due to small variations in the optical alignment of components on the central glass block. Interpolation to a common wavelength scale was achieved using a cubic spline interpolation technique.

Although the photometric response of these instruments is similar, due to this company's rigorous photodiode selection criteria, differences between instruments with long periods between manufacture dates was observed. The photometric response (mean absorbance spectrum of standardistion sample set) of slave and master was ratioed. A comparison of this transfer technique is made against other proposed transfer techniques.

A technique generally used for updating calibration models with new spectra considered to encompass new variables (for example, new cultivars or growing districts) was used to adapt to new instrumental variables. To assess the capabilities of model updating (MU), increasing numbers of Kennard–Stone selected samples were added to the master data and new models generated. The new model was tested on the original slave data set.

All data sets were subjected to the same data pretreatments (mean centring) and predictive modelling (PLS) with the relevant parameters for both predictive model generation (principal components) and standardisation method implementation (number of samples and/or window size) optimised for each. Calibrations generated used equivalent data preteatment methods which were not optimised for any individual set. Therefore, *RMSECV* and *RMSEP* should not be assessed in an individual context. A 'working' calibration would also require attention to the optimisation of data pretreatment techniques.

Results and discussion

The photometric response of the four spectrometers differed in absolute terms (maximum count) and spectrally (wavelength sensitivity), as illustrated by spectra of a white reference (Figure 1). Spectrometers '729' and '730' were purchased together and since the respective serial numbers (also used as the spectrometer identifiers) are sequential it is assumed that they originate from the same production batch. Spectrometers '845' and '846' were purchased at a later date. While output of all spectrometers varied, the most obvious variation occurred between the photometric response of the 700 series spectrometers, relative to the 800 series (30% higher). The obvious output differences between the two

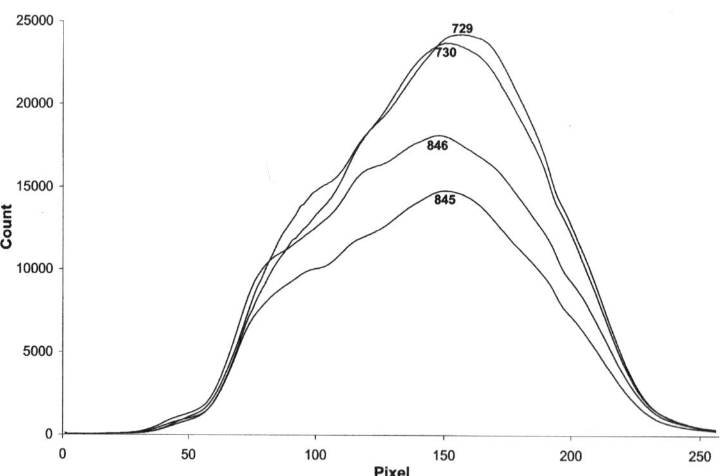

Figure 1. Spectra of the a white reference (teflon tile) acquired with each of four spectrometers used. All operating parameters for the spectrometers (for examples integration time and number of scans) remained constant for the duration of this study. The abscissa is graduated in pixels to highlight the difference in wavelength/pixel allocation between spectrometers as well as difference in photometric response.

Table 2. Performance of calibration transfer process reported in terms *RMSEP* for the prediction of melon SSC using spectra collected on a slave spectrometer and a calibration generated on spectra of the same fruit, collected on a master instrument. Results in bold highlight the standardisation technique with the lowest *RMSEP*. Spectra from the slave (second listed) instrument were transformed to appear as though originating from the master (first listed) instrument spectra. 729, 730, 845, 846 are spectrometer identifiers. The population statistics were = 9.80, *n* = 100 and σ = 0.45.

Spectrometers	729_730	729_845	729_846	730_845	730_846	845_846
RMSEC	0.24	—	—	0.19	—	0.23
RMSECV	0.28	—	—	0.25	—	0.286
Unstandardised	0.50	7.03	8.10	2.51	2.55	0.71
WS	0.65	6.49	7.84	2.68	3.47	0.96
Int + Mod.	4.91	15.11	19.77	11.56	13.21	7.66
DS	0.22	0.28	0.370	0.27	0.43	0.36
PDS	0.41	0.48	0.55	0.60	0.41	0.38
DWPDS	0.40	1.45	0.53	0.46	0.49	0.48
OSC	0.43	0.44	0.47	0.46	0.52	0.55
FIR	0.45	0.45	0.45	0.45	0.45	0.45
WT	0.239	**0.26**	**0.28**	**0.22**	**0.27**	**0.28**
Slope and Bias	0.36	0.80	1.11	3.47	9.913	0.63
WDS (50)	**0.21**	**0.21**	0.31	**0.21**	0.33	0.33
MU(50)	0.22	0.21	**0.30**	0.24	**0.29**	**0.26**

pairs of spectrometers is consistent with the suggestion that these pairs came from different production runs, although the manufacturer (Carl Zeiss GmbH) selects photodetectors (Hamamatsu Q4874) on uniformity to minimise this type of variation. These differences were expected to impact heavily on transferability of calibrations.

Calibrations can be developed using either pixel number or wavelength as the dependent variable. If the pixel number is used, then the standardisation technique must be capable of any misalignment of this variable between spectrometers. Alignment of spectral data from all instruments to a common wavelength scale should overcome this problem, dependent on original wavelength accuracy of the individual instrument. Only the difference in photometric response would remain.

The performance of seven standardisation techniques (SBC, DS, PDS, DWPDS, OSC, FIR, WT), and a wavelength interpolation method with photometric correction were compared using the respective *RMSEP*s (Tables 2). Since WT proved the most successful method in five out of six cases and second most successful in the remaining case, it was compared separately to MU (data presented in the same table) using separately constructed data sets. The sets to be transferred were divided in half, one half used in increasing numbers in the updating process and tested on the unused half (validation set). Of the MU and WT comparison, both performed equally (three of six cases each) , although only one (MU) each proved to be significant, using Fearn's technique. A simple ranking procedure indicated that the relative performance of the techniques to be (best to worst): WT, MU, DS, DWPDS, PDS, FIR, OSC, SBC and Int and Mod.

Of the established standardisation methods, direct standardisation of the wavelet coefficients of the first level decomposition (WT) was demonstrated to be the most efficient for the standardisation of a calibration for the non-invasive assessment of SSC in fresh mandarin fruit samples when used to standardise between MMS1 spectrometers. However, predictive model updating, incorporating 'Kennard–Stone' selected representative spectra of the slave spectrometer, has also been shown to be capable of achieving equally good or better results (in terms of lowest *RMSEP*) with significantly better results in one case. This conclusion is in agreement with our earlier report for calibration transfer for the same instrument for melon spectra.[2]

Model updating has an added advantage over most standardisation techniques of not requiring the measurement of standardisation samples on both spectrometers and allowing the predictive model to evolve to one containing only slave spectra over time. The disadvantage of this method is that a separate model is required for each instrument.

References

1. E. Bouveresse, D.L. Massart and P. Dardenne, *Anal. Chim. Acta* **297,** 405 (1994).
2. C.V. Greensill, P.J. Wolfs, C.H. Spiegelman and K.B. Walsh, *Appl. Spectrosc.* **55(5),** 647 (2001).
3. Y. Wang, D.J. Veltkamp and B.R. Kowalski, *Anal. Chem.* **63,** 2750 (1991).
4. Y. Wang, M.J. Lysaght and B.R. Kowalski, *Anal. Chem.* **64,** 562 (1992).
5. B.M. Wise and N.B. Gallagher, *PLS_Toolbox Version 2.0 for use with MATLAB.* Eigenvector Research Inc., Manson, WA, USA (1998).
6. S. Wold, H. Antii, F. Lindgren and J. Öhman, *Chemom. Intell. Lab. Syst.* **44,** 175 (1998).
7. T. Fearn, *Chemom. Intell. Lab. Syst.* **50,** 47 (2000).
8. B. Walczak, E. Bouveresse and D.L. Massart, *Chemom. Intell. Lab. Syst.* **36,** 41 (1997).
9. R.W. Kennard and L.A. Stone, *Technometrics* **11(1),** 137 (1969).
10. C.V. Greensill and K.B. Walsh, *Appl. Spectrosc.* **54(3),** 426 (2000).
11 T. Fearn, *NIR News* **7(5),** 5 (1996).
12. G.W. Snedecor and W.G. Cochran, *Statistical Methods*, 6th Edn. Iowa State University Press, Iowa, USA (1967).

A new method for mapping the visible-near infrared light levels in fruit

D.G. Fraser,[a] R.B. Jordan,[b] R. Künnemeyer[a] and V.A. McGlone[b]

[a]*University of Waikato, Private Bag 3105, Hamilton, New Zealand*

[b]*Technology Development Group, HortResearch, Private Bag 3123, Hamilton, New Zealand*

Introduction

Little is known about the path that light takes inside an intact fruit and the attenuation it experiences through the different fruit regions. We have developed a probe which is able to directly measure the light distributions in fruit with minimal effect on the distribution being measured. Monte Carlo simulations were able to verify the experimental light measurements made by the probe. Knowing the distribution may enable the selection of more effective near infrared (NIR) spectroscopy modes of measurement (reflectance, interactance or transmission). Indeed the ability of optical techniques to detect internal defects or to estimate the blocking effect of a fruit's stone could be assessed directly if the light distribution profile is known.

We have developed a probe system to explore the light intensity at different points in a fruit with minimal effect from the measurement process. The probe consists of a 400 μm diameter glass fibre encased in a stainless steel tube. This assembly is mounted on a translation stage to enable accurate positioning of the probe tip inside a firmly held fruit. The fruit is illuminated with an 808 nm laser or a white light source to measure the monochromatic or spectral distributions respectively. The light collected by the fibre is relayed to a spectrometer for subsequent analysis. Measuring the transmitted light directly in this fashion is less invasive than cutting away sections of the fruit, which can alter the light distribution.[1,2]

Monte Carlo simulations have been generated using varying absorption, scattering and anisotropy parameters for the fruit tissue and different reflective properties to allow for the boundary conditions of the skin. In these simulations photons are traced through the model using the tissue parameters to randomly determine the step size, path length and scattering angles. Photons are traced until they are fully absorbed or exit the model via the partially reflective skin.

While illuminating a 22.5 mm diameter spot on one shoulder of a Royal Gala apple with an intense tungsten lamp, we inserted the fibre optic

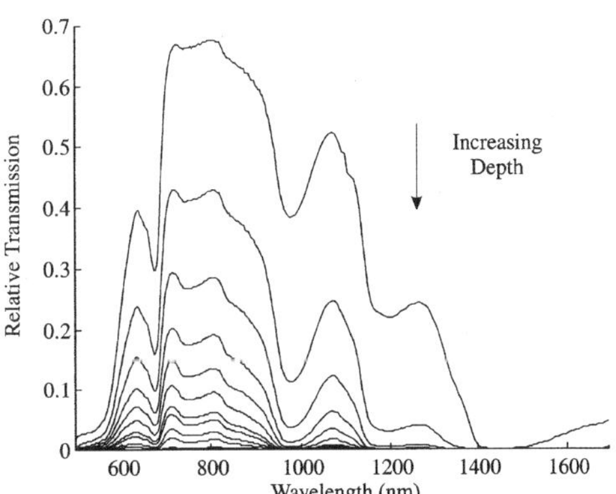

Figure 1. Transmission spectra recorded inside an apple which is illuminated by a white light source; the largest spectrum indicates a position 1 mm from the illuminated surface, each subsequent measurement is 3.175 mm further into the fruit. (Reprinted with permission of Elsevier Science).

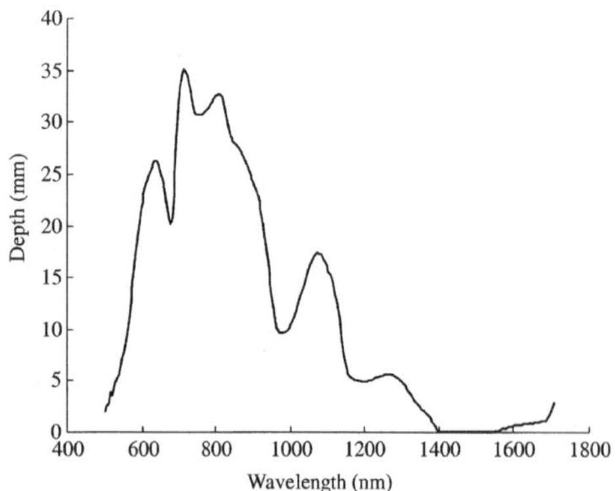

Figure 2. The depth inside an intact apple where the light has been reduced to 1% of the incident intensity. (Reprinted with permission of Elsevier Science).

probe into the apple through the centre from the opposite shoulder towards the illuminated area. Light measurements in the spectral range 500–1690 nm were recorded at incremental depths. The spectra (Figure 1) show chlorophyll absorption around 650 nm and water absorption at approximately 950, 1150, and 1450 nm. From the rate of light extinction at each wavelength, we can compute the 1% depth of penetration (Figure 2). This represents the depth into the apple where the light intensity has been reduced to 100th of the initial intensity.

For NIR transmission where the objective is to sample the internal tissue, the greatest signal-to-noise ratio will be achieved by using wavelengths that are not strongly absorbed by water, specifically those in the 'diagnostic window' 700–900 nm. Diffuse reflectance measurements taken at the wavelengths where there is low light penetration will only contain information describing the near surface of the apple.

By using a diode laser the light distribution throughout the fruit can be mapped. Figure 3 shows the light intensity measured along three paths travelled by the probe into a mandarin and the path loca-

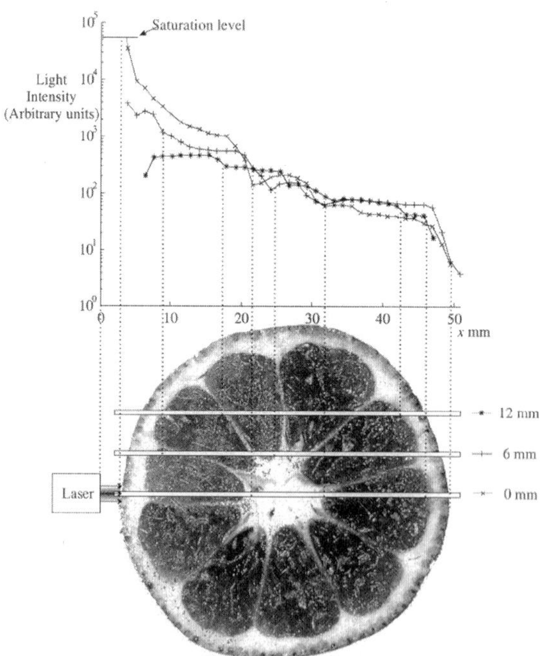

Figure 3. Light distributions through a mandarin.

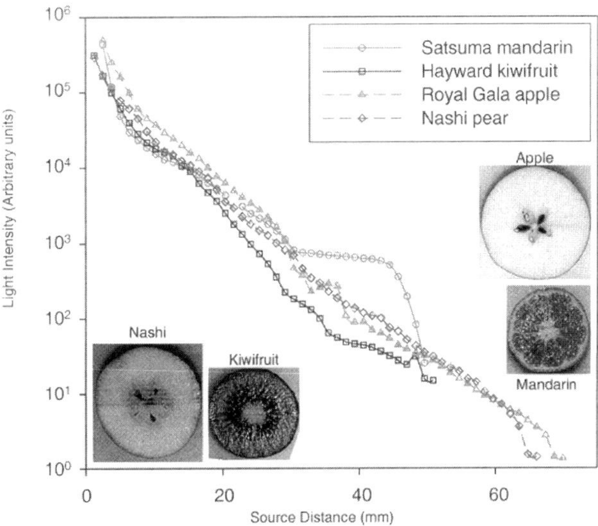

Figure 4. Light extinction curves at 808 nm for four types of fruit.

tions. The skin can be seen to reduce the light level by more than an order of magnitude, confirming the assertion[3] that the mandarin skin is highly absorbent and/or scattering. Perturbations are visible at the core, seeds, segment boundaries and skin interactions.

Figure 4 shows the light extinction curves through the centres of four different fruit parallel to and in line with the input laser beam. For each fruit, the initial power law rate of light attenuation reduces to an exponential rate after a characteristic diffusion distance. In this diffusion region the initially forward oriented light becomes isotropic, the distance over which this occurs depends on the tissue's absorption and scattering properties as well as the degree of scattering anisotropy. It can be seen that different regions in each fruit have different exponential rates of light level extinction. The rate of reduction depends on the tissue scattering and absorption properties as well as on the influence of the fruit skin. The mandarin, in particular, stands out as having an elevated light level in the region distant from the illuminated side. The kiwifruit also showed a marked change between inner and outer pericarp.

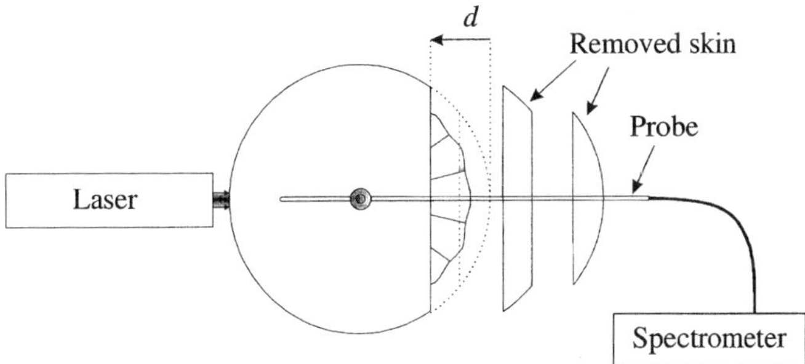

Figure 5. Experimental investigation of the mandarin skin effect on light levels.

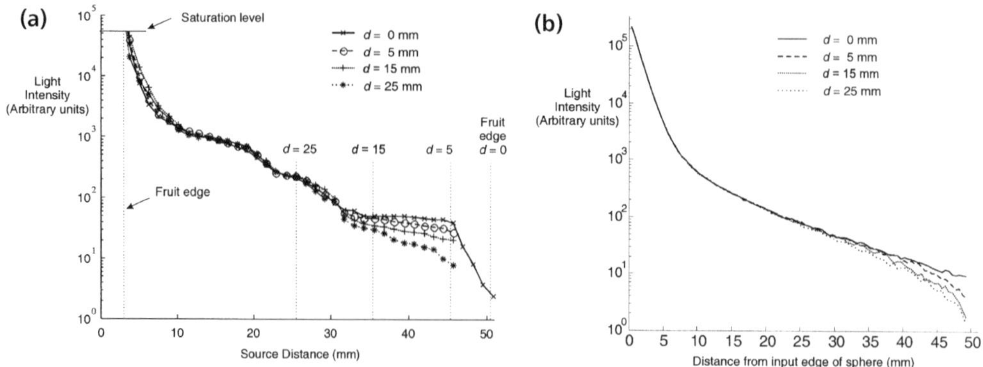

Figure 6. (a) Experimental light extinction curves for a mandarin with incremental amounts of skin removed. (b) Monte Carlo simulation where the equivalent sections of the boundary are removed.

There is evidence that the mandarin skin exhibits significant internal reflectance which reduces the rate of light extinction inside the fruit.[2] By repeatedly inserting the probe through the centre of the fruit towards the source as incremental amounts of skin are removed (Figure 5) the influence of the skin on internal light levels can be seen [Figure 6(a)].

Monte Carlo simulations were able to produce a similar effect [Figure 6(b)]. Our simulations used scattering and absorption coefficients based on the measurements of Cubeddu[4] to define the tissue model and made assumptions about the effect of the skin to define boundary conditions.

These results suggest that boundary conditions cannot be ignored for optical measurements of small fruit, which have low absorption coefficients.

Conclusion

We have developed a probe which is able to directly measure the light distributions in fruit with minimal effect on the distribution being measured.

Monte Carlo simulations were able to help verify the experimental light measurements made by the probe.

References

1. D.G. Fraser, V.A. McGlone, R.B. Jordan and R. Künnemeyer, *Postharvest Biology and Technology* **22(3),** 191 (2001).
2. D.G. Fraser, R.B. Jordan, R. Künnemeyer and V.A. McGlone, *Postharvest Biology and Technology*, in press (2001).
3. S. Kawano, T. Fujiwara and M.J. Iwamoto *Journal Japanese Society Horticultural Science* **62(2),** 465 (1993).
4. R. Cubeddu, C. D'Andrea, A. Pifferi, P. Taroni, A. Torricelli, G. Valentini, C. Dover, D. Johnson, M. Ruiz-Altisent and C. Valero, *Applied Optics* **40(2),** 538 (2001).

The effect of particle size on the determinability of maize composition in reflection mode

Sándor Turza and Mária Váradi

Central Food Research Institute, Herman O. út 15, H-1022 Budapest, Hungary

Introduction

Near infrared (NIR) spectra contains information on the shape, size and probably the surface characteristics of particles.[1] The particles in any kind of food sample have a distribution of sizes. The effect of particle size on NIR spectra has been discussed many times, but its importance was first recognised and documented by Williams.[2,3] For smaller particles, the depth of sample penetration of NIR radiation decreases and the scattering coefficient increases. For a long time it was thought that grinding finely was essential to obtain accurate and precise NIR results but some works have shown that cereal grains are a special case because of the heterogeneous distribution of absorbers through the kernel. Among cereals, the effect of particle size has been investigated mostly for wheat and wheat products. However, this could be equally important in the case of maize. Therefore, our aim was to investigate the effect of different particle sizes on the accuracy of maize composition determination and also that of some pre-treatment methods to eliminate it.

Materials and methods

Samples and chemical analysis

Forty-seven maize samples (300–400 g) were obtained from the Hungrana Starch and Iso-sugar Factory, Szabadegyháza, Hungary. They were all last year's crops. Chemical analyses for moisture, starch, protein and oil were performed at the company's laboratory. The determinations were carried out according to the relevant Hungarian Standards and the results are shown in Table 1.

Sample preparation and NIR spectral recording

The samples were ground to three different particle sizes, 1.3 mm, 1.8 mm and 2.0 mm, respectively, using a Labmill QC-114(Labor-MIM, Hungary) grain grinder and then stored in small glass containers with screw-caps until measurement. The sample cell was a standard powder cuvette and the same one was used for every sample. Spectra were recorded using an NIRSystems 6250 spectrophotometer in the 1100–2498 nm range with 2 nm increments. Every sample spectrum was the average of fifty

Table 1. Basic statistical features of the constituents analysed.

	Starch	Protein	Oil
Sample number	47	47	47
Minimum %	69.05	8.02	3.15
Maximum %	73.95	9.95	5.55
Average %	71.59	8.99	4.27
Variance	1.93	0.27	0.21
Standard deviation %	1.39	0.52	0.46

scans and the cuvette was rotated twice at 120 degrees. Three spectra were taken for all samples, which were then averaged to give one log1/*R* spectrum.

Spectral pre-processing

Four different pre-treatments methods were tried and tested (Figure 1). The first was full MSC and second derivative with different gap sizes, the second was full MSC with a 10 nm boxcar smoothing, followed by second derivative, the third was second derivative only and the last one was 10 nm boxcar smoothing with second derivative. The gap size was set at 10, 20, 30, 40, 50 and 60 nm and the segment size was fixed at a value of 2 nm.

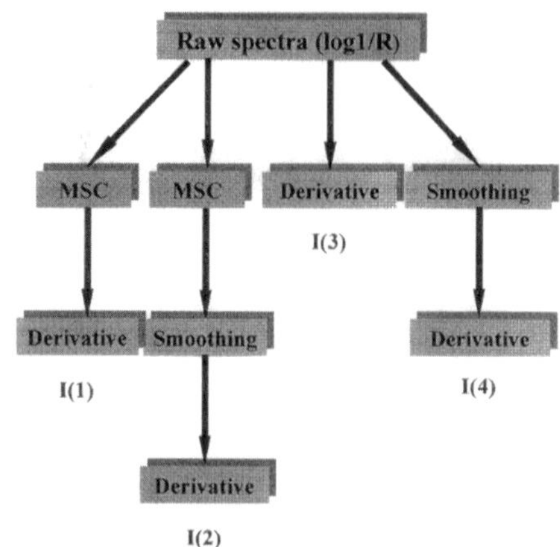

Figure 1. Flowchart of the pre-processing treatments applied.

Qualitative analysis

By taking the average of all 47 spectra for the three mesh sizes after pre-treatments, some wavelengths were selected that are perhaps related to particle size differences. With these selected wavelengths, linear discriminant analysis (LDA) was performed to prove their relevance to particle size.

Calibration and validation procedure

The pre-processed spectra were fed to an MLR algorithm in NSAS. The standard regression option was used with four terms. Linear additive models were calculated, so derivative ratios were not used. As the number of samples was not large enough to have a separate test set, full cross-validation was used. However, there is no possibility in NSAS to carry out cross-validation, so only calibration was done. Those calibrations which had the highest multiple regression coefficient (*R*) and the lowest standard error (*SE*) were selected to be used in Unscrambler. Here, the selected wavelengths were used as inputs for MLR regression, where full cross-validation was applied to test model validity. Regression models were built for starch, protein and oil for the different particle sizes. A so-called "mixture model" was also built, where the 47 samples were made up from the different mesh size groups, i.e. 15 from the 1.3 mm group, 17 from the 1.8 mm group and 15 from the 2.0 mm group, respectively. The difference between the three best "pure" calibrations and the "mixture model" calibrations before outlier removal (same sample number) were tested by a method first described by Pitman.[4]

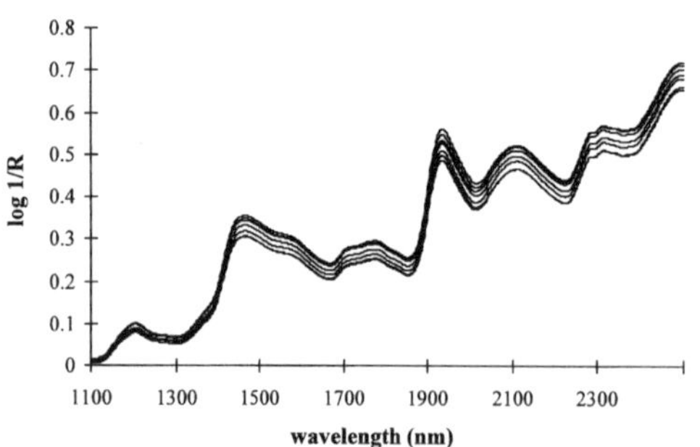

Figure 2. Raw spectra of some maize samples for the 1.8 mm group in the 1100–2498 nm region.

Results

The raw spectra indicated that beside the additive there was a multiplicative type of noise as well. This was manifested through the scissor -like opening of the spectral swarm (Figure 2). When taking the average of all 47 spectra for the three mesh sizes it appeared that this phenomenon was more pronounced in the longer region, in our case from 1932 nm to 2498 nm. When the

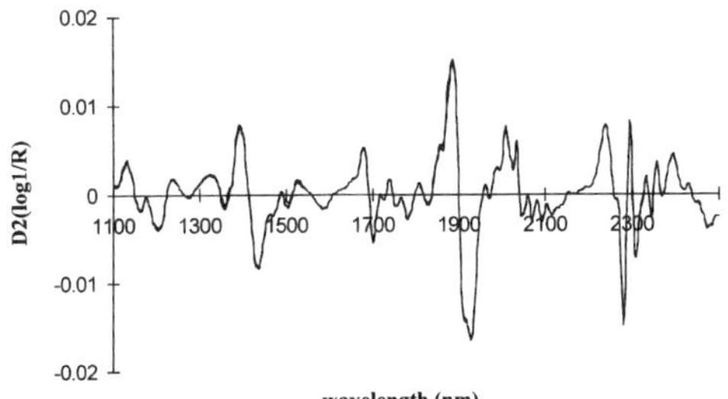

Figure 3. Second derivative spectra of the average of 47 samples for the three mesh size groups.

spectra were preprocessed in some way, the differences disappeared. However, a much closer examination revealed some wavelengths which still exhibit minute differences between different mesh size groups in the appropriate order (Figure 3). With these wavelengths LDA was performed and the results are summarised in Table 2. The wavelength selection was conducted with a 10 nm gap size treated spectra in all four preprocessing ways, for this gap size had the best calibration results in the majority of the cases. The highest classification accuracy was between the 1.3–2.0 mm groups, followed by the 1.3–1.8 mm and the 1.8–2.0 mm groups. In the last case, the proportion of correctly classified samples decreased, especially for the MSC treated spectra. For the first two groups the most dominant wavelength was 1202 nm, which is a relatively weak starch band. For the third group, the 2092 nm proved to be the most important. When the derivative was applied only, a second discriminant variable was necessary to have better classification.

The best regression results are summarised in Table 3. The values are not so good, but it has to be kept in mind that whole maize kernels were used, i.e. the germ was not separated. It is obvious that all results involve MSC treatment, which supports its use under conditions like this. The best figures for starch and protein were attained when the 2.0 mm group was used. For starch, in every pre-processing case, the 2.0 mm group was found to be the best. The oil content was best determined if the 1.3 mm group was selected for regression. The "mixed group" calibrations were not as good as the "pure" ones for starch and oil. The protein determination was not effected by the mixing. The goodness of the models, before outlier removal, was compared and it was found that there were no significant differences between them. After outlier removal the differences changed a little, but not so much that it could effect the significance to a considerable extent. The test method was not applied after outlier removal, as it is only applicable if there are an equal number of samples in each validation set.

Table 2. LDA results of the different mesh size groups with the wavelengths used and reclassification percentages.

	1.3–1.8 mm	1.3-2.0 mm	1.8-2.0 mm
Variables used I(1)	1202 nm 90.4%	1202 nm 93.6%	2092 nm 71.3%
Variables used I(2)	1202 nm 92.6%	1202 nm 93.6%	2092+1202 nm 76.6%
Variables used I(3)	1202 + 1506 nm 80%	1202 + 1506 nm 87.2%	2092 + 1202 nm 75%
Variables used I(4)	1202 nm 86.2%	1202 + 1504 nm 91.5%	2092 nm 79.8%

Table 3. The best calibration and validation results of the "pure" and the "mixed group" samples.

	Rcal	*SEC*	*Rval*	*SEP*	Bias
Starch (2.0 mm), I(2)	0.93	0.51	0.91	0.57	-0.00
Protein (2.0 mm), I(1)	0.93	0.18	0.91	0.20	0.00
Oil (1.3 mm), I(2)	0,88	0.20	0.85	0.23	-0.00
Mixed mesh sizes					
Starch I(1)	0.86	0.66	0.82	0.75	0.00
Protein I(1)	0.93	0.18	0.91	0.21	-0.00
Oil I(1)	0.79	0.24	0.76	0.29	0.00

Conclusions

The combination region is more sensitive to minor differences in particle size. There are some wavelengths that are less affected by the pre-treatments applied. It can also be concluded that the oil content could be best determined using the smallest mesh size group and for starch, the 2.0 mm group seemed to be the best choice. MSC treatment was helpful for every constituent. The small number of samples and the narrow calibration range can be held partially responsible for the relatively poor results. More samples, wider calibration range and more mesh size groups are necessary to prove the validity of results.

References

1. V. Chapelle, J.P. Melcion, P. Robert and D. Bertrand, *Science des Aliments* **9,** 387 (1989).
2. P.C. Williams, *Cereal Science Today* **19,** 280 (1974).
3. P.C. Williams and B.N. Thompson, *Cereal Chemistry* **55,** 1014 (1978).
4. Tom Fearn, *NIR news* **7,** 5 (1996).

Transfer of calibration between on-farm whole grain analysers

Phillip J. Clancy

NIR Technology Australia, 34 Clements Avenue, Bankstown, NSW 2200, Australia

Introduction

On-farm near infrared (NIR) analysers are becoming more accepted around the world. The benefits offered to growers to measure the quality of their wheat, barley, canola, sorghum, soy bean, corn and other crops leads us to believe that on-farm NIR analysers will become a standard piece of farming equipment in the next 20 years.

With such a large potential number of instruments placed in remote locations, the problems of calibration and calibration transfer become very significant. To this end, NIR Technology Australia has been very conscious of the need to make NIR analysers with the ability to transfer calibrations from master to slave instruments.

Calibrations based on partial least squares (PLS) techniques may make this task even more complex since PLS calibrations tend to compensate for instrument and sampling noise. Since these two parameters are commonly instrument-dependent, then a PLS calibration developed on a master instrument may not transfer easily to all slave instruments. Nonetheless, PLS calibrations have proven very successful for the analysis of whole grains and, therefore, a transfer method had to be developed which would allow PLS calibrations to be transferred across a network of instruments.

Procedure

The first step in calibration transfer should always be the standardisation and normalisation of the slave instruments to the master instrument. There are two parameters to be considered, (a) wavelength alignment and (b) photometric response.

Wavelength alignment

The NIR Technology Australia Cropscan 2000G and 2000B Whole Grain Analysers are diode array-based spectrophotometers. They use a flat field spectrograph to project an image of the entrance slit onto a large silicon photodiode array detector. The flat field spectrograph is a concave holographic grating with a spectral dispersion from 720 to 1100 nm. To align the wavelengths from master to slave instruments requires the adjustment of the angle of the grating relative to the incident beam of light. By scanning a sample of wheat grains and using the moisture absorption peak at 967 nm, it is possible to align the gratings between instruments. However to correctly align the gratings, it has been found that a first derivative spectra of wheat grains provides several points to match the spectra and is far more sensitive than the absorbance spectra.

Photometric response

Even though the gratings are produced as identical masters, there is always a mechanical tolerance which effects the efficiency curve of the gratings. As well, the lamp, lenses, detector and sampling compartment have mechanical tolerances which result in differences between the photometric response from instrument to instrument. To correct for the total variance which exists between master and slave instruments, we scan samples on the master and the slave instruments and compare the spectral data. By scanning samples of wheat with absorbances covering the broadest range, i.e. two to four absorbance units, we can look at the presence of skew and shift at each wavelength. Simple linear regression techniques can then be used to estimate the required correction to each wavelength reading in order to make the slave instrument's response the same as the master instrument. We have found that either a slope and bias or simply a slope correction of the photometric output of the pixels on the diode array detector, is sufficient to match instruments.

The Cropscan 2000G and 2000B can use either of these methods. Within the instruments set up files, there is a look-up table which contains the S and B factors (slope and bias). When a sample of grain is scanned, the reading from each pixel is multiplied by the S factor and the B factor is added to give the corrected output. The PLS calibration is then applied to the corrected spectrum to compute the protein and moisture results. Other methods, such as multiplicative scatter correction (MSC), have been tried but showed poorer results.

Analysis

A study was undertaken to prove the viability of this procedure. The master calibration was developed over three years using 425 samples of Australian hard wheats. All samples were scanned on the one instrument, not at one time, but progressively. This calibration set includes as much variation in variety, growing conditions and region as was available. Five samples were scanned at 10°C, 25°C and 45°C and added to the set for sample temperature stabilisation. As well, samples were scanned when the instrument was at 10°C and 45°C and added to the calibration set for instrument temperature stabilisation. Five scans of each sample were collected and used in the PLS calibration. By using the un-averaged scans, we feel that it builds in a tolerance to the sample packing variation. Even though using the five scans increases the standard error of calibration, it reduces the standard error of prediction and provides a more robust calibration. It is also felt that the use of the five scans makes the calibration less instrument-dependent and, therefore, more easily transferred to another instrument. The calibration statistics for this calibration are listed below:

Number of scans:	2125			
Number of PC:	Protein:	12	Moisture:	9
SEC:		0.33%		0.26%

This calibration for hard wheat is used in all instruments sold in Australia and is provided in instruments sold overseas, although local calibrations have been developed in Europe and the USA. To match the instruments to a specific laboratory, each instrument has slope and bias adjustments for each calibration. Generally, five or six samples of wheat from 10% to 15% protein content are used to compute a simple linear slope and bias adjustment. This slope and bias adjustment corrects the final protein and moisture results but does not change the B coefficients of the master calibration nor the photometric response of the instruments.

Five Cropscan 2000G and two Cropscan 2000B instruments were set up using the above procedure for wavelength alignment and photometric response. A set of nine hard wheat samples were then scanned on each instrument. Table 1. shows the prediction data from the seven instruments.

Table 1. Comparison of five Cropscan 2000G and two Cropscan 2000B Whole Grain Analysers against the Master Cropscan 2000G. Wheat(2000) calibration was used on all instruments. Nine samples of Australian hard wheats were scanned on each instrument including the Master instrument.

ID	Prot%	Master Diff.		Cr104 Diff.		Cr105 Diff.		Cr106 Diff.		Cr107 Diff.		Cr 108 Diff.		Cr109BDiff.		CrIIOB Diff.		Average St dev.		Ref – Av
1	13	13.5	−0.5	13.1	−0.1	13.6	−0.6	13.3	−0.3	13.4	−0.4	13.6	−0.6	13.5	−1	13.7	−0.7	13.5	0.19	−0.5
2	14	14,1	−0.1	13.8	0.2	13.9	0.1	14	0	14	0	14	0	14.2	0	14	0	14	0.12	0
3	12.2	12.4	−0.2	12.6	−0.4	12.2	0	12.4	−0.2	11.9	0.3	12.2	0	12.1	0.1	12	0.2	12.1	0.25	0.1
4	10.4	10,1	0.3	10	0.4	9.8	0.1	9.7	−0.2	9.6	0.3	9.7	0.2	9.8	0.1	9.8	0.1	9.9	0.21	0.5
5	13	12.8	0.2	12.7	0.3	12.8	0.2	12.9	0.1	12.7	0.3	12.8	0.2	12.9	0.1	12.4	0.6	12.8	0.16	0.2
6	11.2	11.7	−0.5	11.6	−0.4	11.3	−0.1	11.3	−0.1	11.6	−0.4	11.5	−0.3	11.3	0	11.4	−0.2	11.5	0.19	−0.3
7	11.9	11.7	0.2	11.6	0.3	12.1	−0.2	11.4	0.5	11,9	0	12.2	-0.3	12.1	0	11.8	0.1	11.8	0.27	0.1
8	12.7	119	−0.2	12.7	0	12.5	0.2	12.4	0.3	12.4	0.3	12.5	0.2	12.4	0.3	12.3	0.4	12.4	0.13	0.3
9	9.2	9.7	−0.5	9.4	−0.2	9.1	−0.1	9.7	−0.5	9.1	−0.1	9.2	0	9.3	0	9.6	-0.4	9.5	0.29	−0.3
SEP		0.34		0.31		0.25		0.32		0.28		0.28		0.2		0.41		0.32		

Conclusion

It can be seen that Sample 1 is consistently predicting higher than the reference value. However the remaining samples are predicted well. Nonetheless, all seven instruments predict to the level expected of whole grain analysis.

It should be noted that Cr109 and Cr110 are 2000B models which use a mechanically-driven sample cell with a pathlength of 18 mm. The other five instruments use a manually-loaded cell with a 20 mm pathlength. This illustrates the ability of the above procedure to correct for even 10% variation in effective pathlength.

There are several algorithms for transferring calibrations between instruments sighted in the literature, but the above study shows that a simple linear slope and bias correction can be effective in normalising instruments and thus allowing a single master calibration to be used on all instruments.

Qualification of volatile oils using near infrared spectroscopy and electronic nose

Gabriella Kiskó[a] and Zsolt Seregély[b]

[a]*Szent István University, Faculty of Food Science, H-1118 Budapest, Ménesi út 45, Hungary*

[b]*Metrika R&D Co., H-1119 Budapest, Petzvál J. u. 25, Hungary*

Introduction

People have used natural aromatic plants from the beginning of history. Yet, chemical information related to these same products only began to appear a century ago. The use of natural plants in curing, pharmaceutical and other industries has become increasingly important in recent years. Increased use of bioproducts and the use of natural additives (aromas, colourants etc.) are but two of the more noticeable trends within the food industry. More than 1400 species of aromatic plants produce volatile oils on an industrial scale. The potential use of these ingredients for natural anti-microbial agents has been exploited to a lesser extent. Consumers tend to use and consume bio-products, essential oils as aromas, preservatives, colorants, etc. more than ever before. Attempts have been made to use active ingredients from medicinal and aromatic plants for bio-preservatives rather than resorting to synthetic derivatives.

Quality parameters of volatile oils are usually determined for aroma, taste and other organoleptic criteria. Detailed quality specifications from sensory data are lacking due to fallible human opinions. People panels demand extensive training, are slow to respond and the results are noticeably variable.

Materials and methods

This project is part of an international effort. Its objective was to discriminate and identify seven natural volatile oils obtained from dried plant materials. The oils were prepared by a hydrodistillation process used for the inhibition of food spoilage microorganisms and foodborne pathogens. The investigated plant materials were as follows:

- Thyme (*Thymus vulgaris* L.) marked with "t" in the figures
- Peppermint (*Mentha piperita* L.) marked with "p"
- Dill (*Anethum graveolens* L.) seed marked with "s"
- Dill (*Anethum graveolens* L.) weed marked with "w"
- Cassia (*Cinnamonum cassia* Presl) marked with "c"
- Oregano (*Oreganum vulgare* L. spp. *vulgare*) marked with "v"
- Oregano (*Oreganum vulgare* L. spp. *hirtum*) marked with "h"

Spectra of the samples were recorded with a Spectralyzer (PMC Model 10-25, Switzerland) scanning NIR spectrometer. Log (1/*R*) spectra were recorded over the range 1000–2500 nm in 2 nm steps. Qualitative indices were calculated according to Kaffka and Gyarmati[1,2] using the polar qualification system (PQS) program. The indices were determined in a "quality plane". The index was defined as the centre of the spectrum (of the spectral points) represented in polar co-ordinate system.

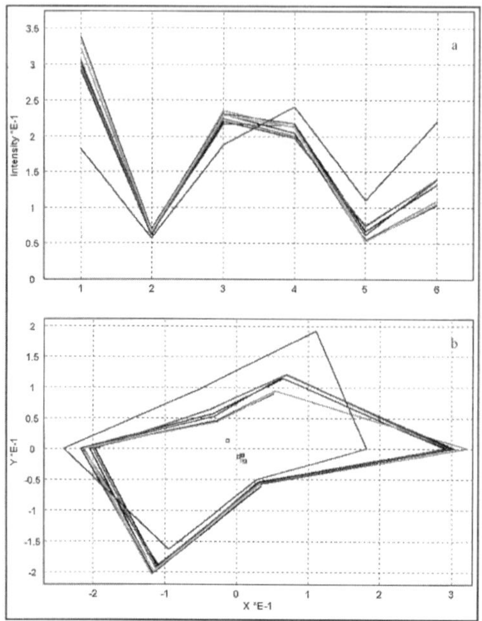

Figure 1. The normalised e-nose data for the seven volatile oils studied in the rectangular (a) and polar (b) coordinates.

"Electronic nose" (or e-nose) is a generic name for an analytical instrument that contains an array of chemical sensors whose outputs are integrated by an advanced signal processing system in order to rapidly identify complex odour mixtures. Current e-nose systems use comparative techniques and produce "fingerprints" of a volatile in a headspace over or around a sample.

SamSelect electronic nose (Daimler Chrysler Aerospace,) was used to take odour measurements. All samples were measured in the standard headspace vials. Headspace auto-sampling were used as a standard and reproducible sampling technique. After normalising the spectral data, principal component analysis (PCA) and a newly developed method, the polar qualification procedure was applied to create classification models from the sensor array data.

The polar qualification system was developed to work with near infrared spectra. The basic principles of the method are described in the relevant literatures.[1–5] A "sequence optimisation" technique was developed to allow PQS to work with non-spectral data. Figure 1(a) shows the normalised sensor data for the seven measured volatile oils represented in Descartes co-ordinate system as a spectrum. Figure 1(b) shows the same data set in polar co-ordinates.

Results and discussion

Principal component analysis (PCA) was used in the classification procedures. E-nose data for the seven volatile oils was measured nine times. Figure 2(a) shows the location of the quality points of the volatile oil samples in the projection plane determined by the first two principal components. As it can be seen, the quality points of thyme and dill weed samples overlap and the identification of all samples cannot be performed using only one projection plane. In such cases, cascade classification must be used. This means that identification is performed in multi-steps. The overlapping samples become

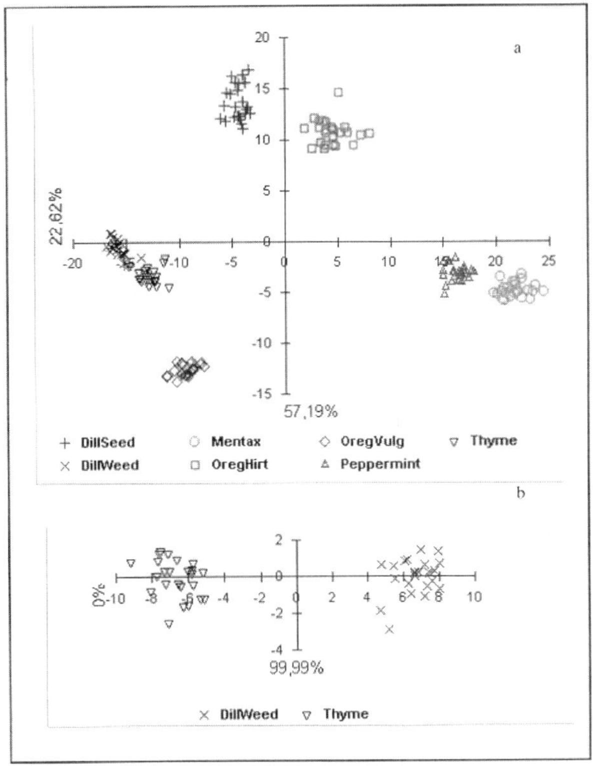

Figure 2. Multi-step discrimination using PCA: (a) location of the quality points for the oil samples in the first projection plane determined by the first two principal components and (b) overlapping quality points of thyme and dill weed samples can be separated in another projection plane.

separated in another projection plane calculated by omitting the already classified samples [Figure 2(b)].

In the second step (sequence optimisation), the PQS technique, designed to work with NIR spectra, was enhanced to be applied to almost any multivariate problems. Figure 3(a) shows the location of the quality points (centre of their normalised data sets drawn as polar spectra) of the measured volatile oils in original data coming directly from the instrument. It can be seen the normalised distances and the sensitivities are high. In addition, the distances are considerably higher among the groups compared to their standard deviations. Separation of all samples was not possible for the quality plane using the original data sequence for the quality points of peppermint. Also, *oreganum hirtum* samples overlap a bit. After applying sequence optimisation to normalised distances, all the investigated oil samples can be identified in one plane [see Figure 3(b)]. The calculated optimal signal sequence was S1 S5 S3 S4 S2 S6.

The cascade identification was also a possible tool in the PQS evaluation. By classifying an unknown sample into separated clusters having overlapping subclasses using a given data sequence, the clusters can be further evaluated using some other data sequence (very difficult to follow) providing full classification.

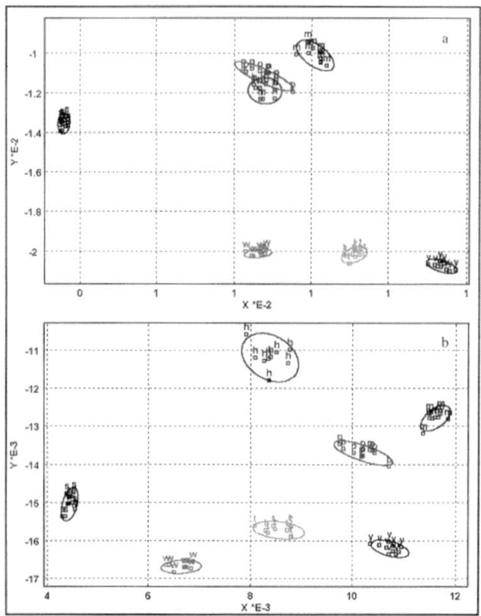

Figure 3. Location of the quality points of the volatile oil samples in (a) original order and (b) after sequence optimisation. The optimal data sequence is (S1 S5 S3 S4 S2 S6).

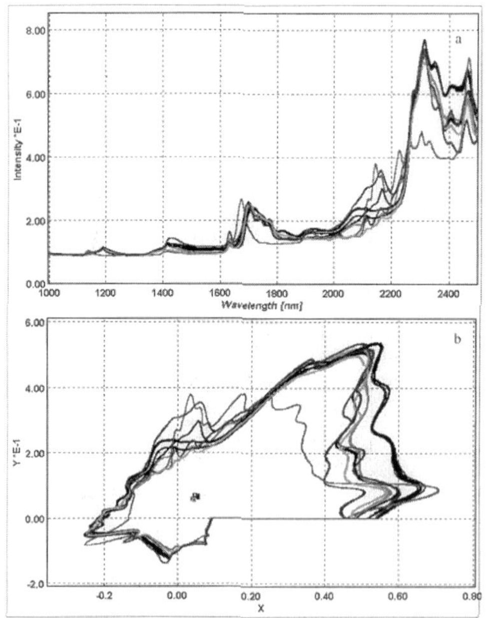

Figure 4. NIR spectra of the volatile oil samples in (a) the rectangular and (b) in the polarco-ordinates using the 1000–2500 nm range.

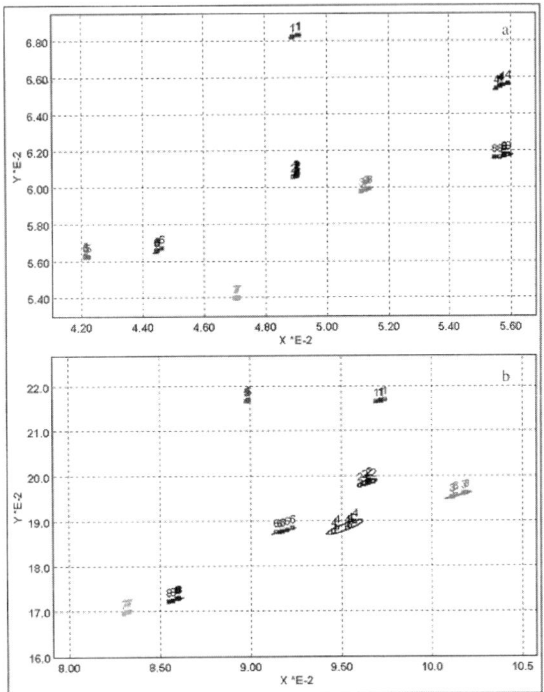

Figure 5. Location of the quality points for the volatile oils using the 1000-2500 nm range (point method) (a) and using the 1408–1436 nm, 1696–1720 nm and 2132–2280 nm ranges (b) The regions among the selected ranges are zero (surface method).

An NIR polar qualification system using wavelength range optimisation was used to create classification models. In Figure 4, the log(1/*R*) spectra of the samples can be seen in (a) rectangular and (b) polar co-ordinates. It was found that well-defined clusters could not be achieved using log (1/*R*) spectra [see Figure 5(a)]. However, after using wavelength range optimisation, clusters could be separated.

The wavelength range optimisation program was repeated several times in order to define the highest optimum range. By determining the best second and third wavelength ranges and by using the whole near infrared region the optimal ranges can be situated opposite to each other resulting in a worse classification. By omitting the ranges among the selected ones the optimal ranges can be collected together, the direction of their shifting effect to the location of the quality point can be turned to the same direction, summarising their effects. The spectrum ranges providing the best separation were as follows: 1408–1436 nm; 1696–1720 nm; 2132–2280 nm, which are associated with oil peaks [see Figure 5(b)].

Conclusion

Both e-nose and NIR data are suitable for identifying the seven volatile oils studied. Sequence optimisation opens new perspectives in the application of PQS. The method offers a rapid, accurate, cheap and simple method for identifying products using different data sets.

Acknowledgement

The authors would like to express their sincere thanks to the participant organisations of the European INCO Copernicus "PLANTCHEM" project (No. EU IC 15 CT97-1000) for their courtesy placing the volatile oil samples at the authors disposal.

This project was supported by a grant from the National Scientific Research Fund (OTKA) No: T 023020 and No: T 032814.

Thanks are due to J. Farkas, G. Bayer and K. Aschenbrenner for their conscientious scientific and technical assistance in performing measurements and data processing.

References

1. K.J. Kaffka and L.S. Gyarmati, *Proceedings of the Third International Conference on Near Infrared Spectroscopy*, Ed by R. Biston and N. Bartiaux-Thill. Agricultural Research Centre Publishing, Gembloux, Belgium, p. 135 (1991).
2. K.J. Kaffka, *Proceedings: International Diffuse Reflectance Spectroscopy Conferences*, Ed by R.A. Taylor. The Council for Near Infrared Spectroscopy. Gaithersburg, Maryland, USA, p. 63 (1992).
3. K.J. Kaffka and L.S. Gyarmati, *Near Infrared Spectroscopy: The Future Waves*, Ed by A.M.C. Davies and P. Williams. NIR Publications, Chichester, UK, p. 209 (1996).
4. K.J. Kaffka and L.S. Gyarmati, *J. Near Infrared Spectroscopy*, **6**, A191 (1998).
5. K.J. Kaffka and Zs. Seregély, *NIR Spectroscopy: Proceedings of the 9th International Conference*, Ed by A.M.C. Davies and R. Giangiacomo. NIR Publications, Chichester, UK, pp. 259–265 (1999).

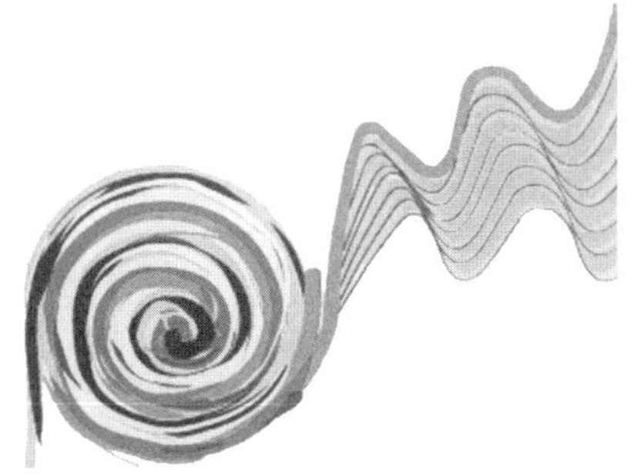

Artificial Neural Networks

Disease diagnosed and described by near infrared spectroscopy

Roumiana N. Tsenkova

Department of Environmental Information and Bio-production Engineering, Kobe University, Faculty of Agriculture, 1-1 Rokkodai, Kobe 657, Japan

Introduction

A mammary gland is made up of tissue, which has the capability of producing a large volume of secretion, i.e. milk, under normal and healthy conditions. When bacteria enter the gland and establish an infection or inflammation it initiates an udder disease called mastitis.[1] It is accompanied by an influx of white cells from the blood stream into the milk, altered secretory function and changes in the volume and composition of secretion. Cell counts in milk are closely associated with inflammation and udder health. Somatic cell count (SCC) has been accepted as the international standard measurement of milk quality in dairy and for mastitis diagnosis.[2] In previous studies near infrared (NIR) spectroscopy has been successfully applied for non-invasive mastitis diagnosis performed by qualitative raw milk spectral analysis[3,4] when an expert's diagnosis has been used as a reference.

The purpose of this investigation was to evaluate the accuracy of mastitis diagnosis based only on near infrared spectra of milk when somatic cell count (SCC) has been used as a standard reference method. Two-dimensional correlation spectroscopy (2-DCOS) was applied for further understanding of the disease on a molecular level.

Materials and methods

Samples

A total of 189 composite milk samples from seven Holstein cows were analysed. The samples were collected for 28 days, consecutively, beginning on 7th day after calving. All cows were fed the same rations, twice daily and always had access to drinking water. The average BW of the cows was 552 kg.

Each milk sample was divided into two subsamples. One was subjected to spectral analysis and the other was analysed for SCC by using the fluoro-opto-electronic method using a Fossomatic 400 (Foss-Electric A/C, Hillerød, Denmark). Somatic cell count standards were used to calibrate the Foss instrument throughout the study. The repeatability coefficients of variation of this method are 4 to 5% for the region between 400,000 and 500,000 cells mL^{-1}and 5 to 10% for the region between 100,000 and 200,000 cells mL^{-1} and over 500,000 cells mL^{-1} (IDF Standard 148A, 1995).[5] Log_{10}SCC was calculated to normalise the SCC distribution. Samples were also analysed for fat, total protein and lactose content[1] by Milko Scan (Foss-Electric A/S, Hillerød, Denmark). Three of the examined cows were healthy, with SCC lower than 137,000 cells mL^{-1}. One cow was mastitic during the entire experimental period. Her measured SCC varied from 204,000 to 11,876,000 cells mL^{-1}. Three cows had mastitic and healthy periods (SCC was between 80,000 and 4,737,000 cells mL^{-1}).

NIR spectra

Near infrared transflectance (T) spectra were obtained using an the InfraAlyzer 500 spectrophotometer, (Bran+Luebbe, Nordestedt, Germany), in terms of optical density log (1/*T*) in a wavelength range from 1100 to 2500 nm. A flow cell with pathlength of 0.2 mm, connected with automated liquid sampling system and taking alternatively milk samples and cleaning solution, was used. Before the spectral analysis, each sample was warmed up to 40°C in a water bath with a temperature control of ± 0.1°C. During the analysis, the same temperature was controlled through the use of an integrated water-jacketed holder of the flow cell connected to the water bath.

NIR analysis

A commercial software program (Pirouette Version 2.6, Infometrics, Inc., Woodinville, WA, USA) was used to process spectral data.

To develop a regression equation for logSCC, spectra were randomly divided into a set of 128 calibration samples and a set of 68 validation samples. Both data sets covered similar ranges of each investigated parameter.

Methods used for preliminary examination of the data included smoothing the spectral data, multiplicative scatter correction, standard normal variance correction, baseline correction and first or second derivative transformation of log (1/*T*) data. The smoothing and derivative transformations were based on the Savitzki–Golay second-order polynomial filter.[6]

Calibration for SCC was performed by partial least square (PLS) regression using the calibration set of samples. Calibration and validation statistics for each regression model included standard error of calibration, coefficient of multiple correlation, standard error of prediction, correlation coefficient between measured and NIR-predicted values. Statistical parameters were used to evaluate the accuracy of NIR for SCC determination.

Classification of milk samples in class "healthy" or class "mastitic" was performed using soft independent modeling of class analogy (SIMCA) and various spectral data pretreatments. Two levels of SCC—200,000 cells mL^{-1} and 300,000 cells mL^{-1}, respectively, were used and compared as thresholds to discriminate between healthy and mastitic cows.

Two-dimensional correlation (2-DCOS) analysis of NIR milk spectra was done to assess the changes in milk composition, which occur simultaneously with the variation of SCC.

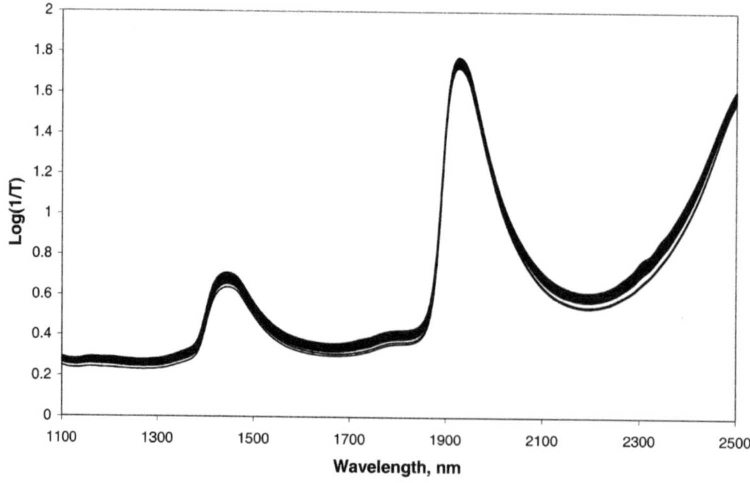

Figure 1. NIR milk spectra, 1 mm cell.

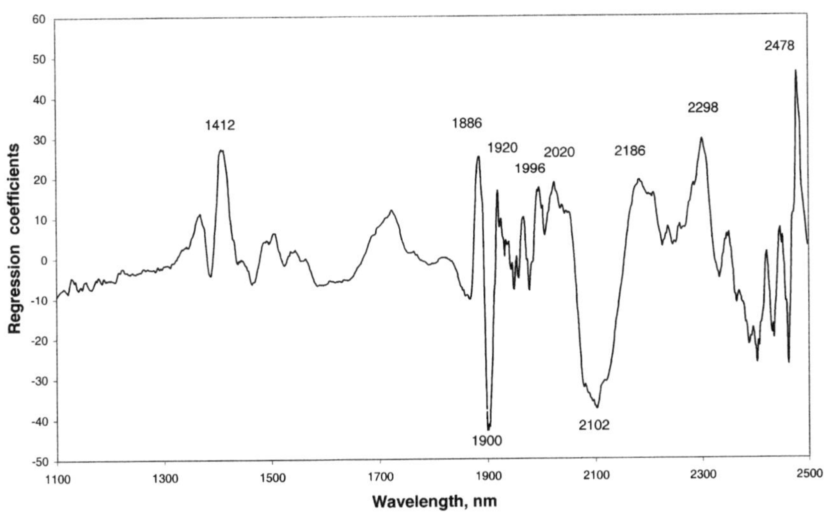

Figure 2. Regression vector for determination of logSCC based on milk spectra.

Results and discussion

A PLS model for prediction of logSCC, based on NIR milk spectra (Figure 1), was developed. The best accuracy of determination for the 1100–2500 nm range was found using smoothed absorbance data and 10 PLS factors. The standard error of prediction for independent validation set of samples was 0.382, correlation coefficient—0.854 and the variation coefficient—7.63%. It has been found that SCC determination by NIR milk spectra was indirect and based on the related changes in milk composition. From the spectral changes and the regression vector (Figure 2), we learned that when mastitis occurred, the most significant factors that simultaneously influenced milk spectra were alteration of milk proteins seen around 2100 nm and changes in ionic concentration of milk that appeared around 1412 nm and 1900 nm. This result was consistent with the milk compositional changes observed in mastitic milk[7] and the results we obtained further when 2-D correlation spectroscopy was applied (Figure 3).

Different thresholds for SCC were set up and SIMCA classification performed on milk spectra was used to find out the most appropriate one to be used in further mastitis diagnosis based on NIR spectra of milk. Two levels of SCC—200,000 cells mL^{-1} and 300,000 cells mL^{-1}, respectively—were set up and compared as thresholds to discriminate between healthy and mastitic cows. The best detection accuracy was found with 200,000 cells mL^{-1} as the threshold for mastitis and smoothed absorbance data:—98% of the milk samples in the calibration set and 87% of the samples in the independent test set were correctly classified (Table 1). When the spectral information was studied it was found that the successful mastitis diagnosis was based on reviling the spectral changes related to the corresponding changes in milk composition.

Two-dimensional correlation analysis of NIR milk spectra was done to assess the changes in milk composition, which occur when somatic cell count (SCC) levels increase with mastitis. The synchronous correlation map revealed that when SCC increases, protein levels increase, while water and lactose levels decrease. Results from the analysis of the asynchronous plot (Figure 3) indicated that changes in water and fat absorption occur before the changes with other milk components. In addition, the same technique was used to assess the changes in milk spectra during a period of time when SCC

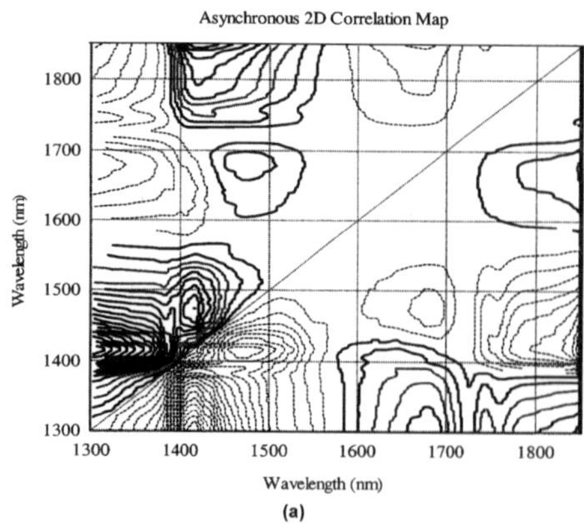

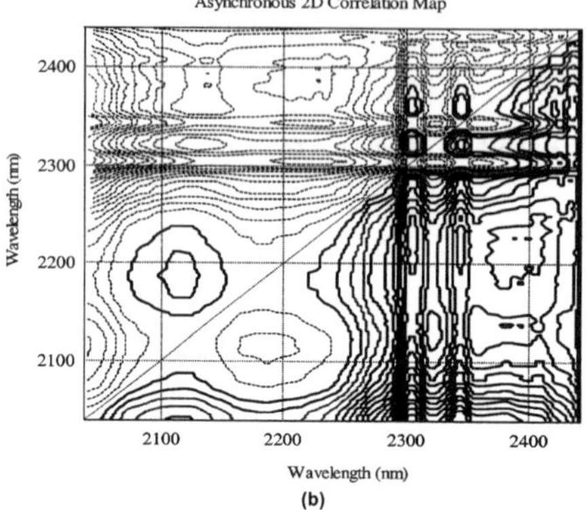

Figure 3. (a) 1300–1900 nm 2-D synchronous and (b) 2000–2450 nm asynchronous correlation map constructed from milk spectra.

levels did not have substantial variations. Results indicated that milk components were in equilibrium and no appreciable change in a given component was observed with respect to another. This was found in both healthy and mastitic animals. However, milk components were found to vary with SCC content regardless of its range. This important finding demonstrates that 2-D correlation analysis may be used to track even subtle changes in milk composition of individual cows, as well as to explain these changes on a molecular level.

Conclusion

NIR spectra of milk subjected to different ways of data mining can provide a fast and accurate alter-

Table 1. SIMCA classification results based on milk near-infrared spectra and two different somatic cell count levels used as thresholds for mastitis diagnosis.

Threshold	PC factors	Calibration set			Validation set		
		Correct classification		Incorrect classification	Correct classification		Incorrect Classification
SCC (cells mL^{-1})	Cal set/val set	*n*	%	kind	*n*	%	Kind
200,000 cells mL^{-1}	11/11	124	98	2 false positive	55	87	4 false negative 2 false positive 2 non classified
300,000 cells mL^{-1}	11/11	117	93	7 false negative 2 false positive	49	78	8 false negative 4 false positive 2 non classified

native to current methods for mastitis diagnosis and a new insight into understanding the disease at a molecular level.

References

1. J.E. Hillerton, *Bulletin of the International Dairy Federation* **No 345,** 4 (1999).
2. Association of Official Analytical Chemists, *Official Methods of analysis,* 15th Edn. AOAC, Arlington, VA, USA (1990).
3. R. Tsenkova, K. Yordanov and Y. Shinde, in *Prospects for Automatic Milking*, Ed by A.H. Ipema, A.C. Lippus, J.H.M. Metz and W. Rossing. PUDOC Scientific Publishers, Wageningen, The Netherlands, p. 185 (1992).
4 R. Tsenkova, S. Atanassova, S. Kawano and K. Toyoda, *J. Animal Sci.,* in press (2001).
5. International IDF Standard 148A, *Milk Enumeration of Somatic Cells*. International Dairy Federation, Brussels, Belgium (1995).
6. A. Savitzky and M.J.E. Golay, *Anal. Chem.* **36,** 1627 (1964).
7. R.J. Harmon, *J. Dairy Sci.* **77,** 2103 (1994).

Advantages of using artificial neural networks techniques for agricultural data

N.B. Büchmann and I.A. Cowe
Foss Tecator AB, Box 70, Höganäs, Sweden

Introduction

The last decade has seen an increase in the use of artificial neural networks (ANN) as a mathematical tool in such diverse fields as medicine (identification of cancers), surveillance (identification of faces from video footage), communications (routing of telephone calls), finance (automated trading on the stock exchange) and biochemistry (identification of the tertiary structure of protein molecules). In this paper, we report some results for prediction of chemical composition of various agricultural products using ANN for both transmittance and reflectance near infrared data.

Use of ANN for ground samples of compound feed measured in transmission

Borggård[1] was the first to use ANN for agricultural data in the development of near infrared calibrations for controlling pig carcass quality. Encouraged by this work, Büchmann proposed the use of ANN as a modelling technique for near infrared transmission measurements of compound feed samples.[2] Three broadly based models are currently available for use with the Foss Infratec1275, one for use with compound feed for pigs and poultry, another for cattle and the third, typical feed ingredients.

The data sets used for these calibrations were extremely large; for example, the pig and poultry model was based on almost 18,000 records, which improved the robustness of the calibrations. Based on an independent prediction set of 2,270 samples, the following accuracy values (*SEP* corrected for bias) were achieved: crude protein: 0.59%; crude fat: 0.32%; moisture: 0.39%; and crude fibre: 0.51%. The accuracy for moisture was slightly disappointing, probably due to the fact that it is often difficult to take spectra immediately before drying, and changes may occur between taking spectra and reference analyses. The *SEP* values for the other parameters met commercially acceptable standards.

As the data for the ANN models were derived from more than ten instruments, instrument variation was, to a large extent, built into the model. Only a simple bias-adjustment is, therefore, required for calibration transfer to new units.

Use of ANN for crop data measured in transmission

The possibility of using ANN with transmission data from cereals for the determination of protein and moisture was first proposed by Büchmann in 1995.[3] Initial models were based on European data only. Later, Australian and North American data were added to the model and extensive testing has demonstrated the models can be applied worldwide. The European ANN models were first used commercially in 1996 by the Swedish Infratec Network. Since then, a number of European and Australian NIT networks have adopted the (now global) ANN models and extensive testing has been conducted in North America.

In an evaluation with 481 US wheat samples, ranging from 8 to 22% protein, the *SEP* was 0.24%. With moisture, the improvement obtained by using ANN instead of PLS was very noticable. In a set of Danish barley samples, ranging from 13% to 26% moisture, the *SEP* for the PLS model was 0.75%, and predictions were non-linear over 18%. Using ANN, the *SEP* was reduced to 0.22% and the model was linear over the entire range. A comprehensive comparison between the PLS and ANN models for Danish crops has been published elsewhere.[4]

A single global ANN model for protein is suitable for wheat, barley and rice, while the moisture model can be used with wheat, barley, rye, oats and triticale. Currently, ANN models are in development for moisture and oil in rapeseed, for moisture and amylose in rice and for moisture and oil in maize. These models are, as yet, still restricted in terms of geographical coverage and cannot be considered completely global at this stage. The data set for cereals now consists of nearly 40,000 records including samples from the last 15 years, obtained from all over the world and measured on Foss Infratec types 1221, 1225, 1229, 1241 and 1275. With such extensive sample variation built into the database, an annual update of the global models is sufficient. After updating, the performance of each new model is then evaluated with independent data sets from each NIT network and network-specific bias adjustments are applied where appropriate. A total integration of ANN models across different networks is, thus, still one step away from completion. This, however, is not due to any shortcomings of the global ANN models themselves; imperfect harmonisation of the reference analyses used at the individual control laboratories is certainly a major influence.

Characteristic differences between Infratec transmission and NIR reflectance spectra

The applications mentioned above all relate to Foss Infratec instruments, where the spectra consist of 100 data points over a narrow wavelength range from 850 nm to 1050 nm measured in transmission. Reflectance spectra from the Foss NIRSystems (NIR) 5000 series consist of 700 data points over a range from 1100 nm to 2500 nm. NIR spectra are thus more complex than those from Infratec instruments and, because of their greater range, tend to display more non-linearity. However, absorbance levels measured in transmission are, typically, in the range between 2 to 4 O.D., where similar samples measured in reflectance rarely exceed 1.2 O.D. Transmission data, therefore, tend to show greater deviation from Lambert–Beer's Law compared with reflectance spectra.

Use of ANN for ground samples of compound feed measured in reflectance

Most of the major Infratec transmission applications are based on models developed centrally, using ANN combined with a Foss Tecator proprietary pre-processing technique. ANN as a calibration tool has never been extensively used for agricultural data measured in reflectance, perhaps because non-Foss expert users developed these applications and, without the proprietary pre-processing technique, results may have been disappointing. Over the last year, however, Foss Tecator and ISI initiated extensive evaluations of modified partial least squares (MPLS) regression, a proprietary ISI technique, and ANN as alternative calibration methodologies for reflectance data from FossNIRSystems instruments. An evaluation of the proprietary ISI technique, LOCAL, was an equal part of the studies, but these results are outside the scope of the current proceedings. One study focused on dry, ground forage data from Australia, Europe and North America. Results relating to this study will be presented separately by Berzaghi[5] and Dardenne[6] at this conference. The second study was carried out on compound feed and feed ingredients of plant origin using historical data from Australia, Europe and North America. Some of the findings with compound feed and feed ingredients will be discussed here.

Historical data derived from Foss NIRSystems instruments trimmed to NIRS5000 format (1100 to 2500 nm) were obtained from Australia, Belgium, Holland, North America and the UK. All of these instruments were standardised to the ISI Master (Ser. No. 1272). The combined modelling set consisted of 6,115 records. All types of feed and feed ingredients were combined into a single ANN model. An independent data set for validation of the models consisting of 530 records of mixed feed collected on an instrument that was not standardised against the ISI Master was obtained from a single Italian site. The *SEP* values for the independent data set were 1.12% for crude protein and 0.79% for crude fibre.

Although these values are not accurate enough to be commercially acceptable, the study does indicate that data from very diverse samples can be combined in broadly-based models. Several factors should be taken into account. Firstly, the independent test sets were not standardised to the ISI master, and hence not to the data set used to derive the model. Secondly, even though the spectra used to derive models were standardised, these standardisation exercises were not always performed using to the same protocol. Finally, there is no certain way to ensure that the different, historical sources of reference data are properly harmonised. The study did, however, confirm that correcting for current laboratory differences by means of a ring test did improve the performance of the resulting models.

Use of ANN for unground samples of compound feed measured in reflectance

Until recently, near infrared measurements of compound feed were typically carried out on samples ground to a 1 mm particle size (reflectance) or 2 mm size (transmission). However, near infrared measurements of such products can be further simplified by omitting grinding altogether and measuring whole feed pellets directly. As part of the calibration methodology study with feed, we compared PLS with ANN on a set of unground samples of feed for pigs, cattle and poultry (both broiler and layer). A separate MPLS model was developed for each of these feed classes, while the ANN model was based on all four classes combined. The ANN calibration set consisted of 6,142 protein, 4,860 fat and 4,967 moisture values. The results, expressed as the *SEP* values averaged over the different feed classes, are shown in Table 1. These results indicate that ANN, on average, offered a 10% improvement over PLS. In addition, the combined ANN model is easier to use and support compared with several individual PLS models. While not as low as the corresponding *SEP* figures for ground samples (data not shown here), the *SEP* values for whole pellets are sufficiently good for practical use (Jos Zegers, personal communication).

Use of ANN for molasses measured in reflectance

ANN has also been studied briefly for use with liquid samples. Purity of molasses is defined as Pol*100 $Brix^{-1}$, where Pol is the optical rotation and Brix the soluble solids. ANN and PLS calibration models developed from approximately 1,000 samples were evaluated using an independent dataset with 80 records. With PLS, the *SEP* for purity was 0.40, too high for commercial use; with ANN, the *SEP* improved to 0.30, approaching the limit for a useful calibration (Monique Seegmans, personal communication). However, further testing with the ANN model indicated that the *SEP* obtained was not reliable, probably because a calibration set of only 1,000 records was lower than the size usually considered necessary for ANN to produce stable models.

Table 1. Accuracy of ANN vs PLS models for unground samples of compound feed.

% improvement in *SEP* of ANN over PLS				
	N	ANN	PLS	% improvement
Protein	710	0.89	1.05	16
Fat	580	0.60	0.62	3
Moisture	546	0.35	0.39	11

Spectral robustness of ANN and PLS models

David Funk has studied the effect of deliberate spectral pertubations on the prediction output of various calibration models for wheat protein (personal communication, extracted from an unpublished presentation given at a GIPSA NIRT Calibration/Standardisation Meeting, May 16–17 2000, Kansas City, Kansas, USA).

The perturbations included absorbance offset, pathlength increase, stray light addition, bandwidth changes, wavelength axis stretch, wavelength axis offset and wavelength stretch offset. The models assessed were six separate official GIPSA PLS calibrations compared with the global ANN model from Foss Tecator (WHPR2121). Funk concluded that the ANN model was less sensitive to most disturbances than the GIPSA PLS calibrations. Thus, the ANN model was more able to cope with instrument differences than the PLS models. In a broader context, the study is an elegent (although indirect) approach allowing a comparison between the behaviour of non-linear and linear calibration methods.

Conclusion

Some distinct features of ANN techniques for agricultural data can be summarised as follows:

- ANN, because it is a non-linear technique, is better than PLS where the range of concentration is large where there is non-linearity in the dataset.
- ANN can combine diverse samples into broadly-based models, simplifying calibration support.
- Although ANN is often vastly superior to PLS when dealing with high absorbance levels in transmission measurements, ANN has also consistently been found to be more accurate than PLS with reflectance data.
- ANN requires very large data sets; real improvements can still be found with more than 20,000 spectra. While such datasets are costly to procure and complicated to compile, they still offer the best solution to the problems of long-term stability and cost-efficient calibration support.
- Enhancement of ANN models only required a few additional samples and existing levels of performance were not upset by the new data.
- Harmonisation of laboratories improved performance of the models, regardless of whether they were based on MPLS or ANN. However, historical datasets can not be reliably harmonised because the records required rarely exist.
- ANN is less sensitive to most spectral pertubations than PLS and can cope better with instrument differences.

Acknowledgements

The molasses results are published with the generous permission of Mrs Monique Steegmans, Tiense Suikerraffinaderij-Orafti, Tienen, Belgium while those for unground compound feed pellets are published with the generous permission of Mr Jos Zegers, Maasweide Laboratory Services, Boxmeer, The Netherlands. The results for spectral robustness are extracted with the generous permission by Mr David Funk, GIPSA (USDA), Kansas City, MO, USA.

References

1. C. Borggaard and A.J. Rasmussen in *Near Infrared Spectroscopy—Bridging the gap between data analysis and NIR applications*, Proc. of the 5th Int. Conf. on NIR Spectrosc., Haugesund, Norway, Ed by K.I. Hildrum, T. Isaksson and T. Næs. Ellis Horwood, Chichester, UK, p. 73 (1992).

2. N.B. Büchmann, in *Leaping ahead with Near Infrared Spectroscopy*, Ed by G.D. Batten, P.C. Flinn. L.A. Welsh and A.B. Blakeney. Royal Australian Chemical Australian Institute, Victoria, Australia, p. 248 (1996).
3. N.B. Büchmann, in: *Near Infrared Spectroscopy: The Future Waves*, Ed by A.M.C. Davies and P. Williams. NIR Publications, Chichester, UK, p. 479 (1996).
4. N.B. Büchmann, H. Josefsson and I.A. Cowe, *Cereal Chemistry* **78(5),** 572 (2002).
5. P. Berzaghi, P.C. Flinn, P. Dardenne, M. Lagerholm, J.S. Shenk, M.O. Westerhaus and I.A. Cowe, *Changing the world with NIR, (*in press).
6. P. Dardenne, I.A. Cowe, P. Berzaghi, P.C. Flinn, M. Lagerholm, J.S. Shenk and M.O. Westerhaus in *Changing the world with NIR*, (in press).

Near infrared spectroscopy for evaluation of apples using the K-mean algorithm

Masahiro Muramatsu,[a] Yoshiyasu Takefuji[a] and Sumio Kawano[b]

[a]*Graduate School of Media and Governance, Keio University 5322 Endo, Fujisawa 252-0861, Japan*

[b]*National Food Research Institute, 2-1-2 Kannondai, Tsukuba 305-8642, Japan*

Introduction

Nondestructive quality evaluation schemes for agricultural products using near infrared (NIR) spectroscopy have been developed since Norris detected the difference in moisture content in grain in 1965.[1] Norris revealed that diffuse reflectance and transmittance spectra of agricultural products contain information about chemical structures because each of the structures has specific absorption properties.[2] This detection is profitable for both producers and consumers, since NIR can predict internal compositions of products. The producers can provide valuable products and consumers can receive high quality foods. Through Norris' contribution, NIR spectroscopy has been used practically and spread widely as an automatic on-line method for evaluating food quality.

A number of studies using NIR spectroscopy have been performed to determine compositions of different types of fruits and vegetables. Most research uses the regression approach and shows high correlations as follows. Birth *et al.* determined dry matter in onions[3] and Dull *et al.* determined soluble solids in cantaloupe melons.[4] Kawano *et al.* analysed sugar content of intact peaches and showed high correlation ($R = 0.97$) between NIR measurements and Brix values which indicate the sweetness of the fruit.[5] Temma and his colleagues revealed high correlations between NIR measurements and constituents of apples including Brix value ($R = 0.94$), sourness ($R = 0.83$) and firmness ($R = 0.75$).[6]

In this paper, a K-mean algorithm is applied to the analysis of intact apples for sugar content. The K-mean algorithm is a clustering algorithm proposed by MacQueen.[7] The clustering is a grouping of data with similar characteristics and is used for various data analyses including spectral computing.[8]

K-mean algorithm

The K-mean algorithm is a clustering method for grouping data with similar characteristics and the steps of the algorithm are as follows.[9]

1. Partition the input vectors into K clusters randomly and compute the central points of the clusters as an initial condition.
2. Assign each vector to the cluster with the nearest central point and update the central points and the clusters.
3. After several iterations, when the change of the cluster becomes small, the programme is terminated.

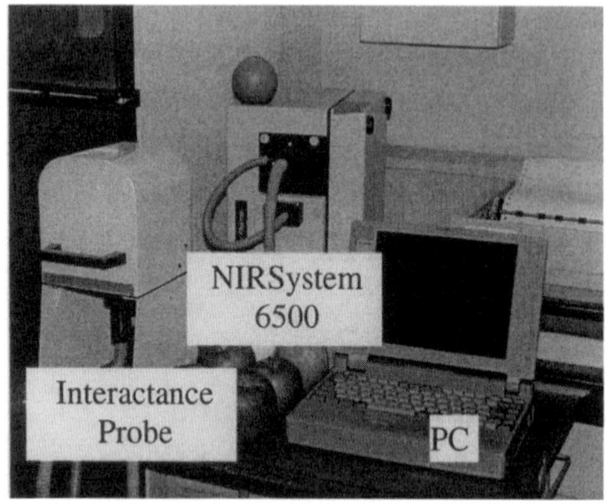

Figure 1. NIR instrument.

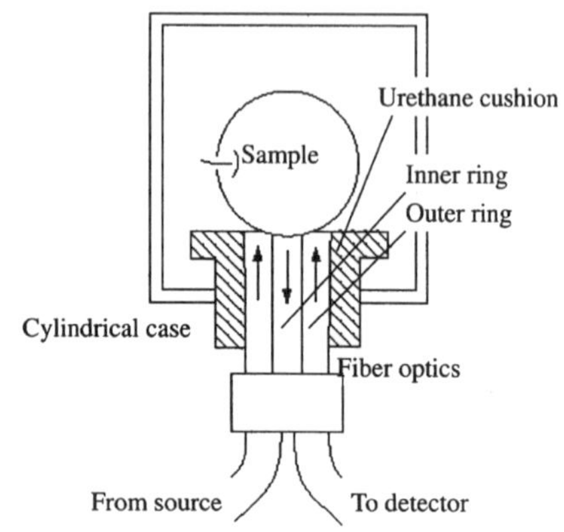

Figure 2. Interactance probe.

Experiments

Spectral acquisition

NIR spectra were acquired with an NIR instrument shown in Figure 1. The instrument has three components. One component is an "Interactance Probe". The fibre-optic probe used has an outer ring illuminator and inner ring receptor (Figure 2). The sample was located on the urethane cushion in a cylindrical case. Another component is an NIR spectrophotometer called NIRSystem 6500. A absorbance of the sample was measured by comparing near infrared energy reflected from the sample with that from the standard reference (8 cm-diameter teflon sphere). Teflon was used as a standard material because it has low absorption. Birth *et al.* also used a Teflon rod as a standard reference. The other component is a PC.

67 intact apples (variety Fuji), cultivated in Aomori prefecture in Japan, were used as the sample. The experiment was performed at 22°C. The spectral data were acquired every 2 nm from 400 to 1100 nm and data from 750 to 1050 nm were used in subseuqent calculations.

Chemical measurement

Reference chemical measurements of Brix values were performed after NIR spectra were acquired. The Brix values are determined using a commercially available refractometer.

Results and discussion

Figure 3 shows the original NIR spectra of the 67 apples. Since NIR spectra of foods are composed of broadbands arising from overlapping absorption, 2nd derivative spectra are used for calculation. With the single regression, the highest correlation coefficient is –0.654 when the wavelength of the spectrum is 904 nm (segment size is 8 nm and gap size is 0 nm). Therefore, the following calculations were performed using the conditions mentioned above. Table 1 is a quotation from *Near Infrared Spectroscopy of Food Analysis*[10] and shows absorbance wavelengths against specific chemical structures.

Table 1. NIR absorbance wavelengths.

Vibration	Wavelength (nm)
C–H str. third overtone	874, 900, 913, 938
O–H str. second overtone	990, 1000
2 × C–H str. + 3 × C–H def.	1015

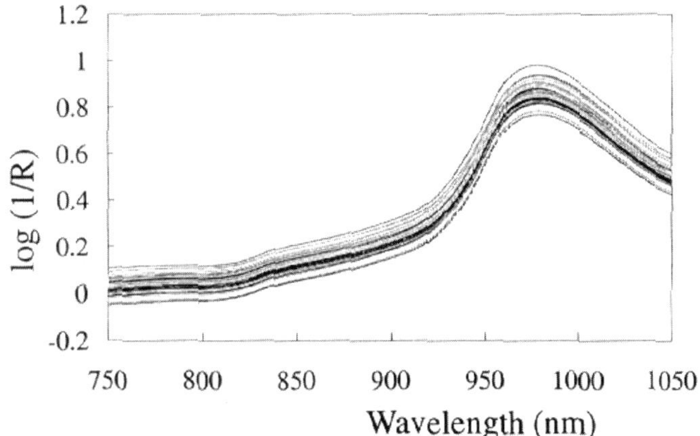

Figure 3. Original spectra of apples (*R*: reflectance).

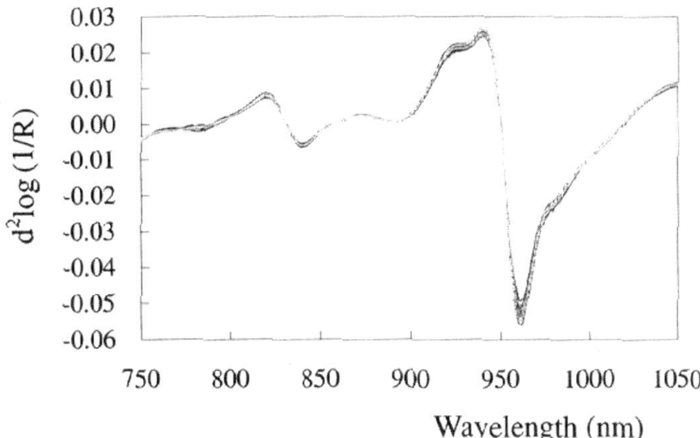

Figure 4. 2nd derivative data (*R*: reflectance).

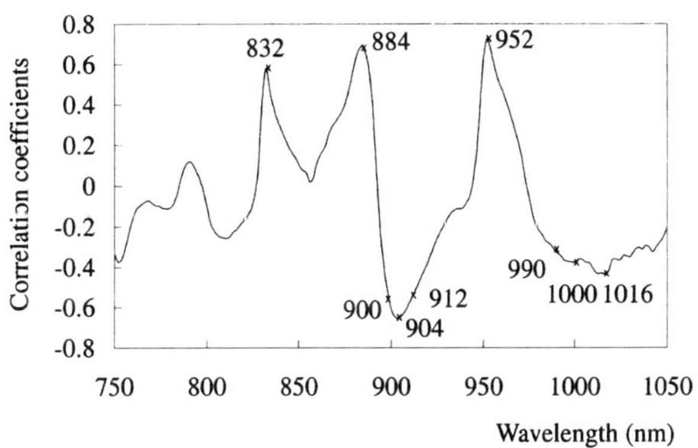

Figure 5. Correlation coefficients between $d^2\log(1/R)$ and Brix value plotted against wavelength.

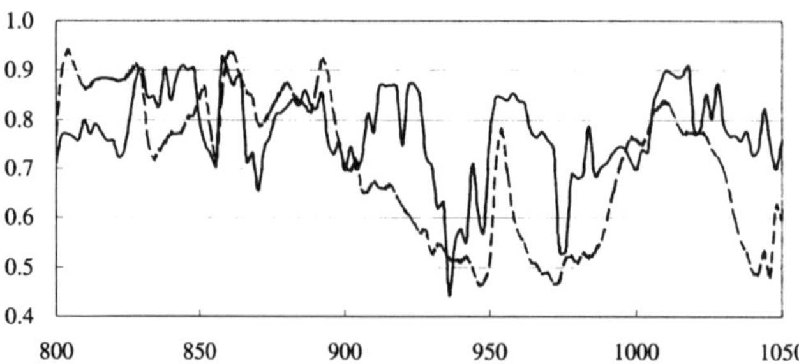

Figure 6. Correlation coefficient between actual and computed Brix value plotted against the second wavelength (the first wavelength: 904 nm.

Figure 4 shows the 2nd derivative spectra and Figure 5 shows correlation coefficients between the Brix values and 2nd derivative spectra which are called "correlation plots". In the case of the 2nd derivative spectra, the correlation plots should show negative peaks at the sugar bands. Therefore, the three wavelengths, 832, 884 and 952 nm, which show high positive correlation coefficients, are not regarded as the absorbance wavelengths.

The K-mean analysis was performed with a 57 sample calibration set and a 10 sample prediction set. The input is the spectra at the two wavelengths and the output is five Brix values which are the mean Brix values of the samples of the five clusters in the calibration set. The number of iterations is 100. In order to estimate the K-mean results, the regression results are measured (Figure 6). Figure 7 shows the plots of the samples in the calibration set. The circles indicate the central points of the cluster and the other marks indicate the samples which belong to the same cluster. The K-mean results show correlations between the actual and predicted Brix values as high as those produced by the regression results. In Figure 6, K-mean results are shown for the best values of correlation coefficients when initial conditions are changed. The highest correlation coefficient of K-mean is 0.939. However, the coefficient is calculated with the spectra at 804 and 904 nm. According to the NIR absorbance wavelengths shown in Table 1, 801 nm is not considered as an NIR absorbance. Then, the highest correlation coeffi-

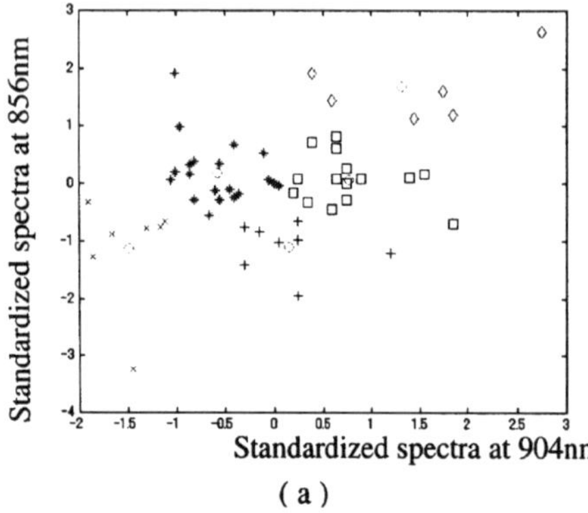

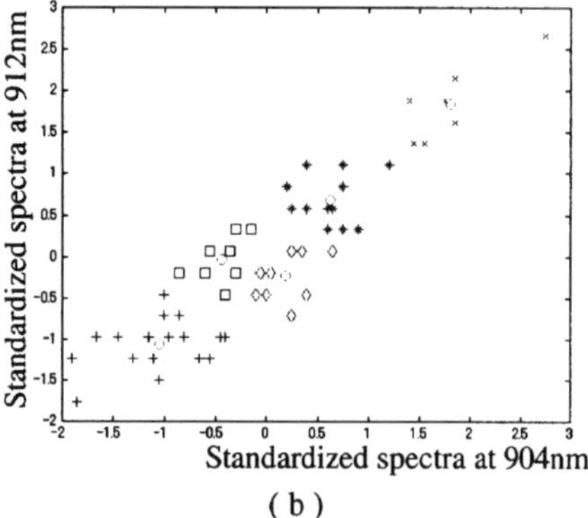

Figure 7. Plots of the sample . The clustering results with the spectra at (a) 856 and 904 nm and (b) 904 and 912 nm. The circle indicates the central point of the cluster and the other marks indicate the samples of each cluster.

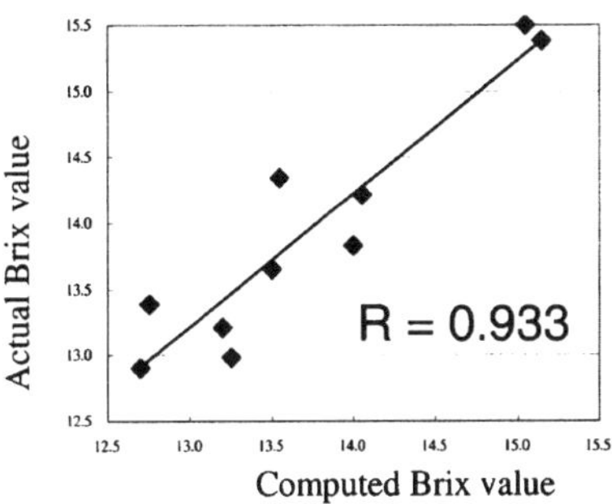

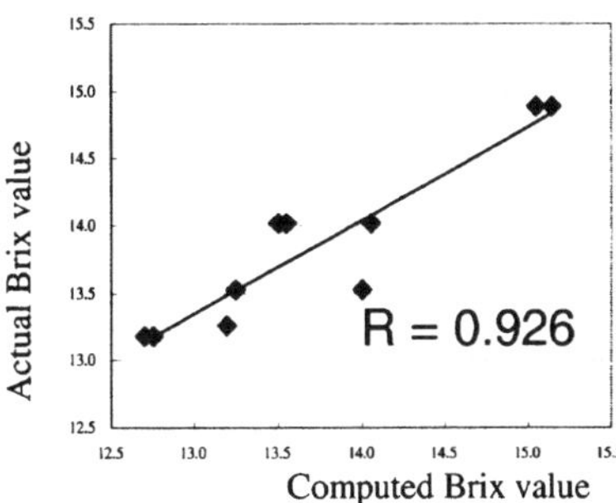

Figure 8. Plot of actual vs predicted Brix value. The regression result is calculated using (a) wavelengths 862 and 904 nm and (b) the K-mean result is calculated using wavelengths 865 and 904 nm.

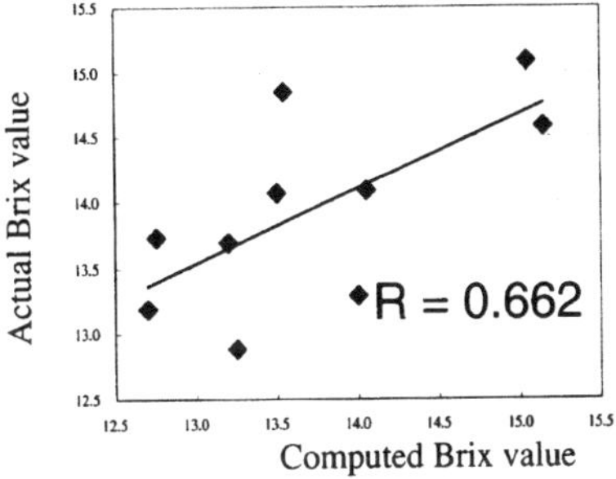

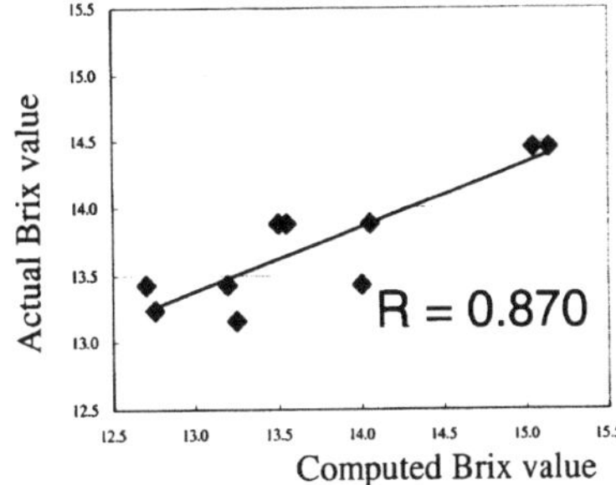

Figure 9. Plot of actual vs predicted Brix value. The results are calculated using wavelengths of 904 and 912 nm for both (a) the regression and (b) the K-mean procedures.

cient related to the absorbance is 0.926 at 865 and 904 nm [Figure 8(b)]. On the other hand, that of regression is 0.933 at 862 and 904 nm [Figure 8(a)].

The difference between K-mean and regression occurs when the spectra at 904 and over 906 nm are used for the analysis. When the spectra at 904 and 912 nm are chosen, the K-mean result shows a higher correlation than the regression (Figure 9). It is noted that 912 nm is considered as a NIR absorbance in Table 1.

Finally, it is confirmed that the K-mean analysis shows high correlations between the actual and computed Brix values corresponding to the selection of the used wavelengths.

Conclusion

In this paper, a new approach analysing near infrared spectroscopy of apples using K-mean algorithm is proposed. The result of K-mean analysis shows correlations between the actual and computed Brix values which are as high as those produced by regression analysis. Moreover, it is confirmed that the K-mean analysis shows high correlations between the actual and computed Brix values corresponding to the selection of the wavelengths used.

References

1. D.R. Massie and K.H. Norris, *Transaction of the ASAE* **8(4),** 598 (1965).
2. P. Chen, *Proc. Int. Conf. On Agric. Machinery Engineering*, **1,** 171 (1996).

3. G.S. Birth, G.G. Dull, W.T. Renfroe and S.J. Kays, *J. Amer. Soc. Hort. Sci.* **110,** 297 (1985).
4. G.G. Dull, G.S. Birth, D.A. Smittle and R.G. Leffler, *J. Food Sci.* **54,** 393 (1989).
5. S. Kawano, H. Watanabe, M. Iwamoto, *J. Japan Soc. Hort. Sci.* **61,** 445 (1992).
6. The Report of Aomori Advanced Industrial Technology Center, (1992).
7. J. MacQueen, *5th Berkeley Symp. Math. Statist. Prob.* **1,** 281 (1967).
8. H.C. Romesburg, *Cluster Analysis for Researchers.* Lifetime Learning Publications, California, USA (1984).
9. L. Kaufman, *Finding groups in data.* Wiley-Interscience, New York, USA (1990).
10. B.G. Osborne and T. Fearn, *Near Infrared Spectroscopy in Food Analysis*, Longman Scientific & Techinical, Harlow, UK (1986).

Application of Benfor's equations to the problem of "seeing through layers"

Georgi P. Krivoshiev,[a] Raina P. Chalucova[a] and Donald J. Dahm[b]

[a]*Institute for Horticulture and Canned Foods – 154, V. Aprilov Blvd, Plovdiv-4000, Bulgaria*

[b]*TechMan Enterprises, 90 Chapin Greene Dr., Ludlow, MA 01056, USA*

Introduction

The V-method for non-destructive measurement of the internal transmittance of objects, was developed in 1996 by G. Krivoshiev.[1] The internal transmittance is obtained by correcting the overall transmittance spectrum with the diffuse reflectance spectrum. The method was used successfully in the sorting of fruits and vegetables according to their internal quality. There was an increase in the accuracy of assessment, because the contribution of the peel to the spectrum was eliminated without the products being peeled physically. Because of this, the method was called the V (virtual) method. (A detailed description of the method was published by Krivoshiev, Chalucova and Moukarev in the journal *LWT*[2] and Krivoshiev reported it during the 10th International Diffuse Reflectance Conference—IDRC-2000 inChambersburg, Pensilvania, USA. The applicability of the V-method for on-line sorting of potato tubers was presented by the same authors in *NIR news*[3] and also published as a poster at IDRC-2000 and an abridged version of the report.)

The method is based on the hypothesis that the objects can be treated as a three-layer structure according to Figure 1—two external layers, A and B and an internal layer, O. The total transmittance for this structure may be modelled using the following approximation

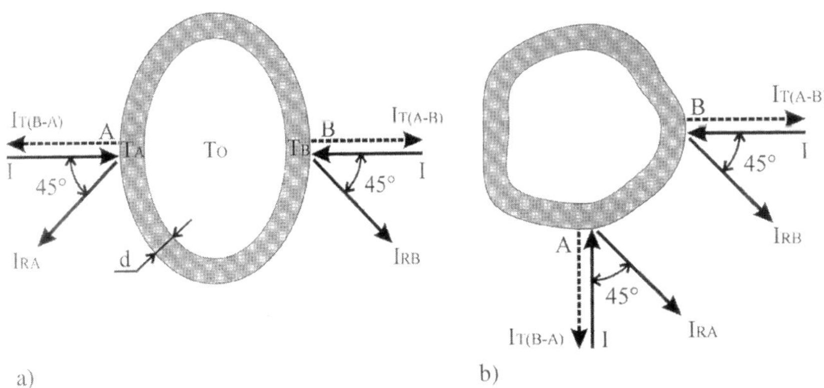

Figure 1. Schemes illustrating the procedure for the nondestructive measurement of the real product flesh transmittance T_O: (a) direct transmittance geometry ($T_{0/180}$) and (b) geometry $T_{0/90}$ — incident and reflected light flux; --- scattered/transmitted/emitted light flux.

$$T = T_A.T_O.T_B \tag{1}$$

which would be exact for plane-parallel layers from a non-scattering material. In formula (1), T is the overall transmittance of the object, and T_A, T_O and T_B are the transmittances of the layers A, B and O, respectively.

Since the external (surface) layers A and B have only one accessible side, their transmittances are measured by using the relationship known as the Kubelka–Munk (K–M) theory.[4]

$$T_{A(B)} = \frac{b}{ashbSd + bchbSd} = \frac{(1 - R_\infty^2)\exp(-Sd)}{1 - R_\infty^2 \exp(-2bSd)} \tag{2}$$

where: R is the diffuse reflectance from a layer with a thickness large enough so that a further thickness increase does not have an effect on the reflectance; Sd is the scattering power (S: scattering coefficient, cm^{-1}; d: average length of the light path, cm);

$$a = \frac{(1 + R_\infty^2)}{2R_\infty} \quad \text{and} \quad b = \frac{(1 - R_\infty^2)}{2R_\infty}$$

Substituting formula (2) in formula (1) and after straightforward transformations, Krivoshiev *et al.* (2000) and Krivoshiev (2000) have proposed two basic models for measuring the internal optical density. The first model is:

$$\begin{aligned} OD_O &= a_0 + a_1 \operatorname{In}(1/T) + a_2\, In\left(1 - R_A^2\right) + a_3 \operatorname{In}\left(1 - R_B^2\right) + \\ &+ a_4 \operatorname{In}\left[1 - R_A \bullet \exp(-2b_A \bullet (Sd)_A\right] + a_5 \operatorname{In}\left[1 - R_B^2 \bullet \exp(-2b_B \bullet (Sd)_B)\right] + \\ &+ a_6 b_A (Sd)_A + a_7 b_B (Sd)_B \end{aligned} \tag{3(a)}$$

The second model

$$OD_O = a_0 + a_1 \operatorname{In}(1/T) + a_2 \operatorname{In}(1/R_A) + a_3 \operatorname{In}(1/R_B) \tag{3(b)}$$

is based on the first one and is obtained by assuming that $R^2\exp(-2bSd) << 1$ and that there is high correlation between the values $(1 - R^2)$ and b on the one hand and $\ln(1/R)$ on the other hand. The models are optimised empirically.

The model [3(a)] contains the scattering power Sd of the surface layers, the value of which is unknown. In the V-method, this difficulty is overcome by setting Sd to a constant "average" value $\overline{Sd}$, which is obtained empirically for a given wave range and for a particular type of object. $\overline{Sd}$ is also included in the coefficients of the model [3(b)]. In the present work, this approximation, as well as that represented by the use of formula (1), is eliminated. The modelling is based not on the K–M theory, but the theory of discontinuum by using equations of Benfor.[5,6]

The hypothesis

Applying the Benfor equations[5,6] to the two-layer structure A + O, shown on Figure 1, we obtain

$$T_{A+O} = \frac{T_A \bullet T_O}{(1 - R_{A\cdot} \bullet R_O} \tag{4}$$

$$R_{A+O} = \frac{T_A^2 R_O}{(1 - R_{A\cdot} \bullet R_O)} + R_A \tag{5}$$

According to formula (5) the transmittance of the surface layer, A, can be expressed by using the reflectances

$$T_A = \sqrt{\frac{(R_{A+O} - R_A)(1 - R_A \bullet R_O)}{R_O}} \tag{6}$$

It is obtained analogously for T_B

$$T_B = \sqrt{\frac{(R_{B+O} - R_B)(1 - R_B \bullet R_O)}{R_O}} \tag{7}$$

The three-layer structure A + O + B can be presented as a two-layer one—(A + O) + B. According to equations (4) and (5) the overall transmittance in the direction A → B will be

$$T = T_{(A+O)+B} = \frac{T_{A+O} \bullet T_B}{(1 - R_{A+O} \bullet R_B)} \tag{8}$$

and the reflectance from side A.

$$R_{(A+B)+B} = R_{A+O} + \frac{T^2_{A+O} \bullet T_B}{(1 - R_{A+O} \bullet R_B)} \tag{9}$$

By replacing formulae (4), (5) and (6) into formula (8) we obtain the overall transmittance.

$$T = \frac{T_A \bullet T_O \bullet T_B}{(1 - R_A \bullet R_O)(1 - R_{A+O} \bullet R_B)} \tag{10}$$

The inclusion of the denominator in formula (10) removes the approximation included in formula (1). However, the values in the denominator cannot be measured non-destructively.

If we replace formulae (6) and (7) into equation (10), the expression of the overall transmittance will be based entirely on the Benfor's equations

$$T = T_O \frac{\sqrt{(R_{A+O} \bullet R_A)(1 - R_A \bullet R_{OA})(R_{B+O} - R_B)(1 - R_B \bullet R_{OB})}}{\sqrt{R_{OA} \bullet R_{OB}(1 - R_A \bullet R_{OA})(1 - R_{A+O} \bullet R_B)}} \tag{11}$$

It is obvious that T_O is a function of seven parameters, from which only three parameters—T, R_{A+O}, R_{B+O}—can be measured instrumentally without the destruction of the object. We assume additionally that:

(a) $R_{A+O} \approx R_{A\infty}$ and $R_{B+O} \approx R_{B\infty}$ because the thickness of the intermediate layer is sufficiently big;

(b) the reflectance R_O of the internal layer O is marked with two indices, A and B, in order to underline that the object inside is not homogenous as a matter of principle, thus the inside reflectance in the directions A → B and B → A is not equal, i.e. $R_{OA} \neq R_{OB}$. Therefore, the theory of discontinuum could be applied only in cases when R_A, R_O and R_B are *a priori* known.

The analysis of equation (5) and the pure physical considerations indicate that if, $R_{A+O} >> R_{OA}T^2_A$, $R_A \approx R_{A+O}$. It seems that for the objects, approximately complying with that requirement (such as fruits, vegetables, eggs, etc.) a reasonable solution for practical utilisation of equation (10) is the transmittances T_A and T_B, may be obtained from the K-M equation (2). By replacing formula (2) for T_A and T_B into formula (10) and by putting $R_A \approx R_{A+O} = R_{A\infty}$, $R_B \approx R_{B+O} = R_{B\infty}$ $R^2e^{-2bSd} << 1$, we obtain

$$T = T_O \frac{(1 - R_A^2)(1 - R_B^2)\exp(-b_A(Sd))_A - b_B(Sd)_B}{(1 - R_A R_{OA})(1 - R_A R_B)} \tag{12}$$

Hence it is easy to obtain the regression model for the determination of the internal optical density

$$OD_O = \text{In}(1/T_O) = a_0 + a_1 \text{ In}(1/T) + a_2 \text{ In}(1 - R_A^2) + a_3 \text{ In}(1 - R_B^2) + \quad [13(a)]$$
$$+ a_4 b_A + a_5 b_B + a_6 \text{ In}\left(1 - R_A \overline{R_{OA}}\right) + a_7 \text{ In}(1 - R_A R_B)$$

R_{OA} is the average reflectance, which is obtained empirically. A simplified version of this model is

$$OD_O = \text{In}(1/T_O) = a_0 + a_1 \text{ In}(1/T) + a_2 \text{ In}(1 - R_A) + a_3 \text{ In}(1 - R_B) + \quad [13(b)]$$
$$+ a_4 \text{In}\left(1 - R_A \overline{R_{OA}}\right) + a_5 \text{ In}(1 - R_A R_B)$$

Conclusion

From the comparison of formulae (3) and (13), it would seem that the last two terms of formula (13) should yield higher accuracy of the OD_O prediction than formula (3), but this is a question that must be answered experimentally for the specific case. It may be presumed that for objects with a great degree of varying internal transmittance it is more suitable to use formula (3) and for objects with a great degree of varying reflectance of the surface layer, formula (13).

Acknowledgement

This Poster Presentation became possible on account of the first prize that a poster presented by the author was awarded at the IDRC-2000 as well as the European Commission financial support for the Research Project QLK1-2000-00455 of the Fifth Framework Program of the EU.

References

1. G. Krivoshiev, *Bulgarian Patent BG No. 62304* (in Bulgarian) (1998).
2. G. Krivoshiev, R. Chalucova and M. Moukarev, *Lebensm.-Wiss. Technol.* **33(5),** 344 (2000).
3. G. Krivoshiev, R. Chalucova and M. Moukarev, *NIR news* **11(5),** 7 (2000).
4. D. Judd and G. Wyszecki, *Color in Business, Science and Industry.* John Wiley & Sons, New York, USA (1975).
5. D. Dahm and K. Dahm, *J. Near Infrared Spectrosc.* **7,** 47 (1999).
6. D. Dahm and K. Dahm, *Appl. Spectrosc.* **53(6),** 647 (1999).

Spectroscopic and chemometric analysis of short-wave near infrared spectra of sugars and fruits

Mirta Golic, Kerry Walsh and Peter Lawson

Plant Sciences Group, Primary Industries Research Center, Central Queensland University, Rockhampton, 4702, Australia

Introduction

Fruit sweetness, as indexed by total soluble solids (TSS), is a key factor in the description of fruit eating quality. Analysis using near infrared (NIR) spectroscopy can be based on pure statistical methods, but knowledge of the basic band assignments for the compounds of interest is advantageous. Previous considerations of this topic have focussed on the 1100–2500 nm spectral region.[1–5] Given the requirement for speed of assessment and the consequent use of Si diode array hardware in fruit sorting applications, a further consideration of band assignments in the short-wave NIR (SW-NIR) and the effect of changes in sample environment is warranted. A combination spectroscopic and chemometric approach has been adopted, following that of Maeda *et al.*[6,7]

Materials and methods

Sucrose, D-glucose and D-fructose were of analytical grade (BDH). Deuterium oxide of 99.93% isotopic purity was obtained from the Australian Nuclear Science and Technology Organisation. NIR spectra (400–2500 nm) were collected in triplicate for each sample, at each temperature (15–50°C, with 5°C increments) using a scanning spectrophotometer (model 6500, NIRSystems Inc, Silver Spring, MD, USA), equipped with a transmittance detector and a transport module with temperature controlling unit. A pathlength of 26 mm and an average of 32 scans were used for spectral acquisition. An empty cuvette was used as a reference. NSAS v. 3.30 software was employed for data collection and the Unscrambler v. 7.6 for the regression analysis. GRAMS32 v. 5.2 and kg-2d softwares were used for the production of the two-dimensional spectra.

Results and discussion

Spectroscopic analysis

NIR spectra of sugar solutions

Water OH groups display strong absorbances at around 750 and 960 nm in the SW-NIR spectral region (third and the second overtone bands of OH stretching vibrations). Sugar OH groups resonate at similar frequencies. The OH peaks of sugar solutions are better distinguished in deuterated water (for sucrose, Figure 1), with peaks at 740, 960 and 980 nm. The broad band at around 910 nm belongs to the third overtone of the sucrose CH stretching vibrations.

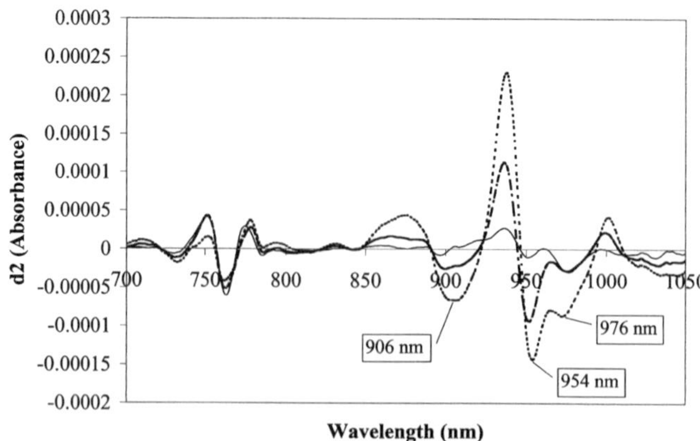

Figure 1. Second derivative spectra of sucrose solutions in deuterated water. — D_2O, –·– 20% sucrose and --- 50% sucrose.

The effect of temperature on the spectra of aqueous sugar solutions

Shifts of the water OH bands towards lower wavelengths upon increasing temperature conditions have been explained by the alteration in the structure of water and degree of hydrogen bonding.[6,8–12] At higher temperatures, singly hydrogen-bonded water molecules predominate. At lower temperatures, higher levels of hydrogen bonding occur, with vibration bands broadened and shifted towards longer wavelengths.

The spectra of the sugar solutions followed the same trend as the spectra of water, with bands at 750, 840, 960 and 985 nm slightly shifting towards shorter wavelengths and narrowing in bandwidth with increasing temperature. These bands appear as temperature-sensitive bands in the 2D spectra (Figure 2).

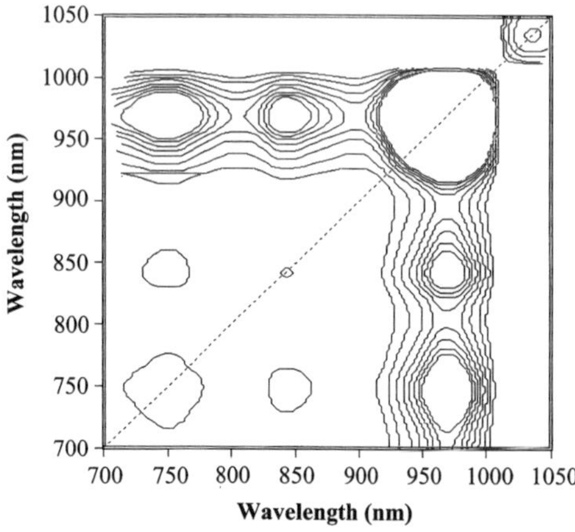

Figure 2. Influence of temperature variation (15–50°C) on the second derivative spectra of 50% sucrose solution.

The effect of concentration on the spectra of aqueous sugar solutions

Increase in sucrose concentration caused a decrease in the intensity of the large 964 nm band (second overtones of water–OH stretching vibrations) and an increase in the intensity of the smaller band at 982 nm (second overtones of sugar–OH stretching vibration) (Figure 3). The position of the 982 nm peak was moved towards higher wavelengths in more concentrated solutions.

Chemometric analysis

Prediction of temperature for aqueous sugar solutions

Temperature of 20% sugar solutions was predicted using the SW-NIR absorbance spectra.

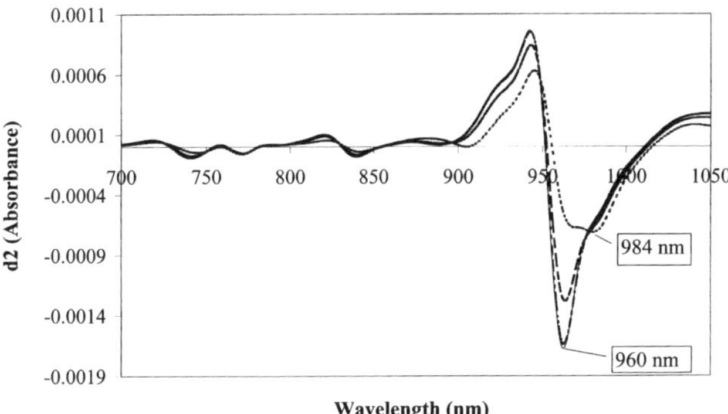

Figure 3. Influence of concentration variation (0, 5, 20 and 50%) on the second derivative spectra of sucrose solutions at 30°C. (— H_2O, –·– 5% sucrose, – – – 20% sucrose, ---- 50% sucrose).

The largest positive PLS1 B-coefficients were placed at 940 and 980 nm, with two other areas of positive coefficients at 750 and 840 nm, consistent with temperature sensitivity of the OH associated vibrations, related to the degree of hydrogen bonding.

Calibration on aqueous sugar concentration

PLS1 calibrations on sugar concentration exhibit high correlation coefficients (for example, at 30°C: R^2 0.989 in all cases, described with *ca* two PLS factors). The largest B-coefficients were placed at about 910 nm, consistent with third overtone bands of sugar–CH vibrations. Other positive B- coefficients were found in the region of 750–820 nm for sucrose and as two partially overlapped sets (750–770 and 790–820 nm) for D-glucose and D-fructose. As the latter regions are affected by temperature, it is to be expected that a calibration for sugar concentration, built across a range of temperatures, will downweight the coefficients in these regions. Indeed, PLS calibrations on concentration for the three sugars (Table 1) developed across a range of sample temperatures (with temperature treated as an x-variable) demonstrated a reliance on the 910 nm band (temperature insensitive CH stretching band). Negative coefficients at 840 nm, present in a single temperature calibration, were not present.

Explaining the SW-NIR spectra of fruit

B-coefficients of the sugars are very similar to those of intact apples, peaches and nectarines (Figure 4) in the 800 950 nm spectral region. This explains why these fruits exhibit excellent calibration and prediction results in the SW-NIR (for example, R^2 for apples (100 fruit, 10 PC), was 0.921, SEC = 0.227, $RMSECV$ = 0.276, SD = 0.99).

Table 1. PLS1 calibration on sugar concentration (absorbance data, 700–1050 nm, temperature added as an *x* variable).

Sugars	#Samples	#PC	R^2	*SEC*	*RMSECV*	Mean	*SD*
Sucrose	429	4	0.993	1.10	1.12	26.48	13.50
D-Glucose	362	3	0.987	1.83	1.86	23.12	16.30
D-Fructos	359	4	0.993	1.29	1.32	23.09	16.30

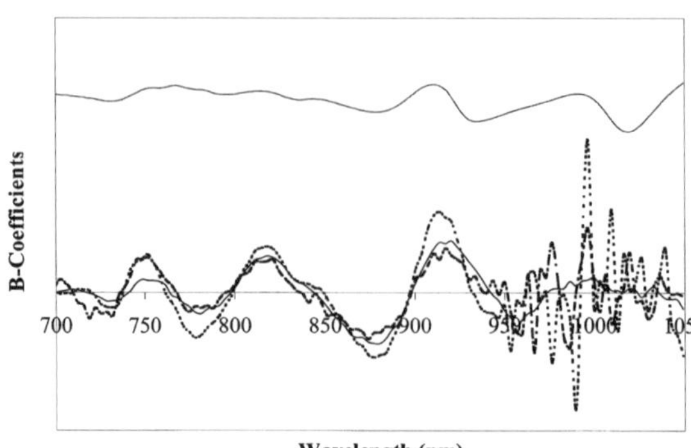

Figure 4. Plot of regression coefficients from PLS calibration on °Brix for apples, peaches and nectarines (absorbance data, 700–1050 nm, temperature added as an *x* variable) (— apples, --- peaches, –-– nectarines). Curve on top represents regression coefficients from PLS calibration on concentration (% w/w) for sucrose solutions, obtained under the same conditions.

Acknowledgements

We thank Dr Yan Wang who has programmed and kindly provided the kg-2d software.

References

1. G.G. Dull and R. Giangiacomo, *J. Food Sci.* **49,** 1601 (1984).
2. R. Giangiacomo, G.G. Dull, *J. Food Sci.* **51,** 679 (1986).
3. J.B. Reeves, III, *J. Near Infrared Spectrosc.* **2,** 199 (1994).
4. J.B. Reeves, III, *J. AOAC Int.* **77,** 814 (1994).
5. J.B. Reeves, III, *Appl. Spectrosc.* **49,** 181 (1995).
6. H. Maeda, Y. Ozaki, M. Tanaka, N. Hayashi and T. Kojima, *J. Near Infrared Spectrosc.* **3,** 191 (1995).
7. H. Maeda, Y. Wang, Y. Ozaki, M. Suzuki, M.A. Czarnecki and M. Iwahashi, *Chem.& Intell. Lab. Systems* **45,** 121 (1999).
8. W.A.P. Luck and W. Ditter, *J. Phys. Chem.* **74,** 3687 (1970).
9. W.C. McCabe, S. Subramanian and H.F. Fisher, *J. Phys. Chem.* **74,** 4360 (1970).
10. H. Abe, T. Kusama, S. Kawano and M. Iwamoto, *Bunko Kenkyu* **44,** 247 (1995).
11. K. Buijs and G.R. Choppin, *J. Chem. Phys.* **39,** 2035 (1963).
12. A.D. Bianco, R. Benes, M. Trinkel, F. Reininger and M. Leitner, In *Proceddings of 9th International Conference on Near Infrared Spectroscopy,* Ed by A.M.C. Davies and R. Giangiacomo. NIR Publications, Chichester, UK, pp. 81–87 (2000).

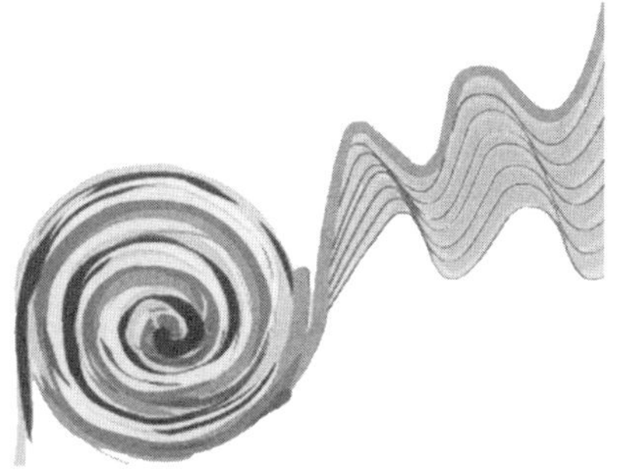

Interpretation of NIR Statistics

Derivatives—a systematic approach to removing variability before applying chemometrics

David W. Hopkins

NIR Consultant, Battle Creek, Michigan, USA.
E-mail: dwh @ voyager.net

Introduction

Derivatives often present the data in spectra in a manner that brings out the information that we are interested in, such as the number and position of bands and their relative intensity and removes unwanted baseline variation. Two major methods of calculating derivative spectra are the Savitzky–Golay method of polynomial curve-fitting and the segment–gap method. Both of these methods can be presented and/or performed as convolution processes. The definition of segment–gap derivatives is clarified by presenting them as convolution functions. A new, normalised form of segment–gap derivative is derived that preserves the basic shapes of the calculated derivatives but brings the results into agreement with the Savitzky–Golay results over a range of parameters in both methods. The effectiveness of the convolution functions in removing the random high frequency noise in the spectra can be evaluated by calculating an index called RSSK/Norm, the square root of the sum of the squares of the convolution coefficients, divided by the normalisation constant. The agreement and utility of the two methods of calculating derivatives is demonstrated by evaluating the results of using derivatives up to the fourth order on the second overtone aromatic CH band of polystyrene, as an example, in the 1680 nm region of the NIR spectrum.

Basics

First, let us consider measurements of y (or Absorbance) measured at equal intervals of x (or wavelength). In reality, a derivative cannot be calculated in the mathematical sense from measurements taken at discrete intervals, because a derivative is defined for continuous functions as the limit of $\Delta y / \Delta x$ as Δx approaches 0. This analytical geometry approach is suggested by the dotted line "tangent" to the curve at 1672 nm in Figure 1. Therefore, the finite difference in y values for a selected difference in x is taken as an approximation to the desired first derivative. Various methods of calculating derivatives include the point-difference method, gap method, Savitzky-Golay[1] method and segment–gap method. (The segment–gap method was called the Norris Method in recent reviews.[2,3] However, in accordance with Karl Norris' wishes, the non-personal reference should be applied.) The first two methods are available in Grams software,[4] and may be considered special cases of the segment–gap method. The Savitzky–Golay method is implemented in many software packages, Grams, Pirouette,[5] Unscrambler[6] and Vision/NSAS software,[7] to name some I have used. There is not much detail written about the segment–gap method outside of the manuals from Foss/NIRSystems, so this method will be described in some detail here.

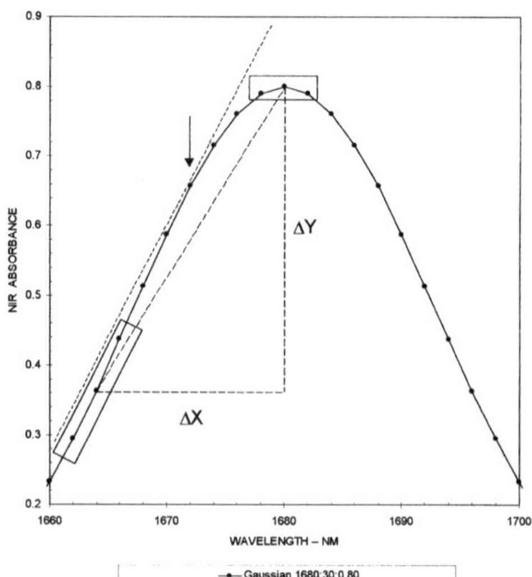

Figure 1. Calculation of a first derivative.

Savitzky–Golay method

Figure 1 illustrates the calculation of a first derivative. In this example, the slope $\Delta y / \Delta x$ is calculated for an interval of nine points. In the Savitzky–Golay method, the data points in the region around a central point (arrow in Figure 1) are fitted to a polynomial and the analytical value of the derivative at

Table 1. Savitzky–Golay method: first derivatives.

Index	Quadratic polynomial fitting				Cubic polynomial fitting			
	5-point	7-point	9-point	11-point	5-point	7-point	9-point	11-point
–5	—	—	—	–5	—	—	—	300
–4	—	—	–4	–4	—	—	86	–294
–3	—	–3	–3	–3	—	22	–142	–532
–2	–2	–2	–2	–2	1	–67	–193	–503
–1	–1	–1	–1	–1	–8	–58	–126	–296
0	0	0	0	0	0	0	0	0
1	1	1	1	1	8	58	126	296
2	2	2	2	2	–1	67	193	503
3	—	3	3	3	—	–22	142	532
4	—	—	4	4	—	—	–86	294
5	—	—	—	5	—	—	—	–300
Norm	10	28	60	110	12	252	1188	5148
RSSK/Norm	0.316	0.189	0.129	0.095	0.950	0.512	0.338	0.246

the mid-point of the interval is taken as the value of the derivative at the wavelength of the central point. Savitzky and Golay[1] showed that the least-squares curve fitting and differentiation could be done in a single convolution operation, simply and elegantly. Convolution involves aligning the values of the convolution function with the values of the spectrum, multiplying the pairs of values, summing the products and dividing the sum by a normalisation constant. The result is entered as the value of the derivative at the central point. This process is then repeated by advancing to the points surrounding the next spectral point, so that a derivative spectrum is determined for the whole spectrum, point by point. They published convolution functions that enable the direct calculation of a smoothed spectrum, or derivatives up to an order of fifth derivative. Their tables contained a number of errors, which were corrected by Steinier *et al.*[8] In Figure 1, a 9-point convolution interval is used to calculate the value at 1672 nm. Values for points in the spectrum are calculated by sliding the convolution interval along the data and calculating the derivative, point by point. The loss of undefined values at the ends of the spectrum is generally not a concern.

Table 1 also contains a parameter for each convolution function called RSSK/Norm, the square root of the sum of the squares of the convolution coefficients, divided by the normalisation constant. Howard Mark[9] derived the relationship (Equation 79, p. 56) for an index of random error, the square root of the sum of squares of the calibration coefficients. This index expresses the sensitivity of calibration equations to the random errors in the spectral data. For results calculated by convolution, I have

Table 2. Segment–gap first derivatives.

Index	Segment, gap values (nanometers)								
	2, 2	2, 6	2, 10	6, 2	6, 6	6, 10	10, 2	10, 6	10, 10
–7	—	—	—	—	—	—	—	—	–1
–6	—	—	—	—	—	—	—	–1	–1
–5	—	—	—	—	—	–1	–1	–1	–1
–4	—	—	—	—	–1	–1	–1	–1	–1
–3	—	—	–1	–1	–1	–1	–1	–1	–1
–2	—	–1	0	–1	–1	0	–1	–1	0
–1	–1	0	0	–1	0	0	–1	0	0
0	0	0	0	0	0	0	0	0	0
1	1	0	0	1	0	0	1	0	0
2	—	1	0	1	1	0	1	1	0
3	—	—	1	1	1	1	1	1	1
4	—	—	—	—	1	1	1	1	1
5	—	—	—	—	—	1	1	1	1
6	—	—	—	—	—	—	—	1	1
7	—	—	—	—	—	—	—	—	1
Norm	1	1	1	3	3	3	5	5	5
RSSK/Norm	1.414	1.414	1.414	0.816	0.816	0.816	0.632	0.632	0.632

called the corresponding term RSSK/Norm. Simplifying the error analysis in this manner assumes that all measurements in the convolution interval have the same error. Thus, the RSSK/Norm represents a figure of merit for the noise reduction factor for the application of the convolution function, because the signals or magnitude of the resulting bands in the convoluted spectra are largely constant, over a range of convolution interval s andes.

Segment–gap method

For slowly changing functions, the derivative can often be approximated by taking the difference in y values for x locations separated by more than 1, up to as many as 15 points. One can see when too large a separation is selected, by observing that the basic shape of the derivative is degraded or changed, or minor or superimposed bands cease to be resolved. For such slowly changing functions, derivative curves with less noise can be obtained by taking the difference of two averages, formed from points surrounding the selected x locations. As a further simplification, the division of the difference in y values, or the difference in y-averages, by the x difference, Δx , is omitted. The term segment indicates the length of the x intervals over which y values are averaged, to obtain the two values that are subtracted to form the estimate of the derivative. The gap is the length of the x interval that separates the two segments that are averaged. This is shown in Figure 1, where a segment of three points and gap of five points is illustrated. The absorbance values in each of the boxes are averaged and the difference is taken as Δy, which is the value of the "derivative", because the division by the Δx term is taken as

Table 3. Normalised segment–gap derivatives: first derivatives.

Index	Point-diff	Segment, gap values (nanometers)								
		2, 2	2, 6	2, 10	6, 2	6, 6	6, 10	10, 2	10, 6	10, 10
–7	—	—	—	—	—	—	—	—	—	–1
–6	—	—	—	—	—	—	—	—	–1	–1
–5	—	—	—	—	—	—	–1	–1	–1	–1
–4	—	—	—	—	—	–1	–1	–1	–1	–1
–3	—	—	—	–1	–1	–1	–1	–1	–1	–1
–2	—	—	–1	0	–1	–1	0	–1	–1	0
–1	—	–1	0	0	–1	0	0	–1	0	0
0	–1	0	0	0	0	0	0	0	0	0
1	1	1	0	0	1	0	0	1	0	0
2	—	—	1	0	1	1	0	1	1	0
3	—	—	—	1	1	1	1	1	1	1
4	—	—	—	—	—	1	1	1	1	1
5	—	—	—	—	—	—	1	1	1	1
6	—	—	—	—	—	—	—	—	1	1
7	—	—	—	—	—	—	—	—	—	1
Norm	1	2	4	6	12	18	24	30	40	50
RSSK/Norm	1.414	0.707	0.354	0.236	0.204	0.136	0.102	0.105	0.079	0.063

constant scale factor that is ignored. Ignoring the division by the Δx term also has the result that derivatives calculated with increasing segments and gaps have greater amplitudes. This feature is a major distinguishing characteristic between segment–gap derivatives and other procedures. The segment–gap derivatives are a very flexible and useful technique for analyising spectra, particularly when the wavelength is sampled at intervals less than the spectral bandwidth of the instrument. Because the segment and gap method is available in the Foss/NIRSystems software that I have used, all further discussion will describe the features of that implementation.

In practice, the length of the segment and gap intervals is specified in actual x units (i.e. nanometers). Furthermore, to maintain wavelength accuracy, the segment and gap intervals are constrained to be an odd number of points, so that averages or results correspond to the midpoints of the in-

Table 4. Normalised segment–gap derivatives: second derivatives.

Index	Segment, gap values (nanometers)								
	2, 2	2, 6	2, 10	2, 14	2, 18	6, 2	6, 6	6, 10	10, 2
–10	—	—	—	—	1	—	—		—
–9	—	—	—	—	0	—	—	1	—
–8	—	—	—	1	0	—	—	1	1
–7	—	—	—	0	0	—	1	1	1
–6	—	—	1	0	0	—	1	0	1
–5	—	—	0	0	0	1	1	0	1
–4	—	1	0	0	0	1	0	0	1
–3	—	0	0	0	0	1	0	0	0
–2	1	0	0	0	0	0	0	0	–2
–1	0	0	0	0	0	–2	–2	–2	–2
0	–2	–2	–2	–2	–2	–2	–2	–2	–2
1	0	0	0	0	0	–2	–2	–2	–2
2	1	0	0	0	0	0	0	0	–2
3	—	0	0	0	0	1	0	0	0
4	—	1	0	0	0	1	0	0	1
5	—	—	0	0	0	1	1	0	1
6	—	—	1	0	0	—	1	0	1
7	—	—	—	0	0	—	1	1	1
8	—	—	—	1	0	—	—	1	1
9	—	—	—	—	0	—	—	1	—
10	—	—	—	—	1	—	—		—
Norm	4	16	36	64	100	48	108	192	180
RSSK/Norm	0.612	0.153	0.068	0.038	0.024	0.088	0.039	0.022	0.030

tervals (rather than allow an error of a half-interval that occurs if even intervals are allowed). In the case where a 2 nm sampling interval is used for spectra, specifying a gap of either 0, 1 or 2 nm (for example) results in the same computation, with a gap of one point. The next higher permitted gap values are 6, 10, 14 nm, etc., due to the requirement of specifying intervals with an odd number of points. Similarly, the permitted segment values are 2, 6, 10, 14 nm, etc., for the same reason. This ambiguity (that deep rounding of the input occurs, with the effect that the identical results are obtained with specified intervals of 3, 4, 5 or 6 nm, for example) is generally invisible to the user, but it is good to keep it in mind when trials for optimal conditions are made.

The calculation of the segment–gap derivatives can readily be expressed as application of convolution functions to the spectral data. The averages in the segments and the gap derivatives can be calculated in a single process of multiplying the spectral values in the convolution interval by factors of 1, summing and dividing by a normalisation factor that performs the averaging within the segments. This normalisation only accounts for the averaging in the segment and the derivative is still not normalised for the length of the convolution interval. The convolution functions for calculating first derivatives with selected combinations of segment and gap are presented in Table 2.

A normalised form of the segment–gap derivatives can be calculated by the application of the same formula that can be deduced from the Savitzky–Golay first derivatives, Equation (1).

$$\text{Norm} \sum_{i=-(n-1)/2}^{(n-1)/2} i * C_i \qquad (1)$$

where i is the index of the set of convolution function integers,
Ci are the convolution function integers and
n is the number of points in the convolution interval.

The convolution functions for calculating first derivatives with selected combinations of segment and gap for the normalised segment–gap derivatives are presented in Table 3.

Second derivatives are frequently used in NIR spectral analysis. Fourth derivatives have been used for UV and Visible spectral analysis,[10,11] but not widely used in NIR analysis. Selected second derivative convolution functions are presented for the normalised segment–gap method in Table 4. The usual segment–gap method convolution functions for the second and fourth derivatives are similar to those given in Table 4, except that the Norm values are 1, 3, 5, etc, depending on the number of points in the segment chosen, as seen already for the first derivatives in Table 2. The normalisation factors for the second, third and fourth derivatives are calculated from equations (2), (3) and (4), which can be deduced from the corresponding Savitzky–Golay tables. For these equations, the index i takes the same range as in Equation (1).

$$\text{Norm} = 1/2\ \Sigma i^2 Ci \qquad (2)$$
$$\text{Norm} = 1/6\ \Sigma i^3 Ci \qquad (3)$$
$$\text{Norm} = 1/24\ \Sigma i^4 Ci \qquad (4)$$

Applications and discussion

The application of various methods for calculating derivatives to a scan of a piece of polystyrene 0.94 mm thick, backed by Spectralon (Labsphere, Inc, North Sutton, NH, USA) and measured in reflectance mode in a Foss/NIRSystems Model 6500 scanning spectrophotometer. The polystyrene was cut from the bottom of a disposable Petri dish (Baxter Healthcare Corp., McGaw Park, IL, USA).

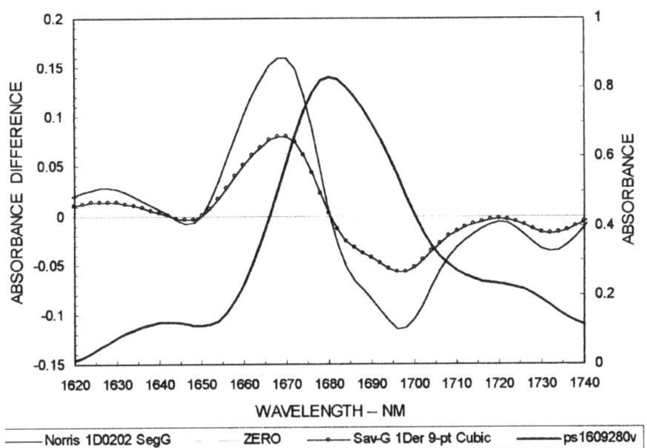

Figure 2. First derivatives of polystyrene, 1620 to 1740 nm region. Bold curve is the original spectrum; the fine solid curve is the segment–gap derivative, calculated with segment and gap of 2 nm; the curve marked with circles is the calculated with the Savitzky–Golay method, 9-point cubic fit.

The results of using these convolution functions in an Excel spreadsheet agree perfectly to the fifth decimal place, with the normal calculations carried out in the Vision and NSAS software (segment–gap derivatives) and Pirouette, Unscrambler and Grams32 (Savitzky–Golay derivatives). The new convolution functions were generated and evaluated in Excel. The results will be presented for the second overtone aromatic CH band of polystyrene as an example. Although the 1680 band appears as a nice, single absorption band, the results demonstrate the ability of the derivatives to identify bands that are highly overlapping. The first derivative results (Figure 2) suggest that this band is actually at least a doublet of very closely situated absorptions, from the two different slopes in the 1670–1695 nm region.

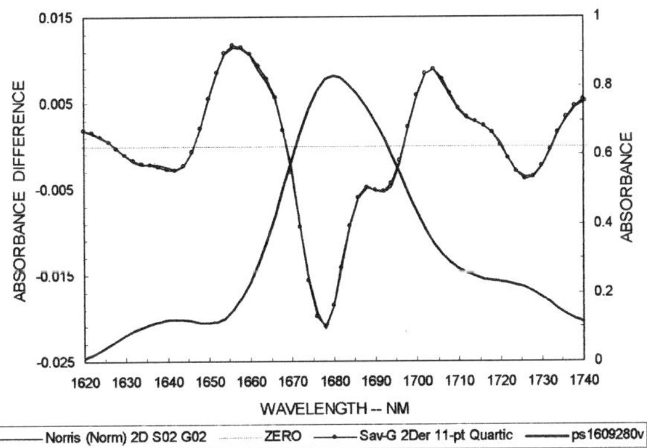

Figure 3. Second derivatives start to resolve two bands in the 1680 nm region. Bold curve is the original spectrum; the fine solid curve is the segment–gap derivative (Normalised), calculated with segment and gap of 2 nm; the curve marked with circles is the calculated with the Savitzky–Golay method, 11-point quartic polynomial fit.

Figure 2 demonstrates a key difference between the segment–gap and Savitzky–Golay results. The segment–gap derivative is of greater magnitude, because no normalisation for the segment and gap length is made. This is usually of no importance when the method is used in a calibration or chemometric analysis; it shows up clearly when results are compared, as in Figure 2. Figure 2 demonstrates that the curve shapes are comparable, differing only by a scale factor.

Normalisation of the segment–gap derivatives as described above, using Equations (1) through (4), brings the segment–gap derivatives into the same scale as the Savitzky–Golay convolution methods. This works because the normalisation constant calculated by Equations (1) through (4) properly determines the effective Δx for the method of calculating Δy determined by the convolution factors, to properly determine the derivative estimator $\Delta y / \Delta x$. Figure 4 compares the results of the segment–gap (normalised) first derivative with segment and gap of 2 nm (only three points in the convolution interval, but RSSK/Norm is 0.707) with the Savitzky–Golay method, 9-point cubic polynomial fit, RSSK/Norm of 0.338); the agreement is so close that the two curves are practically superimposed.

The second derivative suggests the occurrence of bands at 1678 and 1690 nm (Figure 3). This demonstrates the band sharpening and better resolution of complicated spectra that has been noted.[10,11] Looking critically, the asymmetry of the 1680 band can be noted in the raw absorbance scan, suggesting that the band may be more complex than a single absorber. The segment–gap (normalised) second derivative (5-point convolution interval, RSSK/Norm of 0.612) certainly agrees well with the Savitzky–Golay method, 11-point quartic polynomial fit, RSSK/Norm of 0.256).

Third derivative results also indicate the presence of two bands (results not shown). Further, the fourth derivative results show the clearest discrimination of multiple bands in the 1680 nm region of polystyrene; band positions are indicated by the positive peaks in the derivatives (arrows at 1677 and 1694 nm in Figure 4). Clearly the segment–gap (Normalised) method (nine points, RSSK/Norm of 0.523) and the Savitzky–Golay 9-point quartic fit method agree very well (RSSK/Norm of 0.313).

The fourth derivatives (see Figure 4) also suggest the resolution of other bands in the neighborhood of 1680, possibly due to second overtones of –CH– and –CH_2–. There are suggestions of bands at 1632 and 1645 nm that contribute to the low-intensity band observed at about 1642 in the polystyrene

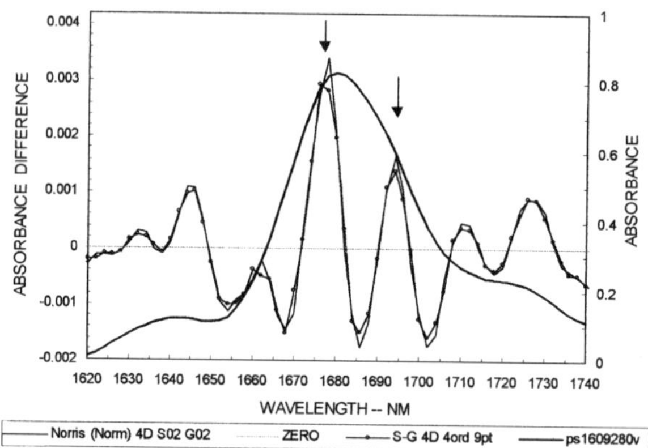

Figure 4. Fourth derivatives of polystyrene clearly indicates two bands in the 1680 region; band positions are indicated by the positive peaks in the derivatives (arrows at 1977 and 1694 nm). Bold curve is the original spectrum; the fine solid curve is the segment–gap Derivative (Normalised), calculated with segment and gap of 2 nm; the curve marked with circles is the calculated with the Savitzky–Golay method, 9-point 4th order polynomial (quartic) fit.

absorbance spectrum. Similarly, the shoulder at the 1720 nm region appears to be due to two underlying bands indicated at 1711 and 1727 nm. The band at 1662 nm is possibly false, an artifact of superposition of side-bands of the adjacent bands. This problem has been noted previously,[10,11] with the suggestion that other methods of spectral characterisation, for example, low temperature spectroscopy or fractionation and purification of various components, should be employed to demonstrate the separation of physically distinct absorbers. Morrey[12] has described the problems that can occur in the application of derivatives to band structure elucidation; he showed that bands can be created in a third-band illusion, or obliterated by the adjacent bands. Such problems need to be recognised in the interpretation of higher order derivatives, but they should not preclude their use.

The selection of which derivative method to use depends significantly on evaluating whether a result is too noisy, or so highly smoothed that details are distorted or lost. It may be necessary to accept a larger amount of noise to obtain good spectral resolution. The trade-off may depend on the object of the analysis. The RSSK/Norm values can offer some guidance in selection.

Conclusions

The segment–gap derivatives have been presented in detail, interpreting the method in terms of convolution. This comparison with the Savitzky–Golay procedures suggests an elegant method to normalise the segment–gap derivatives to obtain consistent signal magnitudes even at various segment and gap selections. The RSSK/Norm values show that even derivatives that do not explicitly involve smoothing achieve signal-to-noise improvement, by the utilisation of gaps of one or more points. The higher order derivatives are useful in interpreting complex spectra and it may be expected that derivatives as high as the fourth order will be useful in chemometric analysis of qualitative and quantitative applications.

The RSSK/Norm values provide a basis for comparing the effectiveness of convolution methods in reducing the propagation of spectral noise. Longer convolution intervals reduce the noise in the results and, typically, choosing a higher order polynomial results in higher noise levels, which can be offset by the ability to utilise longer convolution intervals. However, observation of the results of convolution by any method is necessary, to avoid over-smoothing and loss of resolution of underlying bands. From the results of analysing the polystyrene spectra, it appears that derivative convolution functions with mid-range RSSK/Norm values between approximately 0.3 to 0.7 are most effective for analysing the underlying band structure in the 1680 nm region. Values greater than 1 indicate poor noise rejection, while values less than 0.1 indicate significant loss of band structure at the expense of the noise reduction.

Note added in proof

Upon further presentation and discussion of these calculations, it seems useful to simplify the nomenclature and call the RSSK/Norm values RSSC, root sum of squares of coefficients.

Acknowledgements

This paper reflects my long fascination with the elegance of the convolution process and my hope to bring a little more science to the art of spectral evaluation. I would like to thank Karl Norris, Howard Mark, Richard Kramer and Fred McClure for long discussions on the calculation and interpretation of derivatives.

References

1. A. Savitzky and M.J.E. Golay, *Anal. Chem.* **36(8),** 1627 (1964).
2. D.W. Hopkins, *NIR news* **12(3),** 3 (2001).
3. D.W. Hopkins, *Near Infrared Analysis* **2(1),** 1 (2001).

4. Grams/32 Software is part of a chemometrics package from Thermo Galactic, Salem, NH, USA.
5. Pirouette Software is a chemometrics package from Infometrix, Inc, Woodinville, WA, USA.
6. Unscrambler Software is a chemometrics package from CAMO, Inc, Oslo, Norway.
7. Vision and NSAS are software programs for the collection and analysis of NIR data, available from FOSS/NIRSystems, Silver Spring, MD, USA.
8. J. Steinier, Y. Termonia and J. Deltour, *Anal. Chem.* **44(11),** 1906 (1972).
9. H. Mark, *Principles and Practice of Spectroscopic Calibration.* John Wiley & Sons, New York, USA, p. 170 (1991).
10. W.L. Butler and D.W. Hopkins, *Photochem. Photobiol.* **12,** 439 (1970).
11. W.L. Butler and D.W. Hopkins, *Photochem. Photobiol.* **12,** 451 (1970).
12. J.R. Morrey, *Anal. Chem.* **40(6),** 905 (1968).

A new regularised discriminant analysis; principal discriminant variate method for handling multicollinear data

Jian-Hui Jiang,[a,b] Roumiana Tsenkova,[c] Ru-Qin Yu[b] and Yukihiro Ozaki[a*]

[a]*Department of Chemistry, School of Science and Technology, Kwansei-Gakuin University, Sanda, Hyogo 669-1337, Japan*

[b]*College of Chemistry and Chemical Engineering, Hunan University, Changsha 410082, P. R. China*

[c]*Department of Environmental Information and Bioproduction Engineering, Faculty of Agriculture, Kobe University, Noda-ku, Kobe 657-8501, Japan*

Introduction

In linear discriminant analysis there are two important properties concerning the effectiveness of discriminant function modelling.[1–10] The first is the separability of the discriminant function for different classes. The separability reaches its optimum by maximising the ratio of between-class to within-class variance. The second is the stability of the discriminant function against noise present in the measurement variables. One can optimise the stability by exploring the discriminant variate in a principal variation subspace, i.e. the directions that account for a majority of the total variation of the data. An unstable discriminant function will exhibit inflated variance in the prediction of future unclassified objects and be exposed to a significantly increased risk of erroneous prediction. Therefore, an ideal discriminant function should not only separate different classes with a minimum misclassification rate for the training set but also possess a good stability so that the prediction variance for unclassified objects can be as small as possible. In other words, an optimal classifier should find a balance between the separability and the stability. This is of special significance for multivariate spectroscopy-based classification where multicollinearity always leads to discriminant directions located in low-spread subspaces.

We have developed, recently, a new regularised discriminant analysis technique, the principal discriminant variate (PDV) method, which handles effectively multicollinear data commonly encountered in multivariate spectroscopy-based classification.[11] The motivation behind this method is to seek a sequence of discriminant directions that not only optimise the separability between different classes but also account for a maximised variation present in the data. Three different formulations for the PDV method have been suggested and an effective computing procedure has been proposed for PDV.[11] The PDV method has been applied to near infrared (NIR) spectra of whole blood samples from mastitic and healthy cows to demonstrate the potential of the PDV method in clinical diagnosis of mastitis.[11] Mastitis is an udder inflammation that is a major problem for the global dairy industry.[12] It causes substantial economic losses by decreasing milk production, changing the compositions in milk considerably, reducing milk quality and increasing the risk of early culling of cows. The NIR spectra of the whole blood samples from the mastitic and healthy cows have been clearly discriminated by the PDV method.

Experimental

The preparation of cow blood samples and the instrumentation and experimental procedure for measuring their NIR spectra were described in Reference 11. A total of 162 whole blood samples were prepared and separated into two classes, one associated with the cows suffering from mastitis and the other corresponding to cows not suffering from mastitis. These samples were divided into two sets, the training set and the prediction set. The training set contained 80 samples from Class 1 and 42 samples from Class 2, respectively. The prediction set was composed of the remaining samples, i.e. 25 samples from Class 1 and 15 samples from Class 2, respectively. The algorithm and software used for principal component analysis (PCA), discriminant partial least squares (DPLS), Fisher linear discriminant analysis (FLDA) and soft independent modeling of class analogies (SIMCA) were reported in Reference 11.

Theory and methods

Fisher linear discriminant analysis

In FLDA, one finds a linear combination of the original variables that would maximise the within-class variability relative to the between-class variability.[1,2] Suppose there are N objects $\mathbf{x}_n$ ($n = 1, 2, —, N$) from K classes $\mathbf{C}_k$, ($k = 1, 2, —, K$) with the kth class containing N_k objects, then the between-class, within-class and total covariance matrices are represented as follows:

$$\mathbf{B} = \frac{1}{N}\sum_{k=1}^{K} Nk(\mathbf{m}_k - \mathbf{m})(\mathbf{m}_k - \mathbf{m})^{\mathrm{T}} \tag{1}$$

$$\mathbf{W} = \frac{1}{N}\sum_{k=1}^{K}\sum_{\mathbf{x}_i \in \mathbf{C}_k} (\mathbf{x}_i - \mathbf{m}_k)(\mathbf{x}_i - \mathrm{m}_k)^{\mathrm{T}} \tag{2}$$

$$\mathbf{T} = \frac{1}{N}\sum_{n=1}^{N} (\mathbf{x}_n - \mathbf{m})(\mathbf{x}_n - \mathbf{m})^{\mathrm{T}} \tag{3}$$

$$(\mathbf{T} = \mathbf{B} + \mathbf{W})$$

Here, $\mathbf{m}$ and $\mathbf{m}_k$ are the mean vector of all objects and those from $\mathbf{C}_k$, respectively, as given by

$$\mathbf{m} = \frac{1}{N}\sum_{n=1_k}^{N} \mathbf{x}_n \tag{4}$$

$$\mathbf{m}_k = \frac{1}{N}\sum_{\mathbf{x}_i \in \mathbf{C}_k} \mathbf{x}_i \tag{5}$$

In FLDA, one must maximise over **a** the following formulation;

$$J = \frac{\mathbf{a}^{\mathrm{T}}\mathbf{B}\mathbf{a}}{\mathbf{a}^{\mathrm{T}}\mathbf{W}\mathbf{a}} \tag{6}$$

or equivalently

$$J = \frac{\mathbf{a}^{\mathrm{T}}\mathbf{B}\mathbf{a}}{\mathbf{a}^{\mathrm{T}}\mathbf{T}\mathbf{a}} \tag{7}$$

The drawback of FLDA is that, in multicollinear situations, it is inclined toward overfitting the training data. This would generally result in discriminant variates located in a small-variance subspace, hence showing instability or inflated variance in the prediction of unclassified objects. FLDA maximises the separability without regard to the stability of the classifier, since its formulation does not take account of the variance explained by the discriminant variates. In contrast, PCA aims at seeking the components that represent as much total variation of the data as possible, ignoring the separability of different classes. Therefore, PCA sacrifices the separability to approach maximum stability. Now, one may ask if an integration of PCA and FLDA can result in a discriminant analysis method furnished with favourable separability along with improved stability. This is the motivation behind the proposed PDV method.

Principal discriminant variate method

The PDV method is intended to seek a compromise between PCA and FLDA such that the stability of the method can be improved substantially without significant loss of discriminatory capability. In PDV, one must find a direction, called principal discriminant variate, which maximises the following principal discrimination criterion;

$$J = \frac{\mathbf{a}^{\mathrm{T}}\left[(\lambda\mathbf{B} + (1 = \lambda)\mathbf{T}\right]\mathbf{a}}{\mathbf{a}^{\mathrm{T}}\left[\lambda\mathbf{T} + (1 - \lambda\mathbf{I}\right]\mathbf{a}} \tag{8}$$

where **I** is the identity (unitary) matrix of the same size as the covariance matrix **T** and λ, with a value varied between 0 and 1, is a weight controlling the balance between linear discriminant analysis (LDA) and PCA. By setting $\lambda = 1$, one immediately finds that the principal discrimination criterion (8) is reduced to Eq. (6), the discrimination criterion used in ordinary FLDA. On the other hand, if one sets $\lambda = 0$, the principal discrimination criterion (8) becomes

$$J = \frac{\mathbf{a}^{\mathrm{T}}\mathbf{T}\mathbf{a}}{\mathbf{a}^{\mathrm{T}}\mathbf{a}} \tag{9}$$

which turns out to be the variance criterion maximised in PCA.

If one increases the value of λ, the separability is increased. On the other hand, if the value of λ is decreased, the stability is increased. Thus, λ controls the trade-off between separability and stability.

The optimal value of λ can be taken as a value that yields a minimum error rate found by cross-validation or extra set of test objects. In practical applications, the performance of the PDV method is not so sensitive to the choice of the value of λ so that it is enough to optimise the parameter stepwise. That is, λ can be started from a small value, say 10^{-7}, then increased stepwise by a factor of about 10. For each value of λ one evaluates the misclassification rate on the training set and chooses the largest value that gives the minimum misclassification rate. The reason for this choice is that one can achieve maximised stability without significant loss of separability. Computing procedure for the PDV method was described in detail in Reference 11.

Results and discussion

NIR spectra of whole blood samples from mastitic and healthy cows were almost identical with the spectrum of water. The spectra from two different classes are overlapped severely throughout the whole spectral region.[11] PDV, PCA, DPLS, SIMCA, and FLDA were used to discriminate the NIR spectra of blood samples from two different classes. PCA and DPLS could not discriminate the two kinds of data. SIMCA yielded much better result, but still the discrimination performance was not sufficient. The best discrimination with SIMCA was achieved for the NIR data when the dimensionalities were both set to 20 for Class 1 and 2. The misclassification rate of SIMCA was 5 and 10, respectively, for both the training and prediction sets. It was noted in the plot of FLDA that the training objects from two classes were overfitted and positioned in two clearly separated line segments and the scores of prediction objects were scattered with a large variability so that the prediction objects from different classes yielded scores that overlapped severely.

The PDV method clearly discriminated the NIR spectra of blood samples from mastitic and healthy cows. A clear separation between the scores from different classes was obtained for the training set, but the scores of one prediction object were far-away from the training objects. This indicated that this object was an outliner in the sample set from healthy cows. One could also observe that the scores of the prediction objects were distributed in the neighbourhood of those of the training set, suggesting that the model had a desirable stability.

Conclusion

It was found that the PDV method is quite powerful for the classification of multicollinear data such as NIR spectra. The results with the NIR data of cow-blood samples showed that the PDV method yields superior performance compared to SIMCA, PCA, DPLS and FLDA. This method has opened up a new possibility in NIR spectroscopy-based classification.

References

1. G.J. McLachlan, *Discriminant Analysis and Statistical Pattern Recognition*, John Wiley & Sons, New York, USA (1992).
2. W.J. Krzanowski, *J. Chemom.* **9,** 509 (1995).
3. L. Stahle and S. Wold, *J. Chemom.* **1,** 185 (1987).
4. M. Ortiz, L. Sarabia, C. Symington, F. Santamaria and M. Iniguez, *Analyst* **121,** 1009 (1996).
5. B.K. Alsberg, D.B. Kell and R. Goodacre, *Anal. Chem.* **70,** 4126 (1998).
6. S. Wold, *Pattern Recogn.* **8,** 127 (1976).
7. O.V. Kvalheim, K. Oygard and O.G. Nielsen, *Anal. Chim. Acta* **150,** 145 (1987).
8. P.J. Gemperline, L.D. Webber and F.O. Cox, *Anal. Chem.* **1989,** 138 (1989).
9. I.E. Frank and J.H. Friedman, *J. Chemom.* **3,** 463 (1989).
10. U.G. Indahl, N.S. Sahni, B. Kirkhus and T. Næs, *Chemom. Intell. Lab. Syst.* **49,** 19 (1999).
11. J.-H. Jiang, R. Tsenkova, Y. Wu, R.-Q. Yu. and Y. Ozaki, *application Spectrosc.* **56,** 488 (2002).
12. R. Tsenkova, S. Atanassova, K. Toyoda, Y. Ozaki, K. Itoh and T. Fearn, *J. Dairy Sci.* **82,** 2344 (1999).

Comparison of linear and non-linear near infrared calibration methods using large forage databases

Paolo Berzaghi,[a,b*] Peter C. Flinn,[c] Pierre Dardenne,[d] Martin Lagerholm,[e] John S. Shenk,[f] Mark O. Westerhaus[f] and Ian A. Cowe[e]

[a]*University of Padova, Agripolis, 35020 Legnaro, Italy*

[b]*University of Wisconsin, 1925 Linden Dr., 53706 Madison, WI, USA*

[c]*Agriculture Victoria, Pastoral and Veterinary Institute, Private Bag 105, Hamilton, Victoria 3300, Australia*

[d]*Centre de Recherches Agronomiques de Gembloux – CRAGx, 24 Chaussee de Namur, 5030 Gembloux, Belgium*

[e]*Foss Tecator AB, Box 70, SE-263 21 Höganäs, Sweden*

[f]*Infrasoft International, 109 Sellers Lane, 16870 Port Matilda, PA, USA*

Introduction

Forages represent about 50% of the diet fed to dairy cattle and information about their chemical composition is necessary to correctly balance nutrients in the diet. However, chemical and nutritional composition of forages is highly variable. Major sources of variation include botanical family (for example, legumes *vs* grasses), stage of maturity at harvest, method of conservation (for example, hay *vs* silage) and climatic conditions. As a result of these sources of variation, commercial forage testing labs have been using several different near infrared (NIR) calibrations to cover the analysis of all forages. Type and source of the sample is critical for the selection of the appropriate calibration equation and this information is often missing or incorrect. Forage NIR analysis would be simplified by using few or even only one NIR calibration for all types of forage. However, the large source of variation that the calibration data set must include may cause problems of non-linear relationships between spectral and chemical information resulting in lower accuracy of prediction.

Alternatives to multivariate calibration methods that can handle non-linear relationships are artificial neural network (ANN)[1] and local partial least squares (PLS) calibrations (LOCAL).[2] Although these methods are not new, they have only recently been introduced in practical applications and they were not tested with a large forage database. The aim of this study was to compare the performances of modified PLS (MPLS) calibration to ANN and LOCAL calibrations for the prediction of a large forage data set.

Materials and methods

The study used historical forage data sets (25,977 samples) from Australia, Europe (Belgium, Germany, Italy and Sweden) and North America (Canada and USA). The data sets had already chemistry values relative to moisture (DM), crude protein (CP) and neutral detergent fibre (NDF) content that were obtained from different unharmonised laboratories. Samples spectra were collected during a time span of about ten years with ten different Foss NIRSystems instruments, which were either standardised or not standardised to one master instrument. The spectra were trimmed to a wavelength range between 1100 and 2498 nm.

Two data sets, one standardised (IVAL) and the other not standardised (SVAL) were used as independent validation sets but 10% of both sets were omitted from the validation sets and they were use for later expansion of the calibration database. The remaining samples were combined into one database (n = 21,696), which was split into 75% calibration (CALBASE) and 25% validation (VALBASE).

MPLS equations were developed using WinISI (Infrasoft International LLC, USA). Pre-defined spectral math treatments were first derivative, 4 data points skipping gap and smoothing with SNV-Detrend scatter correction. Local PLS calibrations were also developed under WinISI software. In this case, two settings were defined. The first was decided prior to the trial (LOCAL1), while the second (LOCAL2) was optimised for the prediction of CALBASE. There were also two methods for ANN (ANN1 and ANN2), both developed using Matlab (The Mathworks Inc., USA).

The chemical components in the three validation data sets were predicted with each model derived from CALBASE using the calibration database before and after it was enhanced with 10% of the samples from IVAL and SVAL data sets. Calibration performances were evaluated using standard error of prediction (*SEP*), bias, *SEP* corrected for bias [*SEP*(*C*)], slope and R^2.

Results

Regardless of calibration method, prediction of VALBASE (data not shown) had smaller *SEP*(*C*) and bias values than for IVAL (Table 1) and SVAL (Table 2). This was not surprising as VALBASE was selected from the calibration database and it had a sample population similar to CALBASE, whereas IVAL and SVAL were completely independent validation sets. Part of the problem may be caused by differences in wet chemistry methods, as indicated, for example, by the large bias of DM in SVAL or NDF in IVAL.

None of the models developed before enhancements appeared to be consistently better for the two independent validation sets. However, LOCAL and ANN had lower *SEP* and *SEP*(*C*) than MPLS for all three variables evaluated in VALBASE. This is consistent with previous studies that found LOCAL[3] and ANN[4] were able to handle data sets with large sources of variation.

In most cases, LOCAL and ANN models, but not MPLS, showed considerable improvement in the prediction of IVAL (Table 1) and SVAL (Table 2) after the calibration database had been expanded with the 10% samples of IVAL and SVAL reserved for calibration expansion. The addition of only 439 samples from the two independent sets to the 16272 samples of VALBASE greatly reduced bias, *SEP* and *SEP*(*C*) of LOCAL and ANN of IVAL and SVAL. From a practical point of view, the expansion of a database to predict new forage products will require fewer samples and result in better accuracy using either LOCAL or ANN than using MPLS calibrations.

The effects of sample processing, instrument standardisation and differences in reference procedure were partially confounded in the validation sets, so it was not possible to determine which factors were most important.

Table 1. Prediction performances of the different calibration methods for the independent set from Italy (IVAL).

	Enhancement	*SEP*	*SEP(C)*	Bias	Slope	R^2
DM (*n* = 1885)						
MPLS	Before	1.34	1.33	0.2	1.18	0.79
LOCAL1	Before	1.54	1.44	0.56	1.16	0.74
LOCAL2	Before	1.53	1.43	0.55	1.14	0.74
ANN1	Before	1.38	1.36	0.26	1.09	0.77
ANN2	Before	1.33	1.32	0.23	1.08	0.78
MPLS	After	1.33	1.32	0.17	1.17	0.79
LOCAL1	After	1.12	1.12	0.02	1.06	0.84
LOCAL2	After	1.07	1.07	0.01	1.06	0.85
ANN1	After	1.34	1.34	0.08	1.2	0.79
ANN2	After	1.32	1.32	0.09	1.18	0.79
CP (*n* = 1846)						
MPLS	Before	1.82	1.33	–1.25	0.88	0.96
LOCAL1	Before	2.12	1.52	–1.48	0.87	0.95
LOCAL2	Before	1.91	1.44	–1.26	0.87	0.95
ANN1	Before	2.15	1.54	–1.5	0.85	0.96
ANN2	Before	2	1.41	–1.42	0.87	0.96
MPLS	After	1.74	1.31	–1.14	0.89	0.96
LOCAL1	After	1.26	1.15	–0.51	0.96	0.96
LOCAL2	After	1.14	1.1	–0.3	0.97	0.96
ANN1	After	1.19	1.04	–0.57	0.95	0.97
ANN2	After	1.06	0.99	–0.39	0.97	0.97
NDF (*n* = 1912)						
MPLS	Before	4.62	3.47	3.05	1.02	0.93
LOCAL1	Before	5.53	4.15	3.66	0.99	0.89
LOCAL2	Before	5.36	3.82	3.76	1	0.91
ANN1	Before	4.64	3.56	2.98	1.06	0.93
ANN2	Before	4.98	3.46	3.59	1.05	0.93
MPLS	After	4.2	3.44	2.4	1.04	0.93
LOCAL1	After	3.4	3.2	1.16	1.02	0.94
LOCAL2	After	3.15	2.96	1.08	1.01	0.95
ANN1	After	3.15	3.06	0.77	1.04	0.94
ANN2	After	3.01	2.93	0.71	1.05	0.95

Table 2. Prediction performances of the different calibration methods for the independent set from Sweden (SVAL).

	Enhancement	*SEP*	*SEP(C)*	Bias	Slope	R^2
DM (n = 1861)						
MPLS	Before	3.85	2.41	–3.00	–0.26	0.12
LOCAL1	Before	3.08	2.46	–1.86	–0.15	0.05
LOCAL2	Before	3.20	2.43	–2.08	–0.11	0.03
ANN1	Before	3.30	2.41	–2.26	–0.20	0.07
ANN2	Before	3.52	2.49	–2.48	–0.23	0.11
MPLS	After	2.87	2.35	–1.64	–0.28	0.12
LOCAL1	After	0.90	0.90	–0.08	0.81	0.53
LOCAL2	After	0.90	0.89	–0.09	0.80	0.54
ANN1	After	0.82	0.82	–0.07	0.91	0.59
ANN2	After	0.66	0.66	–0.06	0.98	0.73
CP (n = 1860)						
MPLS	Before	1.01	0.74	0.69	0.97	0.97
LOCAL1	Before	1.34	1.06	0.82	0.98	0.94
LOCAL2	Before	1.56	1.32	0.83	0.96	0.92
ANN1	Before	1.21	0.70	0.99	0.94	0.98
ANN2	Before	1.27	0.72	1.04	0.96	0.98
MPLS	After	0.85	0.74	0.43	0.97	0.97
LOCAL1	After	0.74	0.74	0.00	1.00	0.97
LOCAL2	After	0.72	0.72	–0.01	0.99	0.97
ANN1	After	0.71	0.69	0.12	0.98	0.98
ANN2	After	0.67	0.67	0.09	0.97	0.98
NDF (n = 1660)						
MPLS	Before	2.60	2.39	–1.02	1.06	0.92
LOCAL1	Before	4.46	3.98	–2.02	1.03	0.78
LOCAL2	Before	3.63	3.60	–0.47	1.06	0.82
ANN1	Before	2.90	2.48	–1.49	1.06	0.92
ANN2	Before	2.53	2.53	–0.16	1.07	0.92
MPLS	After	2.27	2.27	0.08	1.04	0.93
LOCAL1	After	2.24	2.23	–0.18	1.03	0.93
LOCAL2	After	2.17	2.16	–0.17	1.04	0.94
ANN1	After	2.20	2.20	–0.13	1.03	0.93

Conclusions

Compared with MPLS, LOCAL and ANN improved accuracy in the predictions of forage samples similar to those in the calibration data set. The accuracy in the prediction of complete independent data sets was unacceptable for all models but LOCAL and ANN were able to reduce *SEP*, bias and *SEP*(*C*) after updates using a small number of samples. LOCAL and ANN were able to manage large sources of variation adding the flexibility of rapid and inexpensive expansion to new forage data sets.

Further work on the development of large databases must address the problems of standardisation of instruments, harmonisation and standardisation of laboratory procedures and, even more importantly, the definition of criteria for the selection of samples used in the creation and updates of the database.

References

1. T. Næs, K. Kvaal, T. Isaksson and C. Miller, *J. Near Infrared Spectrosc.* **1,** 1 (1993).
2. J.S. Shenk, P. Berzaghi and M.O. Westerhaus, *J. Near Infrared Spectrosc.* **5,** 223 (1997).
3. P. Berzaghi, J.S. Shenk and M.O. Westerhaus, *J. Near Infrared Spectrosc.* **8,** 1 (2000).
4. N.B. Buchmann, in *Near Infrared Spectroscopy: The Future Waves*, Ed by A.M.C. Davies and P. Williams. NIR Publications, Chichester, UK, p. 479 (1996).

Calibration transfer from reflectance to interactance–reflectance mode: use of mathematical pretreatments

Víctor M. Fernández-Cabanás,[a] Ana Garrido-Varo[b] and Pierre Dardenne[c]

[a]*Escuela Universitaria de Ingeniería Técnica Agrícola, University of Seville, Ctra Utrera Km. 1, E-41013 Sevilla, Spain. e-mail: victorf@cica.es*

[b]*Escuela Técnica Superior de Ingenieros Agrónomos y Montes, University of Córdoba, Avda Menéndez Pidal s/n, E-14080 Córdoba, Spain*

[c]*Centre de Recherches Agronomiques de Gembloux, Chaussée de Namur 24, B-5030 Gembloux, Belgium*

Introduction

Traditional applications of near infrared (NIR) spectroscopy in agriculture products are being developed in either reflectance or transmittance mode, and calibration transfer from these modes to fibre-optic analysis modes might prove a useful way of saving money and effort. Most methods for calibration transfer between different instruments involve the use of sealed reference cups but, as fibre optic analysis does not use cups, other methods of calibration transfer are required.[1]

Within the framework of the STAFANIR project,[2] highly accurate European calibrations were developed for barley, wheat and maize. These equations were successfully transferred from a master instrument to several satellite instruments of different brands and models. However, all satellite instruments tested were using reflectance analysis.

The primary purpose of this paper was to study the effect of different mathematical pretreatments of log 1/*R* data on the performance of calibration equations obtained for reflectance analysis; a secondary aim was to determine whether use of these mathematical transformations is sufficient to implement the transfer of European calibrations from reflectance to interactance–reflectance mode.

Material and methods

Samples and reference data

Three sets of ground cereals (barley, wheat and maize) were defined as calibration sets. These samples and the reference data were provided by the STAFANIR project.[2] The chemical composition of the calibration sets is shown in Table 1. The validation sets consisted of 92 ground samples of barley, 74 of maize and 76 of wheat.

Table 1. Reference data (expressed as %) of calibration sets.

Set (parameter)	*N*	Mean	Minimum	Maximum	*SD*
Barley (CP)	177	11.86	7.38	17.19	2.83
Wheat (CP)	225	12.16	8.54	17.71	2.51
Maize (Oil)	180	5.27	3.08	9.30	1.49

NIR hardware

Reflectance spectra for calibration sets were obtained on a Foss NIRSystems 6500 monochromator, from 400 to 2498 nm, every 2 nm. Analysis was performed using a spinning module and samples were scanned on standard ring cells.

Validation spectra were scanned at the same time by two scanning monochromators, in reflectance and interactance–reflectance modes. For reflectance analysis, a Foss NIRSystems 6500 without auto-gain detectors, provided with a spinning module, was used. Samples were scanned in standard ring cells. A Foss NIRSystems model 6500 scanning monochromator with auto-gain detectors was used to measure interactance-reflectance spectra from 400 to 2498 nm, at intervals of 2 nm. Analysis

Table 2. Combination of mathematical pretreatments used for calibration development.

Equation	Scatter Correction	Derivative	Equation	Scatter Correction	Derivative
1	None	None	19	DT	2,10,10,1
2	SNV	None	20	SDT	1,5,5,1
3	DT	None	21	SDT	1,10,10,1
4	SDT	None	22	SDT	2,5,5,1
5	Standard MSC	None	23	SDT	2,10,10,1
6	Weighted MSC	None	24	Standard MSC	1,5,5,1
7	Inverse MSC	None	25	Standard MSC	1,10,10,1
8	None	1,5,5,1	26	Standard MSC	2,5,5,1
9	None	1,10,10,1	27	Standard MSC	2,10,10,1
10	None	2,5,5,1	28	Weighted MSC	1,5,5,1
11	None	2,10,10,1	29	Weighted MSC	1,10,10,1
12	SNV	1,5,5,1	30	Weighted MSC	2,5,5,1
13	SNV	1,10,10,1	31	Weighted MSC	2,10,10,1
14	SNV	2,5,5,1	32	Inverse MSC	1,5,5,1
15	SNV	2,10,10,1	33	Inverse MSC	1,10,10,1
16	DT	1,5,5,1	34	Inverse MSC	2,5,5,1
17	DT	1,10,10,1	35	Inverse MSC	2,10,10,1
18	DT	2,5,5,1			

was performed using a fibre optic probe (NR-6775). Three spectra were collected and averaged for each sample at different locations, in order to reduce sampling errors.

NIR software and chemometric treatments

All spectra were manipulated and processed and all calibration equations were obtained, using ISI software NIRS3 ver. 4.0 and WINISI ver. 1.5 (Infrasoft International, Port Matilda, PA, USA).

A total of 35 different calibration equations were obtained for each cereal and chemical parameter (Table 2). All calibrations were obtained for the spectral range 1100–2200 nm, which was found to be free of spectral error for interactance–reflectance analysis in a previous study.[3] The modified partial least squares (MPLS) regression method was used for obtaining equations.[4,5] The mathematical pretreatments used were standard normal variate (SNV),[6] detrending (DT),[6] SNV and DT (SDT), the three versions of multiplicative scatter correction (MSC)[7] included in WINISI software (Normal MSC, Weighted MSC and Inverse MSC) and four different derivative math treatments. The derivative math treatments are referred to by a four-digit notation (a,b,c,d).[8]

NIR equations were evaluated by examining the statistical values obtained for 1-*VR* (coefficient of determination of cross-validation) and standard error of cross-validation (*SECV*). Calibration transfer was evaluated by the statistic R^2 and standard error of differences (*SED*) calculated for the predicted values of the validation samples scanned on the two instruments and in both analysis modes.[9]

Results and discussion

Calibration statistics (R^2 and *SECV*) for the different equations and products are shown in Figures 1 and 2. The 1-*VR* values for CP in barley and wheat and oil in maize ranged between 0.98–0.99, 0.98–0.99 and 0.95–0.97, respectively. The *SECV* values for these samples and chemical parameters were low in all cases, but certain differences are apparent. The *SECV* values for CP in barley and wheat ranged from 0.26 to 0.34 and from 0.25 to 0.36, respectively. Equations obtained for oil in maize yielded a maximum *SECV* of 0.32 and a minimum of 0.27.

From Figures 1 and 2, some trends concerning the effect of the different mathematical pretreatments can be appreciated. DT transformation with any derivative combination (Equations 3, 16–19) and derivatives with no scatter correction (Equations 8–11) yielded the highest *SECV*s values

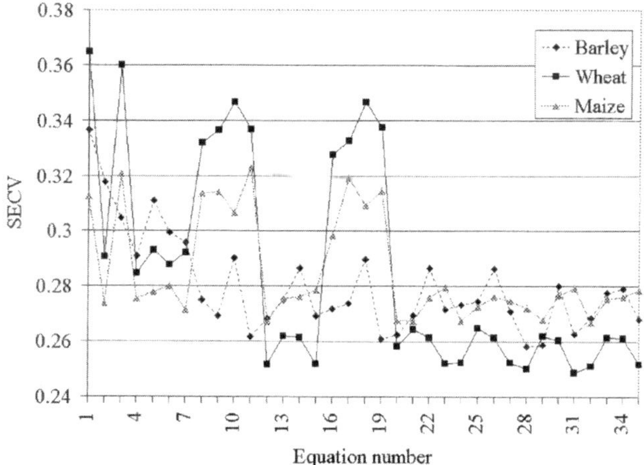

Figure 1. Calibration statistics for all cereals (*SECV*).

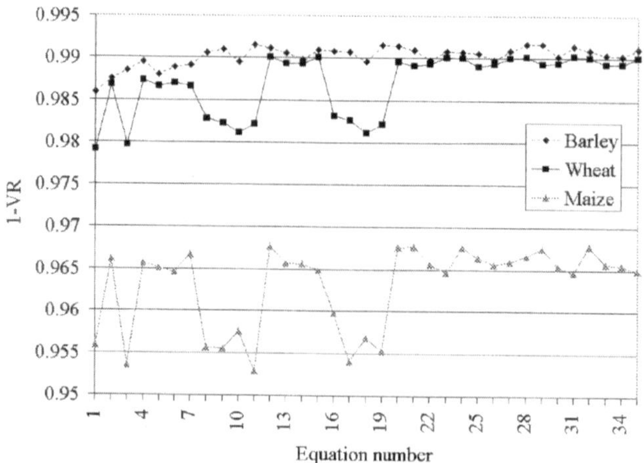

Figure 2. Calibration statistics for all cereals (1-*VR*).

for wheat (0.36, 0.33, 0.33, 0.35, 0.34 and 0.33, 0.34, 0.35, 0.34) and maize calibrations (0.32, 0.30, 0.32, 0.31, 0.31 and 0.31, 0.31, 0.31, 0.32). Best calibrations were obtained for derivatives 1,5,5,1 (all cereals), 2,10,10,1 (barley and wheat) and 1,10,10,1 (maize).

Once obtained, the equations were evaluated by their ability to be transferred from the instrument on which the equations were developed and which uses reflectance analysis and cups, to the second instrument which uses interactance–reflectance and a fibre optic probe.

Figures 3 and 4 illustrate the effects of the application of the mathematical pretreatments. The R^2 values for CP in barley and wheat and for oil in maize ranged from 0.65 to 0.88, from 0.32 to 0.78 and from 0.25 to 0.73, respectively. *SED* values for CP in barley and wheat ranged from 0.40 to 5.95 and

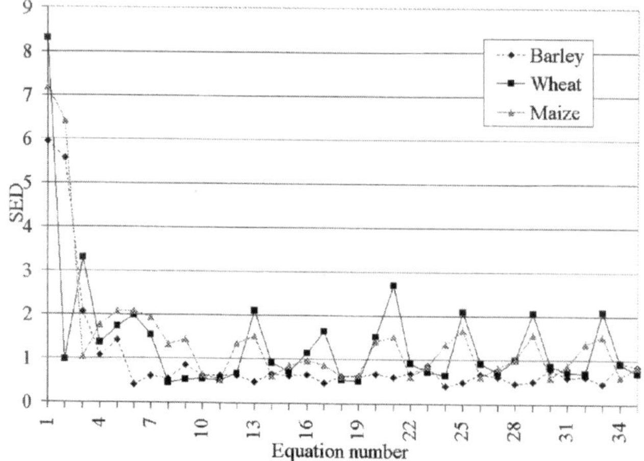

Figure 3. Validation statistics for equation transfer (*SED*).

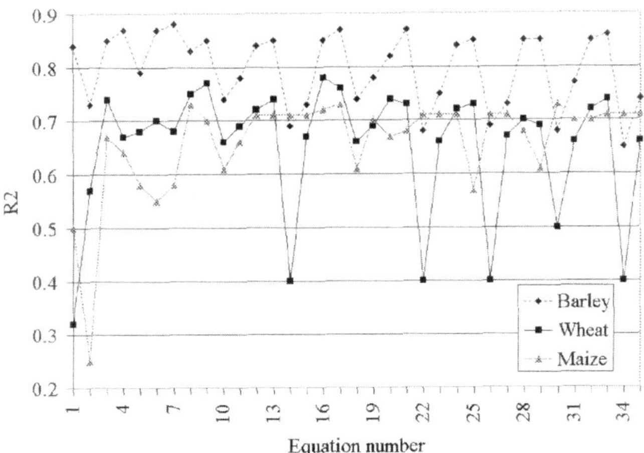

Figure 4. Validation statistics for equation transfer (R^2).

from 0.45 to 8.29, respectively. Validation of oil equations for maize gave *SED* values of between 0.52 and 7.17. From these results it can be appreciated that for some combinations of pretreatments, *SED* values for barley, wheat and maize calibrations are similar to *SECV* values. This suggests that the use of pretreatments in calibration development reduces most spectral differences caused by different instruments and different modes of analysis. This effect is more evident when validation statistics are compared for calibrations based on untreated spectra and for the best combination of mathematical pretreatments. For example, *SED* and R^2 values for Equation 1 (no pretreatments) were 5.95 and 0.84 for the barley set, while for Equation 6 (weighted–MSC) they were 0.40 and 0.87, respectively.

The pretreatments used exerted a major effect on validation results. As Figure 3 shows, higher *SED* values were recorded for combinations with derivative 2,5,5,1 for barley (0.61 to 0.95) and 1,10,10,1 for wheat (1.64 to 2.70) and maize (0.87 to 1.67). In terms of R^2 values, first derivatives gave the highest values for wheat and barley, while derivative 2,5,5,1 produced the poorest results (also for maize, when combined with DT or without scatter correction). Strikingly, the use of derivatives without scatter correction gave good results in terms of *SED* and R^2, confirming the results obtained by Fernández *et al.*,[10] except for maize, where R^2 values proved inadequate for second derivatives and *SED* values were high for first derivatives.

From all the information available, it may be concluded that derivative 1,5,5,1 gave better calibration and validation statistics for the equations transferred for barley and wheat, while derivative 2,5,5,1 produced the worst results. The behaviour of maize calibrations was different; while trends of 1-*VR* and R^2 values in relation to derivatives were similar to the other two cereals, *SED* values were lower for derivative 2,5,5,1 than for 1,5,5,1.

Conclusions

The main conclusion of this study is the importance of derivatives for calibration and equation transference statistics. In general, the effect of scatter corrections alone is not as important as the effect of derivatives. Thus, it is recommended that they be used in conjunction with derivatives, selecting the best combination. The use of pretreatments alone is not sufficient to implement a good calibration transfer, based on the quality criteria proposed by Shenk *et al.*,[9] Further work is in progress to combine the use of mathematical pretreatments with other strategies.

Acknowledgements

This work was carried out using the NIR hardware and software of the Centralised NIR and MIR Spectroscopy Unit (SCAI) of the University of Córdoba (Spain). The authors are grateful to the STAFANIR project for providing spectral and reference data for calibration development.

References

1. T.B. Blank, S.T. Sum and S.D. Brown, *Anal. Chem.* **68,** 2987 (1996).
2. C. Scotter. Final report. European Project "Standardisation of food analysis by NIR-spectroscopy- model NIR analysis of cereals to establish an EU wide network an acceptable CEN and ISO standards. Standard Measurement and Testing Program. IVth Framework Programme. (1999).
3. V.M. Fernández-Cabanás and A. Garrido-Varo, in *Near Infared Spectroscopy: Proceedings of the 9th International Conference, Ed by A.M.C. Davies and R. Giangiacomo. NIR Publications, Chichester, UK, p. 655 (2000).*
4. J.S. Shenk and M.O. Westerhaus, *Crop Sci.* **31,** 6 (1991).
5. H. Martens and T. Næs, *Multivariate Calibration.* John Wiley & Sons, Chichester, UK (1998).
6. R.J. Barnes, M.S. Dhanoa and S.J. Lister, *Appl. Spectrosc.* **43,** 772 (1989).
7. H. Martens, S.A. Jensen and P. Geladí, *in Proceedings of the Nordic Symposium on Applied Statistics.* Stokkand Forlag, Stavanger Publishers, Norway, p. 205 (1983).
8. J.S. Shenk and M.O. Westerhaus, *Routine operation, calibration, development and network system management manual.* NIRSystems, Inc., 12101 Tech Road, Silver Spring, MD 20904. USA (1995).
9. J.S. Shenk, J.J. Workman and M.O. Westerhaus, in *Handbook of Near Infrared Analysis*, Ed by D.A. Burns and E.W. Ciurczak. Marcel Dekker, New York, USA, p. 383 (1992).
10. V.M. Fernández-Cabanás, A.Garrido-Varo and C. Portal-Basurco, *ITEA* **20,** 550 (1999).

Understanding the "H" statistic during routine analysis of animal fats

Juan García-Olmo,[a,b] Ana Garrido-Varo[a] and Emiliano De Pedro[a]

[a]*Department of Animal Production, Faculty of Agriculture and Forestry Engineering, University of Córdoba, PO Box 3048, E-14080 Córdoba, Spain*

[b]*NIR/MIR Unit, SCAI, University of Cordoba, Spain*

Introduction

To ensure the correct usage of a predictive analytical technique such as near infrared (NIR) reflectance spectroscopy, outlier detection should be included as an integral part of instrument operation.[1]

In routine analysis, using NIR reflectance spectroscopy, detection of erroneous or abnormal samples should be based on the spectral information obtained. The calculation of statistics such as the Mahalanobis distance,[2] levarage[3] or the "*H*" statistic,[4] is of great value to detect sample spectra that differ widely from spectra in the calibration set, thus avoiding an extrapolation of the calibration model. In routine analysis, ISI software is especially suitable for this purpose, since it provides both predicted values and the "*H*" statistic of a sample immediately after scanning its NIR spectrum.[4]

However, correct use and interpretation of "*H*" values is crucial, because samples with higher-than-recommended "*H*" values should be sent to the laboratory for reference analysis. Given that reference analysis is usually expensive, it is important to differentiate between real and false "*H*" values in samples that appear to be outliers.

Previous studies have shown that NIR calibration equations, with a precision similar to that of the reference method,[5–7] can be obtained for predicting fatty acids in liquid samples of pig fat. However, over a four-year period spent in obtaining equations, certain uncontrolled variations have been observed in the "*H*" values for predicted samples that hinder the adoption of an outlier detection strategy.

The purpose of this study was to achieve an enhanced understanding of the "*H*" statistic during routine analysis of samples of liquid animal fat.

Material and methods

Samples

Three different validation sample sets were used (see Table 2). Validation set A consisted of the spectra of 20 Iberian pig fat samples obtained in May 1998 not included in the calibration file. Validation set B was the spectra of 30 Iberian pig fat samples obtained in November 1999 included in the calibration file. Validation set C consisted of 150 spectra of one pig fat sample (designated as the fat check sample), which was representative of the mean fatty acid composition values in the calibration file. This check sample had been analysed three times a week from June 1999 to July 2000.

NIR analyses and reference data

Validation sets were analysed by gas chromatography (GC) and NIR in the same way as the calibration set.[7] The percentage by weight of the main fatty acids in liquid fat samples (palmitic acid or C16 : 0, stearic acid or C18 : 0, oleic acid or C18 : 1 and linoleic acid or C18 : 2) was determined by gas chromatography. NIR data were recorded from 400 to 2500 nm using a Foss NIRSystems 6500 scanning monochromator equipped with a spinning module. Samples were analysed by folded transmission using a ring cup with a pathlength of 0.1 mm (ref. IH-03459). Only the NIR range (from 1100 to 2500 nm) was used in the analysis. Spectra were collected and processed by the ISI Ver. 3.11 software (Infrasoft International, Port Matilda, PA, USA).

Monitoring NIR equations

The three validation sets were used to monitor the equations developed earlier and reported in a previous work.[7] Shenk *et al.*[8] have outlined a procedure for monitoring predictions obtained by NIR equations. A monitoring test should be made whenever there is any reason to believe that an equation may not be predicting adequately. This procedure, when applied to a validation sample set ($n > 9$), established limits for bias (0.6 × *SEC*), for unexplained error or *SEP*(*C*) (1.3 × *SEC*) and for the "*H*" statistic ("*H*"< 3). When bias, *SEP*(*C*) or "*H*" statistics of a validation set exceed confidence limits, samples should be sent to a laboratory for reference analysis.

Results and discussion

Table 1 shows fatty acid composition for the three validation sets. Mean values for composition of validation sets A and B and the composition of the sample represented in set C were similar to the mean fatty acid values of the calibration set (Table 2).

Table 1. Weight percentage of fatty acids for validation sets A, B and C.

Constituent	Set A (n = 20)			Set B (n = 30)			Set C (n = 1)
	Mean	*SD*	Range	Mean	*SD*	Range	
C16 : 0	20.13	1.63	17.40–23.40	21.08	1.98	17.90–24.60	20.40
C18 : 0	9.54	1.28	7.60–12.40	10.44	1.66	7.70–14.90	10.00
C18 : 1	53.56	2.70	47.40–58.10	52.40	3.11	46.30–57.50	52.60
C18 : 2	10.11	1.15	8.60–12.40	9.61	1.69	6.90–13.50	10.40

Table 2. Statistics for NIR equations predicting weight percentage of fatty acids in Iberian pig fat.

Constituent	Mean	*SD*	Range	*SECV*	r^2	*SEL*[a]
C16 : 0	21.00	1.39	16.83–25.18	0.26	0.97	0.26
C18 : 0	10.62	1.31	6.68–14.56	0.24	0.97	0.22
C18 : 1	52.24	2.37	45.13–59.36	0.26	0.99	0.25
C18 : 2	9.39	1.30	5.48–13.30	0.15	0.99	0.15

[a]Standard laboratory error calculated from 20 samples analysed in duplicate using the reference method (GC)

Table 3. *SEP*, bias and *SEP*(*C*) values for validation sets A, B and C.

	Set A			Set B			Set C		
	SEP	Bias	*SEP*(*C*)	*SEP*	Bias	*SEP*(*C*)	*SEP*	Bias	*SEP*(*C*)
C16 : 0	0.22	–0.03	0.22	0.29	–0.13	0.26	0.21	–0.01	0.21
C18 : 0	0.21	–0.09	0.20	0.28	0.08	0.27[a]	0.38	0.25[a]	0.29[a]
C18 : 1	0.37	0.29[a]	0.24	0.43	–0.33[a]	0.27	0.25	–0.05	0.24
C18 : 2	0.14	0.03	0.14	0.19	0.12[a]	0.16	0.24	–0.02	0.24[a]

Bias limits: C16 : 0 = 0.14, C18 : 0 = 0.12, C18 : 1 = 0.15 and C18 : 2 = 0.08
SEP(*C*) limits: C16 : 0 = 0.30, C18 : 0 = 0.26, C18 : 1 = 0.31 and C18 : 2 = 0.17
[a]Bias and *SEP*(*C*) values higher than confidence limits

It should be stressed that all validation sets were scanned after taking the spectra of the calibration set. The calibration set was obtained between January 1997 and April 1998, validation set A was collected just after the calibration set (in May 1998) and validation set B was scanned 19 months later (in November 1999). Validation set C was obtained over a period of 13 months (from June 1999 to July 2000), one year after scanning the calibration samples.

Table 3 shows *SEP*, bias and *SEP(C)* values obtained for each fatty acid during routine analysis of the validation sets. As can be seen from this table, *SEP* values were very low and similar to the *SECV* values. In addition, with only a few exceptions, bias and *SEP*(*C*) values for the three data sets were lower than the confidence limits established for each fatty acid. These data confirm the robustness and the high precision of the NIR equations obtained previously.[7]

However, the mean "*H*" values for each validation set differed. This statistic measures the mean distance of validation set spectra to the centroid of the principal component space defined by calibration set spectra. Thus, the "*H*" statistic provides information about the distance between the validation set spectra predicted by calibration equations and the spectra belonging to the calibration set. As Table 4 shows, the "*H*" value for validation set A ("*H*" = 2.63) was below three. Thus, spectra from validation set A can be considered similar to calibration set spectra.

However, the mean "*H*" value for validation set B ("*H*" = 25.44) and Set C ("*H*" = 33.38) were much higher than the maximum value recommended.[8] This means that spectra for validation sets B and C must be considered as outliers, despite their excellent *SEP* values. These anomalous "*H*" statis-

Table 4. Mean "*H*" statistic values for validation sets A, B and C using different principal component spaces.

	"*H*" calculated using different spectra sets	
	Calibration set	Calibration set file + validation set C
Set A	2.63	2.19
Set B	25.44	2.10
Set C	33.38	1.17

tics for validation sets B and C are difficult to explain, since validation set B contained replicate spectra for the same samples scanned during calibration development and Set C had spectra of a single sample, which was similar to the mean spectrum of the calibration set. Validation set C, the check spectra of a sample analysed three times weekly over several months, should reflect the influence of day-by-day variations of the instrument and/or the environment in a fat sample representative of the calibration set.

In order to model these day-by-day instrumental and/or environmental variations, the principal component space used to calculate "*H*" values was reconstructed using not only the calibration set but also validation set C. "*H*" statistics were again calculated by projecting validation sets onto the new principal component space. Mean "*H*" statistic values for validation sets are shown in Table 4. The mean "*H*" value for validation set A was similar using both principal component spaces (2.63 vs. 2.19). However, mean "*H*" statistic values for validation set B and C decreased with the new principal component space (25.44 *vs* 2.10 for validation set B and 33.38 vs 1.17 for validation set C).

When a principal component space taking into spectra of both the calibration set and validation set C is constructed, mean "*H*" statistic values for all validation sets were lower than three. Thus, there is no reason to consider the validation sets as outliers.

Conclusions

Results show that the *"H"* statistic is very useful for detecting outlier fat samples by using NIR spectral information projected onto the principal component space. To construct this multivariate space, spectra of a fat sample that models day-by-day instrumental and/or environmental variations must be included.

Acknowledgments

NIR data were obtained using NIR hardware and software at the NIR/MIR Spectroscopy Unit (SCAI) of the University of Cordoba. GC data were obtained at the Laboratorio Agrario de Córdoba (Junta de Andalucía). Special thanks to Ms Paquita Baena, Mr Antonio López and Mr Alberto Sánchez de Puerta for laboratory assistance.

References

1. H. Martens and T. Næs, *Multivariate Calibration*. John Wiley & Sons, Chichester, UK (1991).
2. R. De Maesschalck, D. Jouan-Rimbaud and D.L. Massart, *Chemom. Intell. Lab. Syst.* **50,** 1 (2000).
3. *Unscrambler User´s guide*, version 5.0. Programme package for multivariate calibration. CAMO A/S, Trondheim, Norway (1993).
4. J.S. Shenk and M.O. Westerhaus, *Routine operation, calibration, development and network system management manual.* NIRSystems Inc., 12101 Tech Road, Silver Spring, MD 20904, USA, (1995).
5. E. De Pedro, A. Garrido, I. Bares, M. Casillas and I. Murray, in *Near infrared spectroscopy Bridging the Gap between Data Analysis and NIR Applications*, Ed by K.I. Hildrum, T. Isaksson, T. Næs and A. Tandberg. Ellis Horwood, Chichester, UK, p. 341 (1992).
6. J. Garcia-Olmo, A. Garrido and E. De Pedro, in *Near Infrared Spectroscopy: Proceedings of the 9th International Conference*, Ed by A.M.C. Davies and R. Giangiacomo. NIR Publications, Chichester, UK, p. 253 (2000).
7. J. Garcia-Olmo, A. Garrido and E. De Pedro, *J. Near Infrared Spectrosc.* **9,** 49 (2001).
8. J.S. Shenk, J.J. Workman and M.O. Westerhaus, in *Handbook of near infrared analysis, Ed by* P.A. Burns and E.W. Ciurczak. Marcel Dekker, NY, USA, p. 383 (1992).

Mastitis diagnostics by near infrared spectra of cow's milk, blood and urine using soft independent modelling of class analogy classification

Roumiana Tsenkova[a*] and Stefka Atanassova[b]

[a]*Faculty of Agriculture, Kobe University, Kobe 657, Japan*

[b]*Department of Physics, Thracian University, Stara Zagora 6000, Bulgaria*

Introduction

Biological fluids, such as milk, blood and urine, contain information specifically related to metabolic and health status of ruminant animals. Some changes in their composition can be attributed to the disease response in the animals. Finding these changes soon after they have occurred could lead to an early diagnosis followed by more effective treatment or even disease prevention.

Mastitis is a disease in a cow's mammary gland, initiated when bacteria enter the gland and establish an infection or inflammation.[1] It causes major problem for the global dairy industry and substantial economic losses from decreasing milk production and reduced milk quality. When mastitis occurs, lactose in milk decreases and lactose in urine increases.[2,3] Alteration in the blood–milk barrier leads to an influx of blood proteins in milk and milk proteins in blood, such as α-lactoalbumin and casein, and to changes in ionic concentration in milk.[4–6] Mastitis is accompanied by an influx of white cells from the blood stream into the milk, altered secretary function and changes in the volume and composition of secretion. Somatic cell count (SCC) has been accepted as the international standard measurement of dairy milk quality and mastitis diagnosis.[7] In previous studies, near infrared (NIR) has been successfully applied for noninvasive mastitis diagnosis performed by qualitative raw milk spectral analysis[8,9] when an expert's diagnosis has been used as a reference. NIR diagnosis has been based on compositional changes in milk caused by mastitis and their respective spectral changes.

In this study, for the first time, it was found that mastitis diagnosis could be performed with equal success when NIR spectra of any biological fluid, such as milk, blood or urine, were classified using SCC in milk as a disease threshold.

Materials and methods

Samples

A total of 112 bulk milk, urine and blood samples from four Holstein cows were analysed. The samples were collected for 28 days, consecutively, starting from 7th day after calving. The milk samples were collected from morning milking. The urine samples were collected before morning milking and stored at –35°C until spectral analysis. The blood samples were collected before morning milking using a catheter inserted into the carotid vein. Heparin was added to blood samples to prevent coagulation. The average BW of the cows was 552 kg.

Each milk sample was divided into two subsamples. One was subjected to spectral analysis and the other was analysed for SCC by fluoro-opto-electronic method using a Fossomatic 400 (Foss-Electric A/C, Hillerød, Denmark). Somatic cell count standards were used to calibrate the Foss instrument throughout the study. The repeatability coefficients of variation of this method are 4 to 5% for the region between 400,000 and 500,000 cells mL^{-1} and 5 to 10% for the regions between 100,000 and 200,000 cells mL^{-1} and over 500,000 cells mL^{-1} (IDF Standard 148A, 1995).[7] The SCC content in milk was used as the indicator for mastitis. One cow was mastitic during the entire experimental period—the measured SCC varied from 204,000 to 11,876 ,000 cells mL^{-1}. Three of the cows had mastitis and healthy periods—SCC was between 80,000 and 437,000 cells mL^{-1}.

SCC values of milk samples were used as the quantitative parameter for respective urine and blood samples collected from the same cow, at the same time.

NIR spectra

Near infrared transflectance (T) spectra of blood and milk samples were obtained using an InfraAlyzer 500 spectrophotometer, (Bran+Luebbe, Nordestedt, Germany), in terms of optical density log (1/*T*) in a wavelength range from 1100 to 2500 nm. A flow cell, with a pathlength of 0.2 mm, connected with an automated liquid sampling system taking, alternatively, milk samples and cleaning solution, was used. Before the spectral analysis, each sample was warmed up to 40°C in a water bath with a temperature control of ± 0.1°C. During the analysis, the same temperature was controlled through the use of an integrated water-jacketed holder of the flow cell connected with the water bath.

Near infrared spectra of urine samples were obtained using an NIRSystem 6500 spectrophotometer (Foss NIRSystems, Silver Spring, MD, USA), using quartz cuvettes with 1 mm sample thickness, in the spectral region from 1100 nm to 2500 nm.

NIR analysis

A commercial program, Pirouette Version 2.6 (Infometrics, Inc., Woodinville, WA, USA), was used for qualitative analysis, i.e. classification of samples. The spectral region from 1100 to 2500 nm was used in calculations for all data sets. Classification of milk samples in class "healthy" or class "mastitic" was done using soft independent modelling of class analogy (SIMCA). The examined methods for data pretreatment included smoothing of the spectral data, first and second derivative transformation of log (1/*T*) data, based on the Savitzki–Golay second-order polynomial filter.[10] All samples were divided into a calibration set (two thirds of the samples) and a test set (one third of the samples). A class variable was assigned for each sample as follows: healthy (class 1)—with SCC lower than 200,000 cells mL^{-1} and mastitic (class 2)—with SCC higher than 200,000 cells mL^{-1}. The calibration set of samples consisted of 75 samples—30 samples from healthy cows and 45 samples from mastitic cows. The test set consisted of 37 samples—17 samples from healthy cows and 20 samples from mastitic cows. SIMCA developed models for each class based on factor analysis, i.e. principal components that describe the variations of the spectral data. Loadings of the principal components for each model were compared.

Results and discussion

SIMCA classification of milk, urine and blood spectra (Figure 1), respectively, based on somatic cell count (SCC) in milk as a threshold for mastitis diagnosis, is presented in Table 1. For the calibration set of samples, SIMCA models (model for samples from healthy cows: class 1 and model for samples from mastitic cows: class 2), correctly classified from 97.33 to 98.67% of milk samples, 97.33 to 98.67% of urine samples and 94.67 to 96.00 % of blood samples. The best results for all data sets were obtained when first derivative spectral data pretreatment was used. Incorrectly classified samples were false negative, false positive or non-classified. Most of them had an SCC close to the defined threshold

Table 1. Results for SIMCA classification based on milk, urine and blood spectra (class 1: samples from cows with milk SCC < 2000,000 cell mL^{-1}, class 2: samples from cows with milk with SCC > 200,000 cell mL^{-1}.

Spectra	Spectral data pretreatment	Calibration set $n = 75$				Test set $n = 37$			
		correct classification		incorrect classification		correct classification		incorrect classification	
		n	%	n	kind	n	%	n	kind
Milk	Smooth	73	97.33	2	1 false negative 1 false positive	26	70.27	11	2 false negative 6 false positive 3 non classified
	First derivative	73	97.33	2	1 false negative 1 false positive	32	86.49	5	1 false negative 4 false positive
	Second derivative	74	98.67	1	1 false negative	25	67.57	12	7 false negative 5 false positive
Urine	Smooth	74	98.67	1	1 false negative	27	72.97	10	3 false negative 4 false positive 3 non classified
	First derivative	73	97.33	2	2 false positive	32	86.49	5	4 false positive 1 non-classified
	Second derivative	74	98.67	1	1 false negative	31	83.78	6	1 false negative 4 false positive 1 nonclassified
Blood	Smooth	7[illegible]	98.67	4	2 false negative 2 false positive	30	81.08	7	4 false negative 3 false positive
	First derivative	72	96.00	3	2 false negative 1 false positive	33	89.19	4	1 false negative 3 false positive
	Second derivative	72	96.00	3	1 false negative 2 false positive	29	78.38	8	2 false negative 6 false positive

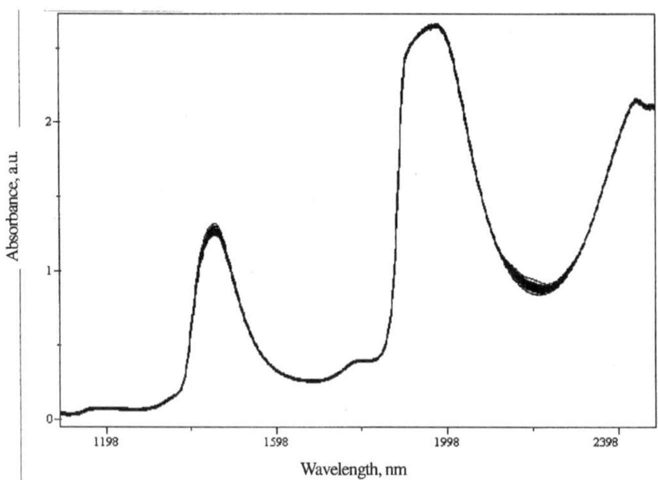

Figure 1. NIR spectra of urine, 1 mm cell.

of 200,000 cell mL^{-1}. There were more false positive samples than false negative ones. This result could be explained by the fact that SCC elevation was a result of the cow's defense mechanism against mastitis and, as such, post infection event was preceded by compositional changes in the biological fluids. These changes were reflected in the cow's milk, urine and blood NIR spectra, respectively. SIMCA models detected these changes earlier and classified the samples as samples from mastitic cows, although their SCC was still low.

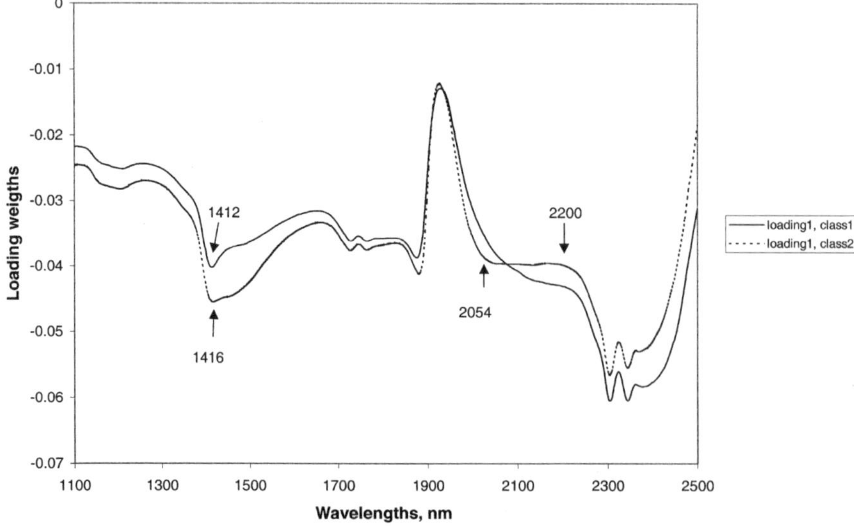

Figure 2. NIR milk spectra: first principal component loading plots of SIMCA models for healthy cows (class 1) and mastitic cows (class 2).—— class 1, - - - - -class 2.

SIMCA models were based on principal components (PC). Their loadings indicated variables (wavelengths) that had the highest contribution in the respective model and loading. Loadings of first PC components included in the SIMCA models were investigated. The first loading describes most of the variations in the related spectral data set. Plots of the first loadings for healthy and mastitic models, based on milk spectra, are presented in Figure 2. The first PC component described 96.88% of the total variations in healthy milk spectra (class 1) and 93.17% of the total variations in mastitic milk spectra (class 2). The respective values of total spectral variation, described by the first PC in the models for class 1 and class 2, were 87.71% and 88.57% for blood samples and 57.47% and 60.15% for urine samples, respectively. Differences in the spectral patterns of the first loadings were analysed in terms of loading weights and shape of the loading at characteristic wavelengths. Significant differences were found in the range between 1412 and 1450 nm for all of the tested biological fluids. For milk spectra, differences in weight and shape of the loadings were observed while for urine and blood spectral shifts in the position of maximum or minimum of the loadings were found, too. NIR absorption in this wavelength range is connected with the O–H absorption of water.[11] Characteristic differences in the loadings of both models, for the investigated fluids, were found in another water absorbance band at around 1910 and 1940 nm, too. It is known that mastitis markedly changes the ionic concentration in milk and blood.[6] Sodium and chloride contents in the milk increase because of the passage from the blood into the milk. Potassium declines because of the passage out of the alveolar lumen between damaged epithelial cells. The changes in blood cause respective changes in urine. Altered ionic concentration in milk, blood and urine exerted influence on water structure due to ion–OH interactions. We speculated that these changes were origininally from the observed differences in the NIR spectra of biological fluids[12] that were used to build up each SIMCA model for classification. PC loadings elucidated significant spectral differences from 2040 to 2200 nm for milk spectra, from 2140 to 2300 nm for blood spectra and from 2150 to 2300 nm, for urine spectra, too. In the region from 2040 to 2300 nm the dominated absorption of the N–H and C–H bands connected with proteins. The combination of O–H deformation/C–O stretching vibration around 2100 nm and of O–H stretching/C–O stretching vibration around 2278 nm were associated with carbohydrate absorption. Mastitis decreases lactose content and alters the types of protein in milk. Casein content decreases while whey protein increases. Alteration of the blood–milk barrier leads to an influx of albumin and other blood serum proteins in milk and lactose and milk proteins in blood, such as α-lactoalbumin. Altered types and concentration of proteins in milk and blood, and changes in lactose content of milk, blood and urine,[2] explained the differences in the loading of the first PC factors for spectra of healthy and mastitic cows in this spectral region.

Conclusion

Cow's mastitis diagnosis, based on SIMCA classification of near-infrared spectra of biological fluids, showed similar performance for milk, blood and urine when somatic cell count (SCC) in milk was used as the disease threshold. The best calibration results were obtained with first derivative spectral transformation. They varied from 97.37% to 98.67% for different fluids.

The success of NIR mastitis diagnostics was defined by the effect on the NIR spectra of biofluid compositional changes caused by mastitis.

Acknowledgments

The authors thank Dr S. Tanabe, Dr F. Terada, Dr T. Hayashi, Dr M. Amari and Dr A Purnomoadi from the National Institute of Animal Industry, Tsukuba, Japan and Dr S. Kawano and Dr J. Chen from the National Food Research Institute, Tsukuba, Japan for the preparation of the milk, blood and urine samples and the NIR measurements. The authors are grateful to GL Science Inc., Tokyo, Japan for providing the Pirouette Version 2.6 software. This work was supported by the Programme for Promotion of Basic Research Activities for Innovative Bioscience (PROBRAIN), Ministry of Agriculture, Fisheries and Food, Japan.

References

1. J.E. Hillerton, *Bulletin of the International Dairy Federation* **No. 345,** 4 (1999).
2. D.E. Shuster, R.J. Harmon, J.A. Jackson and R.W. Hemken, *J. Dairy Sci.* **74,** 3763 (1991).
3. J. Schulz, T. Hanisch, S. Dumke, S. Springer and K. Beck, *Praktische Tierapzt.* **79,** 657 (1998).
4. B. Poutrel, J.P Caffin, P.J. Rainard, *J. Dairy Sci.* **66,** 535 (1983).
5. T.B. McFadden, R.M. Akers, A.V. Capuso, *J. Dairy Sci.* **71,** 826 (1988).
5. R.J. Harmon, *J. Dairy Sci.* **77,** 2101 (1994).
6. International IDF Standard 148A, *Milk Enumeration of Somatic Cells.* International Dairy Federation, Brussels, Belgium (1995).
7. R. Tsenkova, K. Yordanov and Y. Shinde, in *Prospects for Automatic Milking,* Ed by A.H. Ipema, A.C. Lippus, J.H.M. Metz and W. Rossing. PUDOC Science Publishing, Wageningen, The Netherlands, p. 185 (1992).
8. R. Tsenkova, S. Atanassova, S. Kawano and K. Toyoda, *J. Animal Sci,.* in press (2001).
9. A. Savitzky and M.J.E. Golay, *Anal. Chem.* **36,** 1627 (1964).
10. H. Maeda, Y. Ozaki, M. Tanaka, N. Hayashi and T. Kojima, *J. Near Infrared Spectrosc.* **3,** 191 (1995).
11. V.J. Frost and K. Molt, *J. Mol. Struct.* **410/411,** (1997).

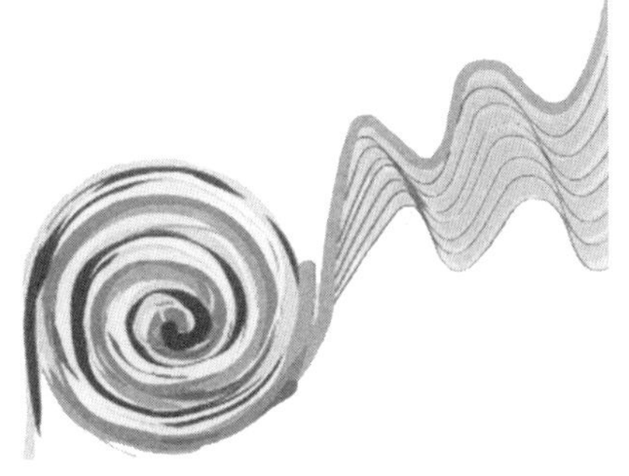

Food

Hand-held near infrared spectrometry: status, trends and futuristic concepts

W. Fred McClure

North Carolina State University, Raleigh, NC 27695-7625, USA

Status

Increased interest in hand-held (HH) technology is reflected in the growing number of HH meters exhibited at PITTCON in just the last five to ten years (see Figure 1). It appears that no instrument manufacturer is exempt from the pressure to place their technology in the hands of customers.

Near infrared (NIR) spectrometers are more compact than they were five years ago. So called *portable* NIR instruments do exist, but truly hand-held NIR (HHNIR) meters have not yet appeared in the market place. The lack of commercial HHNIR meters is due, largely, to three problems: (1) Profit margins for HHNIR meters are small compared with portable units, (2) HHNIR meters furnished with calibrations may require additional resources to maintain and (3) NIR energy sources draw considerable power—too much for battery operation.

It is true that HHNIR meters do not command the profits enjoyed from laboratory and process analysers. Yet, interests in HHNIR meters will not go away. Millions of growers need an instrument to

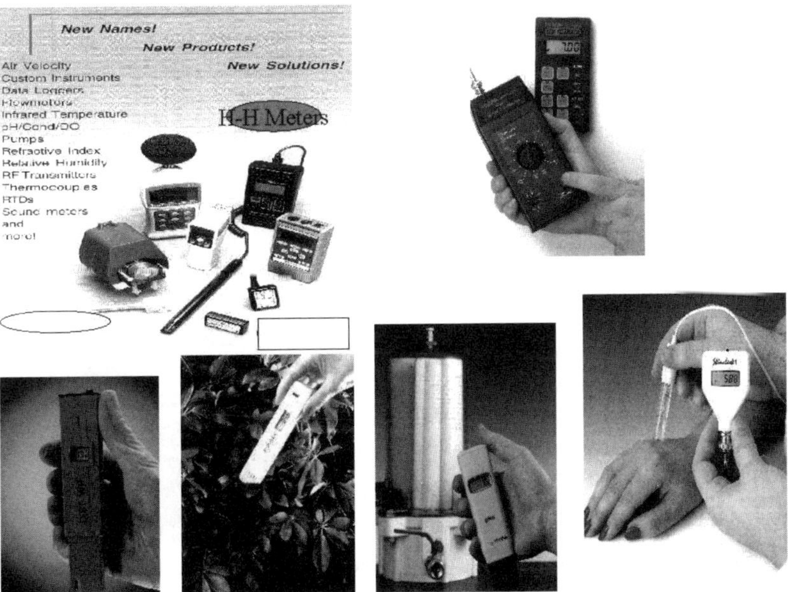

Figure 1. Illustrating the trend in hand-held technology. Hand-held exhibits at the Pittcon meeting have increased several fold in the last five years.

monitor environmental abuse due to the misuse of farm waste and chemical fertilizers. Pharmaceutical manufacturing would like to have HH meters to test incoming raw materials and monitor processes. Clinical medicos are calling for pocket analysers to measure bioparameters related to patient health. This continued interest in HHNIR technology seems to indicate that the number of units required to satisfy needs would tend to offset the profit disparity.

Furnishing instruments with calibrations is not foreign to the NIR instrument laboratory. Infrasoft International does this on a routine basis, calibrations that involve both complex samples and complicated instruments. Research results[1,2] seems to indicate that calibrations for HHNIR meters would be no more of a problem than for laboratory instruments. Data seems to imply that calibration development, transfer and maintenance could be facilitated *via* the world wide web.

Finally, components for HHNIR technology are improving all the time. Recently, with a no-moving parts HHNIR meter costing less than $300 US, McClure and his associates[3] demonstrated that unfiltered light emitting diodes can be used to measure chlorophyll plant tissue and moisture in paper. Morimoto[2] developed a filter-based derivative spectrometer that utilised a low power Halogen lamp. Long life batteries provide adequate power to operate the Gmeter for up to 12 hours of continuous operation. Hence, HHNIR meters can be made to operate for long periods of time.

Though not approved for commercialisation as of this date, several companies are working feverishly to be the first to market a noninvasive hand-held blood–glucose meter.[4] Pushed by an annual $780 million financial incentive from invasive technology, a noninvasive blood-glucose meter would be a welcomed change to the more than 5 million diabetics in the USA alone. One of the first to attack the blood-glucose problem was Rosenthal.[5] His approach was to transmit NIR energy through the index-finger/nail to determine sugar content of the blood in the optical path. That work, conducted by a company called Futrex, was terminated in 1998 due to extenuating circumstances.[6]

Trends

McClure and his associates[2,3,7,8] have developed four NIR meters for measuring: (1) Nicotine and moisture in tobacco (Nmeter), (2) Vanillin and moisture in vanilla beans (Vmeter), (3) chlorophyll in growing plants and moisture in paper (Twmeter) and (4) protein and nitrogen in sugar/protein mixtures and grass tissue (Gmeter), respectively. This article will address the design and performance of these instruments.

Nmeter

Figure 2 shows the Nmeter mounted in a *laboratory caddy*. The Nmeter was initially intended to measure only nicotine in tobacco. However, the design permitted the incorporation of eight filters (1759, 1940, 2139, 2190, 2230, 2250, 2270 and 2310 nm) to the filter wheel, a feature which permitted the expansion of measurements to include total sugars and moisture. The filter-wheel was continuously driven by a stepper-motor. One reading was taken for each filter per revolution of the wheel.

Figure 2. The Nmeter (nicotine meter) mounted in a laboratory caddy. The Laboratory caddy is used when acquiring data in the laboratory for the purpose of developing calibrations.

Software for acquiring filter readings, downloading data to a computer and uploading calibrations to the meter was written in C. With the filter wheel rotating at 60 rpm, twelve spectra (eight readings spectrum^{-1}) were recorded in about 8 s. The twelve readings were averaged to produce an *average spectrum*. Only average spectra were used for developing calibrations and computing the composition of *unknown samples*. The meter weighed 3 kg, a bit heavy for a hand-held device that was to be used continuously on eight hour shifts.

Figure 3. The Vmeter was designed to measure vanillin and moisture in vanilla beans. The meter is shown to the left of the laboratory caddy. The caddy is used to hold the Vmeter when recording data used to develop calibrations.

Vmeter

A picture of the Vmeter is given in Figure 3. It is a filter-based meter much like the Nmeter. It has a filter wheel with seven filters and a reference screen. The filter wheel is rotated with a DC motor at 640 rpm. Filter identification and position is made possible with a decoder-wheel attached to the shaft of the filter wheel. The Vmeter differs from the Nmeter in that it illuminates the sample with two lens-end Halogen lamps and filters the reflected light while the Nmeter filters the illumination and captures a portion of the reflected light. These two lamps draw 0.8 A at 5 volts and have a life of 10,000 hours. Adequate illumination was achieved by running the lamps at 3.4 V, reducing the load on the battery. The Nmeter weighs 2.6 kg without the battery and 2.9 kg with the battery.

The Vmeter was designed to opterate in two modes. In the first mode, the Vmeter is connected to a PC through an RS232 serial port. Due to added functionality, this mode is the preferred mode of operation in a laboratory environment. In the second mode, the Vmeter is disconnected from the computer for *field operation*, after calibrations have been uploaded from the computer to the meter. In this latter mode, meter measurements are displayed on the liquid-crystal display (LCD) and stored in the memory.

TWmeter

The TWmeter,[3] shown in Figure 4, was developed to minimise the cost of HHNIR technology. Dubbed the *TWmeter*, this device was conceived for use by researchers and others in *Third World* countries unable to afford more costly technology found in developed countries. Three light-emitting diodes (LEDs) were selected to measure chlorophyll and moisture in plant tissue. Centre wavelength and emission bandwidth of the LEDs (Sylonex, Inc., Plattsburg, NY, USA) were as follows: (1) 700 nm–100 nm, 880 nm–50 nm and 940 nm–50 nm. The 700 nm filter, near the chlorophyll absorption band of 673 nm, did not exactly correspond to the chlorophyll absorption maximum but the emission did provide illumination at 673 nm. The 940 nm emitter provided illumination at 960 nm, the absorption of water. The 880 nm emitter was chosen as a reference.

Figure 4. The Twmeter was designed to measure chlorophyll and moisture in situ. It has no moving parts and has a parts cost of less than US$300.00.

The TWmeter was powered with four 1.5 V alkaline batteries hooked in series to provide

Table 1. Performance of the Nmeter, Vmeter, TWmeter and Gmeter.

Meter	N	R^2_C [a]	*SEC*	*SEP*	CV_P [b]
Nmeter					
Nicotine	327	0.933	0.226	0.228	8.1
Sugars	327	0.923	1.367	1.373	9.2
Moisture	327	0.728	0.794	0.976	8.1
Vmeter					
Moisture	60	0.977	0.710	0.857	5.8
Vanillin	60	0.865	0.324	0.672	17.4
TWmeter					
Chlorophyll	40	0.847	0.90	0.99[c]	18.1
Moisture	72	0.993	0.90	1.04[c]	1.8
Gmeter					
Protein[d]	60	0.990	2.570	2.740	6.3
Nitrogen[e]	60	0.951	0.581	0.630	17.2

[a]R^2_C = Coefficient of Determination for Calibration
[b]CV_P = *SEP*/Mean
[c]root mean square standard error
[d]in protein/sugar mixture
[e]in dry grass tissue

6.0 V that was regulated to + 5 V to power the meter. Measuring 10 cm wide, 19 cm long and 5 cm high, the TWmeter weighs 364 g and can operate continuously for more than 16 hours without changing batteries. The TWmeter has no moving parts.

Gmeter

A protype of the Gmeter[2] is shown in Figure 5. The Gmeter was designed to measure protein in protein/sugar mixtures and to measure nitrogen in plant tissue. It is unique in that it utilises 2nd derivative calculations from data obtained with narrow-band filters. The derivative is calculated according to a formula derived from the Taylor series expansion of a digital function. The filter wheel in the Gmeter can accommodate up to ten filters, however, only a single-term 2nd derivative utilising three filters has been tested so far.

Only the prototype meter has been built at this time. Calculations indicate that the Gmeter could be reduced to a cylindrical volume with di-

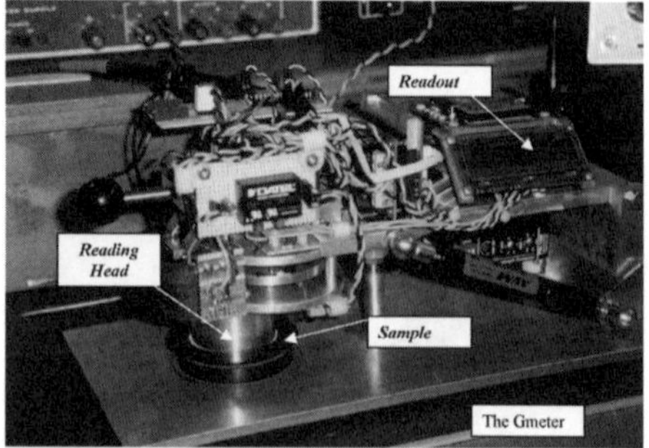

Figure 5. A prototype of the Gmeter mounted in a laboratory caddy. The Gmeter could be packaged in a cylinder 7 cm diameter by 8 cm tall and weigh less than one kilogram. It is the first meter to use filters to generate derivatives.

mensions of 7 cm diamter by 8 cm high and weigh about 1.5 kg. The Gmeter operates in the step-scan mode.

Performance

Performance data for all four meters discussed above are given in Table 1. A general rule-of-thumb, developed from experience in this laboratory, indicates that an application (calibration and prediction) is not very robust unless the coefficient of variation (*CV*) for prediction is less than 10%. Performance of all meters met this performance criteria for at least one assay. Nmeter data resulted in a CV_p of less than 10 for nicotine, sugars and moisture. The Vmeter produced a CV_p of 5.8 for moisture, but its performance for vanillin was less impressive (CV_p = 17.4%). The TWmeter did a good job of measuring moisture, but did not perform well when measuring chlorophyll. However, chlorophyll measurements are quite nonlinear[9] and the data shown is for a linear calibration. The Gmeter was designed to measure protein in protein/sugar mixtures and this calibration was robust. It was not optimised for measuring nitrogen in plant tissue, but the protein calibration did surprisingly well for measuring nitrogen (*CV* = 17.2%). If the filter selection was optimised for measuring nitrogen in plant tissue, the authors are convinced that the calibration would be robust.

Future prospects

Over the last 50 years, we have seen NIR instrumentation evolve into two areas: (1) The laboratory and (2) The process line. Process instruments (located *at-line, in-line* or *on-line* are designed to operate in high-temperature and dusty environments. Laboratory instrument, on the other hand, are designed to function in temperature/humidity-controlled labs.

More recently, managers of both laboratory and process analysers are finding an increasing need to make measurements remote from the line and laboratory. The ever increasing flow of products renders the falible human opinion useless. Objectivity, rendered by HH technology, enables managers to make intelligent decisions concerning the quality of both raw materials and product output.

The demand of HHNIR is here—now! The call for HHNIR will not go away. Problems facing this important contribution to spectroscopy will be overcome. HHNIR meters will appear within five years. More than likely, the first development will be the blood–glucose meter. However, HHNIR to monitor environmental abuse could be the first NIR meter on the market. Who will be the first?

References

1. S. Morimoto, In *Proceedings of 10th Intenational Conference on Near Infrared Spectroscopy*, Ed by A.M.C. Davies and R.K. Cho, Kyongju, Korea, 2001. NIR Publication, Chichester, UK. (2002).
2. S. Morimoto, W.F. McClure and D.L. Stanfield, *Appl. Spectrosc.* **55(1),** 182 (2001).
3. W.F. McClure, *Appl. Spectrosc.*, in preparation.
4. R. Mendosa, *Diabetes Wellness Letter* **November,** 1 (1997).
5. R.D. Rosenthal, *Hand-held near-IR analysis instruments for noninvasive measurement of blood* (1991).
6. B. Rosenthal, *Unbelieveable*. Star Publishing, Hagerstown, MD, USA (1999).
7. W.F. McClure, C.M. Hargrove, M. Zapf and D. Stanfield, *J. Near Infrared Spectrosc.*, in preparation.
8. W.F. McClure and D.L. Stanfield, *J. Near Infrared Spectrosc.*, in preparation.
9. K.K. Katur, A spectrophotometric technique for automatically controlling the flue-cured process (under the direction of W.F. McClure). MS, NC State University, NC, USA (1974).

Thermometrical spectroscopy: temperature programming as a control variable to increase information content from near infrared spectroscopic measurements—characterisation of honey samples

Anthony M.C. Davies

Norwich Near Infrared Consultancy, 75 Intwood Road, Cringleford, Norwich NR4 6AA, UK.

Introduction

Invisible radiation beyond the red end of the visible spectrum was discovered by William Herschel in March 1800.[1] Not surprisingly, Herschel did not understand what he had discovered. He believed that this "heat radiation" was something quite different from light and he suggested the terminology "the thermometrical spectrum" for what we now know as part of the near infrared (NIR) region (which I like to refer to as "the Herschel region", 780–1100 nm[2]) and his suggestion has been long forgotten. In this paper I would like to suggest a new use of this term.

It has been known for a long time that NIR spectra are affected by the temperature of the sample. This is especially true if the sample is present in a liquid form. There have been several studies of the effect of temperature on the spectrum of water[3–5] and Osaki and colleagues have carried out 2D-correlation studies on other liquid systems of organic solvents where temperature is the controlled variable.[6,7] While these experiments have deliberately made measurements at different temperatures, as far as I am aware, the quantitative and qualitative applications of NIR spectroscopy have always seen temperature variation as an interference that has to be managed or avoided. In our work on the characterisation of honey,[8] I deliberately introduced temperature variation. I believed that this would help the discrimination because the expected variation in the sugar composition of the samples would have a varying influence on the spectrum of the water in the sample. The first paper giving the results of the experiment has been accepted for publication in the Journal of Near Infrared Spectroscopy[8] and need not be repeated in detail except to explain how these experiments in "Thermometrical spectroscopy" were performed. Then I will discuss how they might be incorporated in future NIR spectroscopic systems.

Experimental system for honey characterisation

The system used for the characterisation of honey was based on the use of a Foss NIRSystems NIR 6500 spectrometer (Foss NIRSystems, Silver Spring, MD, USA). Honey samples were scanned using a special sample cell prepared for these experiments, which allowed a small sample of honey to be held in a water bath while NIR measurements were made via a fibre optic probe. The cell, was made by cutting out a cell in a solid brass cylinder (60 mm tall by 25 mm diameter) so that a minimum amount of sample was lost in covering the bottom 25 mm of the probe whilst allowing the sample to flow into the measuring cavity of the probe. The warm honey sample, 7–10 g, was placed in the cell and the probe was slowly inserted so that honey would flow into the measuring cavity of the probe. The cell was maintained at one of five chosen temperatures (10, 17, 26, 37 and 50°C) by immersion in a thermostatically controlled water bath, Grant 6G (Cambridge, UK). NIR spectra over the range 1100–2498 nm, were recorded at 2 nm intervals. The cavity, between the end of the probe and mirror was set at a gap of 2 mm giving a path length of 4 mm. Three NIR spectra were recorded at each temperature using the empty cell as the reference measurement for the calculation of transmission (*T*) spectra as log(1/*T*). The recorded spectra were transferred to the hard disk of a personal computer.

Data Analysis

The recorded NIR spectra were transferred to the UNSCRAMBLER (Camo AS, Oslo, Norway) environment for graphical plotting and some preliminary mathematical treatment. Selected portions of the spectra were plotted before and after the calculation of second derivatives using a seven point,

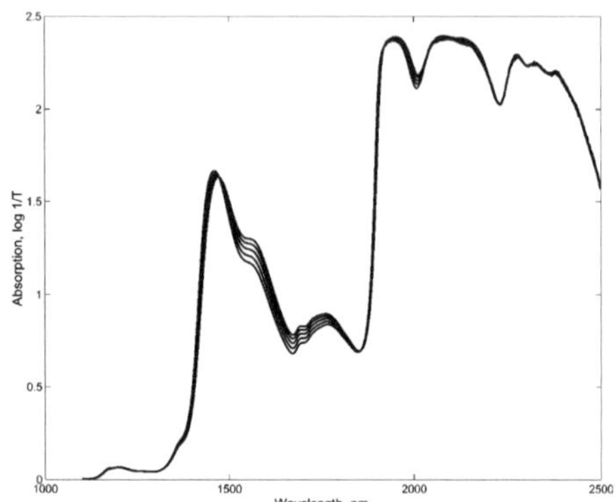

Figure 1. Raw spectra of one honey sample at five temperatures.

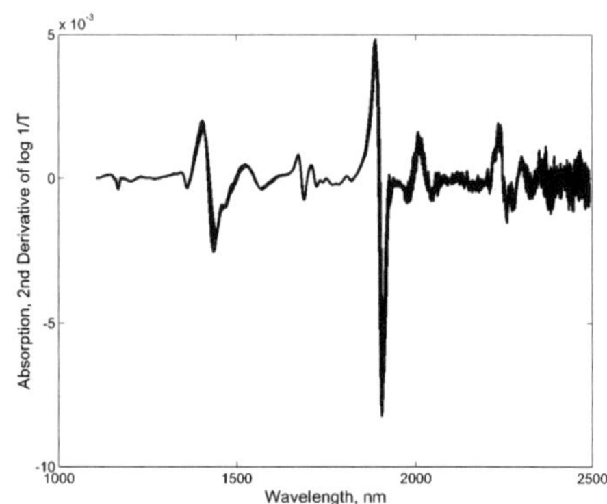

Figure 2. Second Derivative of spectra shown in Figure 1.

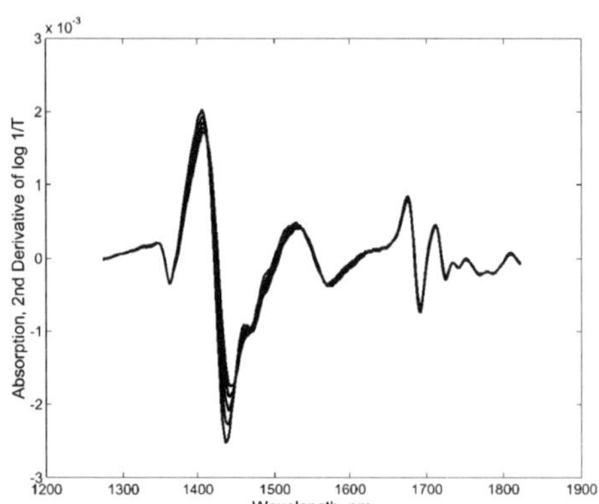

Figure 3. Selected region of second derivative spectra of a honey sample measured at five temperatures.

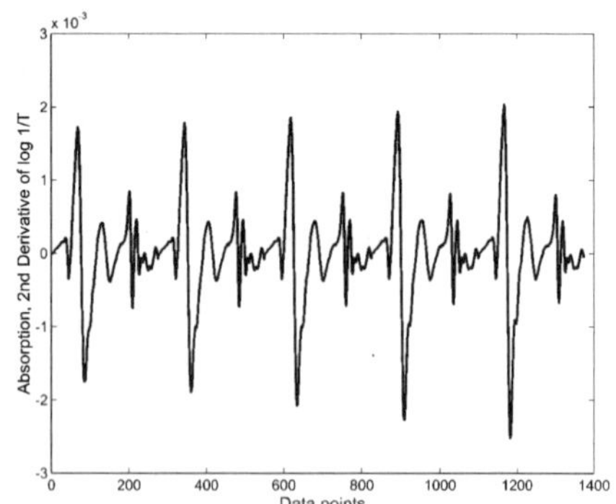

Figure 4. Psuedo spectrum of the selected region of a honey sample measured at five temperatures. The temperature increases from left to right.

second order Savitzky–Golay filter. The data were then transferred to the MATLAB (The MATH WORKS, Natick, MA, USA) environment for subsequent data processing.

The three derivative spectra for each temperature measurement were averaged. For each honey sample, a selected region from 1274 to 1822 nm (275 data points) of each of the averaged derivative spectra for each temperature were combined to make a pseudo spectrum (PS), which combined temperature and NIR measurements of 1375 data points. The processing sequence for a single sample is shown in Figures 1–4. The PSs were then compressed by principal component analysis (PCA) into twenty principal components (PCs).

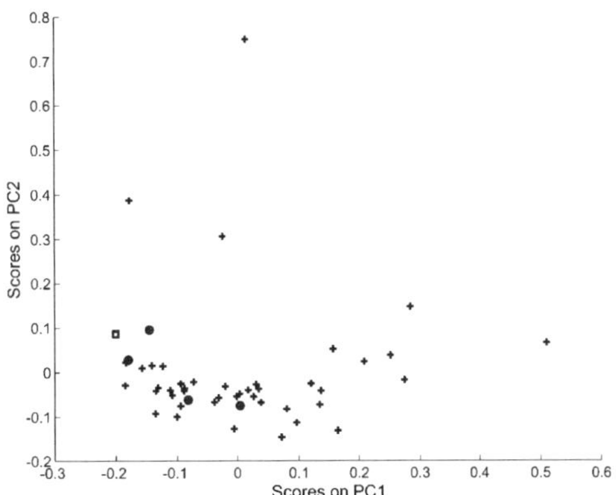

Figure 5. A scores plot of the first two PCs indicating the position of the rape samples (circles). The square symbol indicates the position of a rape sample which was excluded from the PCA.

Canonical variates analysis (CVA) was used to attempt to form groups of honey with five or more members into clusters using 5, 10, 15 or 20 PCs. Cross-validation (leaving out one sample in turn as a test sample, carrying out the calculations on the rest of the samples and then predicting the excluded sample) was used to test each of the 28 members of these groups for membership when they were not used in the computation of the clusters. Figure 5 is a typical scatter plot of the first two PCs for the rapeseed group and Figure 6 is an overview plot of the results of the CVA for the four groups.

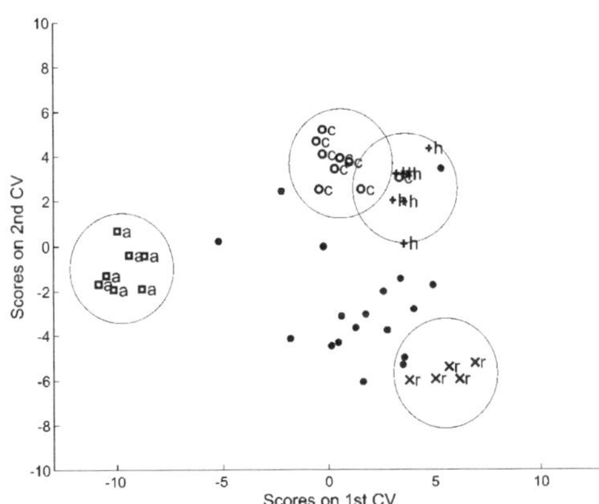

Figure 6. Plot of first two CVs from the CVA of the 28 honeys belonging to one of four groups; a, acacia; c, chestnut; h, heather and r, rape. The unlabelled dots indicate the calculated positions of the twenty honeys which are not members of any of groups in the CVA. The ellipses around the groups are at a confidence level of 0.95.

Discussion

Some preliminary analysis has found evidence that the temperature programming made a contribution to the generally satisfactory results from the honey experiment but this is not required for the discussion of the general idea of introducing deliberate variation in sample temperature. The system used in the honey experiment was very slow to use and impractical for regular application. In the current study it took an average of 45 minutes to make the measurements on one sample, taking three spectra at each of five temperatures. The majority of this time is due to the use of a water bath.

More practical methods are available but the choice depends on the type of spectrometer and the application.

For slow scanning systems (more than one second per scan), the temperature must be at a known stabilised temperature during the scan and this could be achieved by a Peltier effect temperature controlled sample cell. This type of controller can change the controlled temperature much more efficient and would allow measurement of 3 – 4 samples per hour. For fast scanning systems, pulsed heating produced by electrical or by microwaves would be feasible, assuming that the sample is below the minimum temperature of interest.

Application

Slow scan system

- Spectra at a few (5–20) known temperatures
- Gives Added information
  - Improved qualitative analysis
  - Improved quantitative analysis?
  - Search required for most useful temperatures?
  - CARNAC[9]

Rapid scanning system

- Spectra at many temperatures
- New techniques become possible
  - Develop a temperature calibration and an analyte calibration at a given temperature
  - Use the temperature calibration to find the spectrum of the required temperature
  - Use that spectrum to estimate analyte

While thermometrical spectra from slow scanning systems might give some useful information, the possibilities for fast scanning systems are much more interesting. The idea is that during the calibration phase calibrations for the analyte at a given temperature and a temperature calibration would be produced. In use, the thermometrical spectra would be searched to find the spectrum closest to the calibration temperature and this spectrum would be used to predict the analyte concentration of the sample. It would have application wherever sample temperatures cannot be easily control. Biomedical measurements would be an especially interesting application as patient body temperature can introduce serious errors.

It could be said that thermometrical spectroscopy has been dormant for 200 years. In order to end this dormant phase, instrumentation and funding are required for the idea to germinate.

References

1. W.F. Herschel, *Phil. Trans. Royal Society of London*, Part II, 284–292 (1800).
2. A.M.C. Davies, *Appl. Spectrosc.* Also A.M.C Davies. The Herschel Infrared Region in *Proc. Third International Conferenceon Near Infrared Spectroscopy*, Ed by R. Biston and N. Bartiaux-Thill. Agric. Res. Centre Publishing, Gembloux, Belgium, p. 74 (1991).
3. M. Iwamoto, J.Uozumi and K. Nishinari, in *International near infrared diffuse reflectance/transmittance spectroscopy conference*, Budapest, May 1986, Ed By J. Hollo, K.J. Kaaka and J.L. Gonczy. Akademiai Kiado, Budapest, Hungary, p 3 (1987).
4. A. Grant, A.M.C. Davies and T. Bilverstone, *Analyst* **114,** 819 (1989).
5. H. Maeda, Y.Osaki, M. Tanaka, N. Hayashi and T. Kojima, *J. Near Infrared Spectrosc.* **3,** 191 (1995).
6. Y. Ozaki, and I. Noda, *J. Near Infrared Spectrosc.* **4,** 85 (1996).
7. Y.Ozaki, Y. Lin, I. Noda and M.A. Czarnecki, in *Near Infrared Spectroscopy: the future waves*, Ed by A.M.C. Davies and P. Williams. NIR Publications Chichester, UK, p 32 (1996).
8. A.M.C. Davies, B. Radovic, T. Fearn and E. Anklam, , *J. Near Infrared Spectrosc.* **10,** 121 (2002).
9. A.M.C. Davies, in *Near Infrared Spectroscopy: Proceedings of NIR-2001* Ed by A.M.C. Davies and R.K. Cho. NIR Publications Chichester, UK, in press (2002).

Meat speciation using an hierarchical approach and logistic regression

Thorsteinn Arnalds,[a] Tom Fearn[a] and Gerard Downey[b]

[a]*Department of Statistical Science, University College London, London WC1E 6BT, UK*

[b]*Teagasc, The National Food Centre, Dunsinea, Castleknock, Dublin 15, Republic of Ireland*

Introduction

Speciation of fresh, comminuted meat is an important authenticity issue.[1,2] A number of approaches to this problem have been reported, including some based on chemometric analysis of mid- and near infrared spectroscopic data.[3,4] Levels of success achieved in these feasibility studies, which involved discrimination between selected meat species (chicken, turkey, pork, beef and lamb) have been encouraging, although not sufficiently accurate to warrant their immediate use by regulatory agencies or the food industry. Given the obvious practical advantages of spectroscopic techniques, there is a strong interest in the evaluation of alternative chemometric classification strategies to address this issue.

Techniques previously investigated have included factorial discriminant analysis (FDA), *k*-nearest neighbours analysis (K-NN), partial least squares regression (PLSI & PLSII) and soft independent modelling of class analogy (SIMCA). Each of these has its own advantages and disadvantages. In the case of FDA, K-NN and PLSII, the discrimination takes place in a multivariate space defined by all of the classes to be classified. This allows a one-step model development but may present difficulties in distinguishing between all of the different sample types effectively. Additionally, when a new type of, for example, meat needs to be added to the model, it (the model) must be developed *de novo* all over again. For SIMCA, each class of material needs to be modelled separately; addition of a new class is straightforward and quick. PLSI regression requires all of the sample types to be present during model development and necessitates a separate model for each material class.

The work reported in this paper describes two other approaches to general discrimination problems using a dataset previously described.[4] The first approach constructs the classification problem as a hierarchy of binary decisions, the correct solution to each leading to the correct identification of an unknown. The second feature lies in the construction of the decision-making rule applied at each step—this uses a technique called logistic regression. Essentially, logistic regression establishes membership of one or more groups on the basis of a probability function[5] rather than the value of a predicted dummy variable or distance function. It may be applied to two (binary) or more than two (polychotomous) groups[6]—this report considers binary regression only. The hierarchical and logistic regression approaches are applied after factorisation by PCA, PLSI and PLSII.

Materials and methods

Meat samples

Two hundred and thirty (230) homogenised meat samples were utilised in this study. They comprised 55 chicken, 54 turkey, 55 pork, 32 beef and 34 lamb. Chicken and turkey were purchased as breast meat, pork as loin chops, beef as round steak and lamb as side loin chops; all were stored overnight at +4°C following purchase and prior to preparation and spectral collection. Individual samples were cut into cubes of manageable size and homogenised (Robot Coupe SA, Vincennes, France).

Spectral collection

Combined visible and near infrared spectra were collected in reflectance mode using an NIRSystems 6500 instrument (NIRSystems Inc., Maryland, USA) over the wavelength range 400–2500 nm at 2 nm intervals. Spectrophotometer control and spectral file management were performed using NIRS3 software (version 3.10; ISI International, Port Matilda, USA).

Chemometric procedures

The development of FDA, K-NN, SIMCA and PLS regression models for this dataset have been described previously.[4] Logistic regression models were developed from sample scores obtained by principal component analysis or partial least squares. These scores were calculated in MatLab using the PLS_Toolbox 2.0[7] while the logistic regressions were performed in Splus. The sample set was divided (on the basis of alternate samples) into separate calibration development and prediction sets.

Results and discussion

Figure 1 shows the structure of the decision making process. An unknown sample is initially classified as either red meat or white; thereafter, appropriate binary decisions allow its eventual identification. This approach utilises the natural structure of the dataset.

Using PCA factorisation, the results obtained for each of the decision steps and the overall classification success is summarised in Table 1. Initial segregation into red or white meat classes is done on the basis of scores on components 2 and 3; this is achieved with 100% success. Complete discrimination between beef and lamb meats was similarly achieved using sample scores from PCs 5 and 11. In the case of white meat, the first decision is between pork and poultry meats. In this case, a logistic regression model (fitted using stepwise forward regression) comprising four principal components (4,9,15 and 18) proved optimum. In the calibration sample set, one poultry sample was mis-classified as pork while, in the prediction set, two poultry samples were mis-classified as pork and three pork samples as poultry. Discrimination between chicken and turkey samples was problematic as discovered previously.[3,4] A logistic regression model involving three components mis-classified 13 samples overall—three in calibration and ten in prediction. These latter were all chicken.

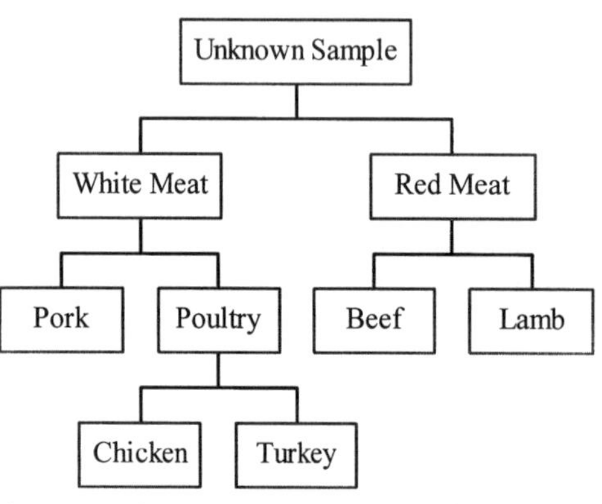

Figure 1. Schematic representation of the hierarchical classification approach.

In the case of PLS factorisation, results are summarised in Tables 2 and 3. As for PCA factorisation, the first two decisions were made most effectively with a linear cut-off rather than a logistic regression model. Subsequent decisions were best made with a logistic regression approach. It can be seen from Table 2, that PLSI factorisation produced perfect classification in the calibration sample sets but that predictive performance with regard to pork vs poultry and especially turkey vs chicken was disappointing. This latter reduced the overall correct classification rate in prediction for the five groups to 78.3%, the lowest of the three factorisation methods. PLSII factorisation produced results which

Table 1. Summary of classification results using PCA factorisation.

			% Correct Classification	
Decision	PCs	Method	Calibration Set	Prediction Set
Red vs white	2, 3	linear cut-off	100	100
Lamb vs beef	5, 11	linear cut-off	100	100
Pork vs poultry	4, 9, 15, 18	logistic	98.8	93.9
Total within 4 groups			*99.1*	*95.7*
Turkey vs chicken	13, 6, 9	logistic	94.4	81.8
Total within 5 groups			*96.5*	*87.0*

Table 2. Summary of classification results using PLSI factorisation.

			% Correct Classification	
Decision	PCs	Method	Calibration Set	Prediction Set
Red vs white	2	point cut-off	100	100
Lamb vs beef	2,7	point cut-off	100	97.0
Pork vs poultry	2, 1 ,4 ,9	logistic	100	90.2
Total within 4 groups			*100*	*92.2*
Turkey vs chicken	3, 4, 6, 15, 16	logistic	100	70.9
Total within 5 groups			*100*	*78.3*

Table 3. Summary of classification results using PLSII factorisation.

			% Correct Classification	
Decision	PCs	Method	Calibration Set	Prediction Set
Red vs white	2	point cut-off	100	100
Lamb vs beef	4	point cut-off	100	87.9
Pork vs poultry	5, 13, 11 ,3	logistic	98.8	93.9
Total within 4 groups			*99.1*	*92.2*
Turkey vs chicken	6, 8, 10, 15	logistic	96.3	85.5
Total within 5 groups			*97.4*	*85.2*

were better than PLSI. Overall, the best performance obtained in this study was using the PCA factorisation approach.

A summary of the results obtained here and those previously reported[3] is shown in Table 4. The PCA factorisation methods which performed best in this work compared favourably with factorial discriminant analysis. The PLSI technique produced the poorest results. In all cases, the models had

Table 4. Summary of % correct classification results using several chemometric techniques.

	Four groups			Five groups		
Technique	Cal	Val	All	Cal	Val	All
FDA[3]	100	95.7	97.8	91.3	86.1	88.7
K-NN[3]	91.2	86.1	88.7	87.0	77.4	82.2
PCA factorisation	99.1	95.7	97.4	96.5	87.0	91.7
PLSI factorisation	100	92.2	96.1	100	78.3	89.1
PLSII factorisation	99.1	92.2	95.7	97.4	85.2	91.3

greatest difficulty distinguishing between chicken and turkey. One of the concerns arising from this work is the high probabilities often associated with mis-classified samples i.e. the estimated probability of a sample belonging to an incorrect class is often close to 1.0 and *vice versa*. This is thought to arise from the intrinsic structure of the dataset being modelled and may indicate a limitation in the utility of logistic regression in this type of application.

Conclusions

The approach to classification described in this report has produced models of comparable accuracy to the best previously published. With regard to the hierarchical approach, it uses the inherent structure in the data and makes the decision-making process transparent. Its success in the classification of red *vs* white and poultry *vs* pork meats is striking. The models produced by logistic regression are not developed using rigorous statistical procedures but on the basis of results obtained and this is a potential weakness. No attempt has been made to optimise the classification results through, for example, data pre-treatment or variable selection. The hierarchical approach has advantages for such optimisations since different sets of variables or data treatments may easily be used for each classification step.

References

1. I.D. Lumley, in *Food Authentication*, Ed by P.R. Ashurst and M.J. Denis. Blackie Academic & Professional, London, UK, pp 108–139 (1996).
2. K.D. Hargin, *Meat Sci.* **43(S),** 277 (1996).
3. H. Rannou and G. Downey, *Analytical Communications* **34,** 401(1997).
4. J. McElhinney, G. Downey and T. Fearn, in *Near Infrared Spectroscopy: Proceedings of the 9th International Conference,* Ed by A.M.C. Davies and R. Giangiacomo. NIR Publications, Chichester, UK, pp. 511–515 (2000).
5. A.J. Dobson, *An Introduction to Generalised Linear Models*. Chapman & Hall/CRC, Boca Raton, USA (1990).
6. C.B. Begg and R. Gray, *Biometrika* **71(1),** 11 (1984).
7. B.M. Wise and N.B. Gallagher, *PLS_Toolbox 2.0* for use with MatLab. Eigenvector Research Inc. (1998).

Near infrared reflectance spectroscopy as an essential tool in food safety programmes: predicting ingredients in commercial compound feed

A. Garrido-Varo,[a] M.D. Pérez-Marín,[a] A. Gómez-Cabrera,[a] J.E. Guerrero,[a] F. De Paz[b] and N. Delgado[b]

[a]*Escuela Técnica Superior de Ingenieros Agrónomos y Montes, University of Córdoba, Avda Menéndez Pidal s/n, E-14080 Córdoba, Spain*

[b]*SAPROGAL, S.A. A Coruña, Spain*

Introduction

A number of EEC directives and decisions lay down rules governing the circulation of raw materials and feed; some of these stress the importance of a detailed statement of the products used in animal feed and set down rules for declaring the feed materials present in a compound product (77/101/EEC, 79/373/EEC, 90/44/EEC, 91/357/EEC, 98/87/EC).

The BSE and dioxin crises have highlighted the need for labelling to include a detailed quantitative statement of feed composition. On 10 January 2000 the Commission submitted a proposal to the Council for a Directive from the European Parliament and the Council, amending Directive 79/373/EEC. The purpose of the proposal is to render compulsory the statement of all the materials included in compound feedingstuffs for production animals, in terms of the percentage by weight, by descending order of weight.[1]

Light microscopy is the method officially used to identify feed materials in mixtures and compound feed. However, this method is not without drawbacks.[2] There is, therefore, an urgent need for swift, economical and accurate analytical techniques that will ensure compliance with the rules mentioned earlier.

Various authors have demonstrated the ability of near infrared (NIR) technology, used either alone or in combination with other techniques (microscopy–NIR), to predict the percentage of different ingredients in mixtures.[3–5] Previous studies have focused on ground samples and have mostly been what are known as viability studies, in that they used a small number of samples and only laboratory- prepared experimental mixtures.

The aim of the present study was to obtain NIR calibration equations for the instant and simultaneous prediction of both chemical composition and percentage of different ingredients in commercial feed analysed in their original form.

Material and methods

Samples

A total of 287 highly-varied compound feed samples were used (cattle, lamb, poultry, pig, ostrich, horse, rabbit, cat and dog feed), as supplied by a Spanish compound feed manufacturer (SAPROGAL SA).

Reference analyses

Samples were analysed using the methods proposed by the AOAC for moisture, H, crude protein, CP, crude fat, F, crude fibre, CF and ashes, ASH.[6] Calibration equations for predicting ingredient proportions were obtained using the formula for each feed as supplied by the manufacturer as reference data.

Instrumentation and software

A Foss NIRSystems model 6500 scanning monochromator, (Foss NIRSystems, Silver Spring, MD, USA) equipped with a transport module, was used to measure reflectance spectra from 400 to 2498 nm every 2 nm. The analysis was carried out using the coarse rectangular transport cell with internal dimensions of 4.1 cm wide, 17.2 long and 1.4 cm deep. The 3.3 cm × 16.4 cm quartz viewing window allows 54.12 cm^2 of the sample surface area to be irradiated. The instrument was set up to read 12 reference scans of the ceramic reference material and the transport speed of the coarse transport cell was set to allow 32 complete wavelength range scans across the full length of the quartz window.[7] Spectra were recorded with the ISI NIRS 3 software ver.3.11 (Infrasoft International, Port Matilda, PA, USA).

Calibration development

Calibration development was performed using WINISI ver. 1.04 (Infrasoft International, Port Matilda, PA, USA). A modified partial least squares method was used to obtain regression equations for all analytical parameters. All equations were obtained using the standard normal variate and detrending method for scatter correction and different derivative mathematical treatments. The statistics used to select the best equations were the coefficient of determination (r^2), the standard error of cross-validation (*SECV*), the *RPD* and the *RER* values [8,9]

Table 1. Calibration statistics for equations obtained for predicting the chemical composition of compound feed (*N* = 287).

	N	Mean	Range	*SD*	*SECV*	r^2	*RPD*	*RER*
H	263	10.664	6.7–13.6	1.324	0.54	0.84	2.45	12.78
CP	252	18.205	12.2–33.6	3.798	0.75	0.96	5.06	15.20
Fat	221	4.438	2.0–9.4	1.445	0.54	0.86	2.68	13.70
CF	248	6.756	2.0–18.6	3.903	0.63	0.97	6.20	26.35
ASH	245	8.401	3.9–16.7	2.196	0.83	0.86	2.65	15.42

Table 2. Calibration statistics for equations obtained for predicting feed ingredient percentages in compound feed (N = 287).

Feed Ingredient	*SECV*	r^2	*RPD*	Feed Ingredient	*SECV*	r^2	*RPD*
% Wheat	8.20	0.43	1.32	% Meat & bone meal	1.65	0.61	1.59
% Barley	8.93	0.57	1.52	%Alfalfa	1.44	0.94	3.92
% Barley +wheat	8.36	0.81	2.28	% Molasses	1.17	0.76	2.06
% Corn	5.37	0.87	2.76	% Total meat meal	0.88	0.98	8.16
% Soybean 44	4.08	0.83	2.43	% Royal palm	0.75	0.95	4.41
% Bran	4.99	0.66	1.71	% $CaCO_3$	0.70	0.67	1.74
% Manioc	4.01	0.51	1.43	% Animal fat	0.61	0.89	2.93
% Gluten meal	2.42	0.93	3.74	% Fish meal	0.35	0.68	1.77
% Lupin	2.13	0.77	2.06	% Sugar beet pulp	0.25	0.90	3.24
% Rice	1.95	0.53	1.47	% Whey	0.13	0.94	4.38
% Sunflower	1.60	0.95	4.67	% Salt	0.09	0.52	1.44
% Poultry meal	1.82	0.89	2.99	% NH_4Cl	0.04	0.78	2.50

Results and discussion

Calibrations for predicting CP and CF displayed (Table 1) excellently predictive ability, judging by the high values obtained for r^2 (0.96 and 0.97) and the low values recorded for *SECV* (0.75% and 0.63%). Calibrations for H, fat and ashes also display an adequate degree of precision, accounting for 84% of variation in moisture and 86% of variation in fat and ashes. Errors (*SECV*) in the calibrations obtained for the intact feeds analysed (Table 1) lie within the usual range of values for feed analysed in ground form.[4,10,11] The fact that the calibrations obtained in the present study using unground feed were similar, in terms of both accuracy and precision, to those obtained by other authors using ground feed may be attributed to the greater product scanning surface, obtained by using the rectangular coarse cell which has a scanning area of 54.12 cm^2 instead of the 28.1 cm^2 of the small ring cup. This difference may be of major importance in NIR analysis of samples as heterogeneous as those analysed here.

Preliminary analysis of the data in Table 2 highlights the excellent predictive ability ($r^2 \geq 0.9$; $RPD \geq 3$) of the calibrations obtained for determining the proportions of alfalfa, sunflower, gluten, sugar beet, royal palm, poultry meal, total meat meal (% meat meal + % poultry meal) and whey. Other equations showing good predictive capacity ($r^2 \geq 0.7$) were those obtained for soybean, maize, molasses, fats and lupin. Calibrations obtained for predicting the proportions of barley, manioc, bran, rice, meat and bone meal, fish meal, calcium carbonate, ammonium chloride and salt displayed an acceptable predictive capacity ($0.5 \leq r^2 \leq 0.7$). Finally, the calibration for wheat at least enabled classification of samples into low, medium and high wheat content.

The equations obtained for predicting meat meal percentages require special comment. Since December 2000, use of these meals in production animal feeds has been banned throughout Europe. Although the equation for predicting meat and bone meal displays poor accuracy, the overall equation (total meat meal) affords excellent precision (r^2 = 0.98, *RPD* = 0.88). The *SECV* values should be interpreted on the basis of the reference method; in this case, the reference method was the percentage of feed material shown in the manufacturer's formula, as obtained by industrial by-weight breakdown, with the attendant errors. The use of other reference methods is likely to enhance the accuracy of equations for predicting meat and bone meal. Indeed, in January 2001, a total of ten European R & D centres, universities, official laboratories and one industrial partner launched a 3-year period of collaboration on a European Project,[12] which has as one of its main objectives the development and validation of new methods (PCR, NIR and NIR reflectance spectro–microscopic techniques) for the rapid, precise and reliable detection and quantification of animal meal in feed.

The equations obtained are still of a preliminary nature and it is hoped that their accuracy and precision will be enhanced through the use of different optimisation strategies. These calibrations are currently being expanded with a larger number of samples (N > 500) and research is also being carried out into various aspects of their development, such as the design of the calibration log, with particular reference to the number of samples to be included for each interval of the range.

Conclusions

NIR technology enables swift and economic compliance with rules relating to the labelling of feed, avoiding both the need for reagents and the production of chemical residues. NIR must, therefore, be considered an essential tool in Food Safety programmes.

Acknowledgements

This work was carried out using NIR hardware and software at the Central NIR and MIR Spectroscopy Unit (SCAI) of the University of Córdoba (Spain). The authors are grateful to the Spanish DGICYT for providing financial support for this study as part of the project CICYT-Feder IFD1997-0990 and would like to thank Mr Antonio López, Mr Alberto Sánchez de Puerta and Ms Isabel Leiva for technical assistance.

References

1. European Commission.Common Position (EC) No 6/2001. O.J.E.C. 2.2.2001, p. C. 36 (2001).
2. J. Jørgensen in Report of the Workshop on Identification of animal ingredients in compound feed focusing on the microscopic method for identification organised by the CEMA group, 25–26th May 1998. DGXII. Contract SMT4-CT97-6524. p. 1 (1998).
3. A. Garrido and V. Fernández in Report of the Workshop on Identification of animal ingredients in compound feed focusing on the microscopic method for identification organised by the CEMA group, 25-26th May 1998. DG XII. Contract SMT4-CT97-6524.Annex X.
4. G. Xicatto, A. Trocino, A. Carazzolo, M. Meurens, L. Maertens and R. Carabaño. *Anim. Feed. Sci. Techn.* **77,** 201(1999).
5. F. Piraux and P. Dardenne in *Near Infrared Spectroscopy: Proceedings of the 9th International Conference,* Ed by A.M.C. Davies and R. Giangiacomo. NIR Publications, Chichester, UK, p. 535 (2000).
6. AOAC. *Official Methods of Analysis of AOAC International.* W. Horwitz. AOAC International. Maryland. USA, Vol I. Chapter 4 (2000).
7. R.S. Park, R.E. Agnew and R.J. Barnes, *J. Near Infrared Spectrosc.* **7,** 2 (1999).
8. J.S. Shenk and M.O. Westerhaus, *Routine operation, calibration, development and network system management manual.* NIRSystems Inc., 12101 Tech Road, Silver Spring, MD 20904, USA (1995).
9. P.C. Williams and D. Sobering in *Near Infrared Spectroscopy: The Future Waves,* Ed by A.M.C. Davies and P.C. Williams. NIR Publications, Chichester, UK, p. 185 (1996).

10. N.B. Büchman in *Leaping Ahead with Near Infrared Spectroscopy,* Ed by G.D. Batten, P. Flinn, L.A. Welsh and A.B. Blakeney. NIR Spectroscopy Group, Royal Australian Chemical Institute, Victoria, Australia, p. 248 (1995).
11. A. Garrido, M.D. Pérez-Marín, J.E. Guerrero. A. Gómez, F. De Paz and N. Delgado, unpublished (2001).
12. P. Dardenne (coord.) in *Strategies and methods to detect and quantify mammalian tissues in feedingstuffs.* Proposal RTD Project. Fith Framework Programme, FP5 (2000).

Assessing and enhancing near infrared calibration robustness for soluble solids content in mandarin fruit

J.A. Guthrie[a] and K.B. Walsh[b]

[a]*Centre for Food Technology, Department of Primary Industries, Rockhampton, Qld 4702, Australia*

[b]*Non-Invasive Assessment Group, Plant Sciences, Central Queensland University, Qld 4702, Australia*

Introduction

Near infrared (NIR) spectroscopy has been applied to the sorting of intact fruit with a high moisture content for constituents such as soluble solids content (SSC) in cantaloupe fruit,[1] sugar content in intact peaches,[2] sugar content, acidity and hardness of intact plum fruit[3] and SCC of intact citrus (mandarin fruit). Commercial application to pack-house fruit sorting lines commenced in Japan in the mid 1990s, for the sorting of sweetness, ripeness and acidity of citrus fruit, apples, pears and peaches at three pieces per second per lane.[5] Commercial application within pack-houses of Western countries is nascent.

The application of NIR technology requires an appreciation of the distribution of the character of interest within the fruit and the absorption and scattering of light through the fruit, in order to design an appropriate optical configuration of light source, detector and fruit (for example References 6 and 7). The robustness of the NIR calibration model must be assessed across populations of fruit differing in, for example, temperature, variety and growing district. Unfortunately, these parameters are not well reported in the literature, with many NIR studies reporting the use of a standard optical design for spectral acquisition and the use of a single harvest population, divided into a calibration set and a validation set. Few studies have explored the issue of validation across populations varying in the locality of harvest, the time of harvest with a given season, or across years. A notable exception is that of Peiris *et al.*[8] who reported calibration validation across three seasons for peaches. A calibration developed in one year predicted poorly on other years, but a combined calibration performed well for validation groups drawn from those years.

In the current study we report on issues related to calibration robustness for intact mandarin fruit assessed for SSC.

Materials and methods

Plant material and SSC analysis

Imperial variety of mandarin were sourced from commercial orchards in Munduberra, Queensland. Fruit were sourced from three separate farms on one day, from three separate harvests over a five-day period from one tree and from one packhouse over three seasons. Fruit were halved, juiced and SSC determined by refractometry (Bellingham and Stanley RMF 320).

Spectroscopy

Spectra were collected using an NIR enhanced Zeiss MMS1 spectrometer and a tungsten halogen light in the optical configuration reported by Greensill and Walsh.[7] Spectra were collected from one side of each fruit, on the equator of the fruit, equidistant from pedicel and stylar ends.

Chemometrics

The software package WinISI (ver.1.04a) was used for all chemometric analysis. Calibration performance was assessed in terms of coefficient of determination (R^2) standard error of prediction (*SEP*), variance ratio (1-*VR*), standard deviation ratio (*SDR*), slope and bias of the validation sets. Further, the criteria of Wortel *et al.*,[9] based on the Taguchi concepts as used in process control, were applied to evaluate model robustness. This approach involved calculation of an average *SEP* and a signal-to-noise statistic (s/n = 20 $\log_{10}$ [mean *SEP* / *SD SEP*]) for the performance of a given model across a range of validation sets.

Table 1. Calibration and validation statistics for a calibration on one population of mandarin SSC, used in prediction of three populations varying in (a) days of harvest, (b) location of harvest and (c) season of harvest.

Fruit population	*SD*	R^2	*SECV*/*SEP*	BIAS
Time				
Cal	0.95	0.90	0.35	
Val				
Day 1	0.73	0.68	0.48	0.173
Day 3	0.72	0.71	0.52	– 0.352
Day 5	0.68	0.55	0.52	0.209
s/n			**26.8**	
Av *SEP*			**0.51**	
Location				
Cal	0.85	0.87	0.353	
Val				
A	0.51	0.55	0.37	– 0.12
B	0.57	0.69	0.41	0.25
C	0.50	0.55	0.52	0.40
s/n			**14.93**	
Av *SEP*			**0.43**	
Seasons				
Cal	0.95	0.84	0.42	
Val				
Year 1	0.96	0.83	0.49	0.24
Year 2	1.05	0.31	2.45	0.40
Year 3	1.05	0.82	3.76	3.73
s/n *SEP*			**2.65**	
Av *SEP*			**2.23**	

Results and discussion

Calibration statistics and B coefficients

Typical MPLS calibration statistics for intact mandarin SSC were: R^2 0.87, *SECV* 0.35, using six principal components, on a population *SD* 0.85, n = 100 (Table 1).

The MPLS B coefficients for the mandarin SSC calibrations contain negative weightings on second derivative spectra around 910 and 850 nm and positive weightings around 880 nm (data not shown). Absorbance at *ca* 910 nm is ascribed to a third overtone stretching of CH bonds (Golic and Walsh, this volume). Absorbance at 880 nm may convey pathlength information. A calibration that does not contain spectroscopically 'relevant' information is likely to be over-fitted to the data and, thus, can be expected to perform poorly when applied to new validation populations.

Calibration validation

A calibration developed from a single population of fruit (100 spectra) was applied to validation sets harvested on different days, different locations and different growing seasons (Table 1). The cause of the decrease in performance of a calibration when applied to a 'new' group presumably reflects change in the physical (optical) properties or the chemical properties (acid, water content) of the fruit. Temperature of the fruit was constant at scanning. Calibration performance across harvest day and location was comparable, as indicated by the mean *SEP* and s/n statistic, while performance was dramatically degraded across seasons. The cause of the dramatic decrease in performance of a calibration when applied to a new season of fruit is not clear and could reflect changes in the instrument used as well as change in the sample (fruit).

To improve calibration performance on a new validation set, a typical strategy involved addition of samples from the new set to the calibration group. The validation sets were divided into two equal groups. One group was retained as a validation set and the other group used for selection of samples for addition to the calibration set. Any validation sample with a GH > 3.0 (calculated on calibration set scores and loadings) was excluded from this process. Several approaches were used in the selection of samples from the validation group for addition to the calibration group, (1) random, (2) selection, on the basis of ascending GH (validation set ordered in ascending order of GH calculated on calibration set scores and loadings, and samples selected at equal GH intervals), (3) selection of the basis of spaced GH (calculated as per 2) and (4) selection on the basis of NH (increasing NH values calculated on calibration set scores and loadings to select increasing numbers of validation set samples, using the ISI 'Expand a Product File with New Spectra' feature). The performance of a calibration developed in one

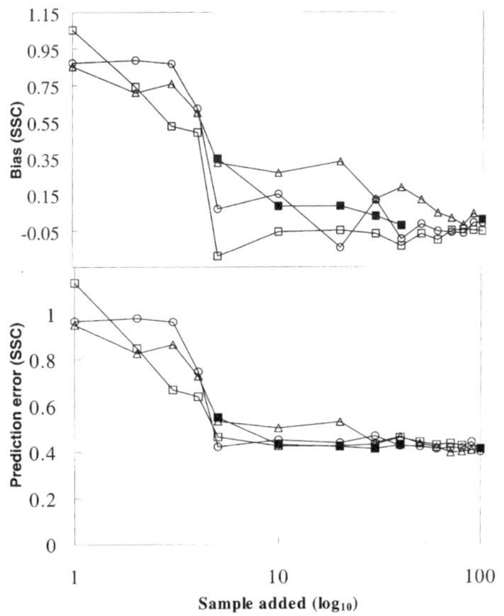

Figure 1. Prediction statistics for SSC of a mandarin validation population (different growing season to calibration population) using three treatments for sample selection from the new season group for addition to the calibration group. Open circle, random selection; open squares, central GH selection; closed squares, spaced GH selection; open triangle, NH selection.

growing season and applied to fruit of a subsequent season was improved in terms of *SEP* and bias as increasing numbers, up to *ca* 10, of 'validation set' samples were added to the calibration set, using any of the three selection approaches (Figure 1). It is surprising that so few fruit were representative of any physical or chemical change in the validation, relative to the calibration, set. In practical terms, we recommend it is sufficient to add data of *ca* 15 fruit to a calibration to update it for use across growing seasons.

Acknowledgements

This work was support by a Citrus Marketing and Development Grant, administered through Horticulture Australia Ltd. Supply of fruit from Gaypak packhouse and Steve Benham of Joey Citrus, Munduberra, is also gratefully acknowledged.

References

1. G. Dull, G. Birth, D. Smittle and R. Leffler, *J. Food Sci.* **54,** 393 (1989).
2. S. Kawano, H. Watanabe and M. Iwamoto, *J. Jpn Soc. Hort. Sci.* **61,** 445 (1991).
3. T. Onda, M. Tsuji and Y. Komiyama, *Journal of the Japanese Society for Food Science and Technology* **41,** 908 (1994).
4. S. Kawano, T. Fujiwara, and M. Iwamoto, *J. Jpn Soc. Hort. Sci.* **62,** 465 (1993).
5. S. Kawano, *NIR news* **5,** 10 (1994).
6. K.B. Walsh, J.A. Guthrie and J. Burney, *Aust. J. Plant Physiol.* **27,** 1175 (2000).
7. C.V. Greensill and K.B. Walsh, *Meas. Sci. Technol.* **11,** 1674 (2000).
8. K.H.S. Peiris, G.G. Dull, R.G. Leffler and S.J. Kays, *J. Am. Soc. Hort. Sci.* **123,** 898 (1998).
9. V.A.L. Wortel, W.G. Hansen and S.C.C. Wiedemann, in *Near Infrared Spectroscopy: Proceedings of 9th International Conference*, Ed by A.M.C. Davies and R. Giangiacomo. NIR Publications, Chichester, West Sussex, UK, p. 267 (2000).

Non-destructive near infrared spectrometers: development of portable fruit quality meters

Susumu Morimoto,* Hitoshi Ishibashi, Toshihiro Takada, Yoshiharu Suzuki, Masayuki Kashu and Ryogo Yamauchi

FSJ-PT, Kubota Corporation, 2-35, Jinmu-Cho, Yao, Osaka, 581-8686, Japan

Introduction

The quality of agricultural products is an important factor for consumers. In Japan, quality is sometimes more important than cost. Quality standards for agricultural products have been established in Japanese markets and it has been determined that price depends on the quality of the product. Usually, the quality of agricultural products is determined in terms of appearance factors, such as shape, colour, size, etc. However, these indices are not always associated with taste, leaving consumers to complain. Today, people are demanding new quality standards related to taste (sweetness, sourness, etc).

Near infrared (NIR) spectroscopy is a proven technique for measuring internal quality of many agricultural products. Since the late 1980s in Japan, on-line NIR graders have been developed and used in grading facilities.[1] Yet, even though these graders are effective, their widespread use has not been realised because they are expensive instruments costing more than $200,000 each.

In 1999, a tabletop fruit quality meter[2] (Fruit Selector: K-FS200, Figure 1) using NIR technology was developed by the Kubota Corporation (Osaka, Japan). This instrument can measure sugar and acid content of fruit or vegetables non-destructively. It is compact and is much cheaper than on-line graders. It has been used widely at grading facilities, wholesalers and supermarkets.

The K-FS200 has enhanced the development of quality standards related to taste. Measurement of taste factors with this instrument provides the consumer with acceptable products, while at the same time, giving the farmer maximum Yen. The disadvantage of this approach is that poor quality products are thrown away.

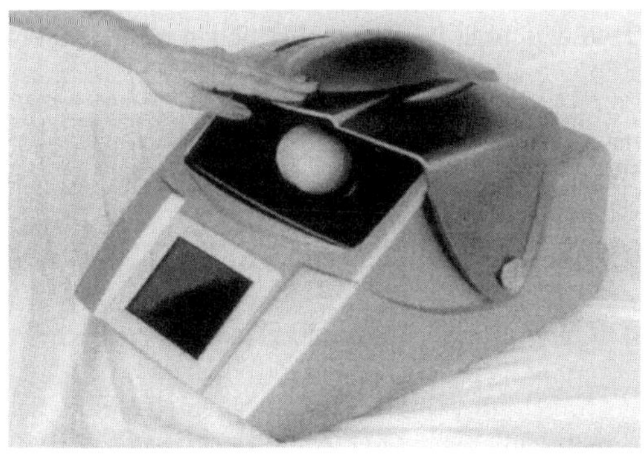

Figure 1. Tabletop fruit quality meter (K-FS200)

Loss of product in grading facilities and market places could be minimised if measurement of taste factors could be made in the field, even before products are harvested. Kubota developed the portable fruit quality meter, Model K-BA100, (shown in Figure 2) to maximise food potential[3] of harvested fruit and vegetables. The portable NIR meter can determine ingredients in the growing product in the field. The farmer can control fertilisers or watering

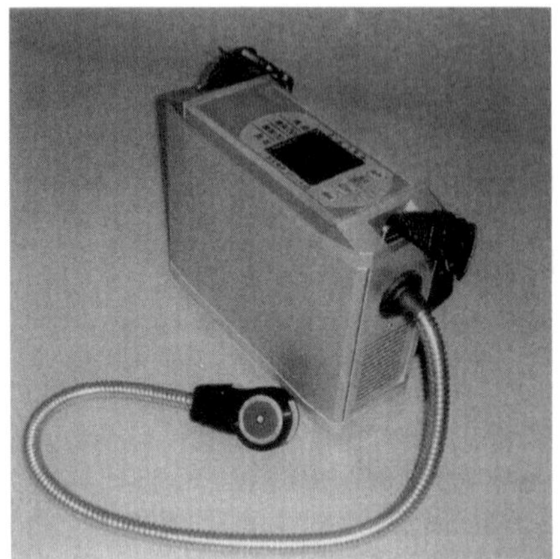

Figure 2. Portable fruit quality meter (K-BA100).

according to information gained with the K-BA100. Thus, quality is improved even before harvest. This relatively new approach results in little or no waste and minimises the use of chemical fertilisers. It is a method of operation that benefits both the grower and consumer and it is environmentally friendly.

This paper discusses the design and performance of the K-BA100 portable NIR instrument.

Features of the K-BA100

Conventional laboratory instruments have been designed to operate in laboratory conditions (25°C, 50% rh) but they are not intended to be moved or carried into the field. On the other hand, the K-BA100 was designed for field use and it can be carried into the field and operated under severe environmental conditions (dusty environment and varying temperature). The features of this instrument are as follows.

(a) non-destructive measurement
- monitoring the quality of the growing sample until harvest
- no waste of samples

(b) stable and precise measurement in the field
- stable and precise under tough conditions such as ambient or sample temperature fluctuation, dusty environment, etc

(c) stand alone and easy operation
- works independently without external equipment and user-friendly design makes operation simple

(d) portability and expandability
- Portable enough to bring to the field (W300 × L 118 × H 240 mm, Weigh 5 kg).
- Optical fibre probe for flexible measurement.
- Development of original calibration model with external computer (RS232C interface and control software).

Key techniques

In this section, we focus and discuss details of two techniques, (a) non-destructive measurement and (b) stable and precise measurement in the field, due to compact design.

Non-destructive measurement

When the fruit sample is measured using NIR, there are some obstacles such as those mentioned below:
- most fruit or vegetables have a high moisture content of more than 80%. NIR light energy tends to be absorbed and weakened by water absorbance
- some fruits, like citrus fruit, have hard and thick peel. The light energy cannot penetrate the sample efficiently

The easiest solution is to use a high-power light source; however, this requires a large battery because of the large energy consumption and a large heat energy that could damage the sample. The following techniques were developed to overcome the problems:

SW (short wavelength)-NIR (500–1000 nm)

SW-NIR has much lower absorption coefficients so water absorbance is reduced and the energy can penetrate into a sample further than longer wavelength NIR (beyond 1100 nm). If the absorption is reduced, the temperature increase will also be reduced.

Interactance method

An optimal design "interactance" probe (Figure 3) transmits the light into the sample and collects the absorbed light more efficiently than "transmittance" or "reflectance" methods.

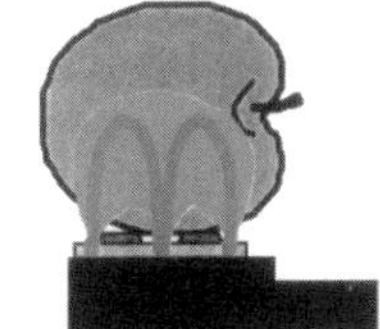

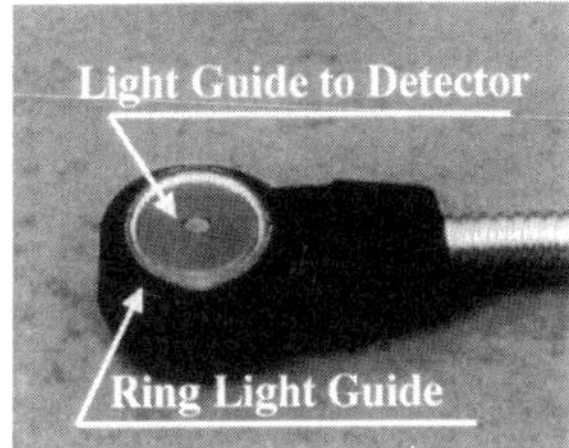

Figure 3. Interactance probe.

Technology for stable and precise measurement in the field

An instrument for field use needs to work in an unstable and dusty environment. In particular, temperature conditions of ambient air and samples are dramatically changed compared with conditions in the lab. The following techniques were developed for the portable NIR meter to allow it to work under those conditions:

Polychrometer with no-moving parts and SW-NIR enhanced detector

The polychrometer (Figure 4) consists of a flat-field concave grating and a linear image sensor (256 pixels). All components are sealed hermetically to prevent dust or shock.

Automatic calibration system (ACS)

ACS (Figure 4) consists of a WC (wavelength calibration) filter and a standard filter to calibrate the wavelength shift and output drift due to temperature variance or change of properties with time. During the operation of the instrument, ACS is working automatically and always maintains the stability of the instrument.

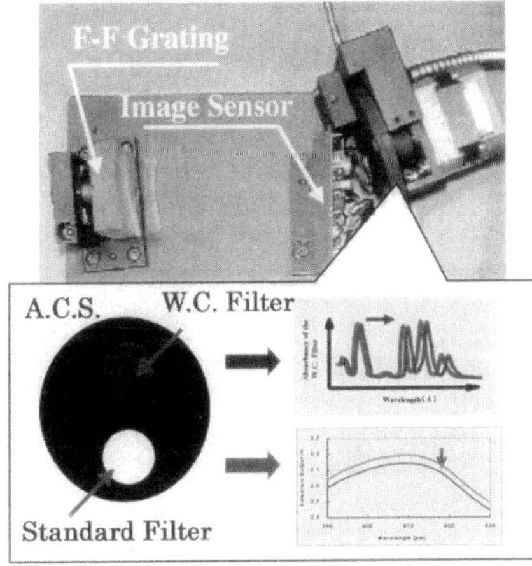

Figure 4. Polychrometer and ACS.

Robust calibration model

According to the nature of NIR absorption, variance of sample temperature affects the prediction value and accuracy in addition to the effect of the environmental temperature on the instrument. The calibration models of the K-BA100 have been developed with samples in various temperature conditions to minimise the effect of temperature variation.

The temperature characteristic of the instrument was tested in a temperature chamber from 5°C to 35°C of ambient air temperature. The fluctuation of predicted sugar content of the mandarin orange was only ± 0.7 Brix% at 13 Brix%. This shows that the instrument works well even under such tough condition.

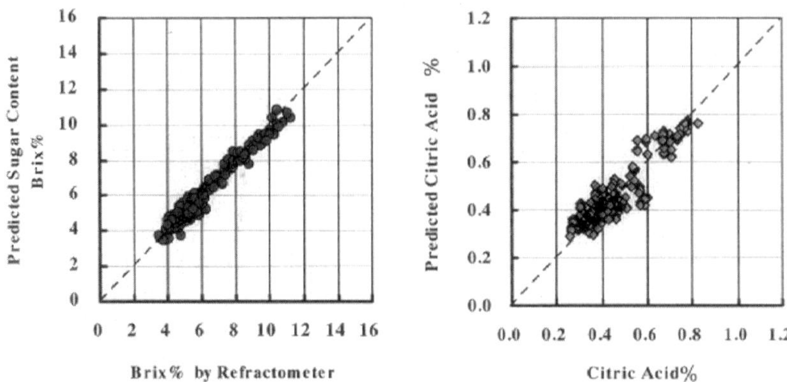

Figure 5. (a) Sugar (R = 0.99, CV = 4.5%) and (b) acidity (R = 0.90, CV = 8.3%) of tomato.

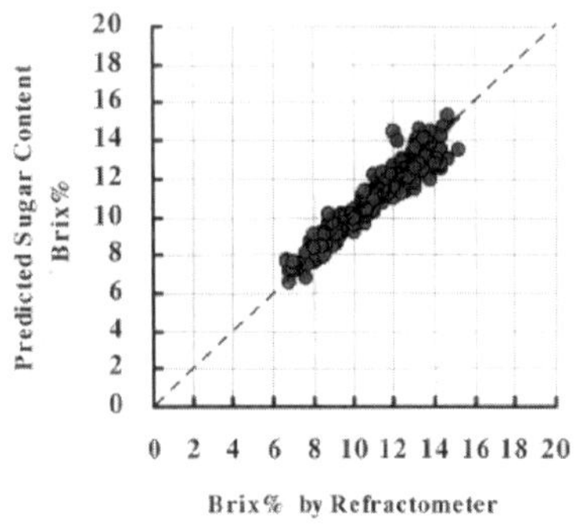

Figure 6. Sugar of citrus fruit (mandarin orange) (R = 0.95, CV = 5.7%).

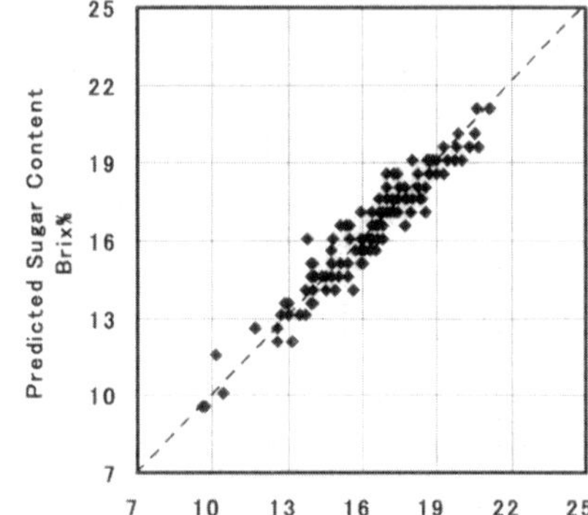

Figure 7. Sugar of grapes (R = 0.92, CV = 4.6%).

Prediction performance

Calibration models for many kinds of fruit were developed. After measurement of spectral data with the K-BA100, the samples were macerated and a refractometer or titratable acidity test was used to measure the sugar content or acid content, respectively. The calibration models for sugar content and acid content were developed using multiple linear regression (MLR) and partial least squares (PLS), respectively. The developed calibrations are listed in Table 1.

Table 1. Calibration models.

Variety	Ingredient
Tomato, apple, mandarin orange	Sugar/acid
Peach, grape, persimmon, pear, melon, strawberry	Sugar

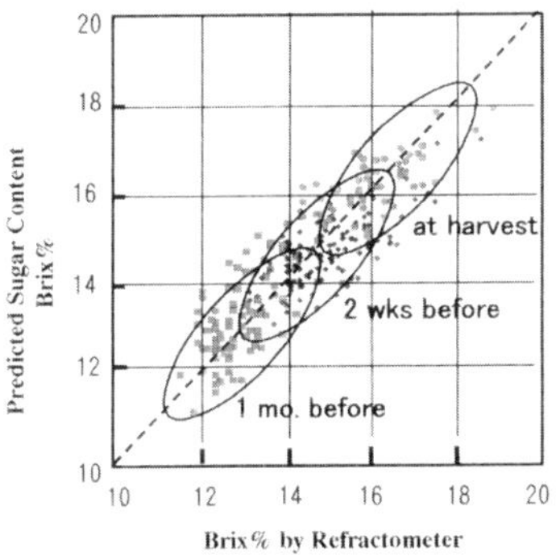

Figure 8. The change in sugar of growing apples on the tree.

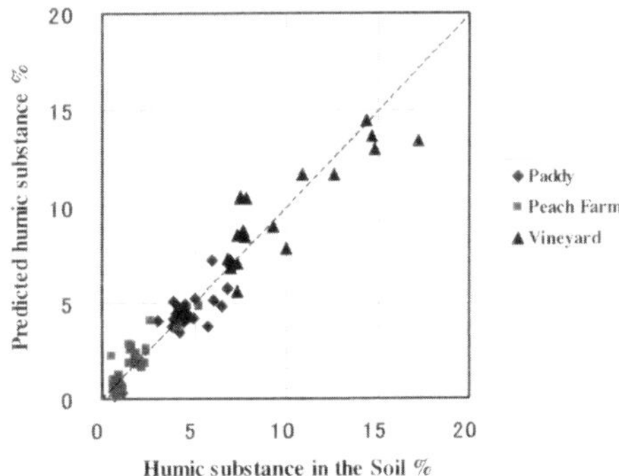

Figure 9. Humic substance in the dried soil (R = 0.96, CV = 11.7%).

Validation results of harvested samples

Some of the prediction results on the harvested samples are illustrated in Figures 5 to 7. Coefficient of variability ($CV = SEP$ / Mean × 100) of sugar content and acid content is less than 6% and 10%, respectively.

Validation results of growing samples

The performances of calibration models were evaluated for growing samples, too. The validation result for apples on the tree at one month before, two weeks before and at just harvested are illustrated in Figure 8. The results show that the instrument predicts the quality of growing products well.

Other applications

The K-BA100 is not only useful for agricultural products but also for other applications. To take one example of the applications, the calibration result of the humic substance in the dried soil[4] is shown in Figure 9.

Conclusion

The K-BA100 portable fruit quality meter using NIR technology has been developed. This instrument can evaluate the taste quality of fruit or vegetables in the field non-destructively.

The portable fruit quality meter is expected to establish new quality standards and production systems. It also promises various applications in the field, at line, on site, etc.

"Changing the world with a portable NIR spectrometer !!"

Reference

1. S. Kawano, in *1999 Edition Annual report on distribution technology of Agricultural products,* p. 73 (1999).
2. S. Morimoto, in *2000 Edition Annual report on distribution technology of Agricultural producst,* p. 152 (2000).
3. W. F. McClure, in *Agricultural and Biological Engineering: New Horizons, New Challenges,* Section 3, Ed by M. Turner. Tynesoft Business Services, Newcastle, UK, pp. 1–9 (1995).
4. K. Takano, in *Proceeding of Kinki-Chugoku agrarian problem symposium.* (2001).

Detection of soy, pea and wheat proteins in milk powder by near infrared spectroscopy

Tiziana M.P. Cattaneo,* Adele Maraboli, Stefania Barzaghi, Roberto Giangiacomo

Istituto Sperimentale Lattiero-Caseario, Via A. Lombardo 11, 26900 Lodi, Italy
E-mail: tcattaneo@ilclodi.it

Introduction

The economic value of milk and milk products is, to a large extent, determined by the fat and protein content. Consequently, the risk of fraud, by replacing them with products of lower value, can be identified. Some of these aspects—such as dilution of milk and milk products, milk fat replacement, milk fat fractionation, milk protein replacement[1,2] and manipulation of the protein composition—have been dealt with in several papers and studies. In this context, the development of a method for the detection of non-milk protein in dairy products is an important step in EU efforts to fight against fraud. Replacement of milk protein with cheaper proteins is usually realised by the addition of vegetable protein isolates such as soy, pea and wheat isolates, on the basis of their cost and their functional properties (yield increase, foaming formation, etc…). There may be a risk that this fraud is carried out in products such as milk powder, yoghurt and processed cheese. Difficulties and limitations encountered by using the methods and techniques developed[3–10] have resulted in the lack of an official method able to detect this kind of fraud. Recently, a research project, financed by the Measurement and Testing Programme of the European Commission,[11] has worked towards a proposed method, based on electrophoretic and/or ELISA principles. However, some questions regarding analytical costs, analysis time and the availability of specific antibodies have directed the research towards faster and cheaper procedures able to provide suitable results.

Near infrared (NIR) spectroscopy, when applied to the dairy field, also proved a fast and feasible technique to determine the chemical composition of different types of dairy products and to detect anomalies in milk and cheese defects.[12–16]

A preliminary approach showed the possibility to detect soy, pea and wheat isolates in milk powder by NIR with satisfactory accuracy. This work aimed to improve the NIR prediction power carrying out separate calibration curves, each designed to detect a single vegetable protein used in milk powder adulteration.

Materials and methods

Samples

Two hundred and twenty-four samples of genuine and adulterated skimmed milk powder containing 0–5% of selected vegetable isolates were used. Genuine (44 samples) and adulterated (1, 2 and 5%) samples were prepared by NIZO (Ede, Wageningen, The Netherlands) in April, May and October 1999 for the purposes mentioned within the EU Project[11] and were strictly controlled for homogeneity,

Table 1. Detection of soy and other proteins by NIR.

Code	Description	% TN	% Protein (TN × 6.25)
A	Soya protein isolate [Supro 500 E]	13.59	85.0
K	Soya protein isolate [Europrod.595]	13.45	84.1
C	Pea protein isolate [Pisane HD]	13.44	84.0
L	Wheat gluten [SWP 100]	12.88	80.5

stability and solubility. Samples were prepared using liquid skimmed milk, applying two different heat treatments: pasteurisation (107 samples) and UHT process (107 samples) using the same spray-drying process. Four commercial vegetable isolates were used: two isolates from soy (A and K; 99 samples), one from pea (C; 62 samples) and one from wheat (L; 63 samples). Table 1 gives a description of vegetable protein isolates. Mixtures at 3 and 4% were made by mixing calculated amount of genuine and adulterated (5%) samples, taking care to accurately blend the two components.

The protein content was measured by the Dumas method.[17] This method was used to determine the protein content of each vegetable isolate, the milk protein content and the protein content of each calibrated mixture before and after the spray-drying process. The percentage of vegetable protein was expressed as [vegetable protein (g) / total protein (g)] * 100].

NIR measurements

NIR spectra were recorded with a holographic grating spectrometer (Bran+Luebbe InfraAlyzer 500, Bran+Luebbe GmbH, Germany) at 1100 to 2500 nm at 4 nm intervals (351 data points).

Light absorption was expressed as log (R^{-1}) values. Measurements were carried out in reflectance mode. The instrument was equipped with a sample cell for solids (code no. 189-0564F, Bran+Luebbe). Samples were analysed in duplicate. Each spectrum was the mean of two spectra, collected rotating the cell by 90°. NIR spectra were collected at room temperature and data were processed by using Sesame Software (Bran+Luebbe). Different data pre-treatments were applied: i.e. raw absorbance data, data normalisation (obtained scaling all measured values between 0 and 1) and first derivative of absorbance values. Pre-treated data were processed by using partial least squares regression (PLSR), using the full spectrum.

Separated calibrations were performed using, respectively, 80, 50 and 50 out of the 99, 62 and 63 samples for soy, pea and wheat adulterated samples. Cross-validation protocol of Sesame software was applied to calculate the optimal number of PLS factors for each calibration and the standard error of prediction (SEP_{CV}). Remaining samples (19 soy, 12 pea, 13 wheat) were used for prediction. The sets were prepared by random selection of samples maintaining roughly constant representation of the addition percentages. Regression coefficients (R^2), *SEE*

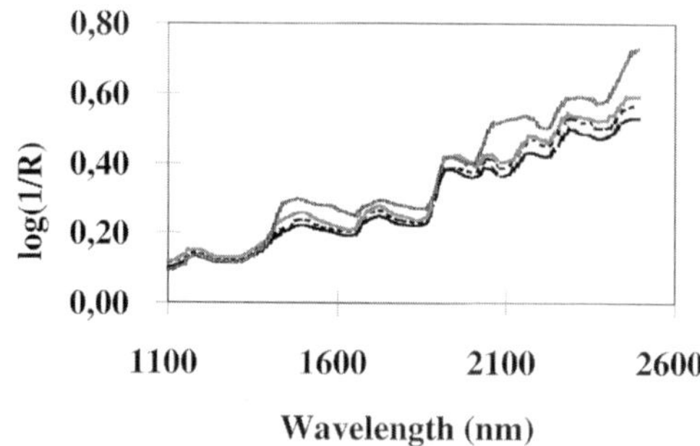

Figure 1. Examples of NIR spectra for genuine skim milk powder (pasteurised ....), soy isolate (A—), pea isolate (C - - -) and wheat isolate (L—).

Table 2. Detection of soy and other proteins by NIR.

Added vegetable protein (*n*)	PLSR factors	*CV* segments	R^2	*SEE*	$RMSEP_{all}$	$RMSEP_{cal}$	$RMSEP_{pred}$	$RMSEP_{cv}$
soy (99)	9	80	0.994	0.193	0.191	0.181	0.229	0.301
pea (62)	9	50	0.997	0.148	0.176	0.133	0.295	0.294
wheat (63)	10	50	0.997	0.152	0.150	0.134	0.198	0.339

(standard error of estimation), *RMSEP* (standard error of prediction) for calibration and prediction sets and $RMSEP_{all}$ (standard error of prediction on whole sets of samples) values were calculated.

Capillary electrophoresis and ELISA procedures

Determination of the content of soy, pea and wheat isolates in milk powder samples was already carried out in the development of the mentioned EU Project,[11] where several collaborative studies were realised among eight European participant laboratories in order to evaluate accuracy and detection limits of both methods. In this study, results obtained by CE and ELISA techniques are compared with NIR predicted values.

Results and discussion

An example of NIR spectra for genuine skimmed milk powder, soy, pea and wheat isolates, collected from 1100 nm to 2500 nm, is shown in Figure 1 as absorbance values *vs* wavelengths.

NIR predictions

The information, collected in a preliminary study,[18] proved the suitability of NIR for detecting vegetable proteins in milk powder and suggested that this method may be used as an accurate screening procedure in routine analyses for quality control.

Soy, pea and wheat protein isolates in milk powder

PLS calibration for soy, pea and wheat protein adulteration was performed using nine factors (F) and 80 cross-validation segments (CVS), 10 F and 50 CVS, 10 F and 50 CVS on first derivative normalised spectra, respectively. NIR prediction results are presented in Table 2. The corresponding correlation coefficients, expressed as R^2, between predicted and actual values (% vegetable protein/total proteins) and number of samples (*n*) are also presented. Figures 2, 3 and 4 show the relationship between the actual percentage of added proteins and NIR predicted values.

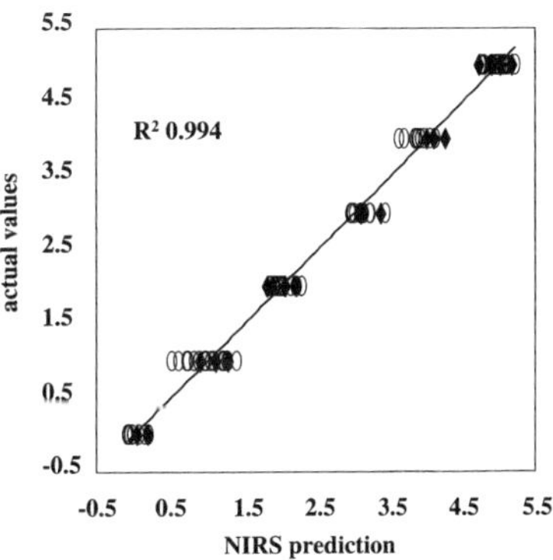

Figure 2. Relationship between the actual percentage of added soy proteins and NIR predicted values by using PLSR [(0) calibration set, (◆) validation set].

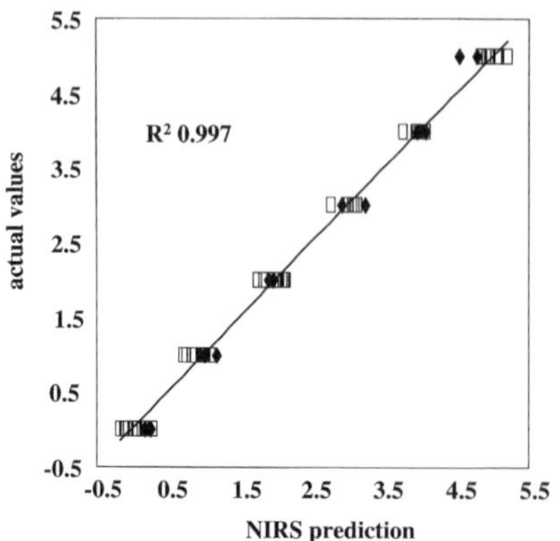

Figure 3. Relationship between the actual percentage of added pea proteins and NIR predicted values by using PLSR [(□) calibration set, (♦) validation set].

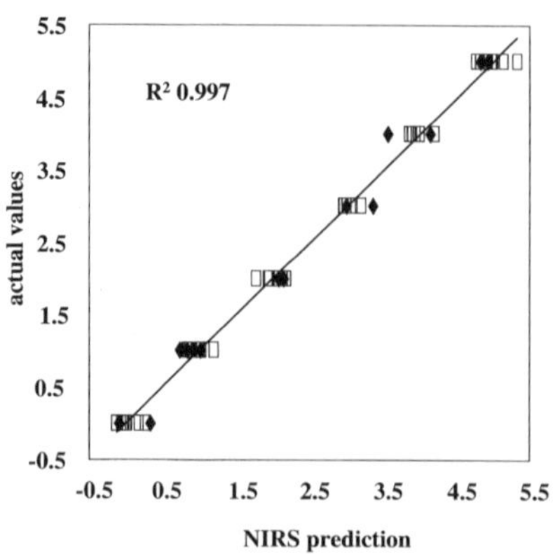

Figure 4. Relationship between the actual percentage of added wheat proteins and NIR predicted values by using PLSR (□) calibration set, (♦) validation set].

NIR predictive power against capillary electrophoresis results

Figures 5 and 6 show the comparison between capillary electrophoresis (CE) results and NIR prediction values. Lines indicating the confidence interval (P = 95%) are also shown. This plot just refers to the sets of pasteurised samples, adulterated with increasing amounts of soy (Figure 5) and pea (Figure 6) isolates. We can note that the variability of data associated with the application of the CE procedure was higher than the NIR variations. In this case, the actual adulteration percentage was known, so NIR could be calibrated against true values, resulting in a precise calibration with satisfactory results in the prediction and cross-validation steps (see Table 2). Conversely, if we needed to calibrate against CE results, NIR prediction would be affected by the standard error associated with the applied reference method, resulting in a higher variability and a worse prediction. Quantitative data

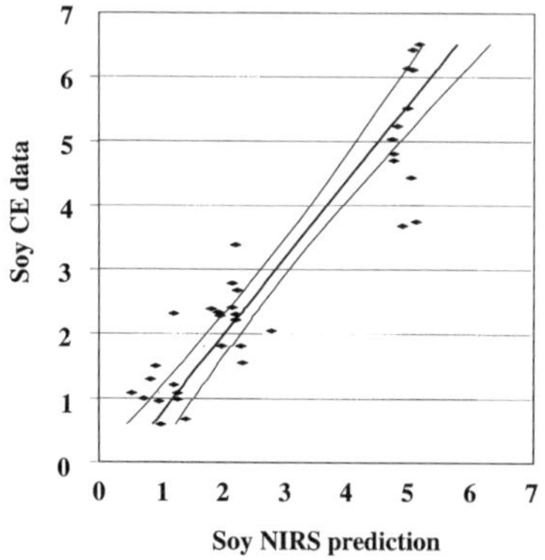

Figure 5. Comparison between capillary electrophoresis (CE) results and NIR prediction values (soy).

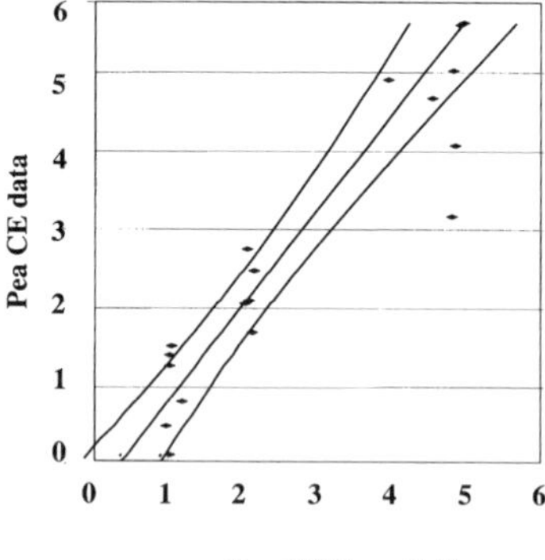

Figure 6. Comparison between capillary electrophoresis (CE) results and NIR prediction values (pea).

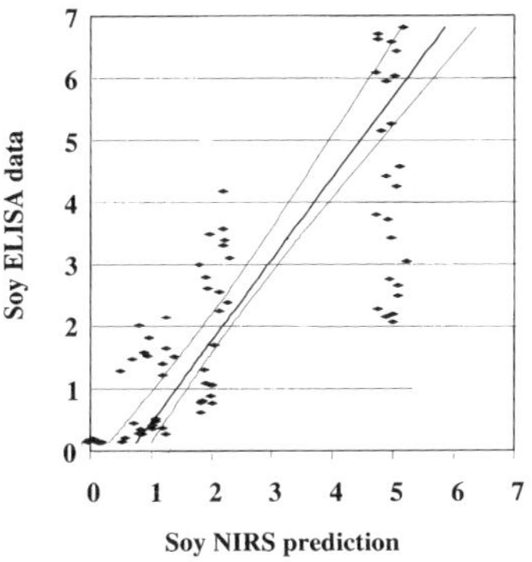

Figure 7. Comparison between ELISA results and NIR prediction values (soy).

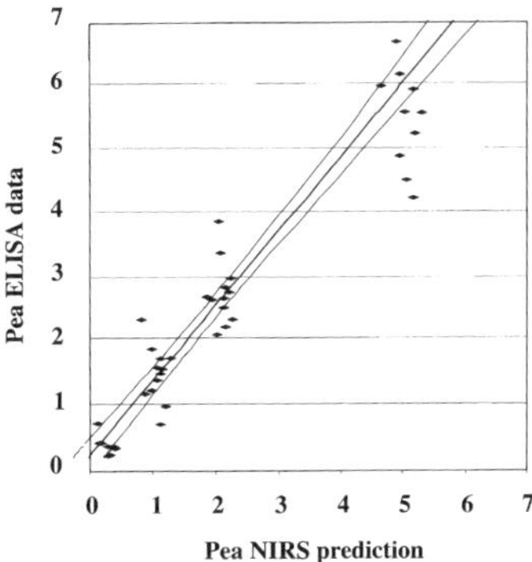

Figure 8. Comparison between ELISA results and NIR prediction values (pea).

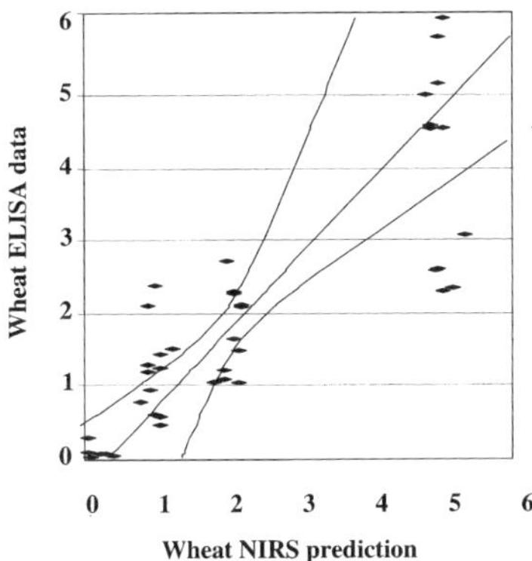

Figure 9. Comparison between ELISA results and NIR prediction values (wheat).

also show the possibility of correctly determining the amount of adulterant proteins in milk powder by using the CE procedure and to discriminate between 1, 2 and 5% additions, but NIR prediction values were characterised by higher accuracy and precision, easier sample preparation and were less time consuming.

NIR predictive power against competitive ELISA data

The comparison between ELISA and NIR results, obtained analysing all sets of samples, is reported in Figures 7–9, which show separated plots for soy, pea and wheat adulterated samples, respectively. Lines indicating the confidence interval (P = 95%) are also shown. A large variability of response and a partial overlapping of quantitative values were found when the ELISA procedure was applied and, also, if genuine samples were, in all cases, correctly classified. Considerations made before about accuracy and precision can be confirmed. We would just point out that the CE and ELISA methods showed satisfactory results, on the basis of data obtained by eight different laboratories, and represented an index of method reproducibility, but NIR prediction seems to be more precise in estimating the actual content of vegetable added proteins, with low *SEE* values.

Conclusions

NIR may have some advantages, such as rapidity, ease of use, no sample preparation and low operator skill required, over existing methods, based on electrophoretic and immunochemical principles.

NIR was able to determine more accurately than the other two techniques the percentage of adulteration in the analysed samples. Furthermore, dedicated calibrations could also help in identifying what type of vegetable protein isolate was added.

A more complete validation of the described NIR procedure needs further investigations, such as: (i) the identification of this kind of adulteration in different skimmed milk powders, (ii) the possibility of identifying the presence of vegetable hydrolysed isolates in milk powder, (iii) the detection of non-milk proteins in yoghurt and processed cheese, (iv) the detection of other vegetable proteins in milk and dairy products; (v) the data collection in order to calculate reproducibility index.

References

1. C.W. Kolar, I.C. Cho and W.L. Valtrous, *J. Am. Oil Chem. Soc.* **56,** 389 (1979).
2. C.V. Morr, *J. Am. Oil Chem. Soc.* **56,** 383 (1979).
3. T.M.P. Cattaneo, A. Feroldi, P.M. Toppino and C. Olieman, *Netherlands Milk & Dairy Journal* **48,** 225 (1994).
4. P.M. Toppino, T.M.P. Cattaneo, M. Casella and V. Denti, in *Ricerche ed innovazioni nell'industria alimentare,* Vol. 1, a cura di S. Porretta. Chiriotti Editori, Pinerolo (TO), p. 234 (1994).
5. T.M.P. Cattaneo, P.M. Toppino and A. Daghetta, in *Proceedings of 24th International Dairy Congress,* Melbourne (Australia) FIL-IDF Eds. (Brussels), p. 103 (1994).
6. K. Shiga, M. Yoshitake, T. Sato and Y. Someya, *Bull. Nat. Inst. Anim. Ind. Japan* **33,** 65 (1978).
7. K. Shimazaki, H. Tojo and K. Sukegawa, *Japanese J. Dairy Food Sci.* **34,** 123 (1985).
8. B.L. Ventling and W.L. Hurley, *J. Food Sci,* **54,** 766 (1988).
9. M.M. Hewedy and C.J. Smith, *Food Hydrocolloids* **5,** 399 (1989).
10. S. Barzaghi, K. Cremonesi and T.M.P. Cattaneo, *Ind. Latte* **35(3/4),** 23 (1999).
11. SMT4-CT97-2205: *Detection of non-milk protein in milk powder.* Final report. EU Project.
12. G. Downey, P. Robert, D. Bertrand and P.M. Kelly, *Appl. Spectrosc.* **44(1),** 150 (1990).
13. R. Giangiacomo, F. Braga and C. Galliena, in *Making Light Work: Advances in Near Infrared Spectroscopy,* Ed by I. Murray and I.A. Cowe. VCH Weinheim, Germany, p. 399 (1991).
14. M.F. Laporte and P. Paquin, *J. Agric. Food Chem.* **47,** 2600 (1999).
15. E.F. Parker, in *Leaping ahead with near infrared spectroscopy,* Ed by G.D. Batten, P.C. Flinn, L.A. Welsh and A.B. Blakeney. NIR Spectroscopy Group, Royal Australian Chemical Institute Publishing, North Melbourne, Victoria, Australia, p. 282 (1995).
16. L.K. Sørensen, and L.K. Snor, In: *Proceedings of 9th Int. Conf. on NIR,* Ed by A.M.C. Davies and R. Giangiacomo. NIR Publications, Chichester, UK, p. 823 (2000).
17. ISO 8968/IDF80-2 - *Milk, Determination of nitrogen content. Part 2: Block digestion method* (Macromethod).
18. A. Maraboli, T.M.P. Cattaneo, R. Giangiacomo, *J. Near Infrared Spectrosc.* **10(1),** 63 (2002).

Measurement of the concentrations of raw material, soya oil and products, mannosyl erythritol lipid, in the fermentation process using near-infrared spectroscopy

Kazuhiro Nakamichi,[a] Ken-ichiro Suehara,[a] Yasuhisa Nakano,[a] Koji Kakugawa,[b] Masahiro Tamai[c] and Takuo Yano[a]

[a]*Department of Information Machines and Interfaces, Faculty of Information Sciences, Hiroshima City University, Ohzukahigashi 3-4-1, Asaminami-ku, Hiroshima 731-3194, Japan*

[b]*Hiroshima Prefectural Institute of Industrial Science and Technology, 3-10-32 Kagamiyama, Higashi-Hiroshima 739-0046, Japan*

[c]*Hiroshima Prefectural Food Technology Research Center, 12-70 Hijiyamahonmachi, Minami-ku, Hiroshima 732-0816, Japan*

Introduction

Nowadays, chemical synthesised surfactant is used in many areas such as detergents, emulsifiers, dispersing agents, coagulant agents, wetting agents, foaming agents, defoaming agents, lubricants, softening agents, microbicides, etc. Although the production cost of chemical synthesised surfactant is low, it has several problems, including high toxicity, low bio-degradability and heavy environmental pollution. However, now the production of biological synthesised surfactant is being studied and developed. Biosurfactant has several advantages, low or harmless toxicity, high bio-degradability and light environmental pollution, although the production cost of it is higher than that of chemical synthesised surfactant.

Yeast, *Kurtzumanomyces* sp. I-11 produces mannosyl erythritol lipid (MEL) from soya oil. Soya oil is classified by triglycerides and is composed of 53% of linoleic acid, 23% of oleic acid, 11% of palmitic acid and 8% of linolenic acid as fatty acid moieties. MEL is the typical amphiphilic compound including both lipophilic and hydrophilic moieties in the molecule and is composed of mannose, erythritol and fatty acids (Figure 1).

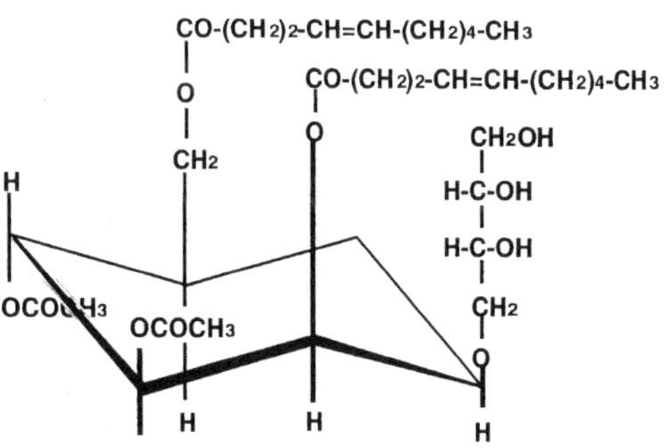

Figure 1. Molecular structure of mannosyl erythritol lipid (MEL).

In the fermentation process, measurement and control of the concentrations of raw materials and products in the culture broth are very important to obtain high productivity and low production cost. Initially, direct measurement of the concentrations of soya oil and MEL in the culture broth were studied. However, no good results have been obtained yet. First, the culture broth was nonhomogeneous. It consisted of air bubbles, water, soya oil, MEL and microorganism cells. It was difficult to make the homogeneous culture broth although the culture broth was mixed at high speed. Second, optical density of the culture broth became high as fermentation developed. As a result, the output of the near infrared (NIR) sensor was saturated. Third, soya oil in the culture broth attached on the inner wall of the cuvette of the spectrophotometer. So it was difficult to obtain the accurate values of the absorbance of the culture broth. However, it will be possible to solve the problem by intermittently washing with organic solvents such as ethyl acetate and acetone.

In this study, soya oil and MEL were extracted from the culture broth with ethyl acetate and the measurement system for the concentrations of soya oil and MEL in the ethyl acetate extract is developed using NIR spectroscopy.

Materials and methods

Microorganism and culture conditions

Five hundred mL of the seed culture of *Kurtzumanomyces* sp. I-11 was added into a 7 L of production medium in a 10 L fermentor (BMS-10PII, Able Co., Shinjuku-ku, Tokyo, Japan). The composition of the production medium contained the following components (gL^{-1}): soya oil variable; NH_4NO_3 0.5; KH_2PO_4 0.4; $MgSO_4{\cdot}7H_2O$ 0.2; yeast extract 1. To measure the concentrations of soya oil and MEL, about 100 mL of the culture broth was drawn out during several runs of the fermentation.

NIR spectroscopy

Four mL of ethyl acetate was added to 4 mL of the culture broth and mixed well. The extraction procedure was repeated three times and the ethyl acetate layer was transferred and combined in a mess flask.The ethyl acetate extract was injected into a cuvette with a 2 mm light path length. After putting the cuvette in the cell holder of an NIR spectrophotometer (NIRS6500SPL, Nireco Co., Hachioji, Tokyo), the absorbance from 400 to 2500 nm was measured at 2 nm intervals. The values of segment and gap size of the calculation of second derivative spectra were 20 and 0 nm, respectively.

To produce calibration equations, multiple linear regression (MLR) using the least-squares method was carried out between the NIR spectral data and the concentrations obtained by the conventional methods, C_{act}, for the calibration sample set.

To select a wavelength used for calibration equation, NIR spectra of MEL, soya oil, mannose, erythritol, ethyl acetate, ethyl ether, 1-butanol and butyric acid were measured. The MEL, soya oil, ethyl acetate, ethyl ether, 1-butanol and butyric acid were injected into the cuvette with a 2 mm light path length and NIR transmittance spectra were measured. The mannose and erythritol powder were packaged in a standard sample cup (Nireco Co.) and NIR reflectance spectra were measured.

Conventional analysis

The concentrations of MEL and soya oil in the ethyl acetate extract of the culture broth were measured using a thin layer chromatography with a flame-ionisation detector (TLC/FID) (Iatroscan newMK-5, Iatron Laboratories Inc., Tokyo, Japan). Microbial cell concentration in the culture broth was measured using the oven drying method.

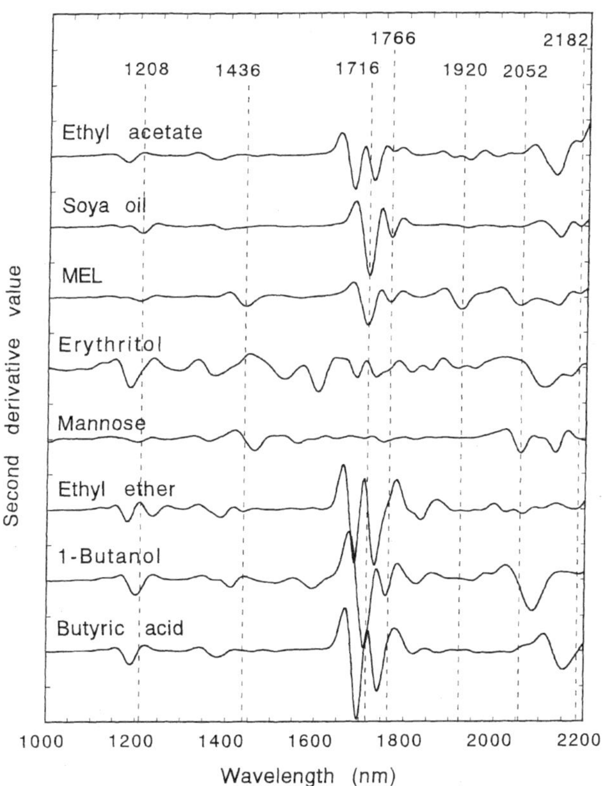

Figure 2. Second derivative spectra of authentic compounds.

Results and discussion

Second derivative NIR spectra of MEL and soya oil

Second derivative NIR spectra of ethyl acetate, MEL, soya oil, mannose, erythritol, ethyl acetate, ethyl ether, 1-butanol and butyric acid are shown in Figure 2. The three significant negative peaks were observed at the wavelengths of 1436, 1920 and 2052 nm on the second derivative NIR spectrum of MEL. The peak at 2052 nm may be caused by a mannose moiety of MEL because the NIR spectrum of mannose powder had a sharp peak at 2055 nm. The peak at around 1436 nm may be caused by ether linkage, C–O–C. Five ether linkages are in the MEL as shown in Figure 1. The peak at around 1436 nm was observed on the spectrum of ethyl ether while the peak was not observed on either spectra of 1-butanol and butyric acid. The peak at around 1920 nm may be caused by a combination band of O–H stretching and O–H deformation.[1] Therefore, the peak at around 1436 or 2050 nm was selected as the first wavelength for MEL calibration.

Four significant negative peaks were observed at 1208, 1716, 1766 and 2182 nm on the spectrum of soya oil and the absorption at these wavelengths was stronger than that of MEL. These negative peaks were mainly caused by the fatty acid moiety. It has been reported that the absorption band at around 1208 nm could be assigned to the second overtone of C–H.[1] The absorption at around 1716 nm

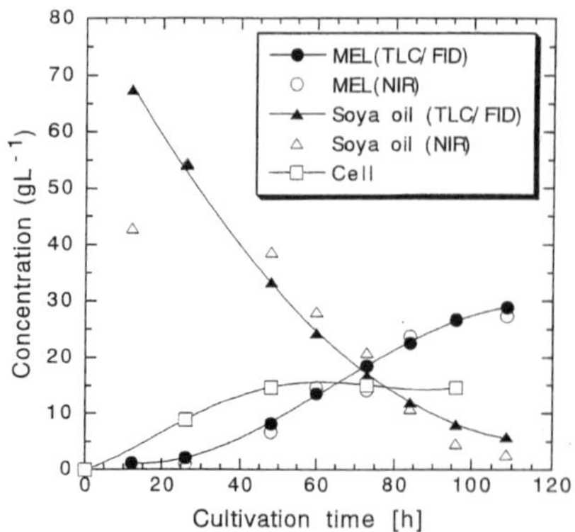

Figure 3. Time courses of the concentrations of soya oil, MEL and microbial cell.

and 1766 nm could be assigned to the first overtone of C–H.[1,2] The absorption at around 2182 nm could be assigned to the combination band of C–H with C=C.[2,3] Therefore, the peak at around 1766 or 2180 nm was selected as the first wavelength for soya oil calibration and the influence of MEL was corrected by using the second wavelength.

Calibration and validation for determining soya oil and MEL in ethyl acetate extract of the culture broth

MLR was conducted on the second derivative spectra and C_{act} for soya oil in ethyl acetate extract to formulate a calibration equation. The following calibration equation of soya oil was obtained.

$$C_{pre,soya\ oil} = -266 - 3680 \cdot A_1 + 2414 \cdot A_2 \quad (1)$$

Here, A_1 and A_2 are the values of second derivative spectra at 2178 and 2090 nm, respectively. The values of regression coefficient (R) and the standard error of calibration (SEC) were 0.974 and 0.77 gL^{-1}, respectively.

Validation of the calibration equation, Equation 1, was carried out. The soya oil concentration in the prediction sample set ($n = 54$) was predicted using Equation 1 and compared with the values of C_{act}. The standard error of prediction (SEP) was 0.56 gL^{-1}. Good agreement between C_{act} and C_{pre} was observed with a correlation coefficient (r) of 0.979.

For the MEL calibration, the following equation was obtained:

$$C_{pre,MEL} = -56 - 5769 \cdot A_1 + 8993 \cdot A_2 \quad (2)$$

Here, A_1 and A_2 are the values of second derivative spectra at 2040 and 1312 nm, respectively. The values of R and SEC were 0.994 and 0.48 gL^{-1}, respectively. As the results of the validation of the calibration equation, Equation 2, SEP was 0.45 gL^{-1}. Excellent agreement between C_{act} and C_{pre} was observed with $r = 0.994$.

Monitoring of MEL and soya oil concentrations in the cultivation

Monitoring of the MEL and soya oil concentrations in the culture broth during the glycolipid fermentation was performed using the calibration equations, Equations 1 and 2. Time courses of the concentrations of MEL, soya oil and microbial cells during the fermentation were shown in Figure 3. As the microorganism growing, MEL was produced to take soya oil into the microorganism. The values of the concentrations of MEL and soya oil measured by the NIR method were in good agreement with those obtained by the conventional method.

Discussions

The TLC/FID method was applied to the measurement of the concentrations of MEL and soya oil in the ethyl acetate extract of the glycolipid fermentation. Although the operational procedure of the TLC/FID method is simple, the time required for the measurement is about 50 min per sample in addition to 30 min for the extraction procedure. However, the operational procedure of NIR was also very simple and the time required for the measurement was only 5 min in addition to 30 min for the extraction procedure. The results of the present study suggest that NIR may be a useful method for the monitoring of the glycolipid fermentation.

Acknowledgment

This work was supported, in part, by a Special Coordination Fund for Promoting Science and Technology (Leading Research Utilising Potential of Regional Science and Technology) of the Science and Technology Agency of the Japanese Government.

References

1. B.G. Osborne, T. Fearn and P.F. Hindle, *Practical NIR spectroscopy with applications in food and beverage analysis*. Longman Scientific & Technical Publishers, Harlow, Essex, UK (1986).
2. T. Sato, S. Kawano and M. Iwamoto, *J. Am. Oil Chem.Soc*. **68,** 827 (1991).
3. J. Hong, S. Yamaoka-Koseki and K. Yasumoto, *Food Sci. Technol. Int*. **2,** 146 (1996).

Detection of physiological processes in wheat using near infrared spectroscopy

Szilveszter Gergely, Éva Scholz and András Salgó

Department of Biochemistry and Food Technology, Budapest University of Technology and Economics, Műegyetem rkp. 3, H-1111 Budapest, Hungary

Introduction

Sequences of biochemical, enzymatic and morphological changes occur during the so-called generative development in seed. To investigate the physiological status of seed, complicated chemical, biochemical, enzymatic, botanical and morphological test methods are needed and the results obtained are often inaccurate and obscure due to extensive biological variations. In order to examine the complex physiological events taking place in plant materials and to understand the biochemical basis of processes, near infrared (NIR) techniques can be applied. NIR is widely used in the quality control of cereals but also can be used as a tool in basic research, breeding programmes, technological process analysis and control and in detection of functional properties of cereal.[1]

According to Batten and Blakeney,[2] Downey[3] and Salgó *et al.*,[4] different physiological processes (ageing, maturation, germination) were followed in cereals using NIR. However, the application of NIR methods were sometimes critical because of the lack of proper reference methods and the fast dynamic changes in the investigated materials.

The aim of the present study was to observe the biochemical physiological changes in intact wheat seeds during seed development (maturation) using near infrared methods. The investigations were focused on the qualitative changes of spectral characteristics and of the main constituents in plant tissues.

Materials and methods

Six different wheat varieties with different harvest time were grown as field trials at the Agricultural Research Institute of the Hungarian Academy of Sciences (ARIHAS), Martonvásár.

The investigated varieties were as follows: GK Öthalom (early), Bánkúti 1201 (early), Jubilejnaja 50 (semi medium), Mv 23 (medium), Fatima (medium), Mv 15 (semi late).

Whole primary ear samples were collected twice or three times weekly from crops of each variety. The sampling began 12 days after flowering (DAF). Because of the high biological variabilities of seed materials six independent ear samples were collected each time from each variety.

The 16 sampling time (1–16 points) covered the whole 41 days long maturation period.

Wheat seed were prepared from ears manually directly after sampling and were scanned in intact form; their moisture content was measured immediately. The remaining fresh samples and the dried materials were frozen (–15°C). From each ear about 40–60 seeds were prepared and the average fresh weight of the samples (mg seed^{-1}) was measured.

Moisture content of wheat samples was determined in triplicates using 105°C and 4 h oven drying method.

Five independent spectroscopic scans were recorded from each sample. A static sample cup was used with a special ring for volume reduction. Scans were collected with an NIRSystems 6500 instru-

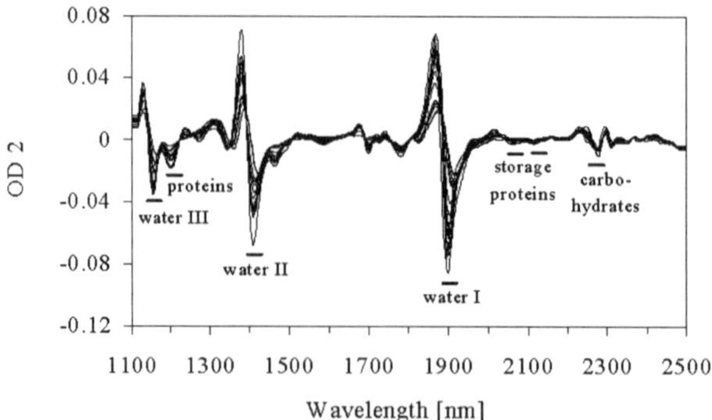

Figure 1. Second derivative spectra of wheat samples taken during maturation.

ment (NIRSystems Inc., Silver Spring, MD, USA) using sample transport module (reflection mode) in the 1100–2500 nm wavelength range, and the raw spectra were transformed, second derivatives (OD 2) were calculated.

Spectral and reference data were processed with the NSAS 3.30 (NIRSystems Inc., Silver Spring, MD, USA) software package.

Results and discussion

A very high level of moisture content was observed in the early lag period of maturation then a proportional but slow decrease was detected (data not shown). An extremely accelerated loss in moisture was measured after 38 DAF. This is the period of maturation where most of the synthesised materials were "dried" into the matrix and the "character" of water was changed significantly.

The changes which took place in wheat seeds were observed in the near infrared spectra. The reflectance spectra of maturing seed samples showed a very high variation in the whole wavelength range. This variation still remained after transformation (calculation of the second derivatives) of raw spectra (Figure 1). Characteristic regions were observed in the spectra where the changes of main constituents can be followed and analysed in details.

The water peak around 1900 nm (water I) was changed dramatically during maturation in all the six varieties. In the early stage of maturation (until 19 DAF) the water peak showed a 15 nm shift to the

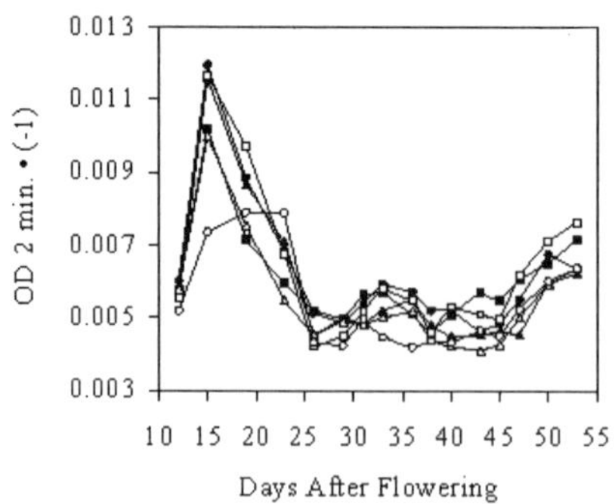

Figure 2. Change of local minimum values of carbohydrate peak (2270–2280 nm) during maturation.

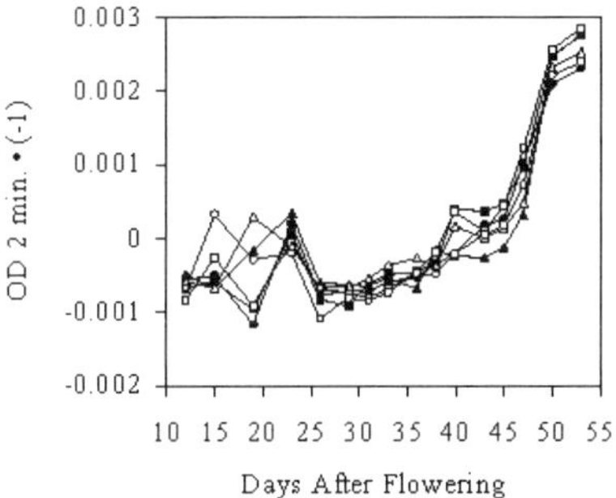

Figure 3. Change of local minimum values of protein peak (2050–2070 nm) during maturation.

lower wavelength value and about a 25% increase in absorbance (decrease in OD 2) showing the higher concentration of "free" type of water in seed. The level of "free" water decreased rapidly in the 23–25 DAF period and at the end of maturation (drying) a characteristic shift (15–20 nm) to a higher wavelength value in the water peak was observed. Similar changes were detected in the region of other water peaks around 1410 nm (water II) and 1155 nm (water III) respectively (data not shown).

The second derivative spectra confirmed a non-linear accumulation of oligo- and polysaccharide (mainly starch) components during maturation, which was concluded from the changes of a characteristic peak in the 2270–2280 nm region. (Local minimum values of the 2nd derivative spectra at 2260–2290 nm were drawn *vs* maturation time.) In the early phase of maturation (until 26 DAF) a huge amount of saccharides were synthesised and converted very fast (Figure 2) and the synthesis of the reserve polysaccharides was started at an accelerated speed. The accumulation of starch was extremely fast in the last period of maturation (45–52 DAF).

Bands of spectral region (2050–2130 nm), which were mainly associated with storage proteins (gliadins and glutenins), have been changed to high extent during maturation. An uncertain "shift" of protein bands (2050–2070 and 2100–2130 nm) until 23 DAF confirmed the formation of different elements (amino acids, peptides) for protein synthesis.[5] In the second phase of maturation (after 26 DAF) a linear and in two phase accelerated increase in the amount of storage proteins were observed (Figure

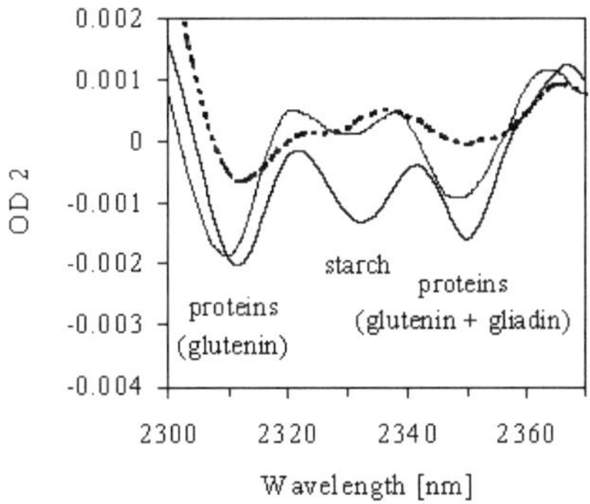

Figure 4. Change of characteristic protein and starch peaks at three different stages of maturation (dashed line = 19 DAF, thin line = 36 DAF, thick line = 53 DAF).

3). Similar results were obtained when the changes of another storage protein band (1190–1210 nm) was analysed.

To see the kinetic of the formation of seed reserves it is essential to follow the changes of peaks of proteins and starch *vs* maturation time (Figure 4). In the early phase of maturation only a limited amount of protein was present. After five days, an accelerated protein synthesis was observed while starch was produced only to a smaller extent. In the final period of maturation the synthesis of starch was very fast.

In order to be able to check the physiological changes or physiological status of seeds quantitatively, calibrations were developed for detecting moisture content, dry weight, nitrogen content and maturation time. Figure 5 summarises the statistical results of calibrations and tests. Moisture content was predicted with 2.42% standard error of prediction (*SEP*) value, which means a 5% coefficient of variation in a broad moisture range. The high correlation coefficient shows that maturation processes (changes in NIR spectra) are highly associated with water content. Calibration and tests carried out for detection of maturation time confirmed that there is information in the NIR spectra which is sensitive enough for predicting the physiological status of wheat seed with *SEP* = 2.4 days value. The coeffi-

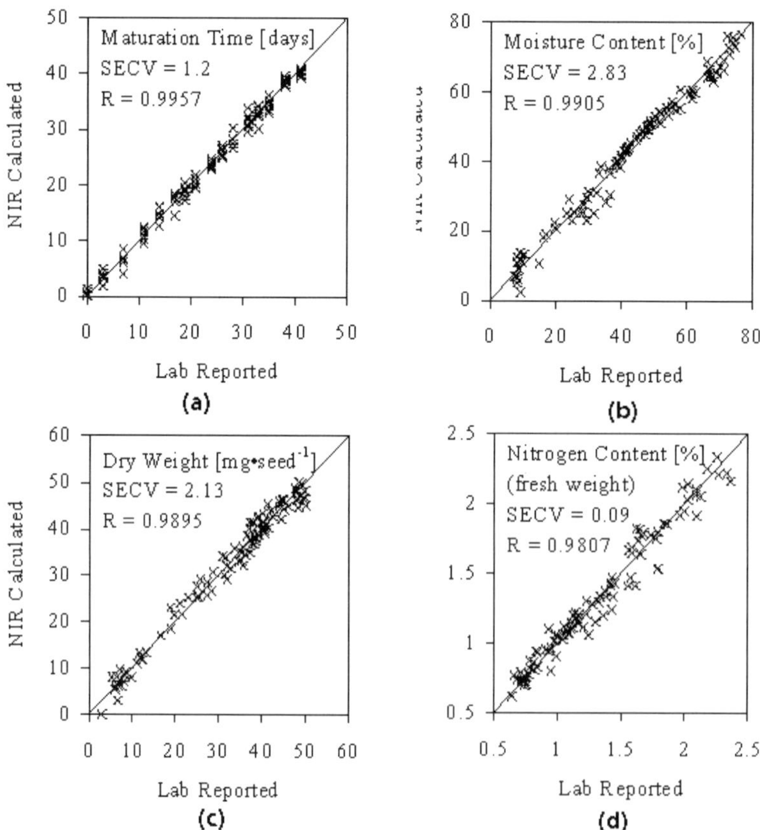

Figure 5. Model equations for determination of (a) maturation time, (b) moisture content, (c) dry weight and (d) nitrogen content.

cient of variation of this measurement was 5–7%, which is very acceptable considering the high biological variability of materials tested.

Conclusion

The seed maturation as a complex metabolic process can be sensitively followed by NIR spectroscopy. The details of NIR spectra are informative concerning the changes of macro components or their fractions and they have plenty of hidden chemical and biochemical information. The kinetic changes of the main constituents can be recognised in the spectra and based on spectroscopic results the physiological status of seed can be predicted quantitatively.

Acknowledgments

Thanks to László Láng (ARIHAS, Martonvásár) for the arrangement of plant trials. This work was supported by National Foundation of Science and Research Hungary (project numbers: T 031902 and A 143).

References

1. B.G. Osborne, *Cereal Foods World* **45,** 11 (2000).
2. G.D. Batten and A.B. Blakeney, in *Near Infrared Spectroscopy: The Future Waves*, Ed by A.M.C. Davies and Phil Williams. NIR Publications, Chichester, UK, p.112 (1995).
3. G. Downey, in *Leaping Ahead with NIR Spectroscopy*, Ed by G.D. Batten, P.C. Flinn, L.A. Welsh and A.B. Blakeney. NIR Spectoscopy Group, Royal Australian Chemical Institute, Victoria, Australia, p. 136 (1995).
4. A. Salgó, Gy. Dely-Szabó and Z. Fábián, in *Leaping Ahead with NIR Spectroscopy*, Ed by G.D. Batten, P.C. Flinn, L.A. Welsh and A.B. Blakeney. NIR Spectoscopy Group, Royal Australian Chemical Institute, Victoria, Australia, p. 506 (1995).
5. A. Salgó and Sz. Gergely, in *Wheat in a Global Environment*, Ed by Z. Bedő and L. Láng. Kluwer Academic Publishers, Dordrecht, Germany, p. 297 (2001).

Near infrared analysis of liquid and dried ewe milk

Nieves Núñez-Sánchez,[a] Ana Garrido-Varo,[a] Juan M. Serradilla[a] and José L. Ares[b]

[a]*Escuela Técnica Superior de Ingenieros Agrónomos y Montes, University of Córdoba, Avda Menéndez Pidal s/n, E-14080 Córdoba, Spain*

[b]*CIFA, Hinojosa del Duque (Córdoba), Department of Agriculture, Junta de Andalucía, Spain*

Introduction

The routine analysis of chemical components of milk is of major importance both for the management of animals in dairy farms and for quality control in dairy industries. A recent review presented by Laporte and Paquin[1] points out that, despite widespread use of near infrared (NIR) spectroscopy in the dairy industry and the general enthusiasm surrounding this technology, there are still problems regarding NIR analysis routines which must be solved before its final implantation in dairy industries and laboratories.

Sample preparation has often been considered as one of the most critical aspects in NIR analysis of liquid products. The main problem associated with their sample preparation is water content because it absorbs most of the infrared radiation and this disturbs the calibration for other constituents. To solve this problem, the dry extract spectroscopy by infrared reflectance (DESIR)[2] system was developed. This method consists of drying a glass fibre filter previously impregnated with the liquid under test. It has been proposed, for goat's milk analysis, to modify the typical drying conditions of this method (70°C, 15 min) to 40°C, 24 h to avoid protein denaturation.[3,4] However, this makes NIR analysis lose one of its main advantages: to provide instantaneous results with little or no sample preparation.

There are other measurement methods appropriate for liquids or semi-liquids, such as transmittance[5] and folded transmission or transflectance.[6–8] In this sense, the prediction results of DESIR and transmittance measurements of prepared liquid foods have been compared by Isaksson *et al.*[9]

The aim of the present study was to compare the accuracy of folded transmission (liquid milk) and reflectance (dried milk) NIR calibration equations to predict quality parameters in ewe's milk.

Materials and methods

Milk samples

A set of 101 ewe's milk samples was used to develop the calibration equations. All of the samples came from individual controls in different lactation in order to obtain maximum seasonal variation. The samples were preserved by adding potassium dichromate, stored at 2–4°C and analysed within 72 hours of collection.

Chemical analysis

Prior to chemical and NIR analyses, milk samples were heated at 40°C, mixed gently in order to achieve uniform dispersion of fatty matter and other components and then left to cool at room temperature.

Milk samples were analysed in duplicate in order to determine the following chemical parameters: total protein (using the colorimetric method described by Bradford),[10] total casein (precipitation of milk casein and subsequent determination of whey protein using the method indicated above), fat (Gerber), total solids (oven drying at 103°C ± 2°C) and somatic cell count-SCC (Fossomatic).

NIR analysis and chemometric treatments

A Foss NIRSystems 6500 SY-I scanning monochromator (400–2500 nm), equipped with a spinning module, was used. All samples were analysed using the two spectroscopic methods to be compared in this study:

Reflectance (R): Small ring cups for solid product analyses were used. One filter per sample was prepared and oven dried at 40°C for 24 h. After one hour in a desiccator, the filters were placed in the cup with the readable side against the quartz window to perform NIR analysis.

Folded transmission (FT): An aluminium reflector 0.1 mm pathlength cam-lock cell for liquid product analyses was used. A sample of 0.85 mL was placed in the cell and the sample was scanned through the quartz window. Two cells per sample were filled and the average spectrum was used in the data analyses.

Folded transmission and reflectance spectra of a milk sample are shown in Figure 1.

Both spectral data collection and chemometric treatment of the data were performed using ISI software (NIRS 3 ver 3.11; Infrasoft International, Port Matilda, PA, USA). MPLS (modified partial least squares) was used for regression purposes; 400–2500 nm and 1100–2500 nm regions (in 2 nm steps) were tested; SNV and Detrending treatments were applied for scatter correction. Several first and second derivative treatments were also evaluated. The methodology followed for the development and evaluation of NIR calibrations is described in different publications.[11–13] The following statistics were used to select the most accurate calibration equations: the standard error of the residuals for the calibration (*SEC*) and for the cross-validation (*SECV*), the coefficient of determination for the calibration

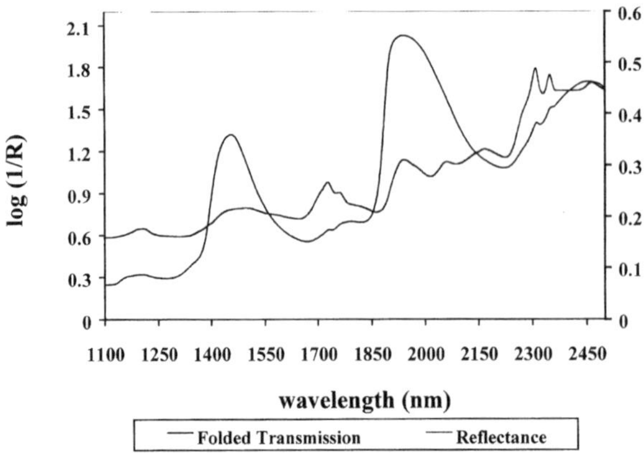

Figure 1. Folded transmission (liquid) and reflectance (dry extract) spectra of one milk sample.

Table 1. Calibration statistics obtained for quantitative analysis of ewe milk in both the Folded Transmission (FT) and Reflectance (R) modes.

Parameter	Mode	Mean	*SD*	*SEC*	*SECV*	r^2	*RPD*	*CV*
Casein	FT	5.49	0.60	0.17	0.21	0.88	2.86	3.77
	R	5.52	0.60	0.13	0.19	0.88	2.95	3.50
Protein	FT	5.95	0.68	0.18	0.19	0.92	3.58	3.21
	R	5.95	0.65	0.11	0.16	0.94	4.06	2.75
Fat	FT	8.06	2.02	0.11	0.14	0.99	14.43	1.71
	R	7.85	1.82	0.21	0.43	0.94	4.23	5.52
Total Solids	FT	17.97	2.23	0.19	0.25	0.99	8.92	1.41
	R	17.78	2.20	0.24	0.34	0.98	6.47	1.91
SCC 10^3	FT	260.63	156.55	25.40	53.11	0.88	20.38	2.95
	R	278.60	169.12	25.39	55.57	0.89	19.95	3.04

(R^2) and for the cross-validation (r^2), the coefficient of variation (*CV*), calculated as ($SECV$ * MEAN^{-1}) × 100 and the *RPD*, calculated as SD * $SECV^{-1}$.

Results and discussion

The results obtained for chemical analysis, calibration and cross-validation of selected calibration equations for each constituent in both NIR analysis modes under study are shown in Table 1.

Protein, fat and total solids calibrations have excellent capacity for quantitative analysis,[12] as their r^2 values are higher than 0.9. Casein r^2 values are slightly lower, but still are high. The equation obtained for somatic cell count (SCC) has adequate accuracy, similar to that obtained for goat's milk by Pérez *et al.*[14] (r^2 = 0.81) and much higher than the model reported by Tsenkova *et al.*[15] in cow's milk (r^2 = 0.35).

In general, both analysis modes present low prediction errors, estimated by the *SECV*, *RPD* and *CV* values. The RPD statistic values were always higher than three, recommended by Williams and Sobering[13] to consider a calibration equation as suitable to use in real conditions of process control.

The accuracy of casein, protein and SCC equations is not affected by sample analysis methods, as they present similar *SECV* values in reflectance and folded transmission modes. Nevertheless, the accuracy of folded transmission fat and total solids equations is significantly higher than the corresponding reflectance equations. Isaksson *et al.*,[9] likewise, reported lower prediction errors with the use of direct measurements (transmittance, 1 mm cuvettes) on liquid foods compared to the DESIR method. *CV* value for folded transmission protein calibration is very similar to those obtained by Albanell *et al.*[7] (4.09%) and by Pascual *et al.*[8] (2.21%) who used an analysis system (transflectance) very similar to the one employed in this study. However, the *CV* value obtained for folded transmission fat and total solids are lower than those obtained by Albanell *et al.*[7] in goat's milk (4.55% vs 3.03%, respectively).

Conclusion

In terms of accuracy and speed of analytical response, NIR analysis of liquid milk (folded transmission) is recommended instead of NIR analysis of dry extract of milk (reflectance) although both analysis modes offer satisfactory results.

Acknowledgements

This study formed part of a PhD founded by a FPI fellowship of the Department of Agriculture, Junta de Andalucia, and was performed using equipment and infrastructure belonging to SCAI (Unidad NIR/MIR), University of Córdoba, (Spain). The authors thank Ms Francisca Baena, of the Animal Production Department (ETSIAM-UCO), for her technical assistance.

References

1. M.-F. Laporte and P. Paquin, *Seminars in Food Analysis* **3,** 173 (1998).
2. G. Alfaro and M. Meurens, *Proceedings of the 2nd International NIRS Conference*, Tsukuba, Japan, p. 204 (1989).
3. E. Díaz, A. Muñoz, A. Alonso and J.M. Serradilla, *J. Near Infrared Spectrosc.* **1,** 141 (1993).
4. C. Angulo, Doctoral Thesis. University of Córdoba (1997).
5. J.W. Hall and K. Chan. *Cheese yield and factors affecting its control,* Ref. SI 9402. International Dairy Federation, p. 230 (1993).
6. R. Franhuizen, in *Handbook of near infrared analysis*, Ed by D.A. Burns and E.W. Ciurczak. Marcel Dekker Inc., New York, USA, p. 609 (1992).
7. E. Albanell, S. Peris, M. Rovai, G. Caja and X. Such, *ITEA* **20-I,** 161 (1999).
8. J.J. Pascual, P. Molina and R. Puchades, in *Near Infrared Spectroscopy: The Future Waves*, Ed by A.M.C. Davies and P. Williams, NIR Publications, Chichester, UK, p. 559 (1996).
9. T. Isaksson, L.E. Jorgenvag and V.H. Segtnan, in *Near Infrared Spectroscopy: Proceedings of the 9th International Conference*, Ed by A.M.C. Davies and R. Giangiacomo. NIR Publications, Chichester, UK, p.139 (2000).
10. M.M. Bradford. *Analytica Biochemistry* **72,** 248 (1976).
11. H. Mark and J. Workman (Eds), *Statistics in Spectroscopy.* Academic Press, Inc. NY, USA (1991).
12. J.S. Shenk and M.O. Westerhaus, in *Near Infrared Spectroscopy: The Future Waves*, Ed by A.M.C. Davies and P.C. Williams. NIR Publications, Chichester, UK, p. 198 (1996).
13. P.C. Williams and D. Sobering, in *Near Infrared Spectroscoopy: The Future Waves*, Ed by A.M.C. Davies and P.C. Williams. NIR Publications, Chichester, UK, p. 185 (1996).
14. M.D. Pérez, A. Garrido, J.M. Serradilla, N. Núñez, J.L. Ares and J. Sánchez, *ITEA* **22-II,** 598 (2001).
15. R. Tsenkova, K. Itoh, J. Himoto and K. Asahida in *Leaping Ahead with Near Infrared Spectroscopy,* Ed by by G.D. Batten, P.C. Flinn, L.A. Welsh and A.B. Blakeney. Royal Australian Chemical Institute, Victoria, Australia, p. 329 (1995).

Characterisation and classification of waxes used in dairy technology by near infrared spectroscopy

Stefania Barzaghi, Claudia Giardina, Tiziana M.P. Cattaneo and Roberto Giangiacomo*

Istituto Sperimentale Lattiero-Caseario, Via A. Lombardo 11, I-26900 Lodi, Italy
E-mail: rgiangiacomo@ilclodi.it

Introduction

Some particular cheeses, like Cheddar, Edam, Gouda and Provolone cheese, are wrapped in a coating of paraffin wax: (i) to prevent development of undesirable micro-organisms;[1–6] (ii) to prevent drying-out during storage; and (iii) to obtain good stability during handling and transportation to customers. Generally speaking, waxes, which are chemically inert and non-biodegradable under ambient conditions, are obtained from the petrochemical industry.[7,8] Previous studies suggested that waxes be replaced with plastic films.[3,5] The main disadvantage in their use may be a migration, caused by direct surface contact,[6] of plasticisers such as adipates and phthalates into cheese. Nowadays biodegradable waxes,[9] which may be produced from microorganisms,[10] may also be used.

The aim of this study was to evaluate the feasibility of near infrared spectroscopy (NIR) to characterise and classify different waxes applied on some Italian cheese types.

Materials and methods

Eight waxes, commonly used at Italian dairies, and four experimental biodegradable waxes, produced from coconut oil by *Pseudomonas putida*,[10] were analysed by an InfraAlyzer 500 (Bran+Luebbe, Norderstedt, Germany) at room temperature.

Preliminary studies were carried out to optimise analysis conditions. Samples of suitable size were obtained by carrying out a melting process at 90°C and cutting pieces of 1.5 cm (height) × 1.5 cm (width) × 0.15 cm (thickness). NIR spectra were collected twice in the range 1100–2500 nm, at 4 nm intervals (351 data points), and data were processed by Sesame Software (Bran+Luebbe) and GRAMS/32AI (Galactic Industries, Salem, NH, USA).

Exploratory analysis was performed using principal component analysis (PCA) on spectra corrected by a standard normal variate algorithm, i.e. each spectrum was normalised by computing and removing the mean value, then scaling by the standard deviation. Discriminant analysis was used to test the qualitative classification power of NIR.

The same waxes were also analysed by FT-IR instrumentation (FT-IR/420-Jasco Europe, Cremella, LC, Italy) coupled with attenuated total reflectance (ATR) in the range 4000–600 cm^{-1} wavenumber.

A two-dimensional correlation between NIR and FT-IR data was generated using the 2D-CORR program (Galactic Industries) to enhance NIR results.

Results and discussion

Preliminary studies were carried out to optimise sample presentation in NIR analysis.

Figure 1 shows an example of exploratory analysis (PCA) of a petrochemical wax, melted at different temperatures (85°C, 90°C, 95°C and 98°C) and cut into pieces of different thicknesses (0.5 mm, 1 mm, 1.5 mm and 2 mm). No significant differences caused by the melting temperature were found. Groups were discriminated according to the sample thickness as a result of absorbance increase. Sample preparation conditions were chosen on the basis of their easiness. In particular, a temperature of 90°C was chosen because (i) this melting temperature is commonly used at factories, (ii) this sample preparation is fast and (iii) it allows possible changes in wax composition due to extreme heat (for example, reaching the point of smoke, although in this case it did not happen) to be avoided.

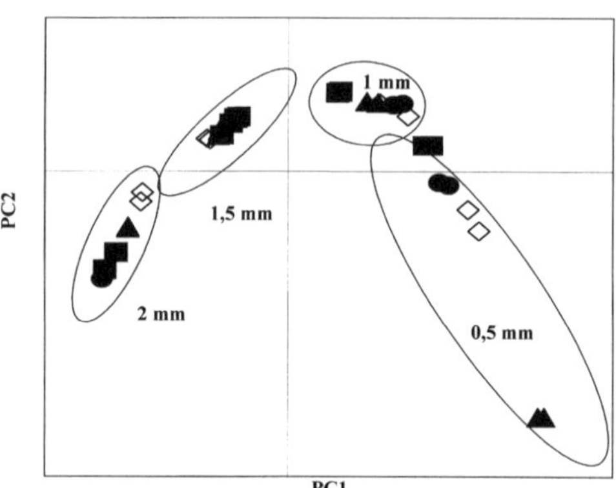

Figure 1. Example of an exploratory analysis (PCA) of a petrochemical wax, melted at different temperatures: (●) 85°C, (■) 90°C, (▲) 95°C and (◇) 98°C.

Variations in the shape of absorbance spectra were related to the nature of samples. Petrochemical waxes showed an opposite trend, as compared with biodegradable waxes in the 2050–2150 nm wavelength regions. Examples of raw spectra are reported in Figure 2.

Exploratory analysis carried out on these spectra (Figure 3) showed a clear separation between the two waxes along the first principal component (PC1). The trend of the first factor loading gave prominence to three principal peaks at 1684 nm, 2136 nm and 2248 nm. These absorbances corresponded to carboxylic acid COOH vibrations, in agreement with literature data.[11] The second principal component (PC2) was able to further discriminate samples within each group and 1220 nm, 1748 nm, 1776 nm and 2252 nm, related to C–H vibrations,[12,13] were found to be the most important wavelengths.

A 2D correlation with FT-IR spectra was performed to carefully recognise the wavelengths involved, to point out other functional groups, possibly covered in the NIR region and to correctly allo-

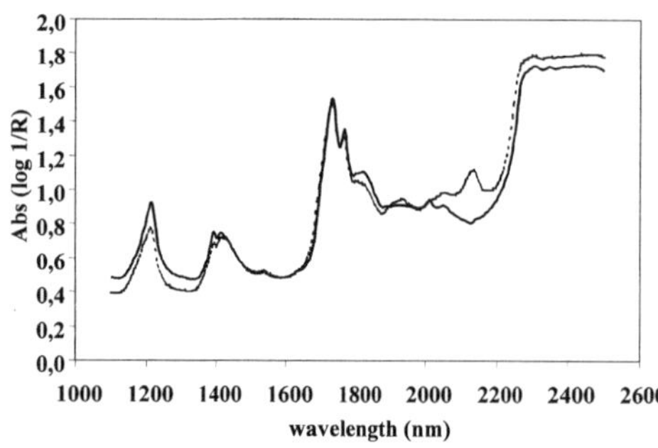

Figure 2. Examples of raw NIR spectra of (—) paraffin and (- - -) biodegradable wax.

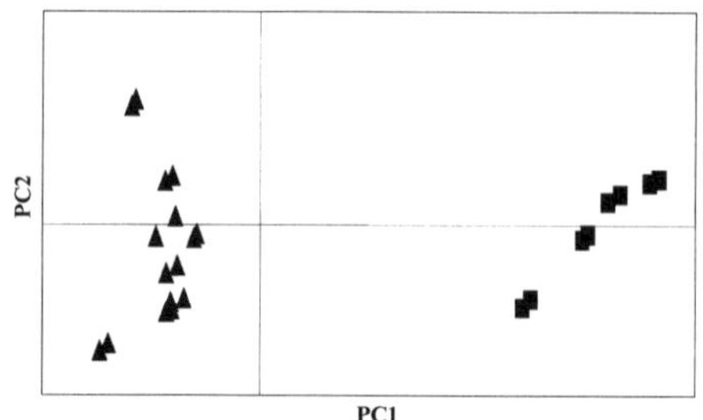

Figure 3. Exploratory analysis (PCA) of (▲) paraffin and (■) biodegradable samples on NIR spectra corrected by an Standard Normal Variate algorithm.

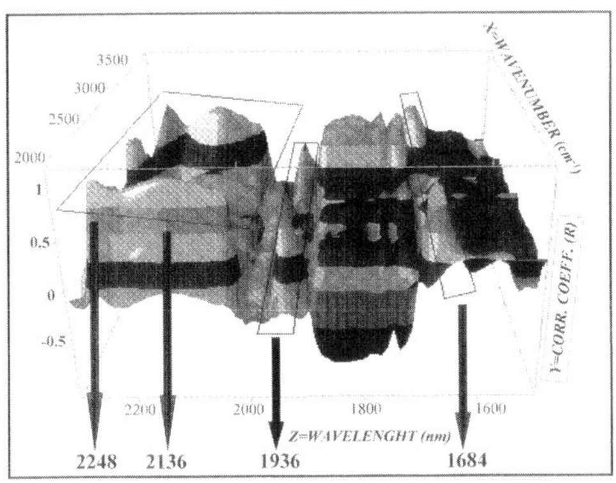

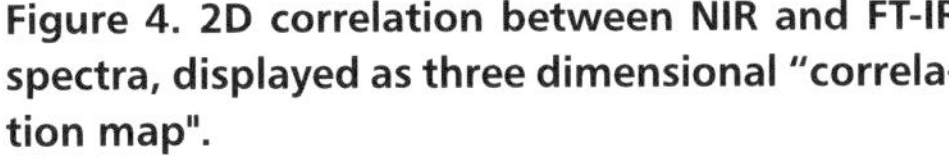
Figure 4. 2D correlation between NIR and FT-IR spectra, displayed as three dimensional "correlation map".

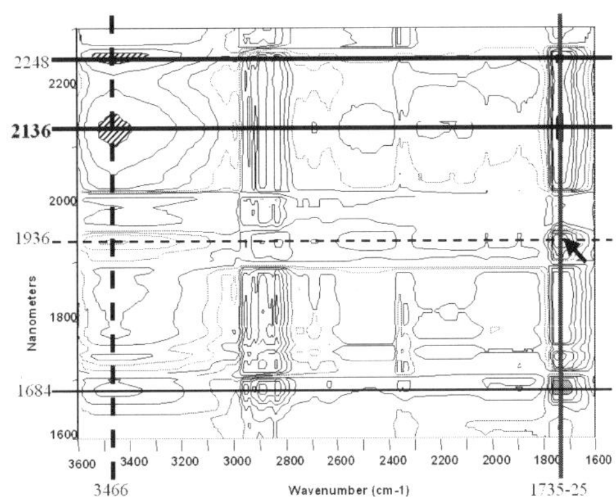

Figure 5. Contour map of bidimensional correlation FT-IR/NIR.

cate some functional groups to a precise chemical class. Figure 4 shows results as a three-dimensional "correlation map": regions of high positive correlation appear as "mountains", where individual bands for the two types of spectra were most similar and regions of negative correlation appear as "valleys" where spectra were most different.

In order to enhance results, a contour map of the bidimensional correlation FT-IR/NIR was plotted (Figure 5). The X dimension (horizontal) is the X matrix spectral axis of FT-IR and the Z dimension (vertical) is the Z matrix spectral axis of NIR.

The maximum value of the correlation coefficient, R, was close to 1 (see black contours under the thick lines). These contours related the interval 1750–1722 cm^{-1} in the mid-IR to two wavelength intervals in the NIR: 2100–2156 and 2234–2252 nm. At 1725 cm^{-1} wavenumber absorption of C=O stretching of carboxylic acids is reported. At 2136 nm the absorption of C–O stretching of long chain fatty acids is attributed. At 2248 nm the absorption is associated with the 2nd overtone stretching of coupled C–O / OH. At 1684 nm a correlation to the 3rd overtone absorption of this group was present with lower values (R = 0.75) of correlation coefficient (see gray contour under the thin line). Wavelengths of 2136 and 2248 nm exhibited further correlations of lower values, i.e. R = 0.5 (see stripped contours under thick-dotted lines), with 3466 cm^{-1} where absorption of O–H stretching of carboxylic acids is reported.[14] The interval corresponding to the absorption of C=O stretching of saturated acyclic esters correlated quite well (R = 0.6) to 1936 nm (see black arrow), ester 2nd overtone C–O stretching.[15]

From these relationships it can be concluded that the NIR wavelengths able to differentiate the chemical compounds giving rise to biowaxes and paraffins were well established. Biowaxes are polymeric chains of fatty acids (3-hydroxy fatty acids) containing the ester group, which permits them to be distinguished from paraffins, i.e. simple hydrocarbon chains with few branches, both in the mid-IR and in the NIR. In addition, they contain, though in lower amounts, natural waxes, other fats and primary alcohols such as stearilic alcohol (saturated C 18). The wavelengths that discriminate waxes (1684 nm, 2136 nm and 2248 nm) were related to COOH group vibrations and the higher correlation coefficient values (R near 1) found, indicated that waxes had a similar behaviour in the near and mid IR regions. Surprisingly, ester C=O stretching did not seem to influence the classification of waxes as expected because biowaxes were made from fatty acid esterification processes.

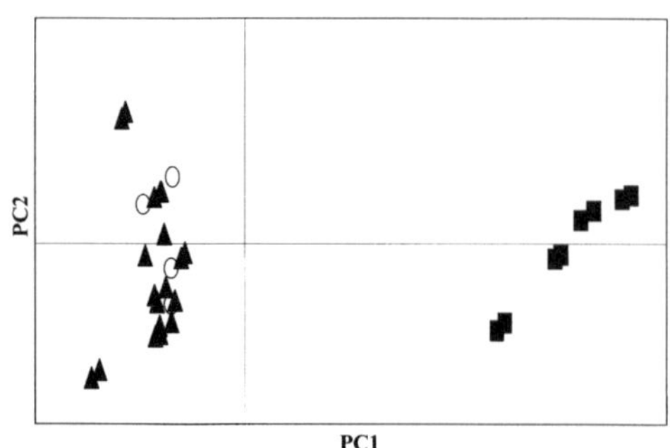

Figure 6. Qualitative classification of unknown samples using discriminant analysis applied on NIR spectra (▲) paraffin, (■) biodegradable samples, and (○) unknown samples.

The discriminant analysis performed on unknown samples (Figure 6) showed that NIR was able to correctly classify waxes using simple criteria (i.e. a qualitative classification), based on the information collected from raw spectra. These data suggested that NIR may be used in routine analysis for the control of raw materials, although further investigations are required to confirm preliminary results.

Conclusions

A two-dimensional correlation between NIR and FT-IR spectra allowed us to explain why the two types of wax were grouped so clearly by NIR on the basis of their chemical compounds.

Interestingly, the correlation NIR/FT-IR permitted us to know the wavelength of absorption exactly, due to ester 2nd overtone C=O stretching. In fact wavelength 1936 nm had a positive correlation with 1740–1735 cm^{-1} (C=O stretching) and a negative correlation with 3466 cm^{-1} (carboxylic acid O–H stretching).

On the basis of its classification power, NIR spectroscopy appeared to be a promising tool when used in routine analysis for a qualitative control of raw materials also on account of a further discrimination within each individual wax group according to the chain length.

References

1. J. Grubhofer, *DMZ, Lebenmittelindustrie-und-Milchwirtschaft* **113(32/33),** 969 (1992).
2. M.G. van den Berg, G. van den Berg and M.A.J.S. van Boekel, *Neth. Milks and Dairy J.* **50(4),** 501 (1996).
3. N. Innocente, L. Cencic and C. Corradini, *Sci. Tecn. Latt. Cas.* **48(2),** 163 (1997).
4. K. Khamrui and G.K. Goyal, *Indian Dairyman.* **50(9),** 25 (1998).
5. K. Khamrui and G.K. Goyal, *J. Food Sci.and Technol. India* **36(2),** 147 (1999).
6. A.E. Goulas, K.I. Anifantaki, D.G. Kokioulis and M.G. Kontominas, *J. Dairy Sci.* **83(8),** 1712 (2000).
7. I.G. Anisimov, T.M. Vyazun, O.I. Soletskii, N.I. Kostin, A.D. Zakharchenko, L.V. Ostrikova, G.G. Shiler, I.A. Rozdov, E.S. Bankaev, R.M. Usmanov and K.I. Utkin, *USSR Patent,* SU 1 613 097 (1990).
8. Paramelt Syntac BV, *Netherlands Patent Application,* NL 8901816 A, NL 8901816 (1991).
9. E.A. Baldwin, M.O. Nisperos, R.D. Hagenmaier and R.A. Baker, *Food Technology* **51(6),** 56 (1997).
10. EU Project BE SE-2659, *Final report* (2001).
11. I. Murray and P.C.Williams, in *Near-Infrared Technology in the Agricultural and Food Industries,* Ed by P. Williams and K. Norris. AACC, St Paul, MN, USA, p. 30 (1987).
12. G. Downey, J. Boussion and D. Beauchêne, *J. Near Infrared Spectrosc.* **2,** 85 (1994).
13. T. Huth-Fehre, R. Feldhoff, F. Kowol, H. Freitag, S. Kuttler, B. Lohwasser and M. Oleimeulen, *J. Near Infrared Spectrosc.* **6,** A7 (1998).
14. R.M. Silverston, G.C. Bassler and T.C. Morril, in *Spectrometric Identification of Organic Compounds*, 5th Edn. John Wiley & Sons, Inc., New York, USA (1991).
15. M. Chen and J. Irudayaraj, *J. Food Sci.* **63(1),** 96 (1998).

Wine quality grading by near infrared spectroscopy

R.G. Dambergs,[a] A. Kambouris,[b] N. Schumacher,[b] I.L. Francis,[a,c] M.B. Esler[a,c] and M. Gishen[a,c]

[a]*The Australian Wine Research Institute, PO Box 197, Glen Osmond, SA 5064, Australia*

[b]*BRL Hardy Limited, PO Box 238, Berri, SA 5343, Australia*

[c]*The Cooperative Research Centre for Viticulture, PO Box 154, Glen Osmond SA 5064, Australia*

Introduction

The ability to accurately assess wine quality is an important part of the wine making process, particularly when allocating batches of wines to styles determined by consumer requirements. Also, grape pricing is often determined by the quality category of the resulting wine—so called "end use" payment. Wine quality, in terms of sensory characteristics, is normally a subjective measure, performed by experienced winemakers, wine competition judges or winetasting panellists. By nature, such assessments can be biased by individual preferences and may be subject to day-to-day variation. An objective quality grading method would therefore be of great assistance.

Flavour compounds are often present in concentrations below the detection limit of near infrared spectroscopy but the more abundant organic compounds offer potential for objective quality grading by this technique.

Materials and methods

Samples were drawn from one of Australia's major wine shows and from BRL Hardy's post-vintage wine quality allocation tastings. Turbid samples were clarified by centrifugation. The samples were scanned in transmission mode with a Foss NIRSystems 6500 spectrometer, over the wavelength range 400–2500 nm, using a 1 mm pathlength.

Data analysis was performed with the Vision chemometrics package (Foss NIRSystems). For the quality allocation samples, the reference values used for calibration development were based on sensory assessments from a panel of winemakers, with a consensus grading into five categories (i.e. discrete values from 1 to 5, with category 1 being the lowest quality grade). For the wine show samples, the average score from a panel of judges was used as the reference value. The scores were based on a standard Australian wine show system with three points for appearance (colour intensity, hue, clarity), seven points for aroma and ten points for palate characteristics, resulting in a combined maximum score of 20 points.

Results and discussion

Wine quality allocation

With samples from wine quality allocation tastings, the best correlations between near infrared (NIR) spectra and tasting data were obtained with dry red wines. Figure 1 shows a correlation plot of the quality grading *vs* the NIR predicted values for a Cabernet Sauvignon dry red wine. The reference

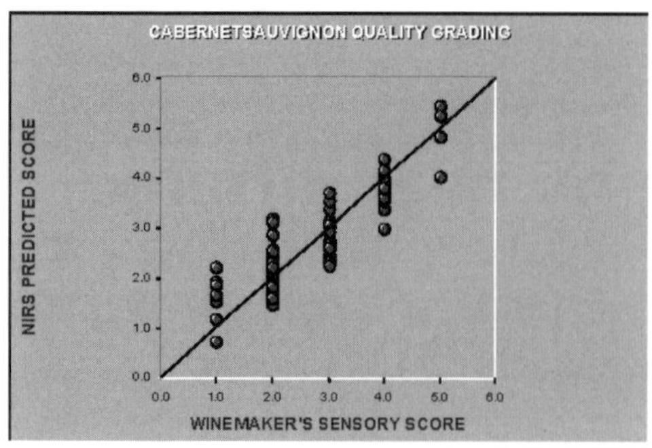

Figure 1. NIR calibration for Cabernet Sauvignon wine quality grading. N = 89. PLS 1st derivative, three factors, R^2 = 0.76, SEC = 0.49 and $SECV$ = 0.55

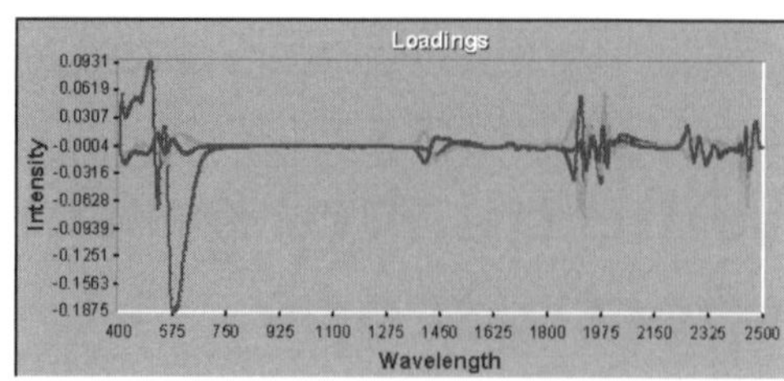

Figure 2. Factor loadings for Cabernet Sauvignon wine quality calibration.

quality scores were discrete values, whereas the NIR values were on a continuous scale: this would have reduced the R^2 value (0.76). Due to this discrete categorisation, the theoretical minimum error is a grade score of 0.5—the *SECV* of the NIR predicted values was 0.6 grade points.

The Cabernet Sauvignon dry red calibration used PLS factor loadings in the wavelengths related to anthocyanins, ethanol and possibly phenolic compounds (Figure 2).[1–3] Only a small number of factors were required and the strongest loadings were in the anthocyanin spectral region (400–700 nm). Anthocyanins are a group of compounds responsible for colour in red wines.[4] Restricting the wavelengths to a region dominated by anthocyanins produced calibrations of similar accuracy to those using the full available wavelength range. Note, however, that this region can also be used to predict pH, another important wine analytical parameter and it has been suggested that NIR prediction of pH may be related to pH-induced shifts in anthocyanin chromophores.[1] One consequence of the NIR predictions being based predominantly on anthocyanin levels may be that a wine could have high anthocyanin levels but may be downgraded because of a major sensory fault, caused by compounds that have a strong sensory effect at very low concentrations (for example, taints of microbial origin). In this situation the NIR predicted value would tend to be higher than the reference value. This appeared to be the case with grade 1 wines in particular (grade 1 being the lowest quality level).

For dry white wines, calibrations were less significant and appeared to be more dependent on the ethanol-related regions of the spectrum (1700–2400 nm), implying that the quality parameters that could be predicted by NIR correlated with fruit maturity.

Wine show grading

The correlations between NIR spectra and sensory data obtained using the wine show samples were less significant in general. The difficulty may have been due to excessive sample matrix variation. With most classes in the show, the samples may span vintages, growing areas and winemaking styles, even though they may be made from only one grape variety. For dry red wines, the best calibrations were obtained with a class of Pinot Noir—a variety that tends to be produced in limited areas in Australia and would represent the least matrix variation. Again, the loadings relied predominantly on anthocyanins.

Strong correlations were also obtained with a tawny port class (Figure 3). The sample set was small but the *SECV* represented a 6% error, relative to the mean score. These were sweet, fortified wines, that were aged for long periods in wooden barrels. During the ageing process, Maillard browning compounds are formed and the water is lost through the barrels in preference to ethanol, producing "concentrated", darkly-coloured wines with high alcohol content. The tawny port quality calibrations

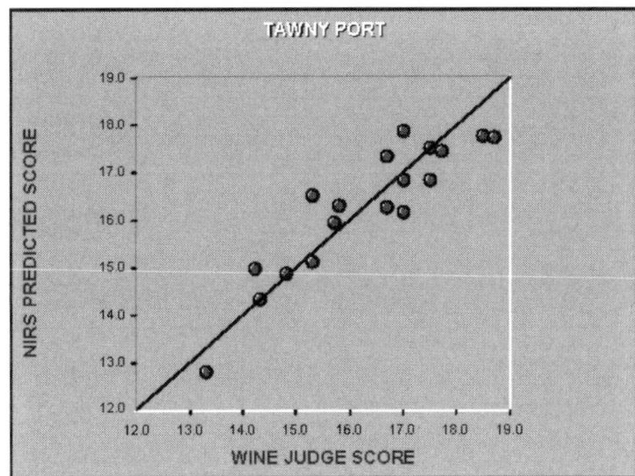

Figure 3. NIR calibration for Tawny port wine show score. $N = 20$. PLS 1st derivative, three factors, $R^2 = 0.84$, $SEC = 0.67$ and $SECV = 0.97$.

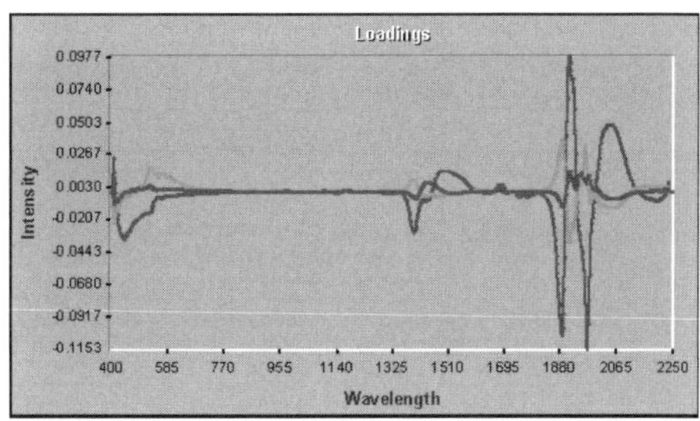

Figure 4. Factor loadings for Tawny port wine show score calibration.

indicated strongest loadings in the water regions of the spectrum, suggesting that "concentration" of the wines was important, whilst the visible and alcohol regions of the spectrum also featured as important factors (Figure 4). It has previously been demonstrated that NIR can predict alcohol and total dry extract in this type of wine[5]—the wine calibrations suggest that these are both important quality parameters.

NIR calibrations based on sensory scores will tend to be difficult to obtain due to variation between individual winetasters and may not pick up compounds at low concentrations, yet with strong sensory properties. Nevertheless, these results warrant further investigation and may provide valuable insight into the main parameters affecting wine quality.

Acknowledgements

The authors thank Peter Dawson and the winemaking section from BRL Hardy and Max Hawke from the Royal Agricultural and Horticultural Society of South Australia for providing samples. This project was funded by the Grape and Wine Research and Development Corporation (Project AWR98/1). Finally the authors thank Peter Høj from the Australian Wine Research Institute for his continuous encouragment and support in regard to NIR research.

References

1. R.G. Dambergs, A. Kambouris, M. Gishen and I.L. Francis, in *Modern Viticulture-meeting market expectations,* Proceedings of the Australian Society of Viticulture and Oenology, Winetitles, Adelaide, South Australia, pp. 45–47 (2000).
2. B.G. Osborne, T. Fearn and P.H. Hindle, in *Practical NIR spectroscopy, with applications in food and beverage analysis*; Longman Scientific and Technical, Harlow, UK, pp. 145–199 (1993).
3. H. Shulz, B. Steuer and H. Krüger, in *Near Infrared Spectroscopy: Proceedings of the 9th International Conference,* Ed by A.M.C. Davies and R. Giangiacomo. NIR Publications, Chichester, UK, pp. 447–453 (2000).
4. T.C. Somers and E. Vérette, in *Modern methods of Plant Analysis, Volume 6: Wine Analysis,* Ed by H.F. Linskens and J.F. Jackson. Springer-Verlag, Berlin, Germany, pp. 219–257 (1988).
5. R. Medrano, S.H. Yan, M. Madoux, V. Baeten and M. Meurens, in *Leaping Ahead with Near Infrared Spectroscopy,* Ed by G.D. Batten, P.C. Flinn, L.A. Welsh and A.B. Blakeney. Royal Australian Chemical Institute, Victoria, Australia, pp. 303–306 (1995).

Measurement of lipid content of compost in the fermentation process using near infrared spectroscopy

Ken-ichiro Suehara, Daisuke Masui, Yasuhisa Nakano and Takuo Yano

Department of Information Machines and Interfaces, Faculty of Information Sciences, Hiroshima City University, Ohzukahigashi 3-4-1, Asaminami-ku, Hiroshima 731-3194, Japan

Introduction

Compost fermentation is one of the key technologies for waste treatment and recycling of the waste matter and residual products from food processing. It is important to measure the compost composition and condition of the fermentation process. In particular, the amount of organic compounds, such as lipids, cellulose and protein, is an important factor in the management of the compost fermentation.

Tofu (soybean-curd) made from soybeans is a traditional food, which is normally eaten in East Asia. tofu refuse, a residual product from tofu processing, has a high moisture and nutrient content and putrefies very quickly with an unpleasant odour. In the past, tofu refuse has been utilised as a feedstuff. However, disposal of this refuse has recently become a serious problem because the amount utilised as feedstuff is decreasing. Because it does not contain toxic or harmful substances, it is very suitable as a compost material and this may be the most appropriate treatment for tofu refuse.

Near infrared (NIR) spectroscopy has already been used for the measurement of cell mass for mushroom cultivation in solid media,[1] measurement and control of moisture content[2] and measurement of carbon and nitrogen content of compost during fermentation.[3] In the present study, the suitability of lipid content as an indicator of compost fermentation of tofu refuse was studied and NIR was used to determine the lipid content of a compost sample during the fermentation process.

Materials and methods

Compost material, fermentation and composter

Fresh tofu refuse, used as a raw material for compost in this study, was supplied by a tofu factory. Compost fermentation was performed in a domestic composter (Model EH4381B-H, Matsushita Electronic Co., Tokyo). The temperature of the compost in the composter was kept between 65–70°C by on-off control of the heater and by mixing with a triple impeller fitted in the composter. To accelerate the fermentation process, compost made from tofu refuse in our laboratory was added as a seed.[2]

The starting mixture placed in the composter consisted of tofu refuse (71% moisture content, 9.65 kg of wet weight), sawdust (7% moisture content, 3.00 kg of wet weight) and the seed (39% moisture content, 200 g of wet weight). During fermentation, the moisture content of the compost was controlled by supplying additional water.[2]

Measurement of lipid, carbon and nitrogen contents

Lipid content of the compost samples was measured by the Soxhlet extraction method. The dried sample (about 10 g, W_1) was extracted with chloroform/methanol (2 : 1 v / v) solvent for 21 h at 75°C using a Soxhlet extractor. The extract in the flask was dried for 24 h at 80°C and the weight of the flask was measured (W_2). The lipid content of the compost sample was calculated using the weight of the flask and the weights W_1 and W_2. Each sample was measured by taking the average value of six measurements.

Carbon and nitrogen content were measured by the Pregie–Dumas combustion method using CN analyzer (MT-700, Yanaco Co., Kyoto, Japan).[3]

Near infrared spectroscopy

Six runs of compost fermentation were performed. During the fermentation process, compost was regularly sampled to produce 95 samples for measuring NIR spectra. These samples were divided at random to make a calibration sample set (n = 60) and a validation sample set (n = 35). The measurement of the NIR spectrum of the compost was carried out by the same procedure described in the previous study.[2] A polyethylene bag, filled with the compost ,was placed in a sample holder after being held at 25°C for 30 min. Reflectance values at wavelengths ranging from 400 to 2500 nm were measured at 2 nm intervals using an NIR spectrometer (NIRS6500SPL, Nireco Co., Hachioji, Tokyo, Japan). All samples were measured in triplicate. To correct for baseline shift in the spectra, second derivatives were calculated from the raw spectra.[2]

Least squares multiple linear regression (MLR) was carried out between the NIR spectral data of the calibration set (n = 60) and the lipid content of the compost obtained by Soxhlet extraction. Calculation of the second derivatives and the regression analysis were carried out using NSAS software supplied by the Nireco Co.

To evaluate the performance of the calibration equation, validation was carried out using a separate sample set (n = 35), which had not been used in the calibration. In addition, time courses of the values of the lipid content during the compost fermentation were predicted using the calibration equation.

NIR spectra of polyethylene bag, dried tofu refuse, residual and extract of Soxhlet extraction

To select a wavelength used for a calibration equation, NIR spectra of polyethylene bags, dried tofu refuse, as well as residual and extract of the Soxhlet extraction of tofu refuse were all measured. Fifteen polyethylene bags were placed in a sample holder and the reflectance measured. The tofu refuse and residual, after Soxhlet extraction, were dried for 12 h at 80°C before the dried samples were packed into a standard sample cup (Nireco Co.) and NIR reflectance spectra were collected. The extracts (in the form of a syrup after removal of the organic solvent) were injected into a cuvette with a 2 mm light path length and measurements taken as NIR transmittance spectra.

Results and discussion

Changes in the carbon, nitrogen and lipid content of compost during fermentation

Figure 1 shows time courses for (a) moisture content, (b) dry weight of the compost in the composter, (c) carbon and nitrogen contents and (d) lipid content of the compost during fermentation. In Figure 1(b), total dry weight of the compost in the composter decreased because compost fermentation is a biodegradation process in which the organic compounds in the compost material were changed to carbon dioxide and ammonia by microorganisms. The nitrogen (protein) in the compost material was decomposed at an early stage of compost fermentation [Figure 1(c)]. On the other hand, lipid content of the compost only begins to decrease in the middle stage of fermentation [Figure 1(d)]. The time course of the lipid content of the compost during fermentation was related to the decrease in total dry weight of the compost in the composter. These phenomena suggested that degradation of the

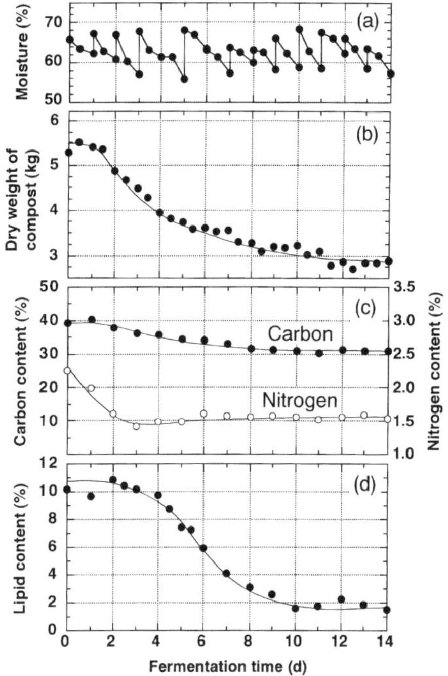

Figure 1. Time courses of (a) moisture content (b) dry weight of compost in the composter, (c) carbon and nitrogen contents and (d) lipid content of the compost during fermentation.

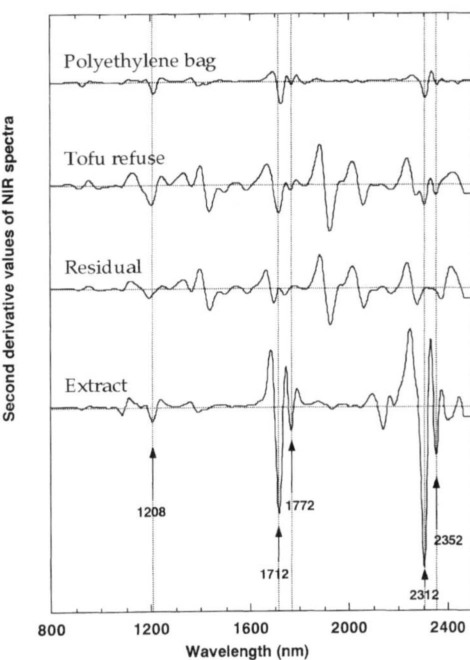

Figure 2. Second derivative NIR spectra of (a) polyethylene bags, (b) tofu refuse, (c) residual after Soxhlet extraction of tofu refuse, and (d) Soxhlet extract of tofu refuse.

lipids in the compost material was slower than that of other biodegradable organic compounds, and the change in lipid content in the compost could be one of the indicators to determine the end point of the thermophilic compost fermentation.

Calibration equation for measuring lipid content

To produce a suitable calibration equation, it is very important to select a wavelength having the absorption assigned to the target compound. Figure 2 shows the second derivative NIR spectra of (a) polyethylene bags, (b) tofu refuse, (c) the residual after Soxhlet extraction of tofu refuse and (d) the Soxhlet extract from tofu refuse. The second derivative NIR spectrum of the Soxhlet extract from tofu refuse had five negative peaks at 1208, 1712, 1772, 2312 and 2352 nm representing absorption of lipids. However, the second derivative NIR spectrum of the polyethylene bag also had the negative peaks at 1214, 1726, 1772, 2312 and 2352 nm.

The wavelength used to formulate a calibration equation of lipid content should, therefore, be selected from 1208, 1712, 1772, 2312 and 2352 nm. MLR was carried out between the second derivative NIR spectral data and the lipid contents of the compost (C_{act}) obtained by the reference method. A calibration was made to predict the carbon content (C_{pre}). The results are shown in Table 1. Where a calibration equation was formulated using only one wavelength, an equation using 1712 nm was better than that any other wavelength. The reason for this was that, while the three absorptions at 1772, 2312

Table 1. Calibration and validation results for prediction of lipid content of compost.

Wavelength (nm)			Calibration (n = 60)		Validation (n = 35)	
λ_1	1_2	λ_3	R (–)	SEC (%)	r (–)	SEP (%)
1208	—	—	0.726	2.17	—	—
1712	—	—	0.971	0.754	0.955	0.902
1772	—	—	0.176	3.10	—	—
2312	—	—	0.543	2.650	—	—
2352	—	—	0.542	2.650	—	—
1208	1712	—	0.975	0.698	0.964	0.815
2312	1712	—	0.974	0.720	0.955	0.902
2352	1712	—	0.975	0.710	0.956	0.897
1208	2312	—	0.888	1.46	—	—
1208	2352	—	0.889	1.45	—	—
1208	2312	2352	0.889	1.45	—	—

and 2352 nm depend mainly on the lipids, the C–H bonds from the polyethylene bags also has absorption bands at these wavelengths.[4,5] The absorption at 1208 nm may relate to the vibration of the C–H stretching second overtone of the lipid.[4] As a result, the wavelengths at 1208 and 1712 nm were selected as the first and second terms in a calibration equation for lipid content. The values obtained for R and SEC were 0.975 and 0.698%, respectively.

$$C_{pre} = -1.11 - 129A_{1208} - 497A_{1712} \tag{1}$$

Validation

The lipid content in the independent validation sample set (n = 35) was predicted using the calibration equation (Equation 1) and compared with the values of C_{act}. A good agreement between the values obtained by the conventional method and those obtained by NIR was observed and the values of r and standard error of prediction (SEP) were 0.964 and 0.815%, respectively (Table 1).

Monitoring of the lipid content during compost fermentation was performed using this equation (Equation 1). In Figure 3, a good agreement between the values obtained by the conventional method and those obtained by NIR was observed during compost fermentation. These results suggest that the calibration equation would be highly suitable for this purpose.

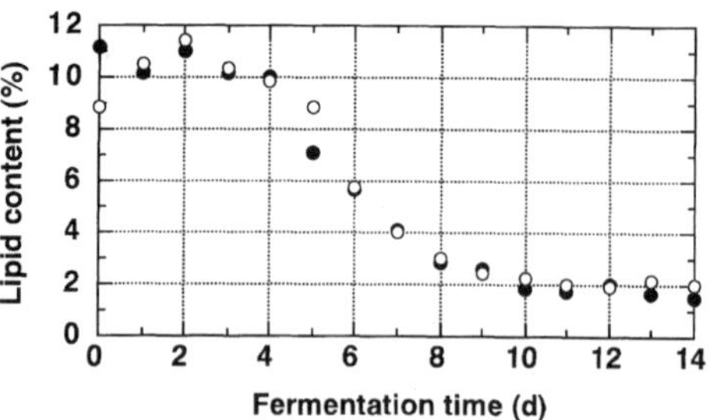

Figure 3. Monitoring of the lipid content during compost fermentation using NIR. Symbols: ●, measurement of the lipid content by the Soxhlet extraction method; ○, predicted value of the lipid content by NIR.

Conclusion

The lipid content of the compost was an important indicator of compost fermentation and could help to detect the end-point of the fermentation process. Furthermore, the validity of the NIR method was shown by applying it to predict the time course of lipid content during compost fermentation. The operational procedure involved in NIR is very simple and nondestructive and the time required for the measurement is but a few minutes. To develop high performance in the composting process, simultaneous measurement of the moisture,[2] carbon,[3] nitrogen[3] and lipid contents of the compost would be a powerful tool.

Acknowledgments

The authors are very grateful to the Eishoku Foods Co. for generously providing fresh tofu refuse. This work was supported by a Hiroshima City University Grant for Special Academic Research (Encouragement for Researchers in 1999 and 2000).

References

1. K. Suehara Y. Nakano and T. Yano, *J. Near Infrared Spectrosc.* **6,** 273 (1998).
2. K. Suehara, Y. Ohta, Y. Nakano and T. Yano, *J. Biosci. Bioeng.* **87,** 769 (1999).
3. K. Suehara Y. Nakano and T. Yano, *J. Near Infrared Spectrosc.* **9,** 35 (2001).
4. B.G. Osborne, T. Fearn and P.F. Hindle, *Practical NIR spectroscopy with applications in food and beverage analysis*. Longman Scientific and Technical, Harlow, UK (1986).
5. J.A. Panford, O.C. Williams and J.M. deMan, *J. Am. Oil Chem. Soc.* **65,** 1627 (1988).

Prediction of physico–chemical and texture characteristics of beef by near infrared transmittance spectroscopy

Mamen Oliván, Begoña de la Roza, Mercedes Mocha and María Jesús Martínez

Servicio Regional de Investigación y Desarrollo Agroalimentario (S.E.R.I.D.A.), Apdo 13, 33.300 Villaviciosa, Asturias, Spain

Introduction

Physico–chemical and texture characteristics of meat determine nutritional, technological and sensory quality. However, the accurate analysis of meat quality requires expensive, laborious and time-consuming analytical methods.

Near infrared (NIR) spectroscopy has become a rapid and effective analytical tool for estimating chemical composition of foods, including meat and meat products.[1–5] It has also been studied for its ability to predict other meat quality attributes such as water holding capacity[6,7] and instrumental and sensory texture.[8–10]

There are some reports on the use of near infrared (NIR) spectroscopy transmittance for determining the composition of ground meat,[11] the composition and shear value of raw beef cuts[8] and the sensory quality of meat sausages.[12] The objective of this work was to evaluate the use of NIR using transmittance for determining physico–chemical characteristics of ground beef, including moisture, fat, protein, total pigment content, water holding capacity and instrumental texture.

Materials and methods

A total of 318 ground beef samples were scanned using an Infratec 1265 Feed Analyzer. The samples were obtained from the *Longissimus thoracis* muscle of the 10th rib of yearling bulls from two local breeds in Northern Spain, "Asturiana de los Valles" and "Asturiana de la Montaña". The meat samples were ground with an electrical chopper, vacuum packed, aged for seven days and frozen at –24°C until the analyses were done.

Moisture content was measured by oven drying at 102°C (ISO 1442-1973), fat content was determined by Soxhlet extraction (ISO 1443-1973) and protein content was estimated from nitrogen content using the Kjeldahl analysis (ISO 937-1978). The total pigment content was determined by using the Hornsey method,[13] showing the results as the absorbance at 512 nm. The water-holding capacity (WHC) was analysed using the method of Grau and Hamm,[14] modified by Sierra.[15]

The instrumental evaluation of texture was conducted on an Instron 1011 with a Warner–Bratzler shearing device. This analysis was performed on a thick chop obtained from the *longissimus* muscle of the 8th rib, aged for seven days and cooked in a hot bath at 75°C for 40 minutes. The results were expressed as maximum load (WB), maximum stress (MS) and toughness.

The NIR transmittance analysis was performed on an Infratec 1265 Feed Analyzer which operated from 850 to 1050 nm at 2 nm intervals. The samples were placed into a glass cup 130 mm in diameter.

Table 1. Physico–chemical and texture characteristics of the sample population.

Component	*n*	mean	range	*SD*
Moisture (%)	114	74.49	70.68 – 76.44	1.42
Fat (%)*	58	2.66	0.56 – 6.65	1.72
Protein (%)*	126	22.43	20.78 – 24.06	0.61
Pigments (absorbance)	318	0.504	0.21 – 0.87	0.13
WHC	318	19.68	6.41 – 28.68	4.43
WB	250	5.01	2.72 – 9.62	1.33
MS	250	5.10	2.67 – 10.59	1.37
Toughness	250	2.49	1.33 – 4.97	0.70

*in fresh matter

The average spectra of each sample was obtained from 15 scan locations and recorded as log 1/*T* (*T* = transmittance).

Calibrations were performed with the WinISI software v. 1.02 (Infrasoft International, Port Matilda, PA, USA) using the modified partial least squares method.[16] To examine the effect of scatter correction or derivation of spectra on the calibration performance, calibrations were calculated with the crude spectra or pretreated with different mathematical treatments like inverse multiplicative scatter correction (inverse MSC), standard normal variate detrending (SNVD) and/or second derivative operation.

Results

Means, standard deviations (*SD*) and ranges for physico–chemical and texture characteristics of samples are presented in Table 1. The crude spectra of meat obtained by transmittance in the region from 850 to 1050 nm showed a wide peak located from 930 to 1000 nm, covering bands mainly related with water (934 to 960 nm and 984 to 996 nm) and fat (962 to 968 nm) (Figure 1).

The results of the calibration statistics are shown in Table 2 (chemical attributes) and Table 3 (texture attributes). Generally, for chemical composition the use of the inverse MSC or SNVD pre-treatments provided better calibration statistics for most of the variables. This is in accordance with the

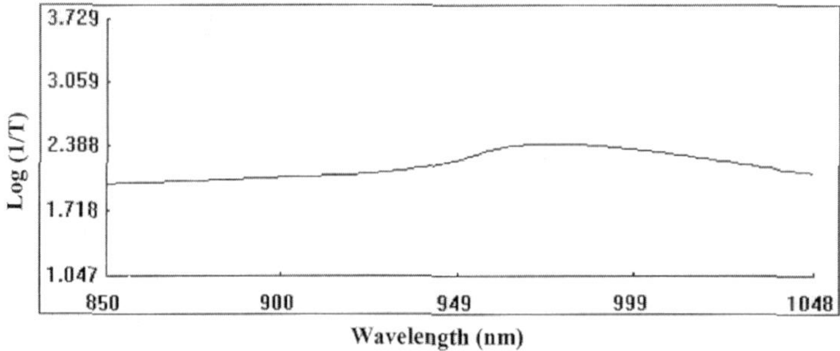

Figure 1. Average crude spectrum of the sample population.

Table 2. Statistics of calibration for chemical composition.

Component	Pretreatment	*SEC*	R^2	*SECV*	r^2	*RER*
Moisture	None	0.409	0.915	0.443	0.901	13.00
	None - D	0.321	0.922	0.364	0.899	15.82
	InvMSC	0.316	0.952	0.340	0.945	16.94
	InvMSC - D	0.317	0.948	0.349	0.938	16.50
	SNVD	0.322	0.951	0.344	0.943	16.74
	SNVD - D	0.293	0.952	0.317	0.944	18.17
Fat	None	0.602	0.880	0.676	0.853	9.00
	None - D	0.352	0.949	0.462	0.913	13.18
	InvMSC	0.424	0.939	0.528	0.909	11.53
	InvMSC - D	0.332	0.961	0.449	0.929	13.56
	SNVD	0.591	0.887	0.682	0.852	8.92
	SNVD - D	0.344	0.958	0.443	0.931	13.75
Protein	None	0.501	0.265	0.505	0.255	6.50
	None - D	0.447	0.378	0.476	0.299	6.89
	InvMSC	0.497	0.269	0.506	0.259	6.48
	InvMSC - D	0.455	0.396	0.481	0.331	6.82
	SNVD	0.445	0.432	0.475	0.353	6.91
	SNVD - D	0.460	0.385	0.483	0.320	6.79
Pigments	None	0.043	0.848	0.045	0.835	14.67
	None - D	0.048	0.829	0.050	0.818	13.20
	InvMSC	0.032	0.918	0.034	0.913	19.41
	InvMSC - D	0.033	0.915	0.034	0.910	19.41
	SNVD	0.032	0.925	0.033	0.922	20.00
	SNVD - D	0.032	0.921	0.034	0.914	19.41

D: 2nd derivative

reports of Ding *et al.*[17] and Ding and Xu,[18] who showed that the SNVD correction improved the classification accuracy of minced meat of beef and kangaroo. The use of the 2nd derivative also improved the predictions, mainly when combined with the SNVD treatment. It is known that the use of derivative

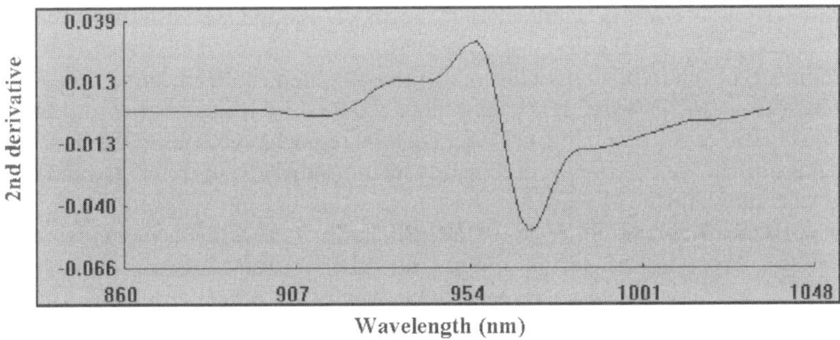

Figure 2. Average spectrum of the sample population after application of scatter correction and 2nd derivative.

Table 3. Statistics of calibration for texture traits.

Component	Pretreatment	*SEC*	R^2	*SECV*	r^2	*RER*
WHC	None	3.124	0.354	3.164	0.340	7.04
	None - D	2.806	0.578	2.935	0.539	7.59
	InvMSC	2.743	0.356	2.861	0.301	7.78
	InvMSC - D	2.797	0.435	2.895	0.395	7.69
	SNVD	2.800	0.360	2.854	0.335	7.80
	SNVD - D	2.801	0.490	2.887	0.463	7.71
WB	None	0.844	0.428	0.885	0.372	7.80
	None - D	0.893	0.362	0.939	0.295	7.35
	InvMSC	1.002	0.306	1.040	0.253	6.63
	InvMSC - D	0.857	0.364	0.898	0.300	7.68
	SNVD	0.928	0.294	0.969	0.233	7.12
	SNVD - D	0.863	0.371	0.904	0.311	7.63
MS	None	0.873	0.405	0.900	0.369	8.80
	None - D	0.889	0.360	0.923	0.309	8.58
	InvMSC	1.098	0.249	1.108	0.233	7.15
	InvMSC - D	0.915	0.298	0.957	0.230	8.28
	SNVD	1.007	0.212	1.021	0.192	7.76
	SNVD - D	0.843	0.367	0.887	0.300	8.93
Toughness	None	0.560	0.265	0.580	0.213	6.28
	None - D	0.493	0.362	0.514	0.308	7.08
	InvMSC	0.560	0.265	0.580	0.213	6.28
	InvMSC - D	0.487	0.306	0.518	0.215	7.03
	SNVD	0.549	0.169	0.557	0.145	6.54
	SNVD - D	0.455	0.386	0.487	0.303	7.47

D: 2nd derivative

treatments not only reduces scattering effects but also increases the resolution of spectrum peaks.[19] In this work, the application of scatter correction and second derivative increased the resolution of peaks and revealed distinct peaks related with moisture (934 to 960 nm and 984 to 996 nm), fat (962 to 968 nm) and pigments (960 nm),[8] as shown in Figure 2. Thus, the use of the SNVD treatment and the second derivative provided the best prediction results for moisture ($r^2 = 0.944$, $SECV = 0.317$) and fat ($r^2 = 0.931$, $SECV = 0.443$) and the application of the SNVD treatment gave the best prediction of total pigments ($r^2 = 0.922$, $SECV = 0.033$).

The coefficients of determination for the total pigment content, both for calibration (R^2=0.925) and cross-validation ($r^2 = 0.922$) were higher than those reported by Mitsumoto *et al.*[8] who obtained a determination coefficient of $R^2 = 0.895$ between the total pigment content and the optical density determined in transmittance mode from beef cuts. The relationship between reference and predicted data for pigment content is shown in Figure 3.

The low coefficients of determination for the estimation of the protein content in meat ($r^2 < 0.35$) must be noticed. Most of the earlier reports on NIR or transmittance prediction for meat composition[2,3,5,8,20] gave lower predictions for protein than for any other constituent. This has been related to analytical errors in the Kjeldahl analysis or a low correlation between the Kjeldahl method which measures nitrogen and the NIR method which measures protein.[2] In this work, the lower prediction of protein could be related to the lack of spectral peaks in the overtone bands for protein (at 910

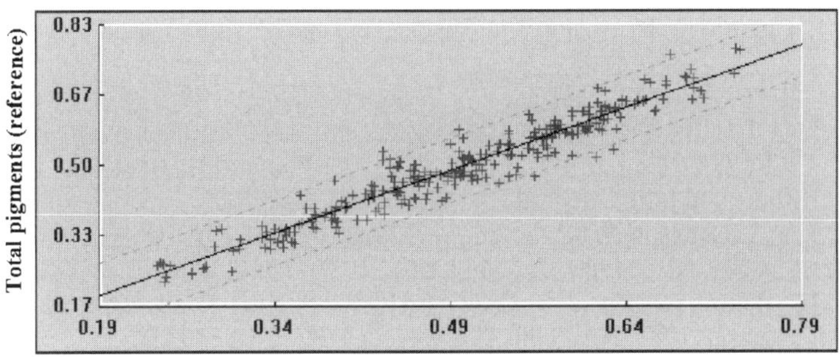

Figure 3. Relationship between reference and predicted data for pigment content.

and 1020 nm)[21] and also due to the relatively large error in the reference method compared with the sample standard deviation for protein, as indicated by Togersen *et al.*[5]

For texture traits, the best estimation was obtained from the crude spectrum (WB and MS) and the 2nd derivative of the crude spectrum (toughness and WHC). The best prediction of the WHC gave better statistics ($r^2 = 0.539$) than the calibrations reported by Brondum *et al.*[7] when determining the WHC of porcine meat estimated with the filter paper method in the NIR range from 802 to 2500 nm ($r = 0.62$).

The prediction of the Instron measurements from the transmittance spectra were low, as expected, when comparing data obtained from different sample matrix. The analyses of the instrumental texture were performed on meat cubes removed from cooked steaks while the transmittance spectra were obtained from fresh ground meat. Probably the best estimation of the instrumental texture of meat should be calculated on the spectrum obtained from meat cuts, as shown by Park *et al.*[9]

Conclusions

The near infrared transmittance from 850 to 1050 nm provided excellent predictions of moisture, fat and total pigment content in ground beef, with r^2 being higher that 0.9. However, the prediction of protein was low, due to the lack of peaks in the bands of protein and the prediction of texture traits (WB, MS, toughness) was poor, due to the low relationship between the texture characteristics of cooked steaks and the transmittance spectra of fresh ground beef.

References

1. W.G. Kruggel, R.A. Field, M.L. Riley, H.D. Radloff and K.M. Horton, *Journal of the Association of Official Analytical Chemists* **64,** 692 (1981).
2. E. Lanza , *J. Food Sci.* **48,** 471 (1983).
3. T. Isaksson, B.N. Nilsen, G. Togersen, R.P. Hammond and K.I. Hildrum, *Meat Sci.* **43,** 245 (1996).
4. R. Sanderson, S.J. Lister, M.S. Dhanoa, R.J. Barnes and C. Thomas, *Anim. Sci.* **65,** 45 (1997).
5. G. Togersen, T. Isaksson, B.N. Nilsen, E.A. Bakker and K.I. Hildrum, *Meat Sci.* **51,** 97 (1999).
6. J.R. Andersen, C. Borggaard and T. Nielsen, *Proceedings ECCAMST meeting.* Roskilde, Denmark (1995).
7. J. Brondum, L. Munck, P. Henckel, A. Karlsson, E. Tornberg and S. Engelsen, *Meat Sci.* **55,** 177 (2000).
8. M. Mitsumoto, S. Maeda, T. Mitsuhashi and S. Ozawa, *J. Food Sci.* **56,** 1493 (1991).

9. B. Park, Y.R. Chen, W.R. Hruschka, S.D. Shackelford and M. Koohmaraie, *J. Anim. Sci.* **76,** 2115 (1998).
10. C.E. Byrne, G. Downey, D.J. Troy and D.J. Buckley, *Meat Sci.* **49,** 399 (1998).
11. W.R. Windham, F.E. Barton, II and K.C. Lawrence, in *"Leaping ahead with near infrared spectroscopy"*, Ed by G.D. Batten, P.C. Flinn, L.A. Welsh and A.B. Blakeney. Royal Australian Chemical Institute, Victoria, Australia, p. 287 (1995).
12. M.R. Ellekjaer, T. Isaksson and R. Solheim, *J. Food Sci.* **59,** 456 (1994).
13. H.C. Hornsey, *J. Sci. Food Agric.* **7,** 534 (1956).
14. R. Grau and R. Hamm, in *"Muscle as Food"*, Ed by P.J. Bechtel. Food Science and Technology. A series of Monograph. 1985. Academic Press. New York. USA (1953).
15. I. Sierra, Instituto de Economía y Producciones Ganaderas del Ebro (IEPGE) **16,** 48 (1973).
16. J.S. Shenk and M.O. Westerhaus, *Analysis of Agricultural and Food Products by Near Infrared Reflectance Spectroscopy.* Monograph. Infrasoft International, Port Matilda, USA.(1993).
17. H.B. Ding, R.J. Xu and D.K. Chan, *J. Sci. Food Agric.* **79,** 1382 (1999).
18. H.B. Ding and R.J. Xu, *J. Food Sci.* **64,** 814 (1999).
19. A.M.C. Davies and A. Grant, *Int. J. Food Sci. Technol.* **22,** 191 (1987).
20. K.I. Hildrum, M.R. Ellekjaer and T. Isaksson, *Meat Focus International* **4,** 156 (1995).
21. B.G. Osborne and T. Fearn, *Near Infrared Spectroscopy in Food Analysis.* Longman Scientific & Technical, Harlow, Essex, UK (1986).

The application of Fourier transform near infrared spectroscopy in the wine industry of South Africa

Anina van Zyl,[a] Marena Manley[a] and Erhard E.H. Wolf[b]

[a]Department of Food Science, University of Stellenbosch, Private Bag X1, Matieland 7602, Stellenbosch, South Africa

[b]Distell, Aan-De-Wagen Road, PO Box 184, Stellenbosch 7599, South Africa

Introduction

During wine production optimum yeast growth healthy alcohol fermentation rates are monitored by the amount of free amino nitrogen (FAN) present in the must.[1] The status of the malolactic fermentation (MLF) in Chardonnay wines must be monitored by determining the degree of conversion of malic to lactic acid.[2] Ethyl carbamate (EC), a suspected carcinogen, is mainly formed during the aging of wine under certain conditions and is restricted by legislation in some countries.[3] It is, therefore, necessary to determine the EC content in wine. Currently these measurements are monitored using expensive, quantitative, time-consuming analytical methods. It would, however, be adequate to use a discriminate screening method as the samples need only to be classified as belonging to a certain class or having reached a specified cut-off point or not. Fourier transform near infrared FT-NIR) spectroscopy can be used as a rapid method to discriminate between different must or wine samples. If the samples are spectroscopically dissimilar, spectral differences can be used.[4] If the spectra are similar, sophisticated techniques such as soft independent modelling by class analogy (SIMCA) can be used. SIMCA takes into consideration both the variability of the spectra of interest and the differences between the spectra.[4]

Objective

The aim of this study was to apply SIMCA on FT-NIR spectra of must and wine to discriminate between the samples in terms of their FAN values, the status of the malolactic fermentation and the level of EC present.

Table 1. Respective classes into which the data were divided for each consituent.

Constituent	Class 1	Class 2	Class 3
FAN (mg N L^{-1})	1 – 800(n = 26)	800 – 2000(n = 71)	
MLF	Not started(n = 18)	Underway(n = 30)	Complete(n = 38)
EC (ppb)	0 – 10(n = 47)	10 – 15(n = 16)	> 15(n = 7)

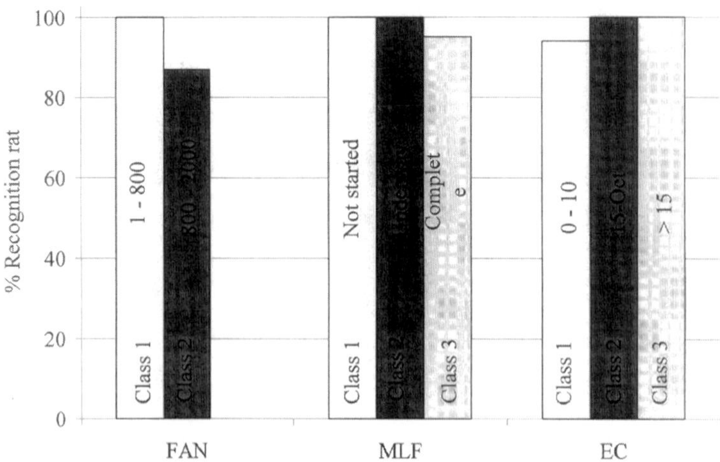

Figure 1. Graphic representation of SIMCA results.

Materials and methods

The FAN content of the must samples was determined spectrophotometrically by means of an auto-analyser.[5] Determinations of the malic and lactic acid content of the wine samples were done by means of high-pressure liquid chromatography (HPLC),[6] while the EC content of the wine samples was determined by means of gas chromagography with mass selective detection (GS/MS) according to the OIV method.[7] Wine and must samples with known FAN (n = 123), malic acid (n = 109), lactic acid (n = 103) and EC (n = 156) values from the 1999 and 2000 harvest seasons were used. Optical absorption spectra for each sample were recorded using the FT-NIR Perkin-Elmer Spectrum IdentiCheck spectrophotometer over the range 10,000 to 4000 cm^{-1} at 16 cm^{-1} resolution in a 0.5 mm path length quartz cuvette. Selected spectra for each constituent were divided into appropriate classes (Table 1) and models were build by performing principal component analysis (PCA) on 2nd derivative spectra using QUANT+ software. SIMCA validation tests were performed on the models to determine the accuracy of the respective models.

Results and discussion

A summary of the results obtained with SIMCA are graphically illustrated in Figure 1. Efficient discrimination between samples has been achieved using SIMCA and FT-NIR can be used successfully as a powerful tool in screening spectroscopically similar samples.

The two models that were created for classes 1 and 2 for the FAN values (Table 1) showed good classification possibilities. The recognition rates were above 87% for both the data sets (Class 1 = 100%, Class 2 = 87%), indicating good separation of each class (Figure 1). Accurate classifications were possible with the three models that were created for the malolactic fermentation (Table 1). Recognition rates of above 95% were reported, indicating good classification of each class (Figure 1). Good classification possibilities were shown with the three models that were created for EC values (Table 2). The recognition rate columns reported 94% for Class 1 and 100% for Classes 2 and 3, respectively, indicating that excellent separation of each class had been achieved.

Conclusion

It would, therefore, be possible to replace expensive, time consuming quantitative analytical methods such as FAN, malic and lactic acid and EC determinations, if not completely, at least to some extent, with FT-NIR spectroscopy.

Acknowledgements

Marais Blom and laboratory personel, Distillers Corporation and the Department of Wine biotechnology, University of Stellenbosch for analyses performed.

References

1. L.F. Bisson, Influence of nitrogen on yeast and fermentation of grapes, in *Proceedings of the International Symposium on Nitrogen in Grapes and Wine.* Seattle, Washington, USA, pp. 78–89 (1991).
2. R.E. Kunkee, *Nature Biotechnology* **15,** 224 (1997).
3. C.E. Butzke and L.F. Bisson, *Ethyl carbamate prevention action manual: 1997.* US Food and Drug Administration. Center for Food Safety and Applied Nutrition (1997).
4. PerkinElmer, *FT-NIR Spectroscopy application notes.* The Perkin Elmer Corporation, Beaconsfield, UK (1998).
5. P.J.A. Vos, E. Crous and L. Swart, Fermentation and the optical nitrogen balance of musts. Wynboer, South Africa, March, pp. 58–63 (1980).
6. A. Schneider, V. Gerbi and M. Redoglia, *Am. J. Enol. Vitic.* **38,** 151 (1987).
7. B.J. Canas, F.L. Joe, G.W. Diachenko and G. Burns, *J. AOAC Int.* **77,** 1530 (1994).

Estimation of total sugar and reducing sugar in molasses using near infrared diffused reflectance spectroscopy

Ranjana Mehrotra, Alka Gupta, Jagdish Tewari and S.P. Varma
Infrared Application Group, National Physical Laboratory, Dr K.S. Krishnan Road, New Delhi 110 012, India

Introduction

Final cane molasses usually contains 30–40% sugar by weight, which is around 10% of the total sugar (non-reducing sugar and reducing sugar) content in cane. The control of this loss is a challenging problem in sugar industries. Also, the ratio of reducing sugars (RS) (glucose and fructose) to non- reducing sugar (sucrose), RS/sucrose, is an important factor which influences the exhaustibility of molasses. Hence, estimation of total sugar and reducing sugar content in molasses is a very important task in sugar refineries. The polarimetric method is commonly employed in sugar industries to determine sugar content in molasses. Although the technique is adequate for sugarcane juice samples under ideal conditions, it has serious limitations in the case of molasses with high a concentration of reducing sugar. Chemical analysis for the estimation of total sugar and reducing sugar in cane molasses is the standard conventional method. However, the complexity of analysis, time required for analysing each sample and the use of hazardous chemicals have made this technique inefficient for process control in any sugar industry. Therefore, a more accurate method is required to separately quantify total sugar and reducing sugar content. Hence, the development of a rapid, inexpensive, physical and also accurate method for sugar determination in molasses will be highly useful.

The near infrared (NIR) spectroscopic technique has emerged over the past several years as an alternative analytical tool for sugar refineries.[1–7] Several types of sugarcane materials with different types of physical or chemical properties have been studied using NIR reflectance as well as transmittance spectroscopy. The work by Berding *et al.*[2] and Clarke *et al.*[8] had pointed out the use of NIR as a realistic method for analysis of sugarcane juice, bagasse and molasses. However, very few attempts to determine the chemical composition of molasses by NIR have been made. In a case study on cane molasses R.A. Pax[9] applied multiple linear regression for quantification of sucrose in molasses using double polarisation data as reference. Large errors of prediction have, however, been reported. It will be of immense value to further explore the potential of near infrared diffused reflectance spectroscopy as a rapid and automated analytical technique for the determination of total sugar and reducing sugar content in molasses.

Materials and methods

The molasses samples were collected during and after the sugarcane season from sugar industry. Care was taken to obtain samples with a wide range of concentration of total sugar and reducing sugar. Each sample was analysed by the standard polarisation method using a digital automatic saccharimeter after clarifying the molasses solution with lead acetate to get the pol value. The samples were also analysed using conventional chemical analysis to find out total sugar and reducing sugar present in the sample. The pol values and chemical values of total sugar and reducing sugar obtained were used as reference to develop the calibration model. An indigenous ELICO NIR system specially designed for the sugar industry was used for taking spectral measurements. The samples were scanned in the region of 1100–2500 nm in diffused reflectance mode. A ceramic block was used as reference.

Chemical analysis

All the samples were analysed chemically for total sugar and reducing sugar content using the standard titration against Fehling's solution. The Fehling's solutions A and B were purchased from the market and used as it was. (1 mL of Fehling's solution equivalent to 0.005 g of invert sugar). Neutral lead acetate solution was prepared by dissolving 100 g lead acetate in 1 L of water. Sodium phosphate–potassium oxalate solution was prepared by dissolving 70 g of di-sodium hydrogen phosphate dodecahydrate and 30 g of potassium oxalate in 1 L water.

Sample preparation

12.5 g of molasses sample was weighed and transferred to a 250 mL volumetric flask with the help of water. 25 mL of lead accetate solution was added to it, volume was made up and the contents were shaken and filtered. 10 mL of sodium phosphate–potassium oxalate mixture was added to 100 mL of this filtrate in a 500 mL volumetric flask. Once again after making up the volume, the contents were shaken and filtered. The solution so obtained was used for determination of total sugar and reducing sugar content. All the samples were titarted against 10 mL of Fehling's solution and the total and reducing sugar in all the samples were calculated as follows.
1 mL of Fehling's solution = 0.005 g invert sugar based on the equation

$$C_{12}H_{22}\,O_{11} + H_2O = 2\;C_6H_{12}\,O_6 \tag{1}$$

342 g of sucrose gives 360 g of invert sugar or 0.00475 g of sucrose gives 0.005 g of invert sugar = 1mL of Fehling's solution. The Fehling factor (invert sugar equivalent to 10 mL of Fehling's solution) = .05

Total sugars (TS) in the molasses samples were calculated according to the formula given as

$$TS\% = \frac{\text{Fehling factor} \times 250 \times 500 \times 100 \times 100}{12.5 \times 100 \times 50 \times \text{Titer volume (mL)}} \tag{2}$$

Reducing sugars (RS) in molasses samples were calculated according to the formula given as:

$$RS\% = \frac{\text{Fehling factor } \times 250 \times 500 \times 100}{12.5 \times 100 \times \text{Titer volume (mL)}} \tag{3}$$

Results and discussion

A commercial program provided with the NIR spectrophotometer, based on GRAMS 386 for multivariate spectral data analysis was used to process the data and to develop the model for pol, total sugar and reducing sugar content in molasses samples. Data were analysed in three different sets. In first set, pol values of the samples were taken as the reference to develop the model. In second the set, reference values were obtained by chemical analysis of the samples. The calibration model was obtained for two components; total sugar and reducing sugar. The wavelength used for calibration was the full spectrum 1100–2500 nm. In the third set, the reference values were again obtained using

Table 1. NIR calibration statistics.

Reference method	Range (nm)	No of samples	Factors	*R*	*RMSD*	No of outliers
Pol	1100–2500	40	5	0.8316	5.0195	
Chemical Analysis	1100–2500	40	6 (TS)	0.9395(TS)	2.6455(TS)	4
	—	—	6(RS)	0.8063(RS)	0.6118(RS)	4
Chemical Analysis	2100–2500	40	6(TS)	0.9556(TS)	1.4833(TS)	2
	—	—	6(RS)	0.9288(RS)	0.3142(RS)	2

R is the correlation coefficient: *RSMD* is the root mean square difference

chemical analysis for total sugar and reducing sugar but the wavelength used for calibration was 2100–2500 nm.

Prior to calibration, spectral data were mean centered. Partial least square (PLS) regression was employed for the calibration. The PLS used both the spectral response and respective reference data for the examined samples to determine PLS factors on the data set. The optimum number of PLS factors used in a model was determined by a cross-validation procedure. One sample was temporarily removed from the calibration set to be used for validation. With the rest of the samples, a PLS model was developed and applied to predict total sugar and reducing sugar content in molasses samples. The results were compared to their respective reference values. This procedure was repeated until the prediction for all the samples were obtained. The optimum number of PLS factors in each set was defined as the one that corresponds to the lowest standard error of cross-validation. The values of the correlation coefficient (*R*) and root mean square difference (*RMSD*) were calculated for each set.

The calibration statistics for all the three cases are summarised in Table 1. Using the pol values as reference, a low correlation coefficient value of 0.8316 was obtained and even the standard error index, root mean square difference was quite high, 5.0195. Poor correlation in calibration based on pol values was not unexpected because polarimetry is not a correct technique for measuring sugar content in molasses which have a substantial concentration of reducing sugars. Hence calibration developed using polarisation values was unsatisfactory and could not be considered suitable for evaluation of sugars in molasses. In the second set, an optimal correlation was obtained with the correlation coefficient value of 0.9395 for total sugar and 0.8063 for reducing sugar when full spectrum was used in developing the calibration model. Although the calibration is acceptable, the error index is a little higher, 2.6455 for total sugar and 0.6118 for reducing sugar. A robust calibration could be developed by selecting the region 2100–2500 nm. The choice was based on the fact that all the relevant combination bands of sucrose, glucose and fructose are located in this region. The total sugar and reducing sugar values obtained by chemical analysis were again used as the reference. High values of correlation coefficients, 0.9556 for TS and 0.9288 for RS with relatively low *RSMD* values, 1.4833 and 0.3142, respectively, could be achieved.

The developed calibration models using the chemical method as a reference were used for prediction of total and reducing sugars in unknown samples of molasses. A prediction set consisting of eight samples was created and values of total sugar and reducing sugar were determined. Table 2 gives the concentration of total and reducing sugars in unknown molasses samples predicted using the full spectral range (1100–2500 nm) calibration model. Similarly, the sugar content in these molasses samples was also determined using range 2100–2500 nm calibration curve. The values are indicated in Table 3. As can be seen, the values obtained in the latter case are more close to the actual values.

Table 2. Value of total sugar and reducing sugar in molasses samples in prediction set (1100–2500).

Chemical	NIR	Bias	Chemical	NIR	Bias
	TS(%)			RS(%)	
35.32	36.17	–0.85	9.90	8.92	0.98
33.69	32.66	1.03	9.94	9.72	0.22
39.91	40.41	–0.50	8.19	8.35	–0.16
41.13	41.98	0.77	8.35	7.73	0.62
41.85	41.48	0.37	8.14	8.40	–0.26
43.98	42.81	1.17	8.16	8.59	–0.43
36.26	32.57	3.69	9.96	10.20	–0.24
47.18	46.02	1.16	8.26	8.61	–0.35

Table 3. Value of total sugar and reducing sugar in molasses samples in prediction set (2100–2500).

Chemical	NIR	Bias	Chemical	NIR	Bias
	TS(%)			RS(%)	
35.32	33.93	1.39	9.90	9.58	0.32
35.85	36.17	–0.32	8.17	8.00	0.17
39.91	39.33	0.58	8.19	8.42	–0.23
41.13	40.71	0.42	8.35	8.52	–0.17
41.85	42.15	–0.30	8.14	8.03	0.11
46.67	45.73	0.94	8.22	8.02	0.20
47.18	47.54	–0.36	8.26	8.54	–0.28
55.96	56.66	–0.70	8.18	8.32	–0.14

Conclusion

The present study shows that near infrared diffused reflectance spectroscopy has considerable potential in rapid estimation of the total sugar and reducing sugar in molasses samples. The approach can be developed into a completely automated on-line technique in a sugar refinery where no sample preparation is required.

References

1. A. Salgo, J. Nagy, E. Miko and I. Bros, *J. Near Infrared Spectrosc.* **6,** 101 (1998).
2. N. Berding, G.A. Brotherton and J.C. Skinner, *Crop Sci.* **31,** 1024 (1991).
3. M.A. Clarke and L.A. Edye, *CITS Tech. Comm. Abs.* (1993).
4. M.A. Clarke, L.A. Edye and W.S. Patout, III, *Proc. Conf. Sugar Process. Res.* 288 (1995).

5. J.H. Meyer, S. Rutherford and K.J. Schaffler, in *Leaping Ahead with NIR Spectroscopy*, Ed by G.D. Batten, P.C. Flinn, L.A. Welsh and A.B. Blakeney. Royal Australian Chemical Institute, Victoria, Australia, p. 204 (1994).
6. G. Vaccari, G. Mantovani and G. Sgualdino, *Proc. Conf. Sugar Process. Res.* **40,** (1990).
7. N. Berding and G.A. Brotherton, *Leaping Ahead with NIR Spectroscopy*, Ed by G.D. Batten, P.C. Flinn, L.A. Welsh and A.B. Blakeney. Royal Australian Chemical Institute, Victoria, Australia, p. 199 (1994).
8. M.A. Clarke, L.A. Edye, C.V. Scott, X.M. Miranda and C. McDonald-Lewis, *Proc. Conf. Sugar Process. Res.* New Orleans, USA, p. 244 (1992).
9. R.A. Pax, *Proc. 15th Conf. Australian Soc. Sugarcane Tech,* p. 202 (1993).

Quick determination of meat colour, metmyoglobin formation and lipid oxidation in beef, pork and chicken by visible and near infrared spectroscopy

Mitsuru Mitsumoto, Keisuke Sasaki, Hitoshi Murakami and Kyouhei Ozutsumi

National Institute of Livestock and Grassland Science, Tsukuba Norindanchi, PO Box 5, Ibaraki-ken 305-0901, Japan

Introduction

Meat becomes brown and rancid during refrigerated storage and display. Colour changes, metmyoglobin formation and lipid oxidation are the most important problems in the transportation/distribution and retail display of meat. The freshness of meat is determined by the senses of vision and smell. Since conventional methods for determining lipid oxidation are time-consuming and destructive (it requires homogenisation of meat with reagents, filtration, time for reaction and reading optical density using spectroscopy), more rapid and nondestructive technical tools are desired. The objective of this work was to evaluate visible and near infrared spectroscopy as an analytical tool for determining meat colour, metmyoglobin formation and lipid oxidation in beef, pork and chicken.

Materials and methods

Semitendinosus and *longissimus thoracis* muscles from six Japanese Black steers, biceps *femoris* and *longissimus* thoracis muscles from twelve LWD crossbred pigs and superficial pectoral muscles from 24 broilers were used. The muscles were sliced into 1 cm-thick steaks. Five cm diameter pieces were cut from these with a template cutter and the weight adjusted to between 19.90 g and 20.10 g. Samples were randomly allotted to the measurement days. Each sample was placed in a disposable 100 ml weigh boat, overwrapped with oxygen-permeable PVC film and displayed under fluorescent lights at 4–5°C for ten days (beef and pork) or four days (chicken). The spectra were measured by an NIRSystems Model 5500 spectrophotometer, using a fibre-optic probe and a scan in the 400–1100 nm range (visible and near infrared region). Data were recorded at 2 nm intervals and ten scans $(10\ sec)^{-1}$ were averaged for every sample. Data obtained were saved as log 1 Re^{-1}, where *Re* is the reflectance energy, and then mathematically transformed to second derivatives to reduce effects of differences in particle size. CIE (Commission Internationale de l'Eclairage) L^*, a^* and b^* values were determined by UV-vis spectrophotometer (Shimadzu UV-2400PC; Shimadzu Corp., Kyoto, Japan) using the integrating sphere unit (Shimadzu ISR-2200). Surface metmyoglobin formations were determined at the same time by the method of Stewart *et al.*[1] 2-Thiobarbituric acid reactive substances (TBARS) were measured to determine lipid oxidation by the method of Witte *et al.*[2] as modified by Mitsumoto *et al.*[3] TBARS values were expressed as mg malonaldehyde (MDA) kg^{-1} meat. A multiple linear regression was used to find the equation which would best fit the data. First, four wavelengths were selected in the calibration model for each trait by means of chemical assignments of absorption bands and stable wavelength region using the near infrared spectral analysis software (NSAS, version 3.18). Secondly, full cross-validation was carried out for prediction using the Unscrambler software package (version 7.01, CAMO ASA, Oslo, Norway). Finally, the best wavelength set was selected using the lowest number of wavelengths, highest multiple correlation of calibration and lowest standard error of prediction by cross-validation as criteria.

Results and discussion

Selected wavelengths and statistical summary of calibration and prediction by full cross-validadion for meat colour, metmyoglobin formation and lipid oxidation in beef, pork and chicken by fibre-optic scans are presented in Tables 1, 2 and 3, respectively.

Beef

The multiple correlation coefficients (R) for a^* and metmyoglobin formation were high ($R = 0.963$ and 0.989, respectively), slightly lower for L^* and TBARS ($R = 0.885$ and 0.891, respectively) and low for b^* ($R = 0.618$). The first selected wavelengths for a^*, metmyoglobin formation and TBARS were in the same region. This could be explained by the fact that the first selected wavelength (624 nm) for metmyoglobin formation was near the maximum absorption band of metmyoglobin (630 nm) and that these values were highly correlated ($P < 0.001$) with each other.

Pork

The multiple correlation coefficients for L^*, a^* and metmyoglobin formation were high ($R = 0.948$ to 0.965) a little lower for TBARS ($R = 0.840$) and lower for b^* ($R = 0.595$).

Chicken

The multiple correlation coefficient for metmyoglobin formation was high ($R = 0.972$) but a little lower for L^*, a^* and b^* ($R = 0.869$ to 0.896) and lower for TBARS ($R = 0.669$). The first selected wavelength for metmyoglobin formation was similar to that in pork.

Table 1. Selected wavelengths and statistical summary of calibration and prediction by full cross-validation for meat colour, metmyoglobin formation (%) and lipid oxidation (TBARS, mg MDA kg^{-1} meat) in beef by fibre-optic scans (n=48).

			Selected wavelengths				Calibration		Prediction by *CV*		
Item		Range	λ_1	λ_2	λ_3	λ_4	*R*	*SE*	*R*	*SE*	Bias
Meat colour	L^*	39.4–55.1	850 nm	606 nm	802 nm	624 nm	0.885	1.65	0.858	1.83	–0.011
	a^*	5.2–22.6	620 nm	744 nm	590 nm	—	0.963	1.14	0.955	1.26	0.004
	b^*	10.4–17.5	554 nm	—	—	—	0.618	1.41	0.578	1.47	–0.012
Metmyoglobin		5.6–89.9	624 nm	1020 nm	—	—	0.989	3.06	0.987	3.29	0.029
TBARS		0.09–2.52	620 nm	978 nm	722 nm	—	0.891	0.325	0.868	0.356	0.003

λ_1, λ_2, λ_3 and λ_4 refer to selected wavelengths for the linear calibration
R: multiple correlation coefficient
SE: standard error
CV: full cross-validation

Table 2. Selected wavelengths and statistical summary of calibration and prediction by full cross-validation for meat colour, metmyoglobin formation (%) and lipid oxidation (TBARS, mg MDA kg^{-1} meat) in pork by fibre-optic scans (n=96).

			Selected wavelengths				Calibration		Prediction by *CV*		
Item		Range	λ_1	λ_2	λ_3	λ_4	*R*	*SE*	*R*	*SE*	Bias
Meat colour	*L**	43.8–62.2	730 nm	892 nm	630 nm	666 nm	0.951	1.35	0.945	1.43	0.009
	*a**	2.0–12.2	588 nm	688 nm	556 nm	—	0.948	0.75	0.943	0.79	0.005
	*b**	8.8–13.6	942 nm	480 nm	—	—	0.595	0.82	0.559	0.85	0.002
Metmyoglobin		4.5–67.9	616 nm	694 nm	744 nm	—	0.965	4.50	0.959	4.84	0.071
TBARS		0.02–0.92	604 nm	934 nm	480 nm	748 nm	0.840	0.106	0.808	0.116	0.000

λ_1, λ_2, λ_3 and λ_4 refer to selected wavelengths for the linear calibration
R: multiple correlation coefficient
SE: standard error
CV: full cross-validation

Table 3. Selected wavelengths and statistical summary of calibration and prediction by full cross-validation for meat colour, metmyoglobin formation (%) and lipid oxidation (TBARS, mg MDA kg^{-1} meat) in chicken by fibre-optic scans (n = 24).

			Selected wavelengths				Calibration		Prediction by *CV*		
Item		Range	λ_1	λ_2	λ_3	λ_4	*R*	*SE*	*R*	*SE*	Bias
Meat colour	*L**	50.2–58.2	572 nm	938 nm	—	—	0.896	0.95	0.867	1.07	- 0.007
	*a**	1.4–5.0	576 nm	462 nm	1018 nm	—	0.869	0.43	0.818	0.50	0.021
	*b**	9.3–15.9	468 nm	712 nm	—	—	0.885	0.70	0.850	0.80	0.018
Metmyoglobin		23.2–64.8	618 nm	998 nm	832 nm	—	0.972	2.44	0.964	2.77	0.098
TBARS		0.28–1.81	626 nm	552 nm	832 nm	—	0.669	0.330	0.521	0.390	0.018

λ_1, λ_2, λ_3 and λ_4 refer to selected wavelengths for the linear calibration
R: multiple correlation coefficient
SE: standard error
CV: full cross-validation

In general, a^* decreases more than b^* during display of meat while L^* does not change. Thus, a^* is a good indicator of the shelf life of meat during display such as metmyoglobin formation and TBARS value. Greene *et al.*[4] reported that consumers would reject beef containing more than 30% to 40% metmyoglobin. Greene and Cumuze[5] reported that a TBARS range of 0.6 to 2.0 mg MDA kg^{-1} was required for inexperienced taste panelists to detect oxidised flavours in meat; Tarladgis *et al.*[6] reported a range of 0.5 to 1.0 for experienced taste panelists. Gray and Pearson[7] reported that consumers were unlikely to detect off-flavors at values below a threshold of about 0.5 mg MDA kg^{-1}. The standard of redness, a^*, was estimated from relationships between a^* and metmyoglobin ($a^* = -0.1924 \times$ metmyoglobin % $+ 21.896$; $r = -0.94$); a^* and TBARS ($a^* = -4.2276 \times$ TBARS $+ 18.9486$; $r = -0.71$). When 30% of metmyoglobin was substituted for the former formula,

a^* was set to 16.1 and a^* was set to 16.4 when 0.6 mg MDA kg^{-1} of TBARS was substituted for the latter formula. Therefore, beef with an a^* value of less than 16 would be rejected by consumers. The degrees of pigment and lipid oxidation in beef can be judged using an a^*, value of 16, 30% of metmyoglobin and 0.6 mg MDA kg^{-1} of TBARS in the above standards. Hence, visible and near infrared spectroscopy can evaluate whether this meat is good for sale or not.

In conclusion, we found that visible and near infrared spectroscopy was a useful tool for quickly determining meat freshness with respect to colour changes (especially, a^* value), metmyoglobin formation and lipid oxidation.

Acknowledgements

We thank Dr Sumio Kawano, National Food Research Institute, Japan, for his advice and help.

References

1. M.R. Stewart, M.W. Zipser and B.W. Watts, *J. Food Sci.* **30,** 464 (1965).
2. V.C. Witte, G.F. Krause and M.E. Bailey, *J. Food Sci.* **35,** 582 (1970).
3. M. Mitsumoto, R.N. Arnold, D.M. Schaefer and R.G. Cassens, *J. Anim. Sci.* **71,** 1812 (1993).
4. B.E. Greene, I. Hsin and M.W. Zipser, *J. Food. Sci.* **36,** 940 (1971).
5. B.E. Greene and T.H. Cumuze, *J. Food Sci.* **47,** 52 (1981).
6. B.G. Tarladgis, B.M. Watts, M.T. Younathan and L. Dugan, *J. Am. Oil Chem. Soc.* **37,** 44 (1960).
7. J.I. Gray and A.M. Pearson, in *Advances in Meat Research, Vol. 3., Restructured meat and poultry products*, Ed by A.M. Pearson and T.R. Dutson. Van Nostrand Reinhold Co., New York, USA, p. 221 (1987).

Prediction of energy content in cereal food products by near infrared reflectance spectroscopy

Sandra E. Kays and Franklin E. Barton, II

United States Department of Agriculture, Agricultural Research Service, Quality Assessment Research Unit, Richard B. Russell Agricultural Research Center, PO Box 5677, Athens, Georgia, 30604-5677, USA

Introduction

Energy content is an important part of the evaluation and marketing of foods in many countries. United States dietary guidelines urge consumers to aim for a healthy body weight by choosing an assortment of foods that includes vegetables, fruits, grains, skimmed milk and fish, lean meat, poultry or beans.[1] The benefits of managing weight and avoiding obesity include a reduced risk for high blood pressure, heart disease, stroke, diabetes and certain forms of cancer.[1] Thus, knowledge of energy content of foods and food portions is one of many important criteria in the selection of foods by consumers.

Methods of measuring the energy value of foods are outlined in the US Code of Federal Regulations.[2] One method is measurement of gross calories by bomb calorimetry with an adjustment for unutilised protein; another is by calculation, using specific factors for the energy values of protein, carbohydrate less the amount of insoluble dietary fibre and total fat. Near infrared (NIR) spectroscopy has been used for the rapid and accurate prediction of nutrients in human foods[3] but very few studies have addressed the use of NIR for the prediction of energy.[4] Previous work has described the use of NIR for the prediction of energy content in feeds for ruminants and monogastric animals.[5,6] The current study explores the potential of NIR for the rapid and accurate determination of gross energy in cereal food products. Subsequent work will address the determination of physiologically available energy.

Materials and methods

Samples and sample preparation

Cereal food products included breakfast cereals, granolas, crackers, brans, flours, unprocessed grains and commercial oat and wheat fibres. The sample set had a wide range in grain types, multi-grain products, fibre, added fat and sugar and additives such as dried fruit, honey, herbs, nuts, cinnamon and cocoa. Samples were ground to < 500 µm in a cyclone mill (Cyclotec 1093, Perstorp Analytical, Silverspring, MD, USA) except for high-fat samples (> 10% fat) which were ground in a coffee mill (Model KSM-2, Braun Inc., Lynnfield, MA, USA).

Spectroscopic analysis

Near infrared reflectance spectra (400–2500 nm) of ground cereal samples were obtained in duplicate with an NIRystems 6500 monochromator (NIRystems, Silver Spring, MD, USA) using cylindrical sample cells (38 × 9 mm). Each sample was scanned 16 times, the data averaged and transformed to log 1/*R*. The duplicate scans of each sample were averaged.

Reference data

For the initial study, energy listed on the nutrition facts label of the products was used as the reference data. Total calories were converted to kcal g^{-1} using label information for serving size. Oxygen bomb calorimetry[7] was used to measure gross energy content of products at the University of Georgia, Poultry Nutrition Laboratory, Athens, GA, USA. Dry matter content of products was determined using a forced air oven at 105°C and gross energy expressed as kcal g^{-1} on a dry weight basis.

Multivariate calibrations

Multivariate analysis was performed using ISI software (NIR3 v. 4.01, ISI International, Port Matilda, PA, USA). For the initial study, with reference data based on nutrition label information, a selection algorithm (SELECT, NIR 3 v. 4.01) was used to select representative samples from the population of 116 spectra. Using a neighbourhood *H* value of 0.6 to define neighbourhoods and principal component analysis, 43 samples were selected and used to develop a modified PLS model for energy prediction. Log 1/*R* spectra were processed using normal multiplicative scatter correction and second derivative processing (gap = 16 nm, smoothing interval = 16 nm) prior to modified PLS. In addition, a modified PLS model was developed for the prediction of gross energy using calorimetry to determine gross energy reference values. One spectral outlier (Mahalanobis distance > 20) was discarded. Log 1/*R* spectra of 128 cereal food products were transformed with normal multiplicative scatter correction and second derivative processing (gap = 16 nm, smoothing interval = 16 nm), prior to modified partial least squares analysis. The number of modified PLS factors used for each model was determined by cross validation. The model for prediction of gross energy was tested using independent validation samples ($n = 58$). Validation samples consisted of additional cereal food products not included in the calibration data set and purchased and scanned at a different time.

Results

Reference analysis

The range in energy content, calculated from product nutrition label values was 1.70–5.00 kcal g^{-1}. After employing the selection algorithm to select representative samples the range was 2.0–4.8 kcal g^{-1} ($n = 43$). The range in gross energy of samples measured by calorimetry was 4.05–5.49 kcal g^{-1} with a method standard error of 0.035 kcal g^{-1}.

Initial model based on product nutrition label values

The initial model used 43 selected, representative samples with reference data for energy calculated from the product nutrition label information (Figure 1). The modified partial least squares model developed used five PLS factors and had a standard error of cross-validation (*SECV*) and multiple coefficient of determination (R^2) of 0.26 kcal g^{-1} and 0.84, respectively. These

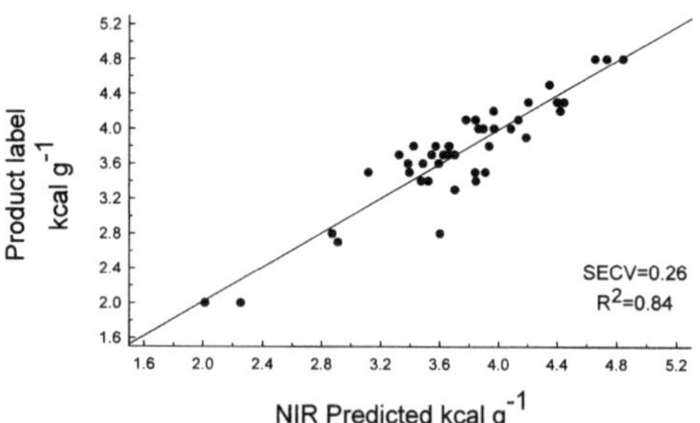

Figure 1. NIR predicted values for energy vs values for energy calculated from product nutrition label information for cereal food products in the calibration data set (*n* = 43 samples selected from a pool of 116 using a selection algorithm).

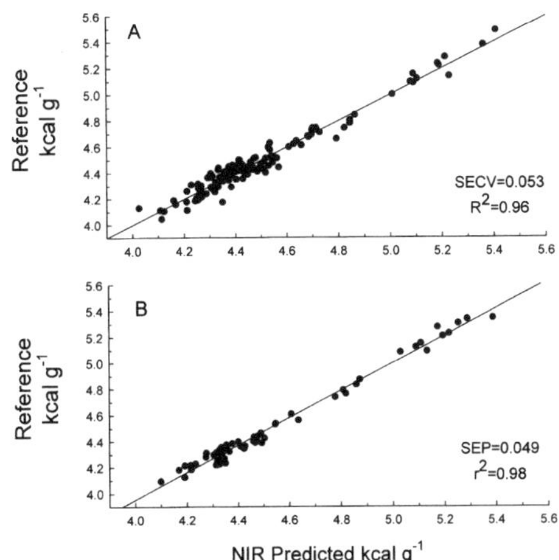

Figure 2. NIR predicted values for gross energy vs gross energy determined by bomb calorimetry for cereal products in the calibration (panel A, *n* = 127) and validation data sets (panel B, *n* = 58).

results indicated promise for NIR reflectance spectroscopy for prediction of energy in diverse cereal food products, therefore, the study was continued measuring the gross energy content using bomb calorimetry.

Calibration for gross energy

A modified PLS model was developed for the prediction of gross energy (Figure 2). The *SECV* was 0.053 kcal g^{-1} and R^2 was 0.96. Seven modified PLS factors were used in the model and described 97.1% of the spectral variation. Sample scores having the highest correlation with gross energy were for factor one and had a Pearson correlation coefficient of 0.92. The PLS loading for factor one had significant absorption peaks correlated to C–H stretch groups in lipids at 1212, 1722, 1764, 2304 and 2346 nm and O–H groups in carbohydrates at 1434 and 2076 nm. Thus, the model appeared to be predominantly influenced by lipids and carbohydrates. The model was used to predict the gross energy of independent validation samples ($n = 58$) with a standard error of performance of 0.049 kcal g^{-1}, coefficient of determination of 0.98, bias of – 0.020 kcal g^{-1} and slope of 1.05.

Discussion

The results indicate that gross energy of a diverse group of cereal products can be predicted accurately and rapidly with NIR reflectance spectroscopy using ground products. The model derived from calorimetry data is far more accurate than that derived from nutrition label values. This is not surprising as label values are not precise. The US Code of Federal Regulations[2] states that the number of calories per serving can be "expressed to the nearest five calorie increment up to and including 50 calories and ten calorie increment above 50 calories".

The gross energy of cereal products is a useful figure in diet selection, however, the actual energy physiologically available is generally less, primarily due to incomplete oxidation of proteins in the human body and indigestibility of fibre. Thus, adjustments in gross energy values for incomplete utilisation of protein and indigestibility of fibre will more accurately reflect food energy available to the body. These values for available energy may be calculated from the predicted values for gross energy, if the protein and insoluble dietary fibre composition of samples is known. Direct and accurate prediction of available energy by NIR spectroscopy will be the subject of subsequent work.

Acknowledgement

Calorimetry work was conducted at the University of Georgia, Poultry Nutrition Laboratory.

References

1. *Dietary Guidelines for Americans,* United States Department of Agriculture, United States Department of Health and Human Services, Home and Garden Bulletin No. 232 (5th Edn) (2000).

2. *Code of Federal Regulations*, FDA, HHS; 21, part 101.9 (2000).
3. B.G. Osborne, T. Fearn and P.H. Hindle, *Practical NIR Spectroscopy with Applications in Food and Beverage Analysis*; Longman Scientific and Technical, Harlow, UK (1993).
4. E. Lanza, *J. Food Sci.* **48,** 471 (1983).
5. D.L. Givens, J.L. De Boever and E.R. Deaville, *Nutrition Research Reviews* **10,** 83 (1997).
6. S. Leeson, E.V. Valdes and C.E.M. de Lange, in *Feed Evaluation: Principles and Practice.* Wageningen, Pers, Stichting, The Netherlands, p. 93 (2000).
7. *Parr Manual No. 120, Oxygen Bomb Calorimetry and Oxygen Bomb Combustion Methods*. Parr Instrument Company, 211 Fifty Third Street, Moline, Illinois, USA (1948).

Use of near infrared spectroscopy to predict oil content components and fatty acid composition in intact olive fruit

L. León-Moreno,[a] A. Garrido-Varo[b] and L. Rallo-Romero[a]

[a]*Plant Production Department, School of Agriculture and Forestry Engineering, University of Cordoba, Avda Menéndez Pidal s/n, PO Box 3048, E-14080 Córdoba, Spain.*
E-mail: ag2lemol@uco.es

[b]*Animal Production Department, School of Agriculture and Forestry Engineering, University of Cordoba, Avda Menéndez Pidal s/n, PO Box 3048, E-14080 Córdoba, Spain*

Introduction

Since 1991 the University of Córdoba has been conducting a breeding programme to obtain new olive cultivars from intraspecific crosses. The objective of this programme is to obtain new early bearing and high-quality cultivars.[1] In the framework of this breeding programme hundreds of samples should be tested every year to increase the chance of getting desirable genotypes. Therefore, fast, inexpensive and accurate methods of analysis are necessary. Conventional laboratory techniques are expensive and time-consuming. Near infrared (NIR) spectroscopy can satisfy the characteristics requested by plant breeders as it offers many advantages such as the simultaneous analysis of many traits and low cost. NIR analysis has enabled plant breeders to quickly select superior genotypes in breeding programmes and this has been one of the most important applications of NIR in agriculture.[2]

The objective of this work was to assess the performance of NIR to estimate oil fruit components (fruit weight, flesh moisture, flesh/stone ratio and oil flesh content in dry weight basis) and fatty acid composition in intact olive fruit.

Material and methods

Genotypes from reciprocal crosses between 'Arbequina', 'Frantoio' and 'Picual' cultivars have been used in this study. A total of 287 samples, each from a single plant, were scanned using a DA-7000 Diode Array vis/NIR Spectrophotometer (Perten Instruments DA 7000 Flexi-Mode), which covers the visible and NIR range from 400–1700 nm.

Intact frozen olive fruit samples were scanned in the "Down-View" mode. The sample is placed in a circular dish and is illuminated from above by light shining directly on the surface of the sample.[3]

All samples were analysed for fatty acid composition by means of gas chromatography (GC) analyses of fatty acid methyl esters. They were prepared following the procedure developed by Garces and Mancha[4] and analysed on a Hewlett Packard gas chromatograph equipped with a flame-ionisation detector. 220 of the 287 samples were analysed for oil fruit components. The oil content was measured by means of nuclear magnetic resonance (NMR).

Calibration development and validation was carried out by Nircal Version 3.0 Software.[5] Partial least squares (PLS) was used to obtain regression equations for all the reference data. The wavelength region used was 900–1500 nm.

Nircal does not use cross-validation so the sample file was randomly split into two files, one containing 70% of the samples (calibration set) and the other containing the remaining 30% of the samples (validation set).

Table 1. Mean, standard deviation and range of variability of oil fruit components (150 samples) and fatty acid composition (201 samples) in the calibration set.

	Mean	Minimum	Maximum	*SD*
Oil fruit components				
Fruit weight (g)	3.8	1.6	7.0	0.9
Flesh/stone ratio	8.0	4.6	12.0	1.5
Flesh moisture (%)	75.8	64.2	86.2	4.2
Oil flesh content (%)	62.1	40.3	74.3	5.7
Fatty acid composition (%)				
Palmitic (C16 : 0)	15.0	9.1	21.5	2.4
Palmitoleic (C16 : 1)	2.8	0.7	7.9	1.2
Stearic (C18 : 0)	1.8	1.1	7.9	0.7
Oleic (C18 : 1)	65.8	43.3	84.7	9.0
Linoleic (C18 : 2)	11.5	1.9	29.7	6.6

Table 2. Mean, standard deviation and range of variability of oil fruit components (74 samples) and fatty acid composition (86 samples) in the validation set.

	Mean	Minimum	Maximum	*SD*
Oil fruit components				
Fruit weight (g)	3.7	2.1	6.4	1.0
Flesh/stone ratio	7.8	4.7	11.7	1.3
Flesh moisture (%)	75.2	63.4	83.9	3.9
Oil flesh content (%)	62.8	51.6	73.2	5.1
Fatty acid composition (%)				
Palmitic (C16 : 0)	15.1	9.0	21.1	2.7
Palmitoleic (C16 : 1)	2.7	0.8	5.7	1.1
Stearic (C18 : 0)	1.8	1.0	4.9	0.6
Oleic (C18 : 1)	65.6	46.0	84.5	9.3
Linoleic (C18 : 2)	12.0	1.7	26.3	6.4

Different data pretreatments (derivative and MSC) were used to improve calibration results. The equation models obtained for each reference data were evaluated following the statistics and rules provided by Nircal, as the standard error of calibration (*SEC*), prediction (*SEP*), coefficient of determination for calibration (R^2) and prediction (r^2).

Results and discussion

Tables 1 and 2 show the oil content components and fatty acid composition of the calibration and validation sets. The breeding programme produces genotypes with a high variability of the characteristics evaluated. Therefore, these samples are very suitable for producing robust NIR analysis. The largest variability was found in flesh moisture, oil flesh content and oleic and linoleic acid content. The available range of variability for the other traits was more limited.

The preliminary results show that calibration for oleic and linoleic acids were highly accurate with coefficients of determination of 0.91 and 0.95 for calibration and of 0.88 and 0.91 for validation (Table 3). Similar results were obtained analysing the fatty acid composition of edible vegetables oils,[6] intact oil seeds such as rapeseed,[7] mustard[8] and other Brassicaceae species.[9] Calibration for palmitoleic and

Table 3. Calibration and prediction statistics in NIRS equations for fatty acid composition.

Fatty acid	Calibration		Prediction	
	R^2	*SEC*	r^2	*SEP*
Palmitic (C16 : 0)	0.85	1.28	0.76	1.75
Palmitoleic (C16 : 1)	0.72	0.83	0.44	1.04
Stearic (C18 : 0)	0.41	0.68	0.14	0.68
Oleic (C18 : 1)	0.94	3.04	0.88	4.40
Linoleic (C18 : 2)	0.95	1.96	0.90	2.73

R^2 = Coefficient of determination for calibration
r^2 = Coefficient of determination for prediction
SEC = Standard error of calibration
SEP = Standard error of prediction

Table 4.- Calibration and prediction statistics in NIRS equations for oil fruit components.

Oil Fruit Components	Calibration		Prediction	
	R^2	*SEC*	r^2	*SEP*
Fruit weight (G)	0.73	0.62	0.72	0.71
Flesh/stone ratio	0.80	0.91	0.68	0.96
Flesh moisture (%)	0.97	0.96	0.94	1.33
Oil flesh content (%)	0.96	1.67	0.91	2.20

R^2 = Coefficient of determination for calibration
r^2 = Coefficient of determination for prediction
SEC = Standard error of calibration
SEP = Standard error of prediction

estearic acids were less accurate, probably because of the narrow range of variability available for these fatty acids.

For the oil fruit components (Table 4), calibration was highly accurate for flesh moisture and oil flesh content in dry weight basis (R^2 and r^2 higher than 0.90) and less accurate for the other characteristics evaluated. These results agree with the ones obtained analysing the oil content in Brassicaceae germplasm,[9] sunflower[10] and the oil and moisture contents in soybean seeds.[11]

Conclusion

The first results obtained indicate that NIR analysis could be an ideal technique to reduce the cost, time and chemical waste necessary to evaluate a large number of olive genotypes. NIR is accurate enough to preselect genotypes for oil content and oleic acid content, two of the most important objectives in our olive tree breeding programme.

References

1. L. Rallo, *Olivae* **59,** 46 (1995).
2. G.D. Batten, *J. Near Infrared Spectrosc.* **6,** 105 (1998).
3. Perten Instruments, Operation manual DA-7000 Flexi-Mode NIR/VIS Spectrophotometer. Perten Instruments (1999).
4. R. Garcés and M. Mancha, *Anal. Biochem.* **211,** 139 (1993).
5. Bühler, Manual for Nircal Version 3.0. Installation, Operation and Tutorial. Bühler AG, Anatec, Switzerland (1998).
6. Y.S. Chen and A.O. Chen, in *Leaping ahead with Near Infrared Spectroscopy*, Ed by G.D. Batten, P.C. Flinn, L.A. Welsh and A.B. Blakeney. Royal Australian Chemical Institute, Victoria, Australia, pp. 316–323 (1995).
7. T.C. Reinhardt, C. Paul and G. Röbbelen, in *Near Infrared Spectroscopy: The Future Waves*, Ed by A.M.C. Davies and P. Williams. NIR Publications, Chichester, UK, pp. 323–327 (1996).
8. L. Velasco, J.M. Fernández-Martínez and A. De Haro, *J. Am. Oil Chem. Soc.* **74,** 1595 (1997).
9. L. Velasco, F.D. Goffman and H.C. Becker, *J. Am. Oil Chem. Soc.* **76,** 25 (1999).
10. B. Pérez-Vich, L. Velasco and J.M. Fernández-Martínez, *J. Am. Oil Chem. Soc.* **75,** 547 (1998).
11. M. Takahashi, M. Hajika, K. Igita and T. Sato, in *Near Infrared Spectroscopy: The Future Waves*, Ed by A.M.C. Davies and P. Williams. NIR Publications, Chichester, UK, pp. 494–497 (1996).

Chemical and microbiological analysis of goat's milk, cheese and whey by near infrared spectroscopy

M.D. Pérez-Marín,[a] A. Garrido-Varo,[a] J.M. Serradilla,[a] N. Núñez,[a] J.L. Ares[b] and J. Sánchez[c]

[a]*Escuela Técnica Superior de Ingenieros Agrónomos y Montes, University of Córdoba, Avda. Menéndez Pidal s/n, E-14080 Córdoba, Spain*

[b]*CIFA, Hinojosa del Duque (Córdoba), Department of Agriculture, Junta de Andalucía, Spain*

[c]*Fromandal, S.A. Lebrija, Sevilla, Spain*

Introduction

Current food legislation requires the dairy industry to perform a number of checks to ensure food safety and product quality. Moreover, the dairy industry's payments to farmers in many cases reflect the bacteriological and nutritional quality of the product. Although most of the required milk analyses can be performed using instrumental methods, these methods rely on the use of expensive equipment (Milkoscan, Fossomatic, Bactoscan) to ensure a thorough analysis.

A number of published studies address the use of near infrared (NIR) spectroscopy in determining the chemical composition of cheese[1–3] and milk from goats,[4–6] sheep[7,8] and cows,[9] although most have been carried out on experimental dairy farms. In contrast, the study reported here used industrial samples, seeking to reproduce real working conditions in the industrial or interprofessional laboratory.

The purpose of the present study was to obtain NIR calibration equations for the determination of quality parameters in goat's milk, cheese and whey.

Material and methods

Experimental material

A total of 123 samples of goat's milk, 109 samples of whey and 190 samples of goat's milk cheese, all from the dairy company Fromandal, S.A., were taken on a weekly basis over one year (January 2000 to January 2001) in order to obtain maximum seasonal variability.

Chemical analysis

Samples were analysed in duplicate to determine, for all three products, the fat content (Gerber for milk and whey, Van Gulik for cheese), total solids (oven-dried at 103 ± 2°C, using marine sand for cheese) and protein content (Kjeldahl). In milk samples, casein (Kjeldahl) and lactose content (Milko-Scan) were also measured. Somatic cell counts (SCC; Fossomatic) and bacterial counts (bacteria mL^{-1}; Bactoscan) were performed.

NIR analyses and chemometric treatment of the data

All NIR spectra were collected using a Foss NIRystems 6500 SY-I scanning monochromator, fitted with a spinning cup, working in reflectance mode in the spectral range 400–2500 nm. Milk and whey measurements were made in folded-transmission gold reflector cups, with a pathlength of 0.1 mm. Two spectra were measured per sample, the mean spectrum being used for subsequent analysis. Cheese samples were analysed unground in small ring cups.

Spectroscopic and chemical data were subjected to chemometric treatment using WinISI ver. 1.04 software.[10] Calibration equations were obtained and evaluated according to Shenk and Westerhaus[10] and Williams and Sobering.[11] MPLS (modified partial least squares) was used for regression purposes; the wavelength range studied was 1100–2500 nm (at 2 nm intervals); SNV and Detrending treatments were applied to correct for scatter. Several first and second derivative treatments were also tested. The following statistical parameters were used to select the best calibration equations: standard error of the calibration set (*SEC*), standard error of cross validation (*SECV*), coefficient of determination for the calibration process (R^2) or the cross-validation process (r^2) and ratios *RPD* ($DT \cdot SECV^{-1}$) and *RER* (Range$\cdot SECV^{-1}$), using bibliography-recommended values of over 3 for *RPD* and over 10 for *RER*.[11]

Results and discussion

The NIR calibration equations obtained afforded a high degree of accuracy in predicting the chemical composition of goat's milk cheese (Table 1), with r^2 values of around 0.9 for all parameters. Similarly, calibration errors were very small and were lower in all cases than those reported (using the DESIR method) by Díaz *et al.*[4] (0.26% for fat, 0.15% for protein, 0.29% for casein and 0.09% for lactose) and by Angulo[5] (0.42% for fat, 0.20% for protein and 0.60% for total solids); error values were also lower than those obtained in liquid samples by Albanell *et al.*[6] (0.24% for fat, 0.18% for protein and 0.34% for total solids).

The equation obtained for the somatic cell count (Table 1), a major health-related parameter, afforded satisfactory precision. Tsenkova *et al.*,[9] in a study of cow's milk, obtained an NIR prediction model for SCC with an r^2 value of 0.35, much lower than the 0.81 obtained here.

The NIR calibration equation obtained for total bacteria in milk (Table 1) accounted for 58% of the variability due to this parameter and thus enables classification of samples into high, medium and low bacteria mL^{-1} content.[12] The value obtained for the calibration set ($n = 93$) can be considered low, given the difficulty of testing this parameter; however, use of a larger calibration set would probably ensure greater accuracy.

Table 1. Calibration statistics obtained for quantitative analysis of goat's milk.

Parameter	Mean	Range	*SD*	*SEC*	*SECV*	r^2	*RPD*	*RER*
Fat	4.99	4.0–6.4	0.701	0.18	0.20	0.92	3.44	11.76
Total solids	13.50	11.7–15.6	0.967	0.19	0.22	0.95	4.40	17.73
Protein	3.76	3.3–4.6	0.277	0.05	0.07	0.94	3.90	17.46
Casein	3.60	3.1–4.3	0.275	0.05	0.07	0.93	3.83	17.27
Lactose	4.37	4.0–4.7	0.158	0.04	0.05	0.89	3.02	12.43
Bacteria $mL^{-1} 10^{-3}$	890.83	20.0–3500	762.0	354.0	499.3	0.58	1.53	6.97
SCC 10^{-3}	2495.10	1343–4168	624.6	169.1	276.9	0.81	2.26	10.20

Table 2. Calibration statistics obtained for quantitative analysis of whey.

Parameter	Mean	Range	*SD*	*SEC*	*SECV*	r^2	*RPD*	*RER*
Fat	0.941	0.70–1.35	0.127	0.07	0.08	0.66	1.69	8.64
Total solids	7.103	6.29–7.84	0.332	0.18	0.19	0.67	1.75	8.16
Protein	1.110	0.88–1.54	0.148	0.05	0.07	0.76	2.11	9.43

Table 3. Calibration statistics obtained for quantitative analysis of goatsmilk cheese.

Parameter	Mean	Range	*SD*	*SEC*	*SECV*	r^2	*RPD*	*RER*
Fat	32.2	28.3–37.0	1.94	0.49	0.57	0.92	3.40	15.3
Total solids	58.6	55.6–66.6	2.07	0.83	0.92	0.80	2.25	12.4
Protein	22.1	19.7–25.5	1.15	0.58	0.63	0.70	1.85	9.3

Equations obtained for whey also afforded satisfactory accuracy (Table 2). Calibrations for fat, total solids and proteins recorded very low *SECV* values and r^2 values—though around 0.7—were normal given the reduced range of variation for the three components analysed; this was also reflected in the values obtained for *RPD* and *RER*.

Table 3 shows calibration statistics obtained for predicting the chemical composition of goat's milk cheese. Predictive performance for the three parameters tested was satisfactory; *SECV* values were lower than those reported by Núñez *et al.*[1] using an interactance–reflectance probe (1.11% for fat, 1.33% for total solids and 0.71% for protein). However, Sorensen *et al.*,[2] in reflectance testing with unground cheese, obtained *SECV* values for fat and total solids lower than those recorded here (0.24% and 0.37%, respectively) while De Santis *et al.*,[3] using an interactance–reflectance probe, obtained an *SEC* value of 1.59% for protein, much higher than that recorded here.

Conclusions

These results confirm the viability of NIR technology for predicting chemical, microbiological and somatic cell count parameters in goat's milk and for predicting chemical composition of goat's milk cheese and whey. The chief benefits for the dairy industry of using this technology rather than other methods of chemical or instrumental analysis are the speed of analysis and, particularly, the versatility it offers; NIR not only measures the quality parameters required in milk analysis but also enables analysis of derived products such as whey and cheese.

Acknowledgements

This study formed part of Project CICYT-Feder IFD1997-0990 and was performed using equipment and infrastructure belonging to SCAI (Unidad NIR/MIR), University of Córdoba (Spain). The authors thank Ms Francisca Baena, Mr Alberto Sánchez de Puerta, Mr Antonio López and Ms Isabel Leiva of the Animal Production Department (ETSIAM-UCO) for technical assistance.

References

1. N. Núñez, A. Garrido, J.M. Serradilla and J.L. Ares, in *Near Infrared Spectroscopy: Proceedings of the 9th International Conference,* Ed by A.M.C. Davies and R. Giangiacomo. NIR Publications, Chichester, UK, p.135 (2000).

2. L.K. Sorensen and L.K. Snor, in *Near Infrared Spectroscopy: Proceedings of the 9th International Conference,* Ed by A.M.C. Davies and R. Giangiacomo. NIR Publications, Chichester, UK, p. 823 (2000).
3. D. De Santis, P. Carlini and R. Massantini, in *Near Infrared Spectroscopy:Proceedings of the 9th International Conference,* Ed by A.M.C. Davies and R. Giangiacomo. NIR Publications, Chichester, UK, p.817 (2000).
4. E. Díaz, A. Muñoz, A. Alonso and J.M. Serradilla, *J. Near Infrared Spectrosc.* **1,** 141 (1993).
5. C. Angulo. *PhD Thesis.* University of Córdoba (1997).
6. E. Albanell, S. Peris, M. Rovai, G. Caja and X. Such. *ITEA,* **20-I,** 161 (1999).
7. J.J. Pascual, P. Molina and R. Puchades, in *Near Infrared Spectroscopy: The Future Waves,* Ed by A.M.C. Davies and P. Williams. NIR Publications, Chichester, UK, p. 559 (1996).
8. N. Núñez, A. Garrido, J.M. Serradilla and J.L.Ares, *ITEA* **22-II**, 595 (2001).
9. R. Tsenkova, K. Itoh, J. Himoto and K. Asahida in *Leaping Ahead with Near Infrared Spectroscopy.* Ed by G.D. Batten, P.C. Flinn, L.A. Walsh and A.B. Blakeney. Royal Australian Chemical Institute, Victoria, Australia, p. 329 (1995).
10. ISI. The complete software solution for routine analysis, robust calibrations, and networking manual. Foss NIRystems/Tecator. Infrasoft International, LLC. Sylver Spring MD, USA (1998).
11. P.C. Williams and D. Sobering, in *Near Infrared Spectroscopy: The Future Waves*, Ed by A.M.C. Davies and P.C. Williams. NIR Publications, Chichester, UK, p. 185 (1996).
12. J.S. Shenk and M.O. Westerhaus, in *Near Infrared Spectroscopy: The Future Waves,* Ed by A.M.C. Davies and P.C. Williams. NIR Publications, Chichester, UK, p. 198 (1996).

Preliminary study on near infrared spectra of retrograde starch

Yoko Terazawa,[a] Takaaki Maekawa[a] and Sumio Kawano[b]
[a]*Tsukuba University, 1-1-1 Tennoudai, Tsukuba 305-0006, Japan*

[b]*National Food Research Institute, 2-1-12 Kannondai, Tsukuba 305-8642, Japan*

Introduction

Gelatinisation and retrogradation are important factors that cause quality deterioration of starch-based foods such as rice. Up to now, research into the mechanism of gelatinisation or retrogradation has been performed by using, say, X-ray diffraction and [1]NMR methods.[2,3] However, a clear explanation of the mechanism has not been given.

Near infrared (NIR) spectroscopy is now widely used for food analysis and also used for basic research on, say, hydrogen bonding related to water in food.[4]

Therefore, in this study, the effect of gelatinisation and retrogradation on NIR spectra was studied and the possibility of their NIR research was examined.

Materials and method

Samples

Wheat starch, commercially supplied by the Wako company, was used to make retrograde starch. First, 10% wheat starch solution was heated under mixing up to 95°C with an amylograph to make gelatinised starch. After that, the gelatinised starch solution was stored at a temperature of –30°C–60°C for several days to make retrograde starch. The retrograde starch solutions were washed three times with ethanol and then dried with acetone.

NIR spectra acquisition

The NIR reflectance spectra (1100–2500 nm) of the samples were measured with an NIRSystems 6500 (Foss NIRSystems, Silver Springs, Maryland, USA) using a spinning sample module and a micro-sample cup for powder samples at room temperature of 25°C. The built-in ceramic tile was used for reference readings.

Chemical analysis

The degree of retrogradation of each sample was indicated as a degree of gelatinisation analysed by the β-amylase-pullulanase (BAP) method.[5]

Data analysis

Data analysis was performed using the NSAS program (Foss NIRSystems, Silver Spring, MD, USA and the Unscrambler program, Version 7.01 (Camo, Trondheim, Norway).

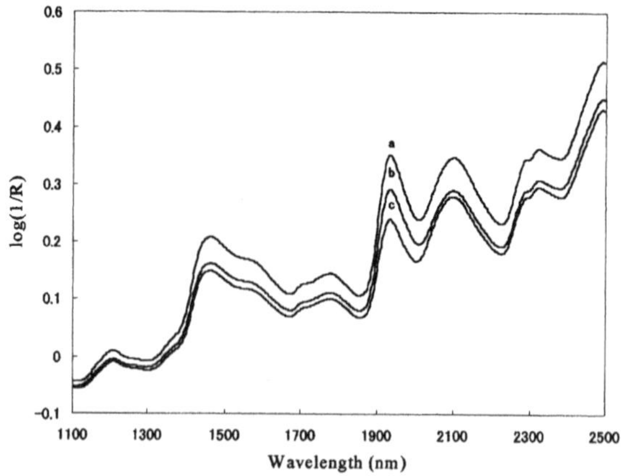

Figure 1. Original spectra of retrograded starch samples with (a) low, (b) medium and (c) high degree of gelatinisation.

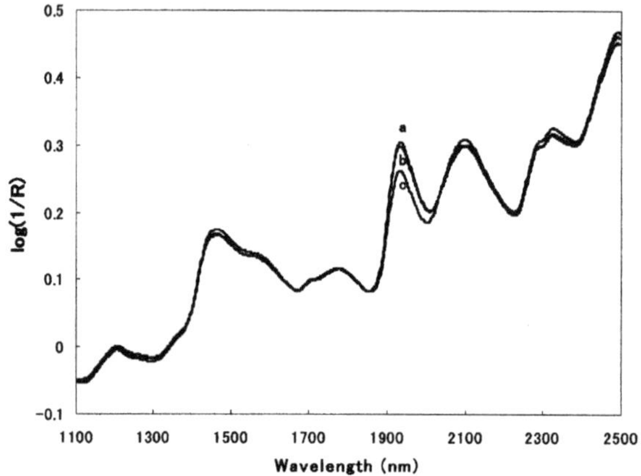

Figure 2. MSC treated spectra of retrograded starch samples with (a) low, (b) medium and (c) high degree of gelatinisation.

Results and discussion

NIR spectra of the retrograded starch

The typical spectra of samples having high (87%), medium (76%) and low (40%) gelatinisation were shown in Figure 1. The spectra were shifted upward depending on the degree of gelatinisation. Pre-processing of the spectra with multiple scatter correction (MSC) treatment removed the base line shift of the spectra (Figure 2). Characteristic changes in spectra due to retro-

Table 1. Chemical data of the degree of gelatinisation of samples used.

N	20
Maximum	93.1
Minimum	30.8
Average	72.6

N: The number of samples

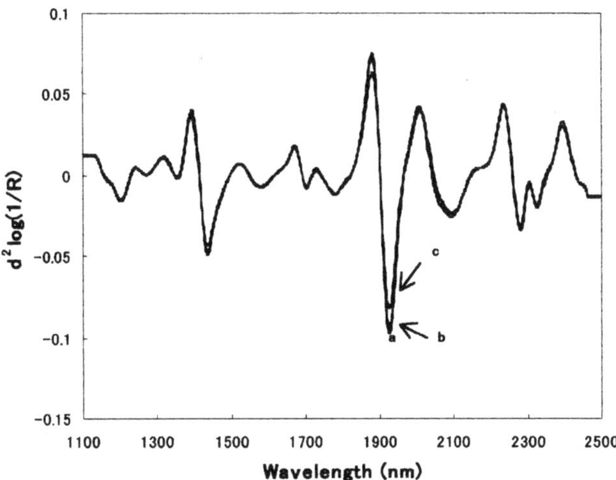

Figure 3. MSC and 2nd derivative treated spectra of retrograded starch samples with (a) low, (b) medium and (c) high degree of gelatinisation.

gradation could be observed at 2100 nm in the MSC and 2nd derivative treated spectra, as shown in Figure 3.

Correlation plots

Correlation plots between MSC-treated spectra and the degree of gelatinisation (DG) (Table 1), and between 2nd derivative values of MSC treated spectra and DG were calculated (Figures 4 and 5). The major negative peaks of 1544 nm and 2258 nm and major positive peaks of 1460 nm, 1602 nm, 1766 nm and 2136 nm could be observed, indicating that NIR absorption at the positive peak wave-

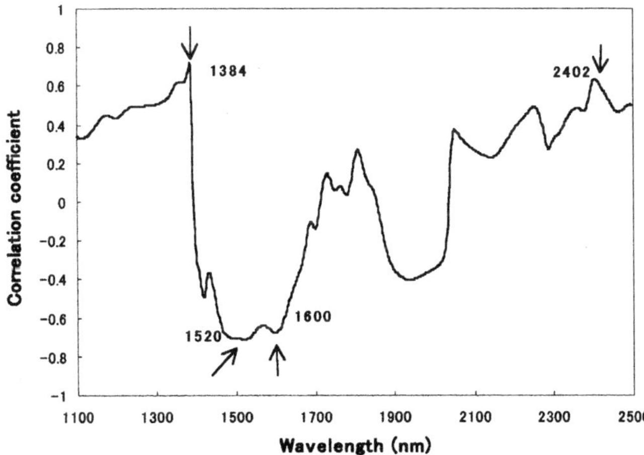

Figure 4. Correlation plots for selecting the first wavelength for the calibration equation developed using MSC treated spectra and degree of gelatinisation.

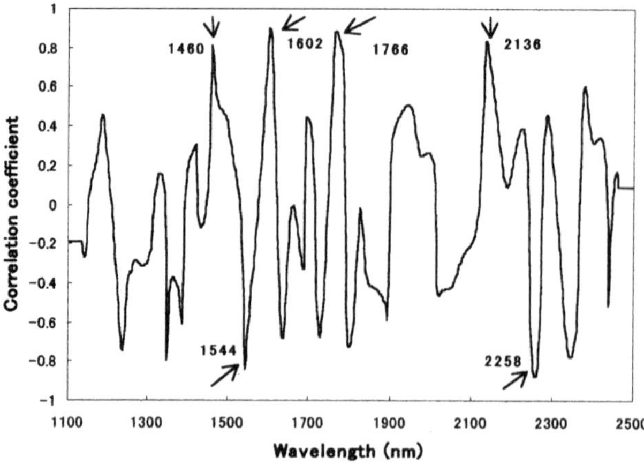

Figure 5. Correlation plots for selecting the first wavelength for the calibration equation developed using 2nd derivative values of MSC treated spectra and degree of gelatinisation.

lengths became weaker while the NIR absorption at the negative peak wavelengths became stronger as the degree of gelatinisation increased.

A simple calibration equation

In order to find wavelengths related to retrogradation, a simple regression base on the 2nd derivative values of MSC treated spectra was performed to obtain a simple calibration equation with two op-

Table 2. Results of making calibraiton equation with two optical terms for determining the degree of gelatinisation.

	Wavelength (nm)	*R*	*SEC*	*SEP*	Bias
MSc	2252, 2158	0.95	5.5	6.5	0.09
	1804, 1596	0.94	6.3	7.5	0.15
	1600, 1688	0.95	6.0	7.1	- 0.01
	1936, 2124	0.94	6.1	7.0	0.03
MSC + D2	1766, 2252	0.96	5.3	6.2	0.04
	1944, 1642	0.97	4.7	5.6	0.06
	1348, 1766	0.96	5.4	6.3	0.11
	1384, 2128	0.96	5.3	6.2	0.05
	1634, 1714	0.97	4.7	5.6	0.04
	1728, 2000	0.96	5.2	6.1	0.20
	2258, 1764	0.96	5.4	6.3	0.12

MSC: Multiplication scatter correction.
MSC + D2: MSC treated and 2nd derivative

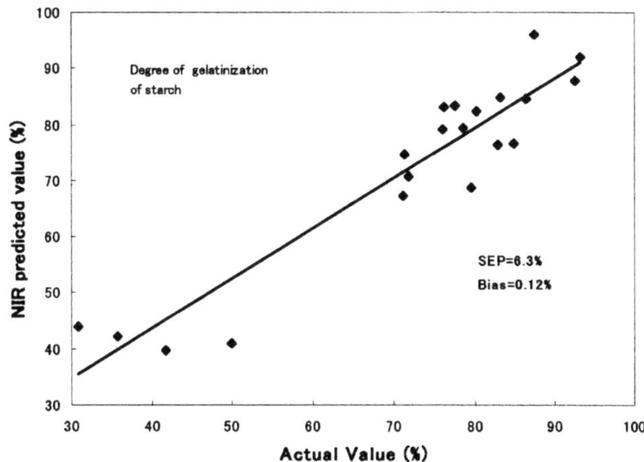

Figure 6. Relationship between actual values and NIR predicted values of the degree of gelatinisation.

tical terms (Table 2). A good calibration equation, which included 2258 nm as the first wavelength and 1764 nm as the second one, was obtained (Figure 6).

Further study is needed to find the assignment of the bands related to retrogradation.

Conclusions

Since a good calibration equation for determining the degree of gelatinisation was obtained, it is concluded that retrogradation of starch contributes to the NIR spectra of the samples.

References

1. A. Yamamoto, *J. Food Sci.* **60,** 1307 (1995).
2. C.E. Mendes Da Silva, C.F. Cialcco, G.E. Barberis, W.M.R. Solano and C. Rettopi, *Cereal Chem.* **73,** 297 (1996).
3. C.H. Teo and C.C. Seow, *Starch* **44,** 288 (1992).
4. M. Iwamoto, S. Kawano and H. Abe, *NIR news* **6(3),** 10 (1995).
5. K. Kainuma, A. Matsunaga, M. Itagawa and S. Kobayashi, *J. Jap. Soc. Starch Sci.* **28,** 235 (1981).

Near infrared spectroscopy applied to *pasta filata* cheese in relation to textural analysis

Tiziana M.P. Cattaneo,* Adele Maraboli and Roberto Giangiacomo
Istituto Sperimentale Lattiero-Caseario, Via A. Lombardo 11, 26900 Lodi, Italy
E-mail: tcattaneo@ilclodi.it

Introduction

Rheological properties of cheese are important factors for the ripening process and basic parameters for quality assessment during shelf-life and trade steps to the consumer. Characteristic curd and cheese texture are associated with different coagulation processes and final products. Cheese texture also influences cutting resistance of the product, spreadability properties and consumption habits.[1]

Objective measurements of rheological properties of cheese are carried out by texturometers,[2–4] based on product deformation caused by the application of mechanical force (compression). Usually, deformation force is applied at constant speed for a defined time on samples kept at constant temperature. Sample preparation and analysis are time consuming. Therefore, a more rapid and sufficiently accurate technique for measuring firmness of cheese is advisable. Near infrared (NIR) spectroscopy is known to be a good technique for the evaluation of cheese components[5] and it has been shown to respond to changes in the state of water in foods.[6] NIR spectroscopy might give information on rheological properties of cheese during ripening and shelf-life, since these properties are related to chemical constituents and water structure. Recently, some studies have been developed to evaluate the feasibility of NIR spectroscopy in predicting physical properties of foods.[7–9]

The present study aimed to seek a correlation between rheological parameters of cheese, such as hardness, determined by a single compression test and data obtained by NIR spectroscopy.

Materials and methods

One hundred samples of *Pasta Filata* cheese, coated with paraffin and biodegradable wax to avoid mechanical damage, were analysed at room temperature during shelf-life at 90 (50 samples) and 120 days (50 samples).

Samples were cut into small cylinders (D = 3.2 cm, height = 1 cm)[10] and cheese pieces were analysed by an InfraAlyser 500 (Bran+Luebbe, Norderstedt, Germany) at 1100 to 2500 nm at 4 nm intervals (351 data points). Spectra were collected as logs R^{-1}. Data were processed by Sesame Software (Bran+Luebbe, Germany) applying the second derivative of absorbance values (gap = 0, segment = 1). Optimisation of the information provided by raw spectra was obtained by applying multivariate analysis to the whole set of processed data using the principal component analysis (PCA).

Textural characteristics were determined by an Instron Universal Testing Machine 4301 (Instron Corporation, Canton, MA, USA) on sample pieces cut into small cylinders (D = 1.7 cm, height = 2 cm), conditioned at 20°C and placed on a flat plate. Hardness was evaluated by a single compression test using a plunger having a plane circular surface (D = 5.8 cm, height = 3.7 cm). This plunger was fixed to the moving crosshead at a speed of 20 mm min^{-1}. A 10 kg load cell was used. Tests were performed by recording the load at 20 mm of compression (max load). Data were processed by Instron series IX Software (Instron Corporation) and expressed as a mean of eight replicates.

Total calibration was performed by applying multiple linear regression (MLR) to the first derivative of normalised absorbance values at seven wavelengths (1160, 1976, 2064, 2080, 2236, 2328 and 2336 nm) and using randomly 70 out of the 100 samples, 35 from each set. The set of 30 remaining samples, 15 at 90 days and 15 at 120 days of shelf-life, was used as a prediction set. Load values, obtained from textural analysis, were plotted against NIR prediction values. Regression coefficient (R^2), *RMSEP* in calibration, *RMSEP* in prediction and $RMSEP_{all}$ were calculated.

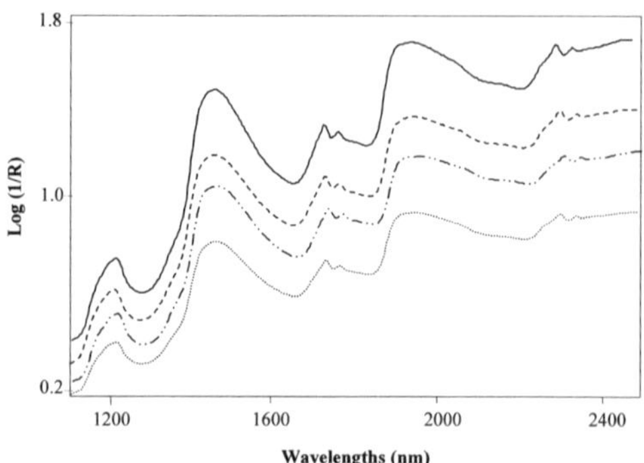

Figure 1. NIR raw spectra for *Pasta Filata* cheese coated with different waxes during shelf-life. (—) paraffin 90 days; (- - -) paraffin 120 days; (-·-·-) biodegradable wax 90 days; (···) biodegradable wax 120 days.

Results and discussion

Figure 1 shows an example of raw NIR spectra for *Pasta Filata* cheese during shelf-life at 90 and 120 days. Since no differences between spectra were apparently pointed out, pre-treatments were performed on data collected. Explorative analysis, carried out by a PCA on second derivative spectra, was able to gather samples according to their shelf-life period. Figure 2 shows results obtained by PCA; the first principal component (PC1) was able to discriminate cheeses according to their shelf-life.

Textural measurements, performed by the Instron, discriminated cheeses according to their hardness, which was measured as applied force.[2] An example of a typical compression curve is shown in Figure 3. Hardness corresponds to the maximum force (max load) exerted on a sample.

Both the type of wax used and the water content affected cheese hardness, in agreement with monitored weight loss data[11] (Figure 4).

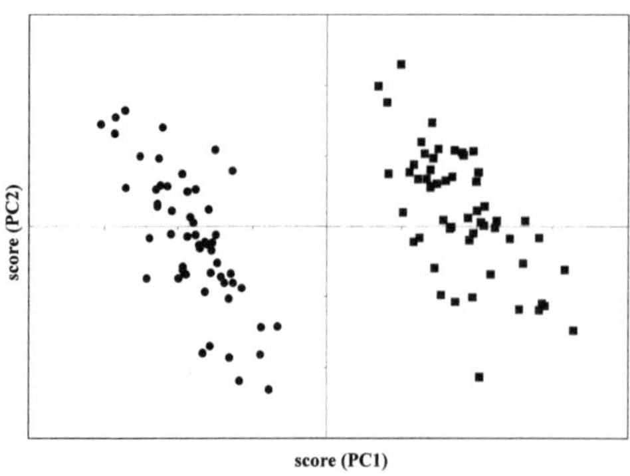

Figure 2. Explorative analysis (PCA) on 2nd derivatives of NIR spectra (gap = 0; segment = 1). (●) 90 days; (■) 120 days.

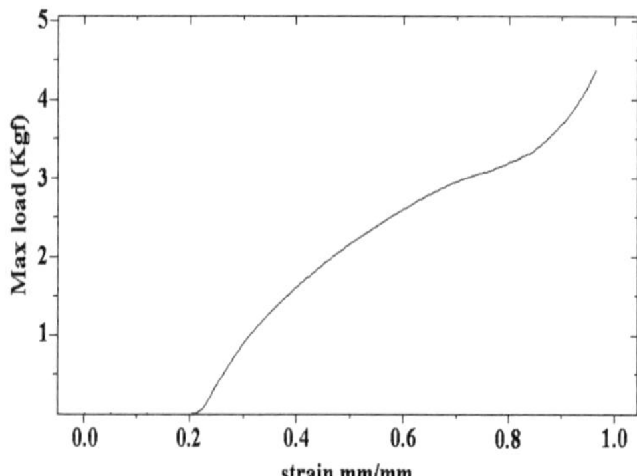

Figure 3. Example of a load-strain curve obtained from a uniaxial compression test on "Pasta Filata" cheese.

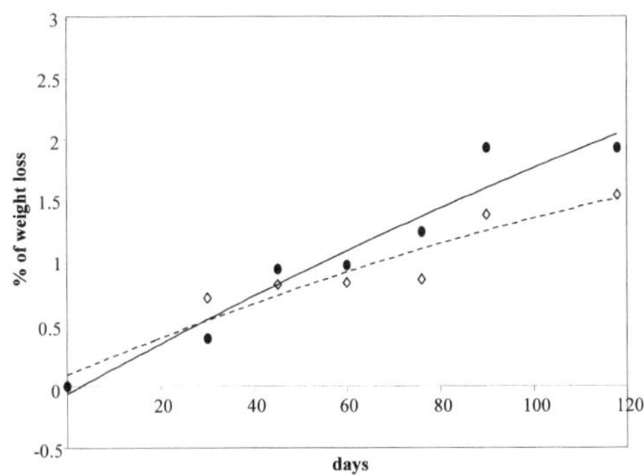

Figure 4. Weight loss rate during shelf-life of cheese. (●) biodegradable wax; (◇) paraffin.

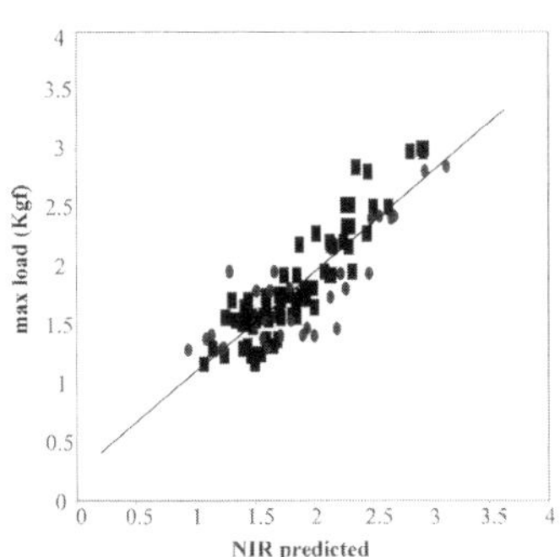

Figure 5. Plot of measured hardness against NIR predicted values for Pasta Filata cheese. (■) calibration set; (●) prediction set.

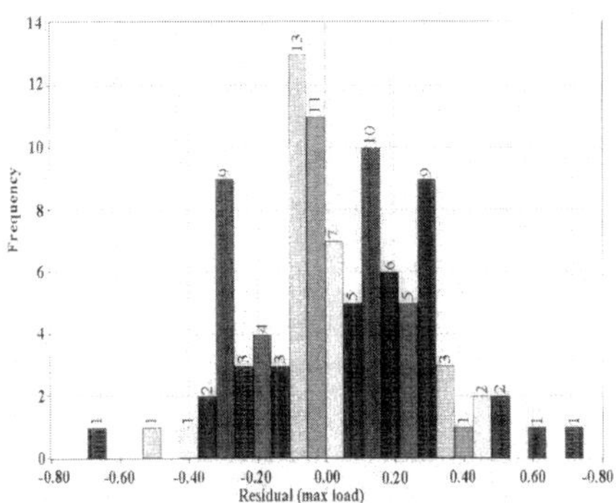

Figure 6. Analysis of residues: prediction and calibration sets.

An MRL calibration model, obtained from the seven wavelengths automatically selected by Sesame software on the first derivative of normalised NIR absorbance values, showed a good relationship between texture measurements (max load) and NIR prediction power. MRL calibration was characterised by $R^2 = 0.916$ and $RMSEP = 0.192$. The prediction set (30 samples) was also characterised by a satisfactory *RMSEP* value (0.345). $RMSEP_{all}$, associated with the whole set of samples, was 0.248. It should be pointed out that the textural analysis, carried out by Instron and used as an objective method for hardness determination, was also characterised by a coefficient of variation (*CV*) of 10% (mean value). Figure 5 shows the NIR spectroscopy prediction power as a plot of hardness values, determined by a texturometer, against the prediction values based on the first derivatives.

From the analysis of residues, calculated on MRL calibration and prediction sets, it can be observed that only 8% of the samples showed higher differences than $\pm$ 0.4 between objective hardness measurements and NIR prediction values (Figure 6). These results confirmed the accuracy of NIR prediction as 92% of residues were found to be close to zero.

Conclusions

NIR spectroscopy proved a useful tool for classifying samples according to their shelf-life period and predicting their textural characteristics, such as hardness. Both the type of wax used and the water content of samples affected cheese hardness, in agreement with monitored weight loss data. Hardness evaluation by NIR spectra was as accurate as that determined by a texturometer, thus confirming that the possible use of NIR spectroscopy for this application may facilitate the routine control of physical characteristics of cheese by a rapid and sufficiently accurate technique.

Acknowledgements

This work was partially supported by EU Project BE SE-2659-1998.

References

1. M. Lucisano, C. Pompei and E. Casiraghi, *J. Food Qualit.* **10,** 73 (1987).
2. A.H. Chen, J.W. Larkin, C.J. Clark and W.E. Irwin, *J. Dairy Sci.* **62,** 901 (1979).
3. E. Casiraghi, M. Lucisano and C. Pompei, *Ital. J. Food Sci.* **1,** 53 (1989).
4. P. Walstra, H. Luyten and T. van Vliet, in *Milk—The vital force, XXII Int. Dairy Congress, De Hague,* D. Reidel Publ. Co., Dordrecht, The Netherlands, p. 159 (1987).
5. L.K. Sørensen and L.K. Snor, in *Near Infrared Spectroscopy: Proceedings of the 9th International Conference,* Ed by A.M.C. Davies and R. Giangiacomo. NIR Publications, Chichester, UK, p. 823 (2000).
6. M. Iwamoto and S. Kawano, in *Making Light Work; Advances in Near Infrared Spectroscopy,* Ed by I. Murray and I.A. Cowe. VCH, Weinheim, Germany, p. 367 (1992).
7. G. Costa, M. Noferini, C. Andreotti and F. Mazzotti, in *Near Infrared Spectroscopy: Proceedings of the 9th International Conference,* Ed by A.M.C. Davies and R. Giangiacomo. NIR Publications, Chichester, UK, p. 863 (2000).
8. K.I. Hildrum and B.N. Nilsen, in *Near Infrared Spectroscopy: Proceedings of the 9th International Conference,* Ed by A.M.C. Davies and R. Giangiacomo. NIR Publications, Chichester, UK, p. 855 (2000).
9. M.R. Sohn and R.K. Cho, in *Near Infrared Spectroscopy: Proceedings of the 9th International Conference,* Ed by A.M.C. Davies and R. Giangiacomo. NIR Publications, Chichester, UK, p. 791 (2000).
10. L.K. Sørensen and R. Jepsen, *Milchwissenschaft.* **53(5),** 263 (1998).
11. EU Project BE SE-2659, *Final report.* (2001).

Discrimination between virgin olive oils from Crete, the Peloponese and other Greek islands using near infrared transflectance spectroscopy

Gerard Downey and Stephen J. Flynn

TEAGASC, The National Food Centre, Dunsinea, Castleknock, Dublin 15, Ireland

Food adulteration is a serious consumer fraud and a potentially dangerous practice. Regulatory authorities and food processors require a rapid, non-destructive test to accurately confirm authenticity in a range of food products and raw materials. Olive oil is a prime target for adulteration either on the basis of the processing treatments used for its extraction (extra virgin vs virgin vs refined oil) or its geographical origin (for example, Greek vs Italian vs Spanish). As part of an investigation into this problem, some preliminary work focused on the ability of near infrared spectroscopy to discriminate between virgin olive oils from geographically-close regions of the Mediterranean, i.e., Crete, the Peloponese and other Greek islands.

A total of 64 oils were collected: 18 from Crete, 28 from the Peloponese and 18 from other Greek islands. Oils were stored at 20°C prior to spectral collection at room temperature (15–18°C). Spectra

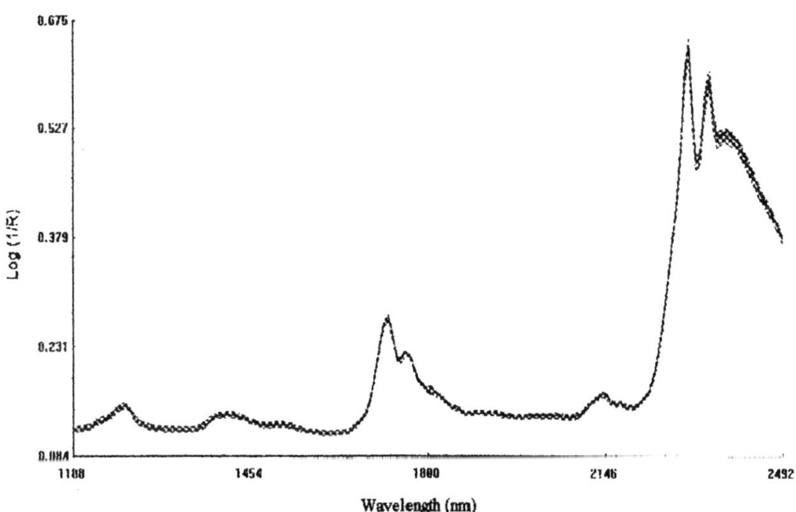

Figure 1. Transflectance spectra of 64 virgin olive oils from Crete (18), the Peloponese (28) and other islands (18).

Figure 2. Camlock cell with gold-plated disc

were recorded between 400 and 2498 nm at 2 nm intervals on a NIRSystems 6500 scanning monochromator (Figure 1). Samples (approximately 0.5 mL) were placed in a camlock reflectance cell with a gold-plated backing plate (Figure 2), producing a sample thickness of 0.1 mm.

Classification into three categories (**C**rete, **P**eloponese and **O**ther) was investigated using factorial discriminant analysis and separate calibration and prediction sample sets (50% of samples in each). Results obtained are shown in Table 1. Best results used spectral data in both the visible and near infrared ranges. Sample scores (Figure 3) reveal the clustering achieved with the best model while the relevant discriminant profiles are shown in Figure 4.

Table 1. Classification results using factorial discriminant analysis.

		% Correct classification	
Spectral range (nm)	Principal components	Calibration set	Prediction set
400–2498	1,3,7,12	96.9	93.9
1100–2498	2,4,5,6	84.4	72.7
400–750	1,3,4	78.1	72.7
1800–2200	4,3,2,13,5,9	96.9	93.9

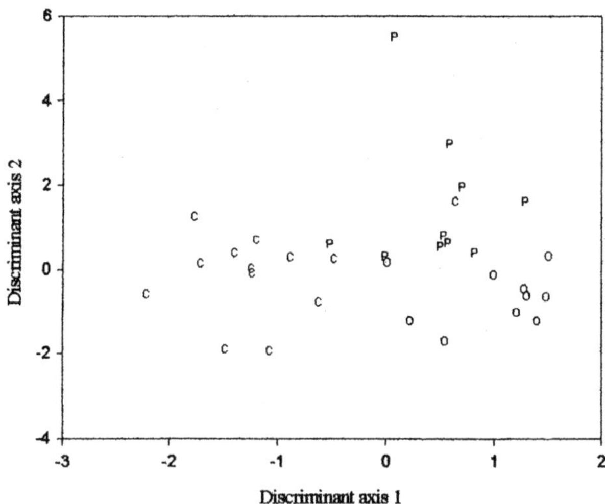

Figure 3. Discriminant scores plot (400–2498 nm).

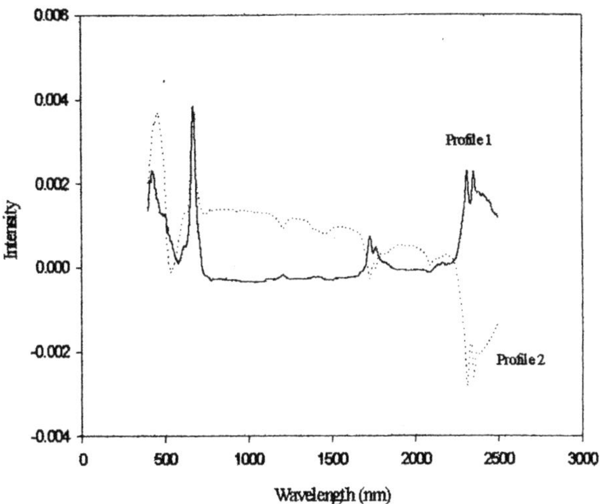

Figure 4. Discriminant profiles.

Single kernel maize analysis by near infrared hyperspectral imaging[a]

R.P. Cogdill, C.R. Hurburgh, Jr, T.C. Jensen and R.W. Jones[b]

Iowa State University, Ames, Iowa , USA

Introduction

As digital imaging equipment and continuously tunable, image-quality optical filters have become available, spectral imaging has developed into a feasible technology. While most applications of near infrared (NIR) spectral imaging (outside remote sensing and microscopy) have been machine vision systems for defect detection and quality control,[1–4] there are many possibilities for applying NIR spectroscopy imaging to two-dimensional, spatial analysis of biological materials.[5–7]

Single-kernel maize analysis is one possible application for NIR spectroscopy hyperspectral imaging. Commercial single-kernel NIR analysers have allowed the segregation of grains based on individual seed quality analysis, rather than on bulk sample properties. However, single-seed NIR spectroscopic analysis predictions are made by single-point sampling of kernel spectral characteristics. To ensure reasonable prediction, such as sampling, must accurately reflect the distribution of spectral characteristics exhibited by the kernel. Hyperspectral imaging spectroscopy (imaging at many wavelengths) could be a more efficient method of analysing spatially heterogeneous materials. To date, no method for calibrating a hyperspectral imaging spectrometer for single kernel grain analysis has been reported.

The objectives of this research were:

1. Develop a technique for creating calibrations for constituents of single maize kernels from NIR hyperspectral image data
2. Evaluate the feasibility of an NIR imaging spectrometer as a tool for single-kernel analysis

Materials and methods

These were met by evaluating methods for handling NIR hyperspectral image data of single kernels of maize to create calibration equations for predicting moisture and oil concentration of single maize kernels.[8] Moisture content was done first, then oil content.

A set of single-kernel samples was imaged with a spectrometer assembled by Dr J.F. McClelland and group, at Ames Laboratory, US Department of Energy (Figure 1). The samples were imaged (512 × 512 pixels) at 5 nm wavelength increments, covering the spectral range 750–1090 nm. Moisture concentration of the kernels was determined gravimetrically by drying for 72 h at 103°C.[9] The moisture calibration reference statistics are summarised in Table 1. Oil concentration of the single-kernel samples was determined gravimetrically, using supercritical fluid extraction.[10] The oil calibration reference data is also summarised in Table 1.

[a]Research supported by the Institute of Physical Research and Technology and the Agriculture and Home Economics Experiment Station, Iowa State University, Ames, IA, project 3261; and by ExSeed, Inc.

[b]Graduate research assistant and Professor, Agricultural and Biosystems Engineering, Iowa State University; Scientist, Center for Non-Destructive Evaluation, and Associate Scientist, Ames Laboratory, U.S. Department of Energy.

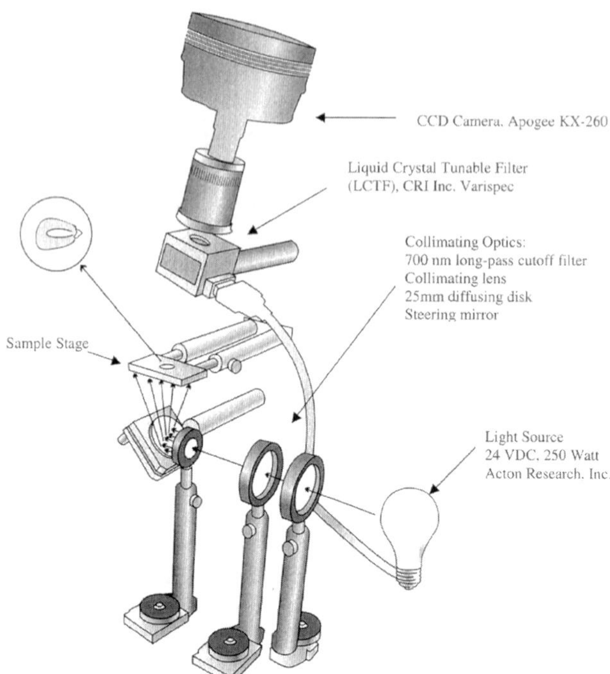

Figure 1. Illustration of hyperspectral imaging spectrometer.

The background (non-kernel) pixels were removed with a binary mask. The remaining image data was standardised (according to equation 1) and rearranged so that the data for every sample was contained in a two-way hypercube array.[11] A single spectrum was extracted for each sample by taking the

Table 1. Moisture and oil calibration reference statistics.

Reference analysis[a]	Calibration	
	Moisture[b]	Oil[c]
Number of samples	452	153
Average	15.66	3.19
Maximum	30.11	12.16
Minimum	9.74	0.26
Spectral outliers removed, %	6.9	5.9
Prediction residual outliers removed, %	1.8	2.9

[a]Reference analysis statistics were calculated after outliers were removed
[b]% moisture concentration (wet basis)
[c]% oil concentrration ('as is' moisture basis)

Table 2. Effect of data pretreatment methods on moisture calibration performance.

	Before Transformation			After Transformation			
Preprocessing Treatment	*SECV* (%)	R^2	lv	*SECV* (%)	R^2	lv	Optimal Transform
Raw Transmittance	1.43	0.783	8	1.21	0.847	18	P = 0.4
SNV	1.32	0.817	16	1.23	0.840	20	P = -0.8
MSC	1.81	0.662	10	1.69	0.703	11	P = 0.2

averaging spectral values by wavelength across pixels. Several data preprocessing treatments were compared:

- raw data (transmittance ratio)
- standard normal variate (SNV)[12]
- multiplicative scatter correction (MSC)[13]
- power transformations ($X = T^P$) of the preprocessed transmittance ratio data were compared to *log(1/T)*
-

$$I_{\lambda,n} = \lfloor C(S_{\lambda,n} - D_n) \div (L_{\lambda,n} - D_n) \rfloor - \lfloor C(E_{\lambda,n} - D_n) \div (L_{\lambda,n} - D_n) \rfloor \quad (1)$$

$I_{\lambda,n}$ = standardised image pixel *n*, at wavelength λ
$S_{\lambda,n}$ = sample image pixel *n*, at wavelength λ
$E_{\lambda,n}$ = empty sample stage image pixel *n*, at wavelength λ
$L_{\lambda,n}$ = light reference image pixel *n*, at wavelength λ
D_n = dark reference image pixel *n*
C = mean brightness level within a 100 × 100-pixel region in the center of ($L_{850,(1:N)}$–$D_{(1:N)}$)

The optimal method of handling the NIR hyperspectral image data was chosen as that which yielded the minimum standard error of 10-block cross-validation (*SECV*). Using the optimal preprocessing method determined in the preceding steps, calibrations for predicting single kernel moisture and oil concentrations were developed.

A discriminant analysis function was also developed to identify genetically modified corn seeds from non-modified seeds. Single maize kernels were classified into two types, CLASS_1 and CLASS_2. A set of 359 samples, 198 CLASS_1 (genetically modified, Cry1 Bt event) and 161 CLASS_2 (non-modified), was analysed. The reference kernels were of diverse origin, both within and between the two classes of samples, in year of harvest, number of hybrids represented and kernel morphology. A discriminant analysis function was created using PLS-DA and cross-validation, where the reference for each sample was either a 1 (CLASS_1) or –1 (CLASS_2).

Results

The comparison of preprocessing methods and transformations are summarised in Table 2. Moisture content was the property used to tune the model. While SNV preprocessing yielded the best calibration performance before the transformation to absorbance was applied, the raw data yielded the greatest calibration performance after the transformation was applied. For all preprocessing methods, a power transformation yielded significantly better calibration performance than the log(1/*T*) transformation. The final calibration achieved a ratio of range to *SECV* of 16.8, which suggests that NIR hyperspectral imaging would be a suitable method for quantitative prediction of moisture concentration in single kernels of maize.[14]

The oil calibration yielded a best *SECV* of 1.49% and R^2 of 0.48, using a PLS model including 11 factors. While the oil calibration performed much more poorly than the moisture calibration, an analysis of variance showed that the random error of the reference chemistry method was significantly greater for the oil calibration and was responsible for 60% of the variance.[8]

The discriminant analysis function achieved a best classification performance during cross- validation of 98.9% correctly classified (0.56% false positive CLASS_1 and 0.56% false negative CLASS_1) using a PLS-DA model with 13 factors. This exceeded previous reports of GMO identification by NIR spectroscopy.[15]

Conclusions

From these results the following conclusions can be drawn:

1. Preprocessing of raw transmittance ratio data before spectral extraction is unnecessary and may detract from calibration performance.
2. The log(1/*T*) transformation was not optimal for adjusting transmittance ratio spectra to a linear model format. In this data, $T^{0.4}$ provided the best linearisation.
3. NIR hyperspectral imaging would be a feasible method for analysing the quality of single kernels of maize.
4. NIR hyperspectral imaging may be an attractive method for discriminating between different classes of complex biological structures.
5. The next and more valuable step will be to develop calibration models that can be applied on a pixel-by-pixel basis, thus predicting the spatial distribution of attributes.

References

1. B.K. Miller and M.J. Delwiche, *ASAE Paper No. 90-6040*. ASAE, St Joseph, MI, USA (1990).
2. V. Bellon, G. Rabatel and C. Guizard, *Food Control* 49 (1992).
3. B.L. Upchurch, J.A. Throop and D.J. Aneshansley, *Transactions of the ASAE* **37(5),** 1571 (1994).
4. B. Park, R. Chen and M. Nguyen, *Journal of Agricultural Engineering Research* **69(4),** 351 (1998).
5. D. Taylor and W.F. McClure in, *Proceedings of the Second International Near Infrared Spectroscopy Conference*, Ed by M. Iwamanot and S. Kawano. Korin Publishing Co., Tokyo, Japan, pp 393–404 (1989).
6. M.D. Evans, C.N. Thai and C.J. Grant, *Trans ASAE* **41(6),** 1845 (1998).
7. J. Sugiyama, Journal of Agriculture and Food Chemistry **47,** 2715 (1999).
8. R.P. Cogdill, *Single kernel maize analysis by near infrared hyperspectral imaging*. MSc. thesis. Iowa State University, Ames, Iowa, USA (2001).
9. ASAE. "Oven temperature and heating period for moisture content determinations," ASAE Method S352.2 DEC92, *ASAE Standards*, 44th Edn. ASAE, St. Joseph MI, USA (1997).
10. AOCS. Supercritical fluid extraction. AOCS methods Am 3-96, *Official Methods and Recommended Practices of the American Oil Chemists Society*, 4th Edn, 3rd printing. AOCS, Champaign, IL, USA (1993 rev.).
11. P. Geladi and H. Grahn, *Multivariate image analysis*. John Wiley & Sons, Inc., New York, NY, USA (1996).
12. R.J. Barnes, M.S. Dhanoa and S.J. Lister, *Appl. Spectrosc.* **43(5),** 772 (1989).
13. H. Martens and T. Naes, *Multivariate calibration*. John Wiley & Sons Ltd, Chichester, UK (1989).

14. "Near infrared Methods: Guidelines for Model Development and Maintenance," *AACC Method 39-00, Approved Methods of the American Association of Cereal Chemists*, 10th Edn. AACC, St. Paul, MN, USA (1999).
15. S. Roussel, C. Hardy, C. Hurburgh and G. Rippke, *Appl. Spectrosc.* **55(10),** 1425 (2001).

Effects of variety and region on near infrared reflectance spectroscopic analysis of quality parameters in red wine grapes

M.B. Esler,[a,c] M. Gishen,[a,c] I.L. Francis,[a,c] R.G. Dambergs,[a] A. Kambouris,[b] W.U. Cynkar[a,c] and D.R. Boehm[a]

[a] *The Australian Wine Research Institute, PO Box 197, Glen Osmond, SA 5064, Australia*

[b] *BRL Hardy Limited, PO Box 238, Berri, SA 5343, Australia*

[c] *The Cooperative Research Centre for Viticulture, PO Box 154, Glen Osmond, SA 5064, Australia*

Introduction

The wine industry requires objective measures of red wine grape quality to determine optimal harvest time, allocate freshly harvested grapes to winery process streams for particular red wine products and determine quality-based payment to grape growers. The practical requirement that these analyses also be rapid and inexpensive currently restricts them to the measurement of TSS (total soluble solids, mainly sugars, in °Brix) by refractometry or hydrometry and pH and acid content by potentiometry and titration. These parameters do not, however, provide comprehensive compositional characterisation for the purpose of winemaking.

Total anthocyanins in red wine grapes

The total concentration of anthocyanin pigments in red wine grapes is believed to be an indicator of potential wine quality and price.[1] However, routine analysis for total anthocyanins is not considered as a practical option by the wine industry because of the high cost and slow turnaround time of this multi-step wet chemical laboratory analysis. The analysis requires homogenisation of about 100 g of grapes, a one-hour solvent extraction of the homogenate, centrifugation of the extract, pH adjustment with three-hour equilibration of the supernatant and, finally, spectrophotometric analysis at 520 nm. Recent work by this group[2,3] has established the capability of near infrared (NIR) spectroscopy to provide rapid and accurate measurement of total anthocyanins as well as the simultaneous measurement of TSS and pH in red wine grapes. For the present study, the only sample processing step still required is homogenisation of the grapes.

Sample collection

In the five weeks leading up to the 1999 harvest, approximately 150 samples of red wine grapes from the single region of the Riverland were collected and then stored frozen. In the lead up to the 2000 harvest, a further 750 samples were collected from commercial vineyards managed by four different wine companies in several different growing regions. Only four varieties of grape account for the overwhelming majority, 98%, of the samples. These were Shiraz (45%), Cabernet Sauvignon (43%), Merlot (7%) and Grenache (3%), approximately representing the relative proportions in which these varieties are grown in Australia.[4] The remaining 2% of samples consisted of the less common grape varieties (Ruby Cabernet, Cabernet Franc, Mataro, Malbec, Petit Verdot and Cinsaut). Similarly, not all regions were sampled to the same extent, the relative proportions being Riverland: 25%, Barossa Valley: 22%, Padthaway: 13%, McLaren Vale: 13%, Langhorne Creek: 11%, Wrattonbully: 9%, Coonawarra: 5% and Mildura, Adelaide Hills and Robe about 1% each. In all, the sample set spans two vintages (1999 and 2000), ten distinct geographical winegrowing regions in South Australia and ten of the varieties of red grape cultivated commercially in Australia.

Analysis

The samples were thawed in batches of about 20, homogenised and scanned in diffuse reflectance mode on a FossNIRSystems 6500 spectrometer and immediately subjected to laboratory analysis the same day using the traditional methods for total anthocyanins, TSS and pH. All of the NIR scanning and the reference method analysis were completed in a period of six months.

Calibrations

All calibrations were generated using partial least squares (PLS) in *The Unscrambler 7.5* (CAMO ASA, Norway). In nearly all cases the calibration initially obtained by the PLS method was improved by using Martens' uncertainty test[5] (or 'jack-knifing') to select the optimal set of wavelengths for inclusion in the calibration. This improvement was manifested as either reduced standard error of cross-validation (*SECV*) or reduced number of PLS factors, or both. The optimal total anthocyanin calibrations used two to four PLS factors and the pH calibrations four to five factors. The calibrations for total anthocyanin and pH required 2nd-derivative preprocessing of the spectra. The TSS calibration used raw reflectance spectra and five PLS factors.

Results and discussion

Figure 1 illustrates the results for total anthocyanin analysis. Figures 1(a) and 1(b) illustrate the correlation of predicted against measured results for the optimal total anthocyanin calibration when all regions and grape varieties (of those sampled) are included. This is referred to as the 'global' calibration and is plotted in Figures 1(a) and 1(b), coded by region and grape variety, respectively. The *SECV* of this calibration, ± 0.14 mg g^{-1}, is equivalent to 12% of the mean total anthocyanin concentration and 6.7% of the observed range of concentrations in the 'global' population. This degree of measurement precision would be of only marginal use to commercial winemakers. In Figure 1(c) the focus of the calibration has been narrowed to include only one variety of grape, Cabernet Sauvignon, but still including all the regions from which that variety was sampled. This 'varietally-localised' calibration has a significantly improved *SECV* of ± 0.106 mg g^{-1}, (9.7% of the mean and 6.2% of the range for that population of samples). Similarly, in Figure 1(d) the calibration has been 'regionally localised' relative to the original 'global' calibration to focus only on a single region, in this case the Riverland, but includes all four varieties of grapes sampled from that region. The *SECV* is better again, ± 0.068 mg g^{-1}, (11% of the mean and 5.1% of the range for that population). Finally, Figure 1(e) illustrates the 'varietally and regionally localised' calibration for Riverland grown Cabernet Sauvignon grapes. The *SECV* of ± 0.054 mg g^{-1} represents 6.9% of the mean and 4.1% of the range of total anthocyanin concentrations for the population of Riverland grown Cabernet Sauvignon grapes and 4.7% of the mean and 2.6% of the range for the original 'global' population. This represents an improvement in measurement precision by a factor of almost three over that obtained in the initial 'global' calibration. This level of precision is likely to be of considerable use to commercial winemakers. The precision of the

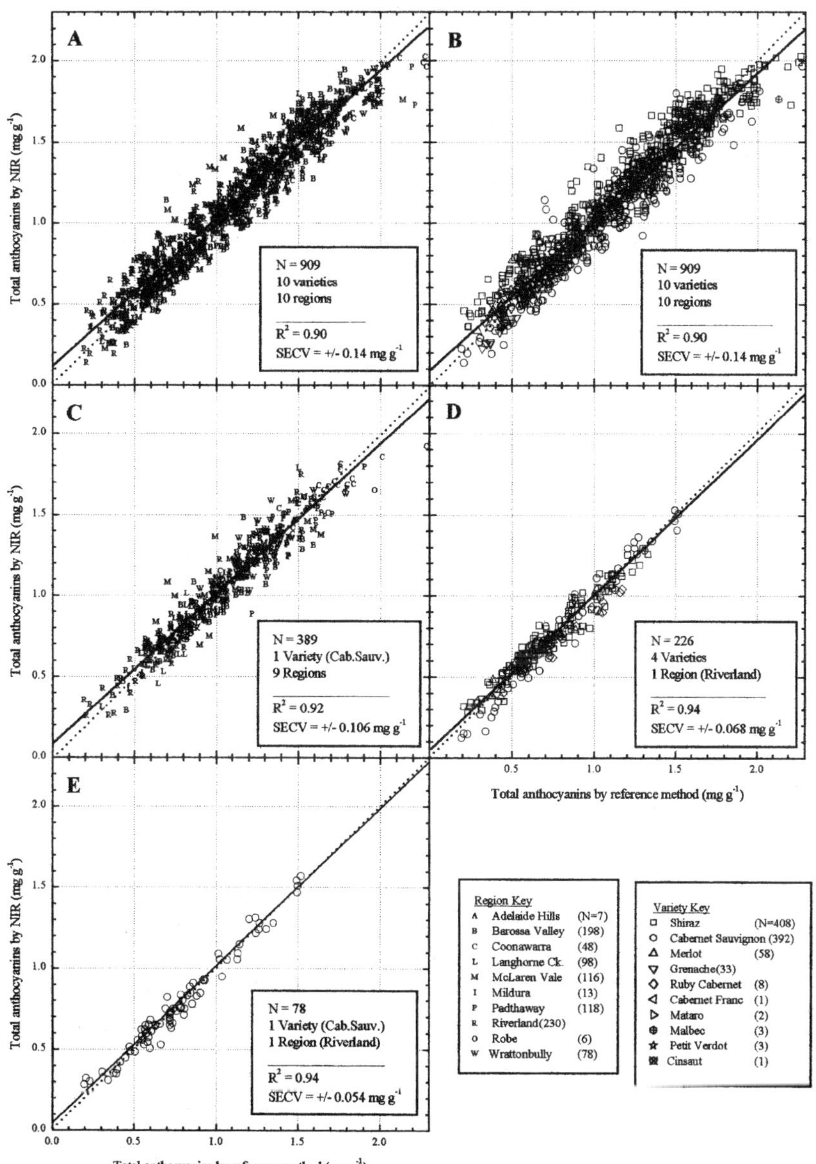

Figure 1. NIR calibrations for total anthocyanin concentration for a population of ~ 900 red wine grape samples and various of its sub-populations. (a) 'Global' calibration; 909 samples across ten growing regions, ten grape varieties and two seasons, plotted by region, (b) Same samples as (a), plotted by grape variety, (c) Subpopulation, localised to a single grape variety only, Cabernet Sauvignon; 389 samples across nine growing regions and two seasons, plotted by region, (d) Subpopulation, localised to a single growing region only, the Riverland; 226 samples across four grape varieties and two seasons, plotted by variety and (e) Subpopulation, localised for both a single grape variety, Cabernet Sauvignon, and a single growing region, the Riverland.

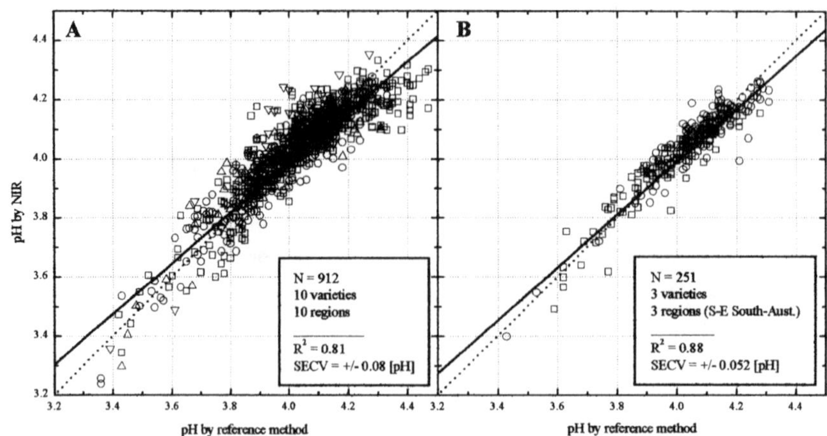

Figure 2. NIR calibrations for pH for a population of ~ 900 red wine grape samples and one of its subpopulations, (a) 'Global' calibration, 912 samples across ten growing regions, ten varieties and two seasons, plotted by variety and (b) Subpopulation, localised to a group of three closely related regions, across three varieties and one season.

reference method for total anthocyanins is estimated to be approximately ± 0.05 mg g^{-1}, so further improvement in the precision of the NIR analysis is unlikely to be seen.

In a similar manner, Figures 2(a) and 2(b) illustrate the improvement in pH measurement precision using NIR spectroscopy on moving from a 'global' to a more 'regionally- and varietally-localised' calibration. It has been suggested that NIR prediction of pH may be related to pH-induced shifts in anthocyanin chromophores and, indeed, much the same region of the spectrum is used for both pH and total anthocyanin prediction. Thus, the same type of improvement seen for both pH and total anthocyanins on going from a global to a more localised calibration is not so surprising. Both pH and total anthocyanin prediction by NIR may be sensitive to the same type of matrix effects due to differing regions and varieties.

In Figure 3, the 'global' calibration for TSS is illustrated, with an *SECV* of ± 0.33°Brix. No significant improvement was observed on 'localising' calibrations for TSS. This may indicate that the NIR measurement of TSS is less sensitive to matrix changes due to different varieties and regions than is the measurement of total anthocyanins and pH. Since the measurement precision of the reference method for TSS is estimated to be approximately ± 0.05°Brix, it is un-

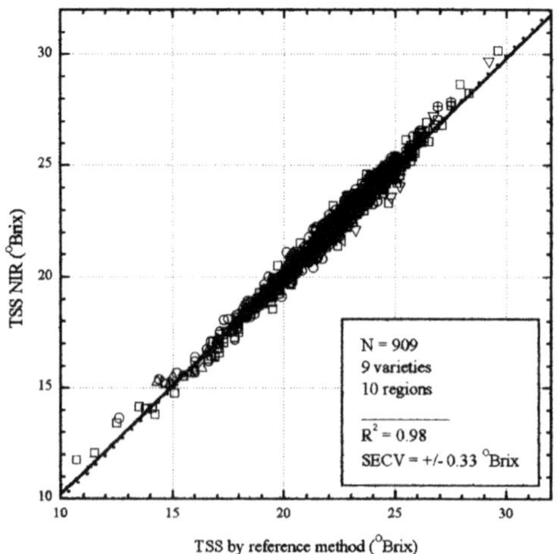

Figure 3. 'Global' calibration for TSS; 909 samples across ten growing regions, nine grape varieties and two seasons.

likely that the NIR prediction of TSS is limited by the reference method precision.

Conclusions

The important red wine grape quality parameters total anthocyanin concentration, TSS and pH can be measured rapidly by NIR spectroscopy. While large "global" calibrations across multiple regions, varieties and seasons provide useful degrees of measurement precision by NIR, calibrations further refined to particular regions, varieties and seasons provide significantly greater precision, at least for total anthocyanin concentration and pH. This study is continuing in 2001 when another ~ 2000 red wine grape samples from a broader range of Australian regions will be analysed, further elucidating regional varietal (and seasonal) effects on optimal calibration design. In addition, as the data set is now growing to an appropriately large size, it is intended to explore the application to it of Shenk's LOCAL algorithm[7] and of artificial neural networks (ANN) for further development of calibrations.

Acknowledgments

The authors thank Peter Høj from the Australian Wine Research Institute for his continuous encouragment and support in regard to NIR research. This research has been funded by the Grape and Wine Research and Development Corporation (Australia) and by the Cooperative Research Centre for Viticulture (Australia).

References

1. I.L. Francis, P.G. Iland, W.U. Cynkar, M Kwiatkowski, P.J. Williams, H. Armstrong, D.G. Botting, R. Gawel and C. Ryan, in *Proceedings–Tenth Australian Wine Industry Technical Conference, Sydney, Australia 1998*, Ed by R.J. Blair, A.N. Sas, P.F. Hayes and P.B. Høj. Australian Wine Research Institute, Adelaide, Australia, pp. 104–108 (1999).
2. M. Gishen and R.G. Dambergs, *Australian Grapegrower and Winemaker,* **414(a),** 43 (1998).
3. M. Gishen, R.G. Dambergs, A. Kambouris, M. Kwiatkowski, W.U. Cynkar, P.B. Høj and I.L Francis, in *Near Infrared Spectroscopy: Proceedings of the 9th International Conference*, Ed by A.M.C. Davies and R. Giangiacomo, NIR Publications, Chichester, UK, pp. 917–920 (2000).
4. *The Australian and New Zealand Wine Industry Directory: 18th Annual Edition 2000*, Winetitles, Adelaide, Australia (2000).
5. The Unscrambler Manual 7.5 Addendum, CAMO ASA, Trondheim, Norway (1999).
6. R.G. Dambergs, A. Kambouris, M. Gishen and I.L. Francis, in *Modern Viticulture- meeting market expectations,* Proceedings of the Australian Society of Viticulture and Oenology, Winetitles, Adelaide, South Australia, pp. 45–47 (2000).
7. J.S. Shenk, P. Berzaghi and M.O. Westerhaus, in *Near Infrared Spectroscopy: Proceedings of the 9th International Conference*, Ed by A.M.C. Davies and R. Giangiacomo, NIR Publications, Chichester, UK, pp. 211–214 (2000).

Near infrared determination of soluble solids in intact melons from a progeny by non-contact mode using a fibre-optic probe

Hidekazu Ito and Nobuko Fukino-Ito

National Institute of Vegetable and Tea Science(NIVTS), Kusawa, Ano, Age, Mie, 514-2392, Japan, e-mail:hito7@affrc.go.jp

Introduction

A previous paper has described the improvement of standard errors of the predicted soluble solids (Brix) in a melon cultivar by non-contact mode using near infrared (NIR) spectroscopy.[1] We then examined mature and immature fruits to provide a broad range of Brix. The objective of this study was to determine the potential of NIR spectroscopy for non-destructive estimation of Brix in melon breeding programmes.

Materials and methods

F_3, F_4, F_5 and F_6 plants, derived from the cross between 'PMAR No.5' and 'Harukei 3', were used as samples. The F_3 and F_4 plants were grown in a plastic greenhouse at NIVTS in 2000, the F_5 and F_6 plants were grown in 2001 and the parents were grown during both years. The harvesting stage of the fruits was judged mainly by external appearance (yellowing of rind and/or formation of abscission layer and so on). The optical absorption spectra were measured using a NIRSystems model 6500 spectrophotometer equipped with a fibre optic probe (Silver Spring, MA, USA). To measure the optical absorption spectra, each fruit was hand-placed with several mm distance apart from the end of the probe so that the blossom end was centred (Non-contact mode).[1] The original spectra were converted to the 2nd derivative spectra. Fruits (n = 31) that were harvested early in 2000 were used as a calibration sample set. A commercial program (NSAS ver. 3.27) was used for multiple linear regression analysis. The other fruits were used as prediction sample sets. Following optical measurement, a piece of tissue was cut out from the blossom end with a cork borer (diameter 18 mm). To obtain its juice, the tissue was comminuted with a grater and centrifuged. Brix of the juice was determined using a temperature compensated refractometer (ATAGO model DBX-55, Japan).

Results

For the calibration sample set (n = 31), *SEC* was 1.27%, with an *MR* of 0.869. The wavelengths selected in the regression equation were 902, 878 and 850 nm [Figure 1(a)]. For the prediction sample sets (n = 40 in 2000, n = 26 in 2001), their *RMS* were 1.09 [Figure 1(b)] and 1.58% [Figure 1(c)] respectively. As mentioned above, Brix in the melons from a progeny could be estimated well.

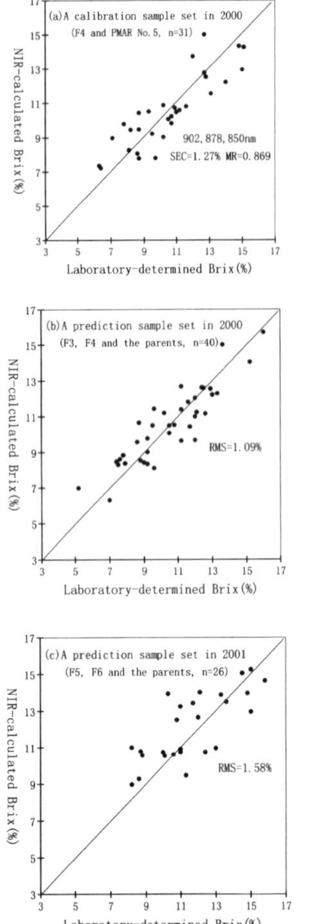

Figure 1. Plots of laboratory-determined Brix vs NIR-calculated Brix for the melons from a progeny.

Discussion

Absorbances at 902, 878 and 850 nm were included in the multiple regression equation as independent variables and the former two wavelengths are key wavelengths for non-destructive Brix determination. Many researchers have selected similar wavelengths for Brix prediction in melons,[1,2] onions,[3] peaches,[4–6] nectarins[6] and pineapples.[7]

Usually, individual cultivar calibrations are successful while predicting Brix in the same cultivar's prediction sample set within the same season. Practically, a calibration equation will be applied to other samples. In this case, the absolute value of a bias sometimes becomes large.[5] In this experiment, fruits that were harvested early in 2000 were used as the calibration sample set. The other fruits were used as the prediction sample sets. The size, weight, netting, surface colour, flesh colour, length of flesh and so on of the melons were more various than those of a melon cultivar (Data not shown). Nevertheless, Brix in the melons could be estimated well (Figure 1).

The previous paper has described improvements in standard errors of the predicted Brix in the melon cultivar by non-contact mode using NIR spectroscopy.[1] We then examined the mature and immature fruits to provide a broad range of Brix. In this experiment, when the results of the non-contact mode were compared with those of the contact mode (usual method), the former mode also improved the latter standard errors in spite of the fruits being almost mature (Data not shown).

Conclusion

NIR spectroscopy could estimate Brix in the melons from a progeny well. 902 and 878 nm, which are the key wavelengths for non-destructive Brix determination, were included in the multiple regression equation as independent variables. Non-contact mode could improve standard errors of the progeny as well as a cultivar compared with contact mode (usual method). It was concluded that NIR spectroscopy could be used for non-destructive measurement of critical selection traits (Brix) in melon breeding programmes.

References

1. H. Ito, K. Ippoushi, K. Azuma and H. Higashio, in *Proceedings of the 9th Intl. Conference*, Ed by A.M.C. Davis and R. Giangiacomo. NIR Publications, Chichester, UK, p. 859 (2000).
2. G.G. Dull, G.S. Birth, D.A. Smittle and R.G. Leffler, *J. Food Sci.* **54(2),** 393 (1989).

3. G.S. Birth, G.G. Dull, W.T. Renfore and S.J. Kays, *J. Amer. Soc. Hort. Sci.* **110(2),** 297 (1985).
4. S. Kawano, H. Watanabe and M. Iwamoto, *J. Japan Soc. Hort. Sci.* **61(2),** 445 (1992).
5. K.H.S. Peiris, G.G. Dull, R.G. Leffler and S.J. Kays, *J. Amer. Soc. Hort. Sci.* **123(5),** 898 (1998).
6. D.C. Slaughter, *Transactions of the ASAE* **38(2),** 617 (1995).
7. J. Guthrie and B. Wedding, *J. Near Infrared Spectrosc.* **6,** 259 (1998).

Application of near infrared spectroscopy for nondestructive evaluation of protein content in ginseng

Guo-lin Lin, Mi-Ryeong Sohn, Eun-Ok Kim and Rae-Kwang Cho
Deptartment of Agricultural Chemistry, Kyungpook National University, Taegu, 702-701, Korea

Introduction

Ginseng has been regarded, for a very long time, as a secret and expensive medicine in the East Asian countries of Korea, China and Japan. Even two thousand years ago, the medical effects of ginseng were recorded in some ancient Chinese medicine books.

Recently, many ginseng producers and consumers in Korea have been damaged because the low-priced Chinese ginsengs have been disguised as the high-priced domestic ones in the market. There are some traditional factors, such as colour, shape and size, to determine the country of origin by the naked eye. However, there is always a difference between individuals. The shape of Chinese peeled ginseng is very similar to that of the Korean one. It is very difficult to distinguish with the naked eye and the result is uncertain. For this reason, it is necessary to distinguish the place of origin and age of ginseng, rapidly and non-destructively. In a previous study we have already reported on the discrimination of origin and age of ginseng by using near infrared (NIR) spectroscopy.

In this work, we attempted to quantify protein content in ginseng by using NIR spectroscopy.

Materials and methods

Sample collection and preparation

Ginseng radix, the root of *Panax ginseng C. A. Meyer*, was studied. A total of 120 samples were used in this study, which consisted of six sets, 4, 5 and 6-year-old samples of Korea ginseng and 6, 7 and 8-year-old samples of Chinese ginseng, respectively. Each sample set was composed of 20 samples. All samples were prepared as a powder using a cyclone mill (Laboratory mill 120, Perten co.) and filtering with a 0.35 mm sieve.

Chemical analysis

The elemental content was measured using an EA1106 (Carlo Erba, Co.) elemental analyser for oxygen and an EA1108 (Carlo Erba, co.) for other elements such as N, C and H, respectively. The protein content was calculated from the nitrogen content (N) using a factor of 6.25. Conditions for EA were as follows: TCD detector, working in 1000°C for EA 1108 and in 1070°C for EA 1106, 0.1 mg of sample weight.

Table 1. Mean value of O, C, H and N content in Korean and Chinese ginseng using an elemental analyser.

Elements	China ginseng (%)	Korea ginseng (%)
O	44.0	45.0
C	39.6	39.7
H	6.2	6.0
N	2.1	1.9
Others	8.1	7.4
Total	100	100

NIR analysis

NIR reflectance spectra were collected over the 1100 to 2500 nm spectral region with an InfraAlyzer 500 (Bran+Luebbe, Germany) equipped with a halogen lamp and PbS detector and data were collected every 2 nm data point intervals. The calibration models were carried out by multiple linear regression (MLR), principal component regression (PCR) and partial least squares (PLS) analysis using IDAS and SESAME software.

Results and discussion

Table 1 shows the results of the elemental analysis of Korean and Chinese ginsengs. The elemental contents of O, C, H and N are similar between both sample sets, respectively. The N content of the Chinese set is 2.1%, which is higher than the value of 1.9% for the Korean samples. Table 2 shows the minimum, maximum and mean value of the protein content of Korean and Chinese sample sets. In the Korean sample sets, the nitrogen content tends to increase according to the year of cultivation. The mean values for protein content were 12.52% for 4-year and 12.56% for 6-year old, respectively. In contrast, Chinese sample sets show the opposite results. The value gradually decreased according to the year of cultivation; 13.56% for 6-year old and 9.98% for 8-year old, respectively. Table 3 shows the calibration and prediction results for determining nitrogen content in ginseng by using NIR. We can find different results according to regression methods used and various spectral pre-treatments. The

Table 2. Protein content of ginseng sample sets by elemental analysis.

Sample sets	Protein Content (%)			
Origin	Cultivation years	Minimum	Maximum	Mean
Korea	4	8.55	15.15	12.52
Korea	5	8.76	17.23	12.56
Korea	6	9.28	24.41	13.45
China	6	9.58	18.69	13.56
China	7	9.41	18.08	12.39
China	8	5.16	14.30	9.98

Table 3. Calibration and prediction results of models developed by various regression method and pre-treatment.

Software	Regression	Pre-treatment	*R*	*SEE* (%)	*RMSEP* (%)	Number of Wavelengths/ factors
IDAS	MLR	Absorbance	0.984	0.506	0.717	8
SESAME	MLR	2nd Deriv.	0.972	0.623	0.596	9
	PCR	2nd Deriv.	0.922	1.17	1.120	6
		Normalisation	0.928	0.974	0.944	6
	PLSR	Absorbance	0.956	0.768	0.741	7
		Normalisation	0.969	0.659	0.630	9
		Smoothing	0.956	0.767	0.741	7
		2nd Deriv.	0.965	0.691	0.664	8

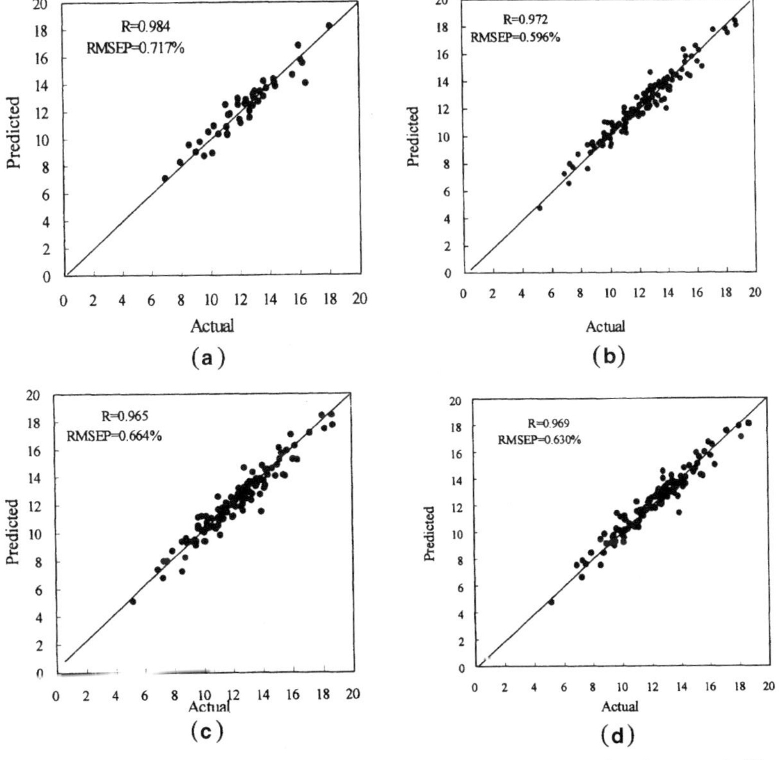

Figure 1. Scatter plots of NIR predicted versus measured protein content in ginseng. Calibration models were developed by different regression and pre-treatment, respectively: (a) MLR (IDAS) and absorbance treatment, (b) MLR (SESAME) and 2nd Derivative, (c) PLSR and 2nd Derivative and (d) PLSR and normalisation.

calibration model by IDAS is a little higher than the one by MLR. When the SESAME software was used, the calibration model was generally accurate in the order of MLR ($R = 0.972$), PLSR ($R = 0.956$–0.969) and PCR ($R = 0.922$–0.928). It was expected that improvement of the calibration model would be obtained by various pre-processing of the spectral data.

For example, *R* and *RMSEP* have the range of 0.9560 – 0.969 and 0.659%–0.767% in calibration models using raw absorbance, normalisation, smoothing and second derivative as the input variables. The best calibration model was obtained using second derivative processing and MLR regression, where *R* was 0.972 and *RMSEP* was 0.596%. The nine wavelengths, 1340, 1412, 1490, 1784, 1862, 1932, 2080, 2204 and 2290 nm were used for this calibration model.

This result indicates that NIR spectroscopy could be used for determine the protein content in ginseng radix with high accuracy. Figure 1 shows the relationship between NIR predicted value and the analysed protein content usind developed MLR calibration models.

Conclusion

The quantification of protein content in ginseng was studied using NIR reflectance spectroscopy and we investigated the effect of regression methods and various pre-processing of the spectral data. The protein content of Korean ginseng tended to increase according to the increase in cultivation years, 4 to 6 years old, but Chinese ginseng showed the opposite result. The accuracy of the calibration model decreased in the order of MLR, PLSR and PCR. The best calibration model was obtained by MLR and a second derivative pre-processing which resulted in 0.972 *R* and *RMSEP* of 0.596% of *RMSEP*.

Reference

1. *An important agreement book of Korea Ginseng*. Korea Ginseng and Tobacco Research Institute (1985).
2. B.H. Han, *Korea J. Pharmacol.* **9(4),** 169 (1978).
3. M.H. Park, *Korean J. Ginseng* **5(2),** 155 (1981).
4. Y.O. Ahn, *Int. J. Cancer Supplement* **10,** 7 (1997).
5. J.D. Park, *Korean J. Ginseng Sci.* **20(4),** 389 (1996).
6. T.K. Yun, *Korean J. Ginseng Science*, **20(2),** 255 (1996).
7. X. Wang, *Anal. Chem.* **71(8),** 1579 (1999).

Discrimination and quantitative analysis of watercore in apple fruit by near infrared transmittance spectroscopy

Eun-Ok Kim,* Mi-Ryeong Sohn, Young-Kil Kwon, Gou-Lin Lin and Rae-Kwang Cho

Department of Agricultural Chemistry, Kyungpook National University, Taegu 702-701, Korea

Introduction

In Korea, most consumers tend to prefer apples including watercore. Apples with watercore are sweeter and more expensive than apples without watercore. However, watercore is a kind of internal disorder[1,2] of apple fruit. It causes the fruit flesh to brown and tissue damage during storage[3] as shown in Figure 1. Finally, these apples will lose their value in the market.[4,5] So, watercore is a very important factor for the storage and sorting industry of apple fruit.[6] However, it is difficult to find apples which have watercore in them. A visual inspection is generally applied to evaluate the amount of watercore by randomly selection of apple fruits. Therefore, a rapid, accurate and non-destructive method is needed.

Previous studies reported on some methods such as X-ray imaging,[7,8] camera imaging,[9] machine vision,[10] magnetic resonance imaging[11] and near infrared (NIR) reflectance spectroscopy[12] for watercore. However, these methods did not obtain high accuracy.

In this study, we attempted the discrimination and quantification of watercore in apple fruit by using NIR transmittance spectroscopy.

Materials and methods

Apple fruits

Apple fruits (*Fuji* variety) were colleted for the 1999 harvest season in Kyungpook prefecture, Korea.

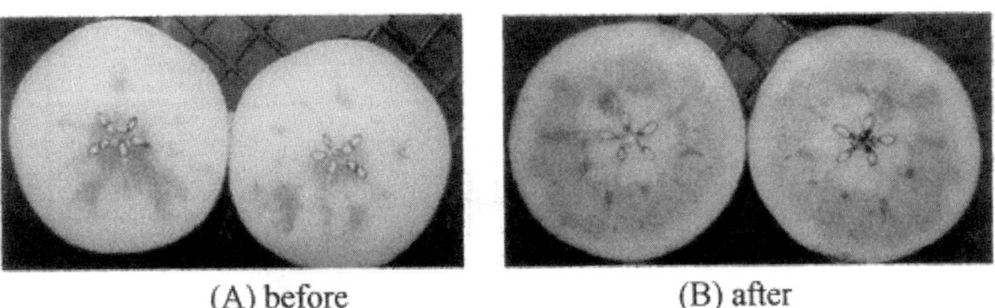

(A) before (B) after

Figure 1. Internal change of apple fruit having watercore (a) before and (b) after storage.

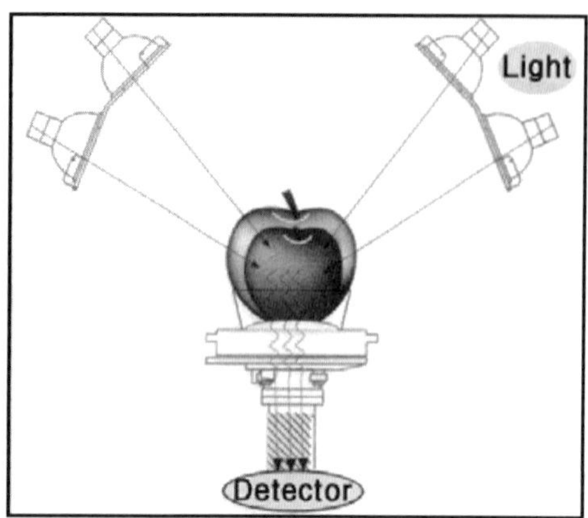

Figure 2. Schematic for measuring NIR spectra of apple fruits.

Spectral analysis

The apparatus used in this study is shown in Figure 2. For the measurement of watercore, the apple fruit was illuminated by four 10 V halogen lamps (Philips Inc., The Netherlands). NIR transmittance spectra were collected using an AH-4240 instrument (American Holographic. Inc). The spectrometer provided 528 data points from 578 nm to 1106 nm. Reference spectra were obtained by measuring a polystyrene sphere. Partial least squares regression (PLSR) was performed using the Unscrambler (Camo Asa, USA). The accuracy of the calibration model was evaluated using the multiple correlation coefficient (R) and root mean square error of prediction ($RMSEP$).

Measurement of watercore content

After measuring NIR spectra, samples were sliced at intervals of 10% thickness according to the cross section. Each slice was scanned by a scanner and then saved onto a computer. Each image was converted to gray scale and treated on the threshold with an image treatment program (Image-Pro PLUS, Media Cybernetics, USA). The process is as shown in Figure 3. The amount of watercore in each apple slice was measured using the threshold image and the watercore content was calculated by the following equation.

$$\text{Watercore content (\%)} = \frac{\text{Total watercore area}}{\text{Total apple area}} \times 100$$

Results and discussion

Figure 4 shows the raw transmittance NIR spectra of apple samples. The spectral difference was found at 732 nm and 820 nm. The high-watercore apples tended to show a higher transmittance values than those of medium- or no-watercore apples. Figure 5(a) shows raw transmission spectra of 75 apple

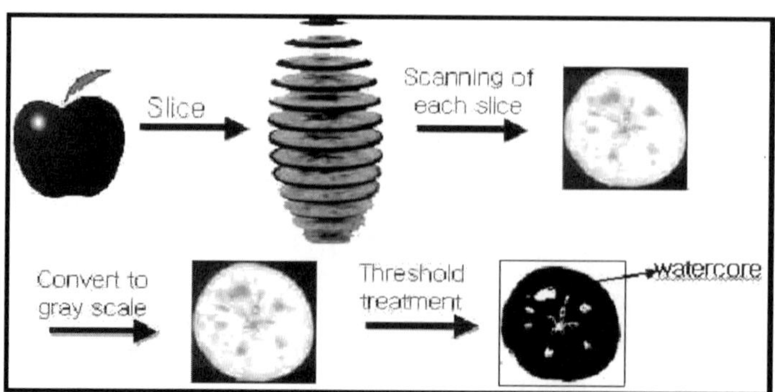

Figure 3. Processing for measuring of watercore content.

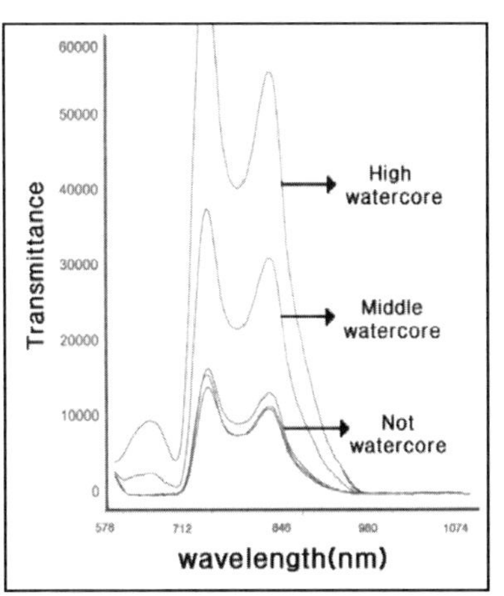

Figure 4. NIR raw transmittance spectra of apple fruits.

samples having a wide range of watercore content. First and second derivatives of spectra were used for an ideal data analysis. Figure 5(b) shows second derivative spectra of apple samples.

Table 1 shows the result of the PLSR calibration model and prediction for the quantification of watercore in apple fruit for each data pretreatment. The correlation coefficient (*R*) and *RMSEP* were 0.959 and 1.379% for raw spectra, respectively. The model was improved after pre-processing of spectral data. For the first derivative model, *R* and *RMSEP* were 0.990 and 0.675% for the calibration set and 0.982 and 0.930%.

The second derivative model resulted in 0.975 of *R* and 1.082% of *RMSEP*. The first derviative was more effective than the second derivative for determining watercore in apples. Fig-

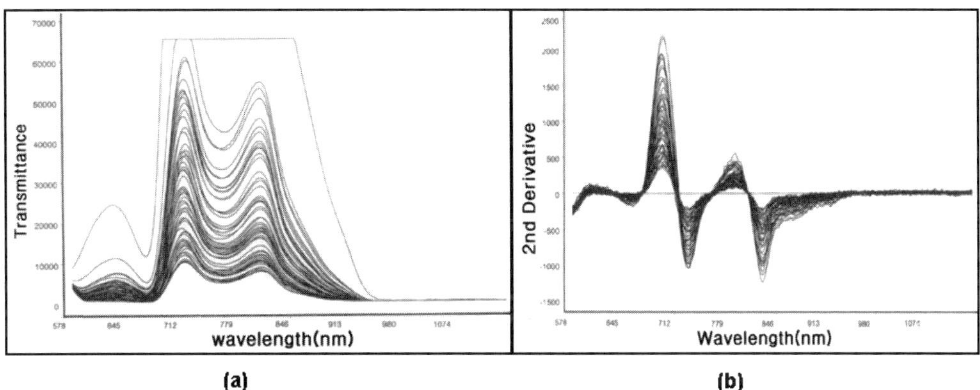

(a) (b)

Figure 5. (a) Raw and (b) second derivative NIR spectra of 75 apple fruit samples.

Table 1. Result of PLSR calibration and prediction for determining watercore content in apple fruit.

Data treatment	Wavelength region	Sample set	Sample no.	*R*	*RMSEP*(%)
Raw	578–1106 nm	Calibration	45	0.978	1.015
		Validation	30	0.959	1.379
First derivative	578–1106 nm	Calibration	45	0.990	0.675
		Validation	30	0.982	0.930
Second derivative	578–1106 nm	Calibration	45	0.983	0.886
		Validation	30	0.975	1.082

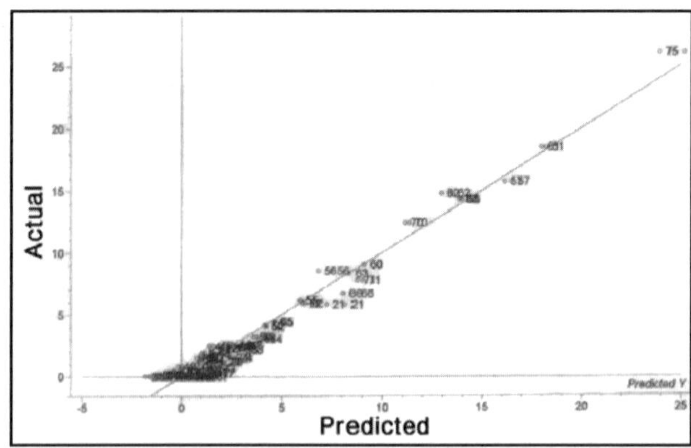

Figure 6. Scatter plots of actual and NIR predicted watercore contents in apple fruit.

ure 6 shows the relationship between actual watercore content and NIR predicted using the 1st derivative PLS model.

Conclusion

The possibility of non-destructive evaluation of watercore in *Fuji* apple fruit using NIR transmittance spectroscopy was investigated and the calibration model was developed for quantification of watercore content. The PLSR model using the first derivative of the spectral data resulted in acceptable error level. The correlation coefficient (*R*) and root mean square error of prediction (*RMSEP*) were 0.990 and 0.930%, respectively. This result indicates that NIR transmittance spectroscopy could be used for discrimination and quantification of watercore in apple fruit.

References

1. I. Ferguson, R. Volz and A. Woolf, *Postharvest Biology and Technology* **15,** 255 (1999).
2. F.R. Harker, C.B. Watkins, P.L. Brookfield and M.J. Miller, *J. Amer. Soc. Hort. Sci.* **124(2),** 166 (1999).
3. R.K. Volz, W.V. Biasi, J.A. Grant and E.J. Mitcham, *Postharvest Biology and Technology* **13,** 97 (1998).
4. Y.C. Hung, Y.Y. Hao, E.W. Tollner and B.L. Upchurch, *Transactions of the ASAE* **37(4),** 1249 (1994).
5. H. Yamada, H. Ohmura, C. Arai and M. Terui, *J. Amer. Soc. Hort. Sci.* **119(6),** 1208 (1994).
6. M.A. Shahin, E.W. Tollner, M.D. Evans and H.R. Arabnis, *Transactions of the ASAE* **42(6),** 1889 (1999).
7. T.F. Schatzski, R.P. Haff, R. Young and I. Can, *Transactions of the ASAE,* **40(5),** 1407 (1997).
8. E.W. Tollner, Y.C. Hung, B.L. Upchurch, *Transactions of the ASAE* **35(6),** 1921 (1992).
9. J.A. Throop, D.J. Aneshansley and B.L. Upchurch, *Transactions of the ASAE* **37(3),** 873 (1994).
10. B.L. Upchurch and J.A. Throop, *Transactions of the ASAE* **37(2),** 483 (1994).
11. C.J. Clark, J.S. MacFall and R.L. Bieleski, *Scientia Horticulture* **73,** 213 (1998).
12. M.-R. Sohn, Doctor thesis of Kyungpook National University (1999).

Measurement of quality properties of honey by reflectance spectra

C.H. Choi[a], W.J. Yang[a], J.H. Sohn[b] and J.H. Kim[c]

[a]*Dept. of Bio-Mechatronic Engineering, Sungkyunkwan University, Suwon 440-746, Korea*

[b]*Korean Bee Product's Research Institute, Suwon 440-150, Korea*

[c]*Korean Food Research Institute, Songnam 463-420, Korea*

Introduction

Honey is the nectar and saccharine exudations of plants that have been gathered, modified and stored in the comb by honeybees. It is the only sweetening material that can be stored and used as it is produced in nature.[1] Honey consists essentially of different sugars, predominantly glucose and fructose. Besides these, honey also contains protein, ash, amino acids, enzymes, organic acids, mineral substances, pollen and oligosaccharides.[2]

There has been a considerable lack of information concerning nutritional aspects of honey. Consumer groups are demanding an accurate description of the nutrient content and its quality. The composition of honey depends upon the floral sources, the composition of the nectar, the climate and the differences in processing.[3] The quality of honey is determined by the composition of sugar, moisture, ash, protein, hydroxymethyl furfural (HMF), proline, PH, diastase activity etc.[4]

The objectives of this study were to develop models to predict quality properties of Korean honeys by visible and near infared (NIR) spectroscopic techniques.

Materials and methods

Two kinds of honey from acacia and polyfloral sources were tested in this study. The honeys were harvested in the spring of 2000 and stored in the storage facility at 20°C during experiments. A total of 394 samples of honey were collected from the Korean Bee Product's Research Institute, Suwon, Korea. Reflectance spectra, moisture content, ash, invert sugar, sucrose, fructose/glucose (F/G) ratio, hydroxymethyl furfural (HMF) and carbon isotope (C12/C13) ratio of honey were measured. Honey samples were left at room temperature when their measurements were made. Temperatures of honey samples were not measured or controlled during measurements.

A digital refractormeter (Atago Co, Japan) was used to measure the moisture content of honey. Ash in honey was determined by the AOAC method.[5] Invert sugar, sucrose, and HMF in honey were measured by high performance liquid chromatography (HPLC) following the Korean Food Code methods.[6] Isotope Mass (Integra-CN, Europa, UK) was used to measure C12/C13 ratio of honey and Pee Dee Belemnite (PDB) was used as the reference.

An NIR spectrophotometer, equipped with a single-beam scanning monochromator (NIRSystems, Model 6500, USA) and a horizontal set-up module, was used to collect reflectance data from the honey. The reflectance spectra were measured in wavelength ranges of 400–2,498 nm at 2 nm intervals. Thirty-two repetitive scans were averaged, transformed to log(1/Reflectance) and were then stored in a microcomputer file, forming one spectrum per measurement.

Honey samples were divided into a calibration set and a validation set. Samples were ranked by values of invert sugar and each set was selected by increasing rank, evenly. Half the total samples were selected for the calibration set and the other 50% were reserved for the validation set. The calibration set was used during model development and the validation set was used to predict quality properties from unknown spectra.

The method of partial least square (PLS) analysis was used to determine the quality properties of honeys. A unique set of PLS loading vectors (factors) was developed. Multiplicative scatter correction (MSC) pre-treatment was applied to all spectra to minimise sample-to-sample light scatter differences. A commercial software package, the Unscrambler (CAMO ASA, Norway), was used to perform the PLS analyses. For each constituent, up to 20 factors were examined. Cross-validation was performed during model development, where calibration samples were removed, one at a time, from the calibration set. The standard error of calibration (*SEC*) was considered to determine the optimal number of factors during calibration. On completion of the calibration, the model was used to predict quality properties of honey from the validation set. Model performance was reported as the correlation coefficient (r), the standard error of prediction (*SEP*) and the average difference between measured and predicted values (bias).

Results and discussion

The minimum, maximum and average values of quality properties of the tested honey are listed in Table 1. The average moisture content of the honey was 19.9% with the range from 17.0 to 27.3%. Moisture content is an important constituent to determine the quality of honey, because it bears direct relation to undesired fermentation. About 90% of tested samples had less than 21.0% moisture content. The ash in the honey ranged from 0.05 to 0.22% with average value of 0.12%. All samples satisfied the Food Code requirements for ash content.

The sugar content is the most important constituents for honey quality. The average invert sugar in honey was 68.4% with the range from 55.7 to 77.3%. About 91% of samples exceeded 65.0% of invert sugar. The average sucrose content of the honey was 5.7% with a wide range from 2.2 to 15.4%. The Food Code requirement for sucrose is less than 7.0%, and 92% of samples were less than limit. The F/G ratio ranged from 1.04 to 2.04 with average value of 1.27.

The average HMF in honeys was 14.4 mg kg^{-1} with the range from 10.0 to 24.9 mg kg^{-1}. The most samples (99%) met the requirement of the honey standards regarding HMF content. The carbon isotope (C12/C13) ratio of honey showed some indication of quality of grading. The C12/C13 ratio

Table 1. Quality properties of honey tested.

	N	Average	Maximum	Minimum
Moisture (%)	394	19.9	27.3	17.0
Ash (%)	393	0.121	0.22	0.05
Invert sugar (%)	394	68.4	77.3	55.7
Sucrose (%)	394	5.7	15.4	2.2
F/G ratio	394	1.27	2.04	1.04
HMF (mg kg^{-1})	208	14.4	24.9	10.0
C12/C13 ratio	359	–19.1	–11.2	–28.3

Table 2. Results of PLS Calibration for honey.

	Wavelengths (nm)	Factor	Correlation	SEC
Moisture (%)	1100 ~ 2200	5	0.985	0.297
Ash (%)	1400 ~ 1800	11	0.873	0.802
Invert sugar (%)	1100 ~ 1300, 1600 ~ 1800	6	0.959	0.794
Sucrose (%)	1100 ~ 1300, 1600 ~ 1800	7	0.966	0.440
F/G ratio	1100 ~ 1300	8	0.988	0.033
HMF (mg kg^{-1})	1100 ~ 1300	8	0.802	2.420
C12/C13 ratio	1100 ~ 1300, 1400 ~ 1800, 1900 ~ 2200	12	0.968	0.092

Table 3. Results of PLS Validation for honey.

	Correlation	SEP	Bias
Moisture (%)	0.973	0.390	0.057
Ash (%)	0.900	0.012	0.000792
Invert sugar (%)	0.942	0.862	0.022
Sucrose (%)	0.952	0.456	–0.035
F/G ratio	0.967	0.042	0.000619
HMF (mg kg^{-1})	0.628	3.320	0.961
C12/C13 Ratio	0.948	1.067	–0.036

ranged from –28.3 to –11.2 with an average value of –19.1. It was found that only 19% of samples were less than –23.0 of C12/C13 ratio.

The PLS analyses showed good correlation between reflectance spectra and quality properties of the honey. As shown in Tables 2 and 3, the PLS model using raw spectra without pre-processing

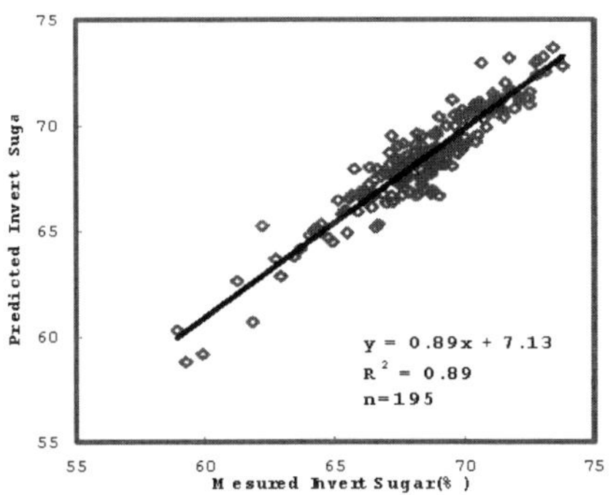

Figure 1. Comparison of actual and predicted values of invert sugar of honey.

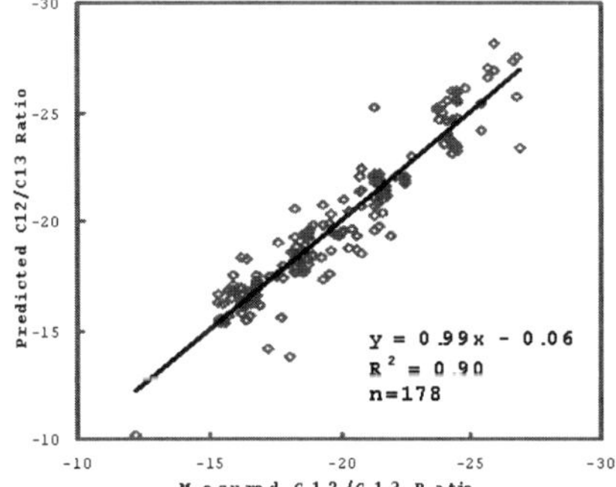

Figure 2. Comparison of actual and predicted values of C12/C13 ratio of honey.

showed the best results in the NIR ranges, but some constituents used many factors. It means that models may include systematic noise to predict the constituents from unknown honey samples. The PLS model for moisture content showed the best performance by using five factors. The moisture content model had a correlation coefficient of 0.973 and an *SEP* of 0.390%. The PLS model for invert sugar had a correlation of determination (R^2) of 0.89 and an *SEP* of 0.862%, as shown in Figure 1. The PLS model for C12/C13 ratio had a coefficient of determination (R^2) of 0.90 and an *SEP* of 1.067, as shown in Figure 2.

The PLS model, using raw reflectance spectra, showed good performance to predict moisture content, ash, invert sugar, sucrose, F/G ratio and C12/C13 ratio of honey in the wavelength range of 1100–2200 nm. However, the PLS analysis was not good enough to predict the hydroxymethyl furfural (HMF) content of the honey. The HMF model had a correlation coefficient of 0.628 and an *SEP* of 3.32 mg kg^{-1}, as shown in Table3.

References

1. J.W. White, Jr, M.L. Riethof, M.H. Subers and I. Kushnir, *Technical Bull.* US Department of Agriculture, Washington, DC, USA, 1261 (1979).
2. H.G. Chang, M.K. Han and J.G. Kim, *Korean J. Food Sci. Technol.* **20(5),** 631 (1988).
3. D.C. Lee, S.Y. Lee, S.H. Cha, Y.S. Choi and H.I. Rhee, *Korean J. Food Sci. Technol.* **29(6),** 1082 (1997).
4. E.S. Kim and C.O. Rhee, *J. Korean Soc. Food Sci. Nutr.* **25(4),** 672 (1996).
5. AOAC, $14^{th.}$ Edn. Association of Official Analytical Chemists. Washington, DC, USA, p. 589 (1984).
6. Korea Food Industry Association, Food Code. 501 (1994).

Qualifying food flavours with near infrared spectroscopy and chemosensor-arrays (electronic nose)

Zsolt Seregély[a] and Károly Kaffka[b]

[a]*Metrika R & D Co., H-1119 Budapest, Petzvál J. u. 25. Hungary*

[b]*Szent István University, Faculty of Food Science, H-1118 Budapest, Ménesi út 45, Hungary*

Introduction

The origin and useage of food flavours and aromas prior to written civilisation remain a mystery. However, the existence of flavour production on an industrial scale is thought to date from the beginning of the 19th century. Flavours are connected to the food industry mainly through confectionery products, liquors and other distilled products and is interwoven with the production of perfumes. As the knowledge of chemistry expanded, natural products were synthesised, others were altered and new products were created. In modern times, product quality can be defined, to some extent, in terms of composition. Numerous regulations exist which restrict the use of synthetic materials. Yet, to understand food quality, researchers keep coming back to sensory tests, tests that depend on fallible human opinion. Training humans to make subjective judgements of quality is very expensive. Consequently, researchers are constantly searching for objective instrumental techniques for determining quality.

The term "Electronic Nose" (or, e-nose) is a generic name for an analytical instrument that profiles headspace volatiles over or around a sample. The technology is based on an array of chemical sensors whose outputs are integrated by advanced signal processing to rapidly identify complex aromatic mixtures.

Near infrared (NIR) spectroscopy and e-nose technology were used in this study to classify food flavours. E-nose data was related to NIR data with the help of chemometric methods that involved "sequence optimisation" and the polar qualification system (PQS).[1–5]

Materials and methods

The aims of this work were to discriminate and identify eight artificial flavours used in the food industry and to discriminate the same aromas originating from different production circles. (The previous sentence is not understandable in english) The investigated food flavours were as follows: cacao, mint, hazelnut, melon, egg liqueur, toffee, vanilla and green apple.

A Spectralyzer, Model 1025, spectrometer was used to record all NIR spectra over the range 1000–2500 nm at 2 nm intervals. A SamSelect chemosensor array (Daimler Chrysler Aerospace RST Rostock, Germany) was used to record the e-nose data. The e-nose instrument incorporated six individual quartz crystal sensors coated with different gas-sensitive materials. The adsorption of volatile molecules on the sensor surfaces caused mass changes resulting in frequency changes of the oscillators. The frequency changes of the six crystals gave six data (one point in the multidimensional space) per measurement from the che3mosensor array. During our experiments we repeated the measurements 15 times per sample (15 points per sample). After normalising the data, (by dividing each sensor signal with the average of the signal coming from the six sensors of the array) principal component analysis (PCA) was used, in conjunction with PQS optimisation techniques, to create classification models from the sensor array data.

PQS was developed to work with NIR spectra. The basic principles of the method are described in the relevant literatures.[1–5] PQS requires the determination of "quality points" defined as the centre of the spectra represented in polar co-ordinates. While the sequence of the data are naturally determined using NIR spectroscopy, there are several multivariate tasks where the order of the data can be changed. The goal is to determine the optimal data sequence with respect to a given classification (giving the best distinction between two samples using their respective quality points). Sequence optimisation was developed to do just that.

Figure 1 shows normalised data from the sensor array for the eight food aromas studied. The data is shown in (a) rectangular and (b) polar coordinates.

The effectiveness of classification was expressed numerically by calculating the normalised distance and sensitivity.[5] The normalised distance is the absolute distance (distance betweent he centre of the clusters) divided by the sum of the absolute distance and the sum of the standard deviations of the quality point coordinates of the investigates samples. By optimising the normalised distance, the calculated data sequence provides the smallest standard deviations of the quality points of the investigated sample groups relative to their distance.

In the first step of our statistical evaluation (carried out on the e-nose data), PCA was used for classification. The eight food aromas were measured six times. Figure 2(a) shows the location of the quality points of the samples in the projection plane determined by the first two principal components. As

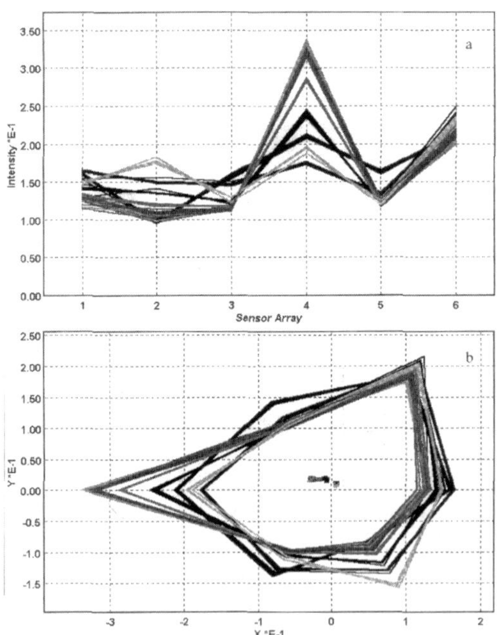

Figure 1. The normalised data of the sensor signals of the investigated eight food flavours measured by electronic nose represented as a spectra in the rectangular (a) and in the polar (b) co-ordinate system.

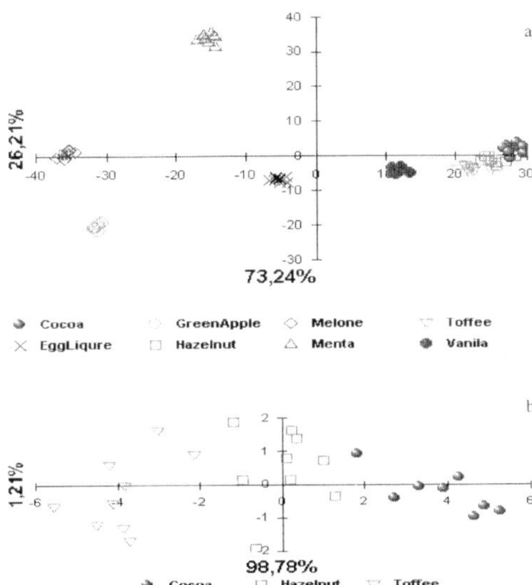

Figure 2. Multi-step discrimination using PCA. (a) The location of the quality points of the eight flavour samples in the first projection plane determined by the first two principal components. (b) The overlapping quality points of cacao, toffee and hazelnut samples can be well separated in an other projection plane.

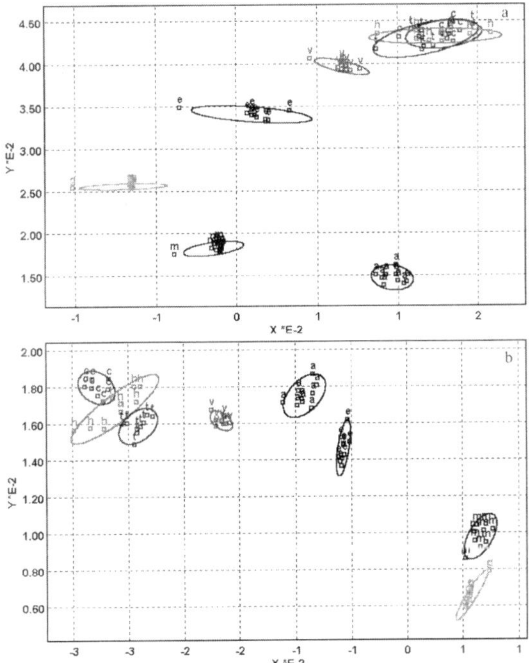

Figure 3. The location of the quality points of the investigated eight food flavour samples in (a) original order and (b) after sequence optimisation. The optimal data sequence is (S1 S5 S3 S4 S2 S6).

can be seen, the quality points of cacao, hazelnut and toffee samples are overlapping and the total identification of all samples can not be performed using only one projection plane. In such cases, multi-step classification can be used to perform the identification. The overlapping samples can be well separated in another projection plane calculated by omitting the already classified samples [Figure 2(b)].

During the second stage of our evaluation work, by defining the term of "sequence optimisation", the PQS technique—used to work with NIR spectra—was further developed and generalised as an evaluation method for almost any multivariate problems. By representing the data sets in the polar co-ordinate system, some data can be situated opposite each other balancing their shifting effect to the location of the quality points. As was mentioned, the aim of "sequence optimisation" is the determination of the optimal data sequence providing the best separation (highest normalised distance and sensitivity).

Figure 3(a) shows the location of the quality points (centre of their normalised data sets drawn as polar spectra) of the measured food flavours, in original data sequence, coming directly from the instrument. As can be seen, the normalised distances and the sensitivities are high. The distances are considerably higher among the groups compared with their standard deviations, but the separation of all samples can not be observed in the quality plane using the original data sequence, as the quality points of cacao, toffee and hazelnut samples are also overlapping here. After sequence optimisation to

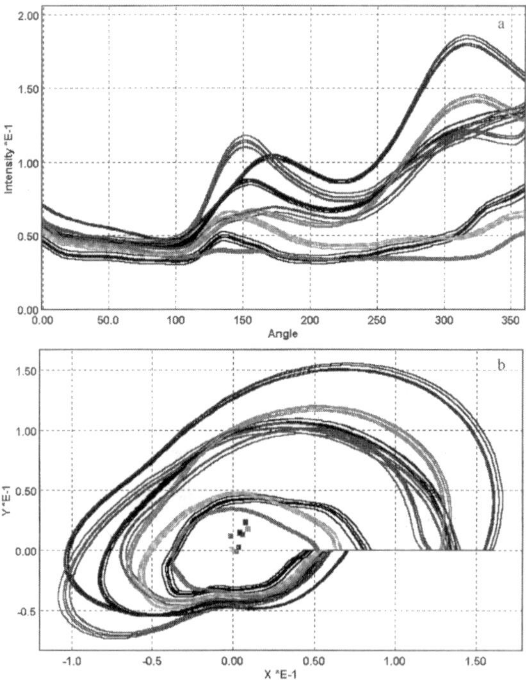

Figure 4. The NIR spectra of the investigated eight food flavour samples in the rectangular (a) and in the polar (b) co-ordinate system using the 1772–2128 nm wavelength range of the spectra. The non-selected points are omitted.

normalised distance, all the investigated flavour samples can be identified in one plane [Figure 3(b)]. The calculated optimal signal sequence is S1 S5 S3 S4 S2 S6.

Passing over to the NIR measurements, the optimal range of the NIR spectra of the investigated flavour samples providing the best separation is 1772–2128 nm. Figure 4 shows the optimal range of the NIR spectra of the food flavour samples in the (a) rectangular and (b) polar co-ordinate system. As was already published in our previous works, the goal of the used wavelength range optimisation is to determine that wavelength range (that part) of the spectrum which gives the best distinction of two sam-

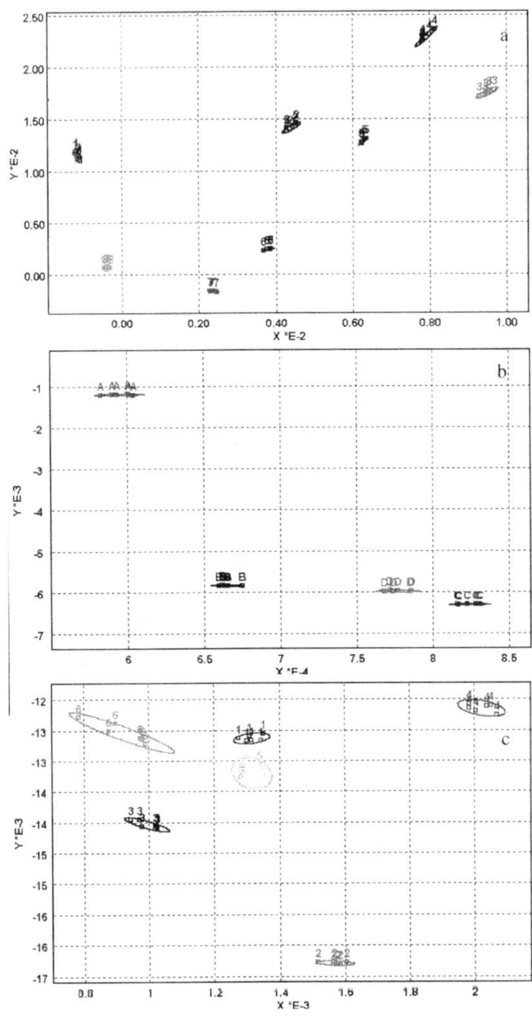

Figure 5. The location of the quality points of (a) the investigated eight food flavour samples using the 1772–2128 nm range, (b) the four lemon flavour samples using the 1628–1662 nm range and (c) the six vanilla flavour using the 1608–1984 nm range of their NIR spectra. The non-selected points are omitted.

ples according to one selected chemical or physical property using their quality points. The criterion of the "best distinction" must, of course, be defined. The maximum of the "normalised distance" or the "sensitivity" are the possible criteria of the optimum, with the help of which the best distinction could be determined. The results of the tasks formulated in our objectives are summarised in Figure 5. Figure 5(a) shows the location of the quality points of the eight flavour samples also investigated by electronic nose and classified in Figures 2 and 3. As can be seen in Figure 5 [four lemon groups (b), six vanilla groups (c)] even the products with the different production circles could be separated well. The used wavelength ranges providing the best separation are based on the normalised distance and the sensitivity.

Conclusion

The sensor signal response of the electronic nose, combined with principal component analysis or polar qualification system as well as NIR spectroscopy, are suitable technologies for discriminating and identifying the eight food aromas. The sequence optimisation opens new perspectives in the application of PQS, offering a rapid, accurate, cheap and simple method for qualifying or identifying products, using their different data sets.

Acknowledgement

The authors would like to express their sincere thanks to Sweet Point Ltd and Döhler Aroma Hungary Ltd for their courtesy in placing the food flavour samples at the author's disposal.

Supported by the National Scientific Research Fund (OTKA) Nos: T 023020 and No: T 032814.

References

1. K.J. Kaffka and L.S. Gyarmati, " *Proceedings of the Third International Conference on Near Infrared Spectroscopy*, Ed by R. Biston and N. Bartiaux-Thill. Agricultural Research Centre Publishing, Gembloux, Belgium, p. 135 (1991).
2. K.J. Kaffka, *Proceedings: International Diffuse Reflectance Spectroscopy Conferences*, Ed by R.A. Taylor. The Council for Near Infrared Spectroscopy. Gaithersburg, MD, USA, p. 63 (1992).
3. K.J. Kaffka and L.S. Gyarmati, *Near Infrared Spectroscopy: The Future Waves*, Ed by A.M.C. Davies and P. William. NIR Publications, Chichester, UK, p. 209 (1996).
4. K.J. Kaffka and L.S. Gyarmati, *J. Near Infrared Spectrosc.* **6(A),** 191 (1998).
5. K.J. Kaffka and Zs. Seregély, *NIR Spectroscopy: Proceedings of the 9th International Conference*, Ed by A.M.C. Davies and R. Giangiacomo. NIR Publications, Chichester UK, pp. 259–265 (1999).

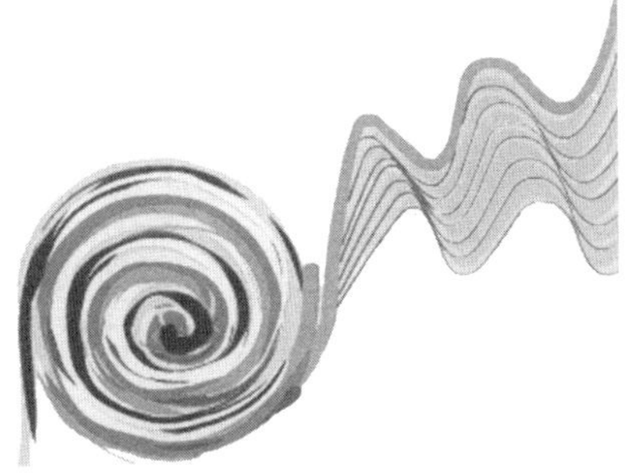

Precision Agriculture

Near infrared technology for soil analysis with implications for precision agriculture

Bo Stenberg,[a] Anders Jonsson[a,b] and Thomas Börjesson[b]

[a]Swedish University of Agricultural Sciences, Department of Agricultural Research, Skara, PO Box 234, SE-532 23 Skara, Sweden
E-mail: bo.stenberg@jvsk.slu.se

[B]svenska Lantmännen, R&D, Östra hamnen, SE-531 87 Lidköping, Sweden

Introduction

The general goal with the concept of precision agriculture is to apply inputs where they best fill their purpose. Adjustment of inputs are to be made as precisely as possible after what is required by the soil and crop potentials on a high spatial resolution. Consequently, precision agriculture is also often called site-specific agriculture. The yield variations within a field could be considerable: several tons of cereals per ha.[1] Nutrient requirement will vary correspondingly. Thus, site-specific inputs of nutrients will save resources and the impact on the environment could be minimised without lowering total yields or putting product quality at risk.[2]

Regulation of field inputs "on the run" has been made possible by the geographical position system (GPS)-technology, which gives the farmer his exact real time positioning in the field. However, the real challenge is to provide a reliable basis for decision-making. To support high spatial resolution, extensive sampling and analysis is required for many soil and plant characteristics. To be able to regulate the inputs of fertiliser, lime, etc. in relation to soil and crop needs on a high spatial resolution, rational soil analyses strategies are needed. The rapidity, minimal sample preparation requirement and potential for direct field measurements make NIR spectroscopy highly interesting as a future soil sensor.

Together, the contents and quality of clay and organic matter (SOM) in soil regulate or influence most soil properties and processes.[3] Therefore, soil clay and SOM are two parameters with prime interest in soil analyses packages and in precision agriculture. In such agriculture systems they are useful to describe soil-based variations in agricultural fields. However, the analyses do not support adjustments of agricultural practice directly, but are often used in pedotransfer functions.[4] The NIR-spectra has been shown to hold information about both clay[5–10] and SOM.[11–14] However, prediction results are variable and, in most cases, the variation of soil types used has been limited. Recently, promising results have also been published for NIR-based predictions of crop uptake of nitrogen.[15,16]

In this paper the performance of NIR calibrations for clay and SOM matter contents on a large set of samples, covering most of the variation of agricultural mineral soils in Sweden, are tested. In addition the stability of NIR calibrations for N-uptake in winter wheat over years and across fields has been studied.

Table 1. General characteristics of data sets for calibration of *clay* and soil organic matter (*SOM*).

Sample set, country	Sample set type	No. of samples	*SOM* %	*Clay* content %
Ekhaga, SE	2 ha field, grid sampled	52	6.2–14.4	~ 60
Ultuna, SE	Long term field experiment, organic amendments	60	1.7–6.9	~ 40
Askov, DK	Field experiment, N-fertiliser	36	2.1–3.3	~ 11
Jyndevad, DK	Field experiment, N-fertiliser	27	1.5–2.6	~ 4
Ribbingsberg, SE	15 ha field, stratified sampling	33	2.8–4.3	7–26
Swedish farmsites	25 m^2 sample plots	2750	0.9–2	0–70
Danish farmsites	Field experiment areas	50	1.5–4.6	3–22

Experimental

Soil samples

For clay and soil organic matter content (SOM) a sample set of 2750 top soils sampled to represent all Swedish agricultural areas was used (Swedish farmsites). The soils have previously been analysed for numerous properties.[17–19] For clay content all soils where used, but for SOM soils with more than 12% organic matter (7% organic carbon) where excluded. For the SOM studies the Swedish farmsite data were evaluated together with six smaller sample sets: three Danish used by permission of Dr. Ingrid Kaag-Tomsen, Foulum Research Station, DIAS, Denmark, and four Swedish. All seven data sets are described in Table 1.

For crop uptake of mineralised soil N, two adjacent fields (I and II, 10 and 15 ha) were sampled at unfertilised 25 m^2 plots distributed on each field. Field I was sampled in 1997 at 15 plots and in 1998 and 1999 at 20 plots. Field II was sampled in 2000 at 20 plots. The fields are located about 100 km NE of Gothenburg. Winter wheat was cropped all years. The fields were very variable, especially in SOM (Table 2).

Reference methods

Soil texture was analysed with the pipette method according to Gee & Bauder,[20] after the organic matter and carbonates had been removed. The content of clay; < 0.002 mm, silt; 0.002–0.06 mm and

Table 2. Soil organic matter in samples from the unfertilised experimental plots 1997–2000.

Year / Field number	Min %	Max %	Mean %	Median %	St. dev. %
1997 / I	3.1	22.4	8.8	4.8	6.4
1998 / I	2.6	19.4	6.9	5.2	4.8
1999 / I	2.8	17.2	6.7	5.0	5.0
2000 / II	2.2	10.8	4.3	3.6	2.4

sand; 0.06–2 mm was determined. SOM was calculated by multiplying organic carbon with 1.72. Organic carbon was analysed on a LECO CNS 700 after removing the carbonates from the soil samples.

Crop uptake of N was measured as the total N content in above-ground plant parts in August, just before harvest. Four 0.25 m^2 squares were sampled in each plot.

NIR measurements

All soils were air dried and crushed to pass a 2 mm screen before analysis. Every fourth wavelength between 1100 and 2500 nm was recorded on a Bran+Lubbe InfraAlyzer 500.

Data analysis

All data was analysed with Unscrambler 7.6. PLS was used for calibrations and only validation results are shown, either from cross-validation (leave-n-out) or test sets. NIR-spectra were smoothed by second order, seven points, Savitzky–Golay. For clay and SOM calibrations, the first derivative was calculated and for crop uptake of N, baseline corrections were made.

Results and discussion

Clay content

For clay content calibrations 25% of the 2750 samples were left out for validation. The validation result is shown in Figure 1(a). This global model levels off at about 25–30% clay, and there are a number of stray samples. In an attempt to improve performance, six regional models consisting of adjacent districts were also calibrated. For each region 25% of the samples were left out as test sets. This regional classification was accomplished according to three criteria: It should follow existing county borders, it should be represented by at least 200 samples to allow for test set validation and the regions should be as large as possible without loss of model performance. The results are shown in Figure 1(b). This classification significantly improved the results, probably due to smaller variations in soil type. The performance of the regional strategy is probably at the limit of what is possible considering that the error of the reference method is at least 10%. In addition, NIR and the pipette methods do not utilise the same features of the clay minerals. Clay, according to the pipette method, is defined as particles of a size equivalent to a sphere diameter of 2 µm or less. NIR, supposedly, measures the chemically-active surfaces of clay minerals, largely through bound water.[21] The aggregate size distribution, which to a large extent reflect the presence of clay, also influences the spectra.[8] In addition, different clay miner-

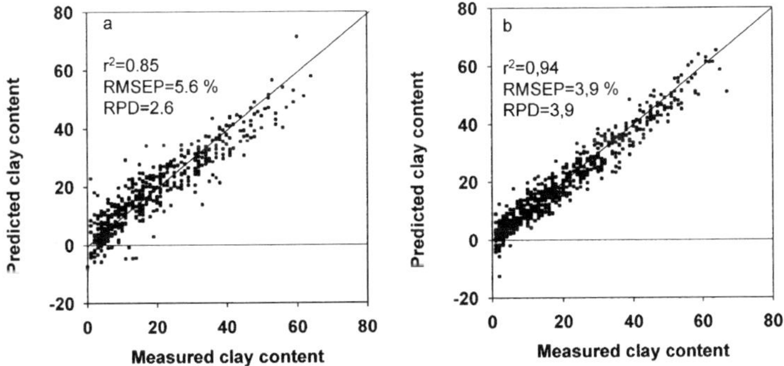

Figure 1. Validation samples for Swedish farmsite data; (a) a global NIR model and (b) six regional NIR models.

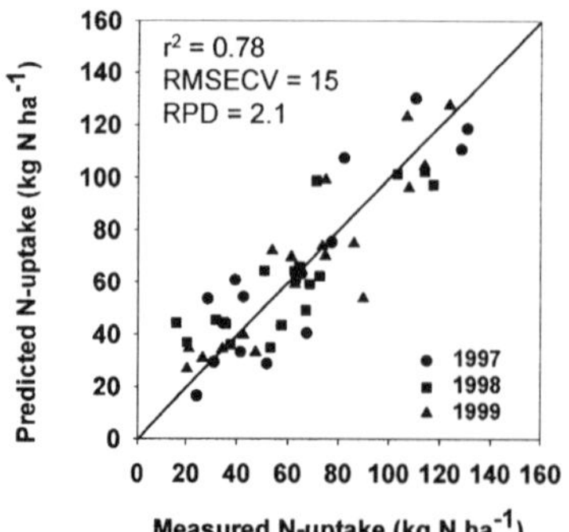

Figure 2. Cross-validation results of NIR predicted *vs* measured N-uptake in Field I 1997–1999.

als give different spectra,. Thus, soils within the same texture class could have very different spectra depending on geological history, agricultural practice and distribution of particle sizes smaller than 2 μm.

Soil organic matter

A global model (Swedish farmsites) for SOM did not perform satisfactorily (Table 3) and division of the Swedish farmsites set into regional classes as for clay content did not improve the results (data not shown). A comparison of smaller data sets from single fields or field experiments indicated that SOM could be better predicted if the models were restricted to soils with a fairly high clay content or with a limited texture variability (Tables 1 and 3). To test this the Swedish farmsites set was divided into classes depending on clay content: 0–15%, 15–30%, 30–45% and > 45%. As can be seen in Table 4, performance increased with increasing clay content. Especially for the samples with the lowest clay contents, the model was poor. This class made up almost half of the total population. Therefore, this class was subdivided into three classes depending on sand content: 0–30%, 30–60% and > 60%. For the samples with

Table 3. Validated performance of *SOM* calibrations.

Sample set, Country	Validation method	r^2	*RMSEP/RMSECV* %	*RPD*
Ekhaga, SE	Leave-one-out	0.92	0.15	8.7
Ultuna, SE	Leave-one-out	0.97	0.14	5.8
Askov, DK	Leave-one-out	0.69	0.09	1.8
Jyndevad, DK	Leave-one-out	0.68	0.09	1.8
Ribbingsberg, SE	Leave-one-out	0.61	0.17	1.6
Swedish farmsites	Test set (25%)	0.46	1.24	1.4
Danish farmsites	Leave-one-out	0.60	0.26	1.5

Table 4. Validated performance of *SOM* calibrations of Swedish farmsites divided in classes after *clay* content. Cross-validation with 50 segments.

Class	*N*	Range % *SOM*	r^2	*RMSECV* %	RPD
0–15 %	1231	0.7–12	0.44	1.3	1.3
15–30%	775	1.6–12	0.80	0.7	2.3
30–45%	408	1.4–12	0.84	0.6	2.5
45–70%	182	2–12	0.86	0.5	3.5

Table 5. N-uptake in over ground plant parts 1997–2000 at time of harvest 1997–2000.

Year/Field no.	Min Kg N ha^{-1}	Max Kg N ha^{-1}	Mean Kg N ha^{-1}	Median Kg N ha^{-1}	St. dev. Kg N ha^{-1}
1997 / I	24.4	130.6	63.6	51.8	35.7
1998 / I	16.1	117.4	61.7	62.8	28.8
1999 / I	20.5	123.5	67.1	68.1	31.7
2000 / II	34.0	103.7	59.0	52.6	22.8

0–30% sand, this procedure improved the results to the level of the most clayey ones (*RMSEP* = 0.41, *RPD* = 3.1), suggesting that the large number of soils with a high sand content in the Swedish farmsites data set disturb the possibilities for good calibrations. The *RMSECV* found for these data sets corresponded well to that which can be found in the literature. However, it is difficult to make a detailed comparison as the variability in soil type and geographical origin usually are smaller than in our investigation.

Crop uptake of nitrogen

The N-uptake in above ground plant parts at the time of harvest was similar in all four years (1997–2000). However, in Field II (2000) the range and variation over the field was smaller (Table 5). This corresponds to the smaller degree of variation of organic carbon in Field II (Table 2). A composite cross-validated NIR model for 1997 to 1999 performed satisfactorily well without shifts between years (Figure 2). N-uptake 1999, predicted by an NIR-model calibrated on 1997–1998 samples, showed a small bias [Figure 3(a)]. A small bias was also evident for a corresponding model using organic carbon as the predictor [Figure 3(b)]. The similar procedure was used to predict the N-uptake in Field II (2000) with models calibrated on Field I (1997–1999) samples (Figure 4). In this case a bias was only seen for the organic carbon based model. The *RMSEP* and *RPD* were also significantly better for the NIR model. This difference between organic carbon and NIR as the predictors supports that the information in the NIR spectra represents a variety of properties of the soil matrix, as suggested by Börjesson *et al.*[15] That the bias generally remains across years is natural, as weather-conditioned differences for crops in different years could not be expected to influence the NIR-spectra.

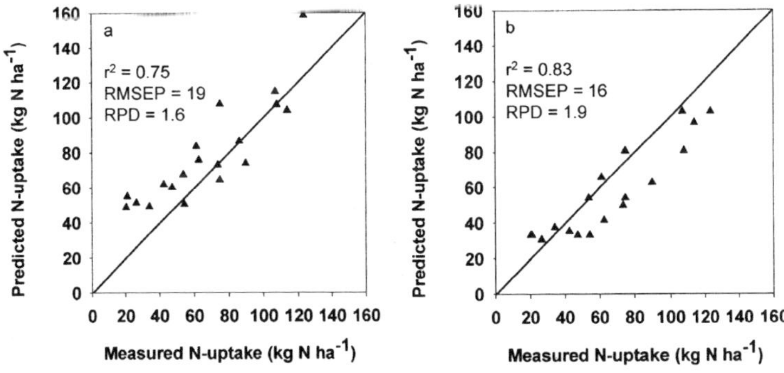

Figure 3. Predicted *vs* measured N-uptake in Field I 1999 by models calibrated on data from Field I 1997–1998. The models were based on (a) NIR spectra and (b) SOM data.

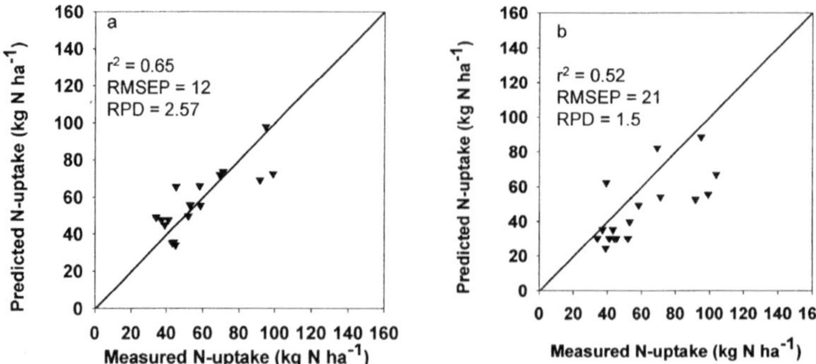

Figure 4. Predicted *vs* measured N-uptake in Field II 2000 by models calibrated on data from Field I 1997–1999. The models were based on (a) NIR spectra and (b) SOM data.

References

1. L. Thylén in *Precision Agriculture '97*, Ed by J.V. Stafford. BIOS Scientific Publishers, Sheffield, UK, pp. 345–350 (1997).
2. P. Robert in *Precision Agriculture '99*, Ed by J.V. Stafford. SCI, Sheffield, UK, pp. 19–33 (1999).
3. B. Stenberg, M. Pell and L. Torstensson, *Ambio* **27(1),** 9 (1998).
4. J. Bouma, *Advances in Soil Science* **9,** 177 (1989).
5. B. Stenberg, E. Nordkvist and L. Salomonsson, *Soil Sci.* **159(2),** 109 (1995).
6. A.H. Al-Abbas, P.H. Swain and M.F. Baumgardner, *Soil Sci.* **114(6),** (1972).
7. E. Ben Dor and A. Banin, *Soil Sci. Soc. Am. J.* **59(2),** 364 (1995).
8. S.A. Bowers and R.J. Hanks, *Soil Sci.* **100(2),** 130 (1965).
9. D.F. Malley, P.D. Martin, L.M. McClintock, L. Yesmin, R.G. Eilers and P. Haluschak, in *Near Infrared Spectroscopy: Proceedings of the 9th International Conference*, Verona, Italy, Ed by A.M.C. Davies and R. Giangiacomo. NIR Publications, Chichester, UK, pp. 579–585 (2000).
10. J.B. Reeves, III, G.W. McCarty and J.J. Meisinger, *J. Near Infrared Spectrosc.* **7(3),** 179 (1999).
11. R.C. Dalal and R.J. Henry, *Soil Sci. Soc. Am. J.* **50,** 120 (1986).
12. C.-W. Chang, D.A. Laird, M.J. Mausbach and J.C.R. Hurburgh, *Soil Sci. Am. J.* **65,** 480 (2001).
13. P. Krishnan, D.J. Alexander, B. Butler and J.W. Hummel, *Soil Sci. Am. J.* **44,** 1282 (1980).
14. M.J. Morra, M.H. Hall and L.L. Freeborn, *Soil Sci. Am. J.* 65**,** 288 (1991).
15. T. Börjesson, B. Stenberg, B. Lindén and A. Jonsson, *Plant and Soil* **214,** 75 (1999).
16. B. Dunn, G. Beecher, G. Batten and A. Blakeney, in *Near Infrared Spectroscopy: Proceedings of 9th International Conference*, Verona, Italy, Ed by A.M.C. Davies and R. Giangiacomo. NIR Publications, Chichester, UK, pp. 565–568 (2000).
17. J. Eriksson, A. Andersson and R. Andersson, **Report 4778**. Swedish EPA, Stockholm, Sweden (1997).
18. J. Eriksson, A. Andersson and R. Andersson, **Report 4955**. Swedish EPA, Stockholm, Sweden (1997).
19. J. Eriksson, A. Andersson and R. Andersson, **Report 45062**. Swedish EPA, Stockholm, Sweden (1997).
20. G.W. Gee and J.W. Bauder in *Physical and mineralogical methods*, Ed by A. Klute. Soil Science Society of America, Madison, USA, pp. 383–411 (1986).
21. J.D. Lindberg and D.G. Snyder, *American Mineralogist* **57,** 485 (1972).

Estimation of clear wood properties by near infrared spectroscopy

Laurence R. Schimleck,[a] Robert Evans,[a] Jugo Ilic[a] and A. Colin Matheson[b]

[1]*CSIRO Forestry and Forest Products. Private Bag 10, Clayton South MDC, Victoria 3169, Australia.*

[2]*CSIRO Forestry and Forest Products. PO Box E4008, Kingston, ACT 2604, Australia*

Introduction

Rapid, cost-effective methods of measuring wood quality are extremely important to tree improvement programmes where it is necessary to test large numbers of trees. Non-destructive sampling of a forest can be achieved by using increment cores generally taken at breast height. At CSIRO Forestry and Forest Products (CSIRO-FFP) methods for the rapid, non-destructive measurement of wood properties (SilviScan-1 and -2 and acoustic methods) and wood chemistry (near infrared spectroscopy) based on increment cores and other samples have been developed.

Recently, a series of experiments have been conducted at CSIRO-FFP that have used near infrared (NIR) spectroscopy to estimate several economically important wood properties such as density, wood stiffness (longitudinal modulus of elasticity—E_L) and microfibril angle (MFA). Experiments were conducted on a hardwood and a softwood species, the two species combined and samples representing a number of different species from around the world, with the aim of developing calibrations that could be applied to increment core samples. In this paper the results of these experiments are reported.

Materials and methods

Sample selection

Eucalyptus delegatensis: 50 dried clear wood (defect-free) samples of mature *E. delegatensis* trees from native forest in East Gippsland, Victoria, Australia.

Pinus radiata: 50 dried clear wood samples of mature *P. radiata* trees from Tallaganda, Canberra, Australia. Schimleck *et al.*[1,2] describes the origin of the *E. delegatensis* and *P. radiata* samples in detail.

Mixed species: 59 samples that had an extremely wide range of wood properties and included many commercially important timber species from around the world. Schimleck *et al.*[3] gives details of the wood properties of these species.

Wood properties

Dried clear wood samples were equilibrated to approximately 12% moisture content and cut to 20 mm × 20 mm transversely and 300 mm longitudinally. Dimensions and masses of these samples were used to calculate their average air-dry densities (D_{stick}). Dynamic elastic modulus was determined using the natural frequency of vibration along the fibre direction.[4] After measurement of E_L, a small strip (2 mm tangentially, 7 mm longitudinally, ~ 20 mm radially) was cut from one end of each sample for MFA analysis (by scanning x-ray diffractometry) on SilviScan-2[5] and NIR analysis. Dimensions and masses of the SilviScan-2 test samples were used to calculate their average air-dry densities (D_{strip}).

Near infrared spectroscopy

NIR diffuse reflectance spectra were obtained from the radial/longitudinal face of each sample using a NIRSystems Inc. Model 5000 scanning spectrophotometer. Samples were held in a custom-made holder.[1] A mask was used to ensure that a constant area was tested. The spectra were collected at 2 nm intervals over the wavelength range 1100–2500 nm. One spectrum was obtained per sample. Second derivative spectra were used for the development of calibrations. A segment width of 10 nm and a gap width of 20 nm were used for the conversion.

Calibration

Calibrations were developed for each wood property using partial least squares (PLS) regression. The *E. delegatensis* and *P. radiata* sample sets were divided, at random, into calibration and prediction sets as follows.
E. delegatensis: calibration set 70 samples, prediction set 34 samples.
P. radiata: calibration set 70 samples, prediction set 34 samples.
Hardwood–softwood (*E. delegatensis* + *P. radiata*): calibration set 140 samples, prediction set 68 samples.
Mixed species: all samples were used to develop calibrations that were then tested on selected samples from the *E. delegatensis* (50 samples) and *P. radiata* (49) sample sets. One *P. radiata* sample was omitted due to fungal staining.

NSAS software (version 3.52)[6] was used to develop calibrations using four cross-validation segments and a maximum of ten factors. In most cases the number of factors recommended by the software was used. For the *E. delegatensis* D_{strip} calibration, eight factors were recommended but only five factors are reported, as the extra factors marginally improved the standard error of prediction. For the mixed species set the calibrations reported gave the best results in prediction on the *E. delegatensis* and *P. radiata* samples.

Calibration statistics

The measure of how well a calibration fits the data is the standard error of calibration (*SEC*). The measure of how well the calibration predicts the trait of interest for a set of unknown samples that are different from the calibration set is given by the standard error of prediction (*SEP*).

Results and discussion

Individual species

Good calibrations were obtained for each physical property examined using NIR spectra obtained from the *E. delegatensis* and *P. radiata* samples (Table 1). Coefficients of determination (R^2) ranging from 0.77 for MFA to 0.93 for D_{stick} were obtained for *E. delegatensis* and R^2 ranging from 0.69 for D_{stick} to 0.94 for D_{strip} were obtained for *P. radiata*. *SEP* results were similar to the *SEC*'s, indicating that calibrations based on NIR spectra of solid wood can be used to rapidly predict solid wood properties in a prediction set.

E. delegatensis and *P. radiata*

The calibrations obtained for each solid wood property, using the combined *E. delegatensis* and *P. radiata* sample set, were also good (Table 2) with R^2 similar to those of the individual species. The

Table 1. Summary statistics of solid wood property calibrations developed for the *E. delegatensis* and the *P. radiata* sample sets. D_{stick} = air-dry density of the stick samples, D_{strip} = air-dry density of the strip samples, E_L = longitudinal modulus of elasticity of the stick samples, MFA = average microfibril angle of the strip samples.

Species	Parameter	Factors	R^2 (calib.^{n})	*SEC*	R^2 (pred.^{n})	*SEP*
E. delegatensis	D_{stick} (kg m^3)	7	0.93	25.9	0.92	30.6
	D_{strip} (kg m^3)	5	0.92	26.4	0.91	31.9
	E_L (GPa)*	4	0.90	1.5	0.88	1.6
	MFA (deg)	3	0.77	1.3	0.74	1.7
P. radiata	D_{stick} (kg m^3)	2	0.69	22.9	0.71*	21.8
	D_{strip} (kg m^3)	7	0.94	12.5	0.85	17.3
	E_L (GPa)	2	0.82	1.0	0.75	1.1
	MFA (deg)	3	0.78	1.9	0.64	2.1

*Note 69 samples were used to develop the *E. delegatensis* E_L calibration and two samples were deleted from the *P. radiata* prediction set when D_{stick} was estimated

Table 2. Summary statistics of solid wood property calibrations developed for the *E. delegatensis* and *P. radiata* sample set.

Parameter	Factors	R^2 (calib^n)	*SEC*	R^2 (pred^n)	*SEP*
D_{stick} (kg m^3)	8	0.97	27.0	0.95*	30.8
D_{strip} (kg m^3)	7	0.97	28.3	0.95	31.4
E_L (GPa)	6	0.92	1.5	0.89	1.5
MFA (deg)	8	0.86	1.6	0.71	1.9

*Note two samples were deleted from the prediction set when D_{stick} was estimated

SEC's were similar to those obtained for *E. delegatensis*. These results suggest that it may be possible to develop general calibrations based on samples from a number of species of a single genus or samples from a number of different genera, even if the species exhibit large differences in wood chemistry and anatomy.

Mixed species

Good calibrations were obtained for density and E_L using the mixed species sample set (Table 3). R^2 were slightly lower than those determined using individual species and standard errors were higher. The wide range in the mixed species set has contributed to the large standard error. The calibration developed for MFA had a lower R^2 than was obtained using the *E. delegatensis* and *P. radiata* sets.

The mixed species calibrations provided good estimates of density (stick and strip) and E_L when applied to the *E. delegatensis* and *P. radiata* sets. Predictions of E_L for the *P. radiata* samples were in close agreement with the reference data (Figure 1). Predicted E_L for the *E. delegatensis* set had an *SEP* much higher than the mixed species *SEC*. Six *E. delegatensis* samples had E_L values beyond the range of the mixed species calibration (1.7 to 23.8 GPa) and were poorly predicted. If these samples were re-

Table 3. Summary statistics for the mixed species calibrations and of predictions made by these calibrations on the *E. delegatensis* and *P. radiata* sample sets.

	Mixed species calibration			*E. delegatensis*		*P. radiata*	
Parameter	# of factors	R^2	*SEC*	R^2	*SEP*	R^2	*SEP*
D_{stick} (kg m^3)	4	0.82	76.2	0.72	65.8	0.67	39.4
D_{strip} (kg m^3)	4	0.83	74.0	0.75	65.4	0.79	40.0
E_L (GPa)	7	0.79	2.14	0.76	3.00	0.75	1.18
MFA (deg)	5	0.52	2.53	0.54	2.19	0.63	3.10

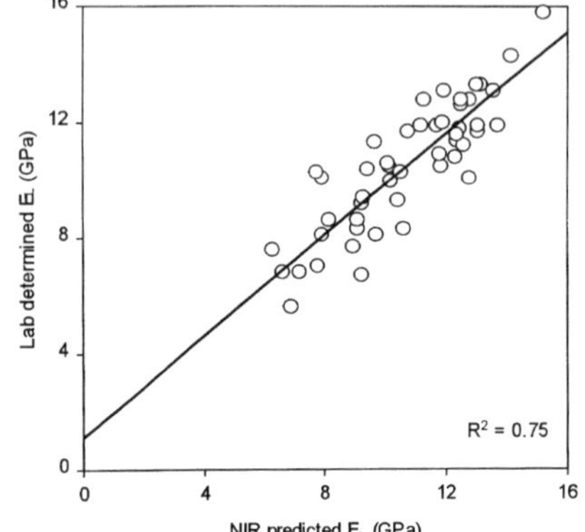

Figure 1. Relationship between laboratory determined E_L and NIR predicted E_L for the *P. radiata* samples. Predictions were made using the seven-factor mixed species E_L calibration.

moved, the *SEP* fell to 2.14 GPa (R^2 = 0.75). The results indicate that mixed species calibrations, encompassing wide variations in wood anatomy, chemistry and physical properties, can be used for ranking trees.

The selected mixed species samples represent an extreme range and it is unlikely that such variation would be required in practice. For most applications, samples from a small number of species would suffice. It is expected that refinement of calibrations through appropriate sample selection would provide improved calibration statistics, owing to less variation in the calibration set and more accurate predictions of the physical properties of test samples.

An important aspect in the success of this work was that samples used for D_{stick} and E_L determination were well represented by the strips used for calibration. For the *E. delegatensis* samples the R^2 for the linear relationship between D_{stick} and D_{strip} was 0.98. For *P. radiata* the R^2 was lower (0.75) but three samples had strip and stick densities that were quite different; exclusion of these samples improved the R^2 to 0.88 (two of these samples were removed from the prediction set for the estimation of D_{stick}). For the mixed species set the R^2 was 0.99.

The results indicate that NIR spectroscopy may be calibrated to estimate several wood properties. The calibrations could be used to estimate the properties of core samples, which would be particularly important to tree improvement programmes. Spectra collected from the prepared surface of cores could be used to provide pith to bark profiles or average properties.

Conclusions

Experiments reported in this paper indicated that useful calibrations for several solid wood properties can be developed using NIR spectroscopy, when appropriately characterised samples are available. NIR spectroscopy offers a rapid and non-destructive alternative to traditional methods of analysis and is applicable to large-scale, non-destructive forest resource assessment and to tree breeding and silvicultural programmes.

Acknowledgments

The authors would like to thank Sharee Stringer and Jenny Carr for their technical assistance.

References

1. L.R. Schimleck, R. Evans and J. Ilic, *Can. J. For. Res.* **31(10),** 1671(2001).
2. L.R. Schimleck, R. Evans and A.C. Matheson, *J. Wood Sci.* **48(2),** 132 (2002).
3. L.R. Schimleck, R. Evans, J. Ilic and A.C. Matheson, *IAWA journal* **22(4),** 415 (2001).
4. J. Ilic, *Holz Roh Werkst.***59,** 169 (2001).
5. R. Evans, *Appita J.* **52(4),** 283 (1999).
6. Anon, *Manual for NIR spectral analysis software,* NIRSystems, Inc., Silver Spring, Maryland, USA (1990).

Analysis of mixed feeds and their components using near infrared spectroscopy

Peter Tillmann,[a] Hartmut Horst,[b] Jürgen Danier,[c] Peter Dieterle[d] and Petra Philipps[e]

[a]*VDLUFA, Am Versuchsfeld 13, D-34128 Kassel, Germany*

[b]*HLDGN, Am Versuchsfeld 13, D-34128 Kassel, Germany*

[c]*HVA, D-85350 Freising, Germany*

[d]*LUFA Rheinland-Pfalz, Obere Langgasse 40, D-67346 Speyer, Germany*

[e]*Untersuchungszentrum Bonn-Roleber LUFA, Siebengebirgsstr. 200, D-53229 Bonn, Germany*

Introduction

Mixed feeds and their components are a very diverse matrix compared with other agricultural products where near infrared (NIR) spectroscopy has classically been applied. On a database of mixed feeds and their components (n = 2.130), universal partial least squares (PLS) calibrations were developed. The results from validation (five sets with n = 180, 130, 124, 41 and 96) show the potential of the calibrations and their limitations. Crude protein, crude fibre, crude fat, starch and sugar were predicted on a dry matter basis with standard errors of prediction (*SEP*) of 1.0%, 1.0%, 0.5%, 1.5–2.0% and 1%, respectively. Gas production was predicted with *SEP* of 2–2.5 mL for a sample of 0.2 g (dry matter basis). Ash content of 15% and more in several mixed feeds or components, as well as rare components limit the use of NIR for routine analyses.

The objective was to show the potential and limitations of using NIR for a heterogeneous matrix as mixed feeds and their components.

Material and methods

Over a three-year period, feed samples were collected from traded feeds (mixed feeds and raw components). On the basis of preliminary results, pet food and feeds with an ash content above 15% (for example, fish meal and laying hen feeds) were excluded from the study. The data set consisted of 2130 samples, from which three validation sets were separated (all samples from 1998: (n = 180), a random selection of samples from 1999 (n – 130) and 2000 (n – 124). The characteristics of samples in the calibration set are shown in Table 1. Furthermore, two additional data sets consisting of mixed cow feeds (n = 41) and mainly pig feeds (n = 96) were collected in different regions in Germany,

All samples were ground and were scanned on an NIRSystems 5000 (Foss, Denmark). Calibrations were developed using ISI software[1] with standard settings (modified partial least squares, multiplicative scatter correction, 1,4,4,1, cross-validation). Reference analyses were done according to

Table 1. Samples in the calibration set.

	Number
Grains	317
Concentrates for cows	242
Pig feed	222
Oilseed meals, cakes	183
Piglet feed	143
Concentrates for fattening cattle	110
Legumes	107
Concentrates for horse	99
Milling by-products	80
Calve feed	71
Concentrates for sheep	39
Poultry feed	33
Others	42

VDLUFA (1997)[2] for crude protein (XP, Kjeldahl), crude fibre (XF, Weende analysis), crude fat (XL, Soxhlet extraction), starch (XS, Ewers), sugar (XZ, Luff-Schoorl) and gas production (GP)[3] in three different laboratories. The unbalanced distribution in the calibration set is shown in Table 2. Due to the targeted use of the samples, not all constituents were analysed on all samples, i.e. starch and sugar were carried out mainly on pig feeds while gas production was measured only on concentrates for cows.

Results and discussion

The calibrations allow the determination, on a dry matter basis, of crude protein (XP), crude fibre (XF) and sugar (XZ) with *SEP*s of approximately 1.0%. Crude fat (XL) was predicted with an *SEP* of 0.5%, starch (XS) with an *SEP* of 1.5–2.0%, while gas production (GP) can be estimated with an *SEP* of 2–2.5 mL per 0.2 g (see Table 3).

Table 2. Distribution of calibraiton samples.

Constituent	*n*	Mean	Range	*SD*
XP	1635	20.6	3.9–64.9	10.9
XF	1051	8.7	0.3–32.5	4.3
XL	748	4.2	1.0–26.3	2.0
XS	692	43.2	0.8–79.9	19.8
XZ	452	6.9	0.7–31.5	3.4
GP	253	56.0	32.5–75.3	6.9

Table 3. Validation of the mixed feed calibration with five validation sets (*SEP*).

Data set/ Constituent	1998	1999	2000	R-P	NRW
XP	1.42	0.98	1.00	0.89	1.1
XF	1.21	0.63	1.22	1.14	0.9
XL	0.54	0.45	0.59	0.41	0.3
XS	1.44	1.36	1.60	—	2.2
XZ	0.87	0.98	0.62	—	0.9
GP	3.23	2.50	2.11	1.79	—

Table 3(a). Composition of validation sets.

Data Set	1998	1999	2000	R–P	NRW
Concentrates					
Cows	41	24	22	41	2
Fattening cattle	17	15	11	—	10
Sheep	18	2	2	—	—
Horse	14	13	15	—	—
Other types					
Calf feed	9	10	10	—	—
Piglet feed	19	18	19	—	1
Pig Feed	24	23	20	—	64
Poultry feed	—	—	—	—	11
Soybean & other oil seed meals	14	4	5	—	—
Milling by-products	14	11	16	—	—
Miscellaneous	10	10	4	—	8

The results for crude protein and crude fat compare favourably with results reported by Berzaghi *et al.*[4] for PLS calibrations for feeds. The best LOCAL procedure (*SEP* 0.8 for XP and 0.4 for XL) reported in the same study were not matched in this study.

Cizmar[5] reported *SEP* for mixed pig and poultry feeds of 0.5%, 0.6%, 0.3% and 1.4% for crude protein, crude fibre, crude fat and starch, respectively. In the present study, samples were collected from mixed feeds as well as raw materials. In the validation sets, about 20% of the samples were raw products.

Conclusions

These calibrations permit the estimation of energy values for the production of mixed feeds and for trading feeds. The NEL of mixed cow feed can be estimated with an *SEP* of 0.15 MJ. For pig feed, the energy estimation has a *SEP* of 0.4 MJ ME, due to the increased *SEP* of starch estimation. All conclusions are based on the official German energy estimation scheme.[6]

References

1. ISI, *WinISI II. Software manual, Version 1.00.* Infrasoft International, LLC (1998).
2. VDLUFA, *Methodenbuch Band III. Die chemische Untersuchung von Futtermitteln.* VDLUFA-Verlag, Darmstadt, Germany (1997).
3. K.H. Menke, L. Raab, A. Salewski, H. Steingass, D. Fritz and W. Schneider, *J. Agric. Sci. Camb.* **93,** 217 (1979).
4. P. Berzaghi, J.S. Shenk and M.O. Westerhaus, *J. Near Infrared Spectrosc.* **8(1),** (2000).
5. D. Cizmar, oral presentation, 29–30 May, ALVA Tagung, Wolfpassing, Austria (2001).
6. GfE, *Proc. Soc. Nutr. Physiol.* **5,** 53 (1996).

Prediction of organic compositions in manure by near infrared reflectance spectroscopy

Masahiro Amari, Yasuyuki Fukumoto and Ryozo Takada

National Institute of Livestock and Grassland Research Science, National Agricultural Research Organisation, Tsukuba Norindanchi, POB 5, Ibaraki 305-0901, Japan

Introduction

Many organic materials are included in the excreta of livestock. These materials are important resources for the production of organic manure and for improving soil quality which are still not being used effectively. Therefore, it is suggested that one standard factor for quality evaluation of manure made from excreta of livestock should be established. The subject of this study is to develop a rapid and accurate analytical method to analyse the organic composition of manure made from excreta of livestock and to establish the quality evaluation method based on some compositions measured by near infrared (NIR) reflectance spectroscopy.

Materials and methods

Samples

Samples were collected from each batch of manure every week from 1st to 7th week of the processing period. A total of 15 samples were used, including the source of the samples. The manure was made from 560 kg of swine feces and 60 kg of sawdust, with a moisture content of 65% RH. The manure was made in the two ways; one by using transparent cover sheets to expose it to sunlight and the other by using black cover sheets to shield it from the light.

Chemical analysis

The 15 samples of manure were analysed for moisture, nitrogen and organic matter (OM) content by an approximate analysis method.[1] Fibrous fractions were also measured using the enzymatic method[2,3] and the detergent method.[4,5] The enzymatic method was used for determining organic cell wall (OCW) and fibres with low digestibility (Organic b fraction: Ob) content, while the detergent method was used for the determination of the neutral detergent fibre (NDF), acid detergent fibre (ADF) and acid detergent lignin (ADL) content. Biological oxygenation demand (BOD) was analysed by the core-metre method.[6]

NIR reflectance analysis

Spectra of samples from 400 nm to 2500 nm were measured with an NIR instrument, Model 6500 (NIRSystems). Calibration equations were developed using eight manure samples and the rest of the samples were used to test the developed calibration equations.

Table 1. Minimum, maximum and ranges of chemical compositions of samples.

	Min.–Max.	Ranges
NDF (DM%)	44.3–55.2	10.9
ADF (DM%)	32.7–41.5	8.8
ADL (DM%)	12.9–21.2	8.3
OCW (DM%)	54.5–65.7	11.2
Ob (DM%)	53.2–62.9	9.7
BOD ($\times 10^4$ mg kg^{-1})	0.0–19.9	19.0

Table 2. Relationship between BOD value and chemical compositions value in manure.

	r	Se
NDF	–0.887	3.70
ADF	–0.909	2.53
ADL	–0.988	1.25
OCW	–0.955	2.37
Ob	–0.789	4.92

Results and discussion

Chemical composition

Table 1 shows the chemical composition of manure. The values of the major components in manure were 54.5–65.7% for OCW, 12.9–21.2% for ADL and 0–19.9 × 10^{-5} mg kg^{-1} for BOD, respectively. During the four weeks of processing, the amount of ADF and ADL increased, while the amount of NDF and OCW decreased. After this period each composition reached a constant value. The BOD rapidly decreased until the third week and then it decreased slowly. The relationship between the BOD and each composition of manure is shown in Table 2 and Figure 1. The correlation coefficients (*r*) between BOD and ADL, ADF or OCW content were negatively high and ranged from –0.909 to –0.988, i.e. manure having a large amount of ADL tended to show a lower BOD. In other words, the BOD decreased as the ADL increased according to the progression of the decomposition of manure by fermentation. The remarkably high negative correlation between BOD and ADL can be attributed to the

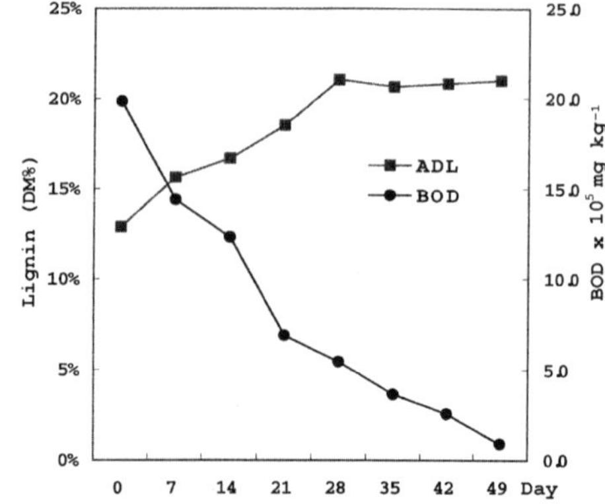

Figure 1. Development of the amount of ADL and BOD during the making of manure.

Table 3. Results of calibration and validation for determining chemical composition in manure.

	Wavelength (nm)	Calibration		Validation		
		r	*SEC*	*r*	*SEP*	*RPD*
NDF	1570	0.971	0.96	0.955	1.03	4.12
ADF	1766, 484,1546	0.983	0.61	0.979	0.60	4.90
ADL	1770	0.996	0.29	0.994	0.39	7.59
OCW	492	0.976	1.07	0.783	2.62	1.46
Ob	2118, 512	0.943	1.07	0.959	0.97	3.35
Nitrogen	2386, 596,1558	0.998	0.08	0.914	0.28	2.71
ASH	566	0.988	0.91	0.866	3.05	1.57
BOD	468	0.997	0.58	0.928	2.44	3.04

decomposed quantity of hard materials such as lignin. The same phenomenon may account for the high correlation between the BOD and OCW as well as ADF.

These facts suggest that the ADL content in manure is one of the factors which shows the degree of fermentation and, therefore, the ADL content can be used to estimate the BOD of manure.

NIR refelectance analysis

Calibration

Table 3 shows calibration results for determining various components in manure, i.e. the correlation coefficient (*r*) and standard error of calibration (*SEC*). The *r* and *SEC* values were 0.996 and 0.29

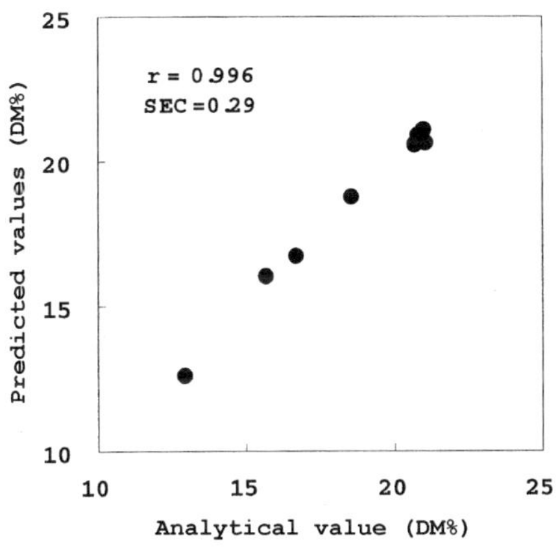

Figure 2. Relationship between the laboratory analysis and predicted content using an multiple reference data employing one wavelength.

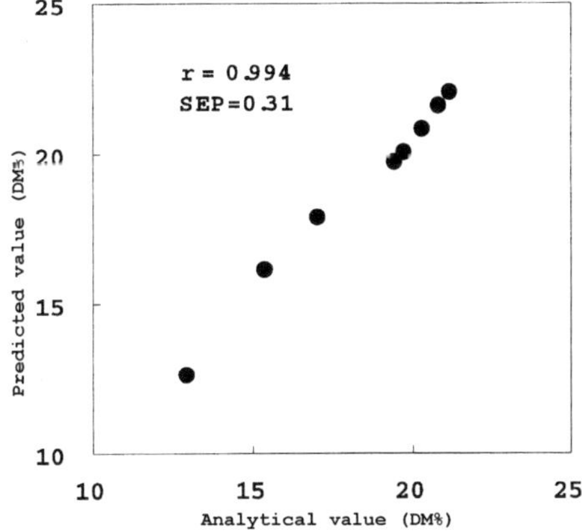

Figure 3. Relationship between the laboratory analysis and predicted content using an NIR calibration of ADL.

for ADL (Figure 2), 0.971 and 0.96 for NDF, 0.983 and 0.61 for ADF, 0.976 and 1.07 for OCW, 0.943 and 1.07 for Ob, 0.997 and 0.58 for BOD, respectively.

Validation of calibration

Table 3 and Figure 3 show validation results for determining various components in manure, i.e. the correlation coefficient, standard error of prediction (*SEP*) and the ratio of standard deviation of the reference data in the prediction sample set to *SEP* (*RPD*). The *r*, *SEP* and *RPD* values were 0.99, 0.39 and 7.59 for ADL, 0.96, 1.03 and 4.12 for NDF, 0.98, 0.60 and 4.90 for ADF, 0.96, 0.97 and 3.35 for Ob and 0.93, 2.44 and 3.04 for BOD, respectively. The results indicated that organic compositions in manure samples are highly predictable using NIR.

References

1. H. Morimoto, *Experimental method of animal nutrition*, (in Japanese). Yokendo, Tokyo, Japan, pp. 280–304 (1971).
2. A. Abe, S. Horii and K. Kameoka, *J. Anim. Sci.* **48,** 1483 (1997).
3. A. Abe and S. Horii, *J. Jpn Grassl. Sci.* (in Japanese with English summary) **25,** 70 (1979).
4. P.J. Van Soest, *J. Assoc. Off. Anal. Chem.* **46,** 825 (1963).
5. P.J. Van Soest and R.H. Wine, *J. Assoc. Off. Anal. Chem.* **50,** 50 (1967).
6. Association for Utilization of Sewage Sludge, *J. Jpn Sewage Works Assoc.* (in Japanese) Tokyo, 146 (1996).

Prediction by near infrared spectroscopy of the content of white clover (*Trifolium repens*) and perennial ryegrass (*Lolium perenne*) in fresh or dry mixtures made up from pure botanical samples

Antonio Blanco, Daniel Alomar and Rita Fuchslocher

Institute of Animal Production, Faculty of Agricultural Sciences, Austral de Chile University, PO Box 567, Valdivia, Chile

Introduction

The contribution of different species to the forage dry mass is a relevant attribute of sward condition.[1] This trait, usually estimated by hand separation, is a tedious, time- and labour-consuming procedure.[2] Near infrared (NIR) spectroscopy has the potential for predicting botanical composition of swards, but most of the work has been carried out taking the spectra of dry samples. The aim of this work was to evaluate the feasibility of developing NIR regression models for predicting the content of white clover and perennial ryegrass in fresh or dry mixtures prepared from pure samples of both species.

Materials and methods

Samples of pure stands of white clover and perennial ryegrass were taken weekly for five weeks at Valdivia, tenth region of Chile, from November 21 to December 19, 2000. Fresh forage was chopped to *ca* 1 cm and manually homogenised. Mixtures were prepared which contained from 0 to 100% of each species, at increments of 10% on a fresh basis, resulting in 55 mixed samples (11 mixtures, 5 cuts).

Spectra were taken in reflectance (400–2500 nm) from fresh samples located in polyethylene bags, inserted in a large rectangular forage cell, placed in a transport module attached to a scanning monochromator (6500, NIRSystems Inc, Silver Spring, MD, USA). Then samples were dried (65°C, 48 h) and ground (Wiley mill, 1 mm screen) and spectra taken in small ring cells (5 cm diameter, 1 cm deep, with a quartz window of 35 mm diameter) placed in a spinning sample module.

The percentage of each species (as contribution to dry weight) in any sample (fresh or dry) was considered as the variable to be predicted.

Calibration models (modified partial least squares) were developed with WinISI II software (Foss NIRSystems) testing different math treatments (differentiation order, subtraction gap, smooth interval), with or without scatter correction of the spectra (*SNV* + Detrend). Equations were tested and selected according to the standard error (*SECV*) and coefficient of explained variance (1-*VR*) of the cross-validation process.

Results and discussion

The spectra for all fresh and dry samples showed differences in absorption bands along the visible and NIR regions. In order to explore species-related absorption bands, spectra from 100% white clover and 100% perennial ryegrass samples of the five cuts, were plotted, either in the fresh [Figure 1(a)] or dry [Figure 1(b)] state. Species-related absorption bands are suggested for both groups, as can be seen at 1450 nm in Figure 1(a), where all clover samples are placed above the ryegrass samples which are located below and in a less scattered fashion (insert). This could be caused by O–H stretching associated to water or starch, which can be expected in higher concentrations as energy storage in legumes in contrast to grasses, which are more adapted to store nonstarch, nonstructural sugars as energy reserves. In Figure 1(b), the insert enhances different absorption bands for clover and ryegrass at 2180 nm, probably caused by differences in protein content.

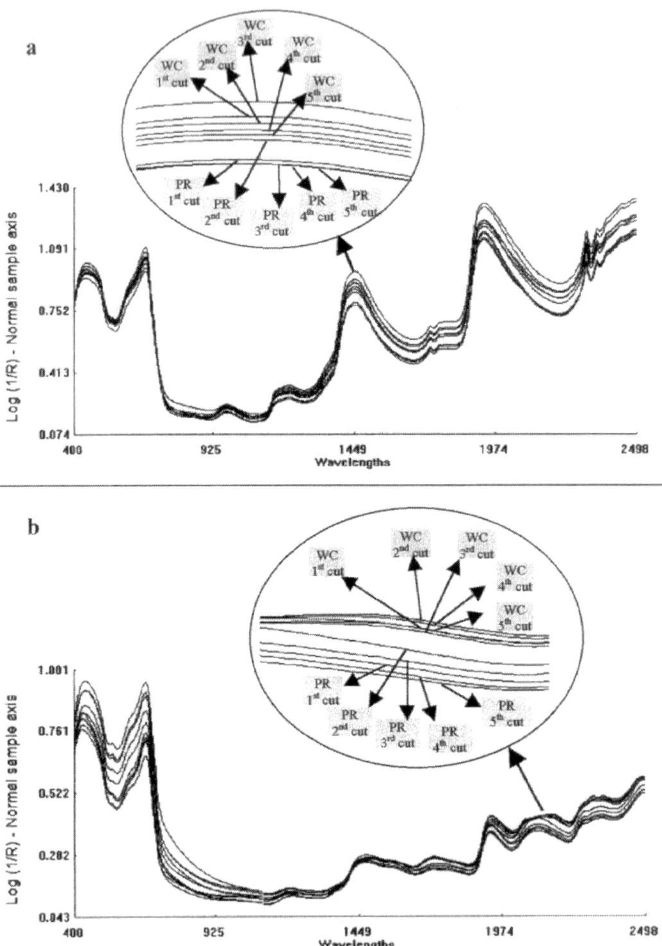

Figure 1. Spectra of (a) fresh and (b) dry samples of pure white clover (WC) and pure perennial ryegrass (PR).

Table 1. Selected equations with different math treatments, ranked by derivative order.

Variable	Terms	*SECV*	1-*VR*	s.d. *SECV*$^{-1}$	Math	Scatter
Fresh clover or ryegrass	6	6.92	0.953	4.64	0–0–5	Snv+det
	6	6.09	0.964	5.27	1–15–15	snv+det
	7	6.05	0.964	5.32	2–10–10	none
	5	5.87	0.966	5.48	3–10–10	snv+det
Dry clover	6	5.30	0.972	5.99	0–0–5	Snv+det
	7	3.61	0.986	8.40	1–5–5	none
	6	3.73	0.985	8.12	2–10–10	none
	6	3.34	0.988	9.07	3–10–10	none
Dry ryegrass	5	5.30	0.972	6.00	0–0–5	Snv+det
	7	3.60	0.986	8.43	1–5–5	none
	6	3.72	0.985	8.15	2–10–10	none
	6	3.36	0.988	9.02	3–10–10	none

SECV: standard error of cross-validation
1-*VR*: coefficient of determination (proportion of explained variance) of cross-validation
s.d.: standard deviation of reference data
Math: mathematical treatment where first number is subtraction order, second is subtraction gap in data points and the third is smooth interval in data points
Scatter: standard normal variate (*SNV*) and detrend (det), treatments were either applied to correct light scattering and spectra curvature, or not applied (none)

The best equations for predicting the percentage of each species on fresh or dry samples (Table 1, Figure 2) show 1-*VR* values over 0.95 and *SECV* that, although not small, if contrasted to mean reference values (*ca* 50%), represent a small fraction of the standard deviation (s.d.) of the reference data. It is generally accepted that if the ratio of s.d. to standard error of performance (*SEP*) is higher than three,

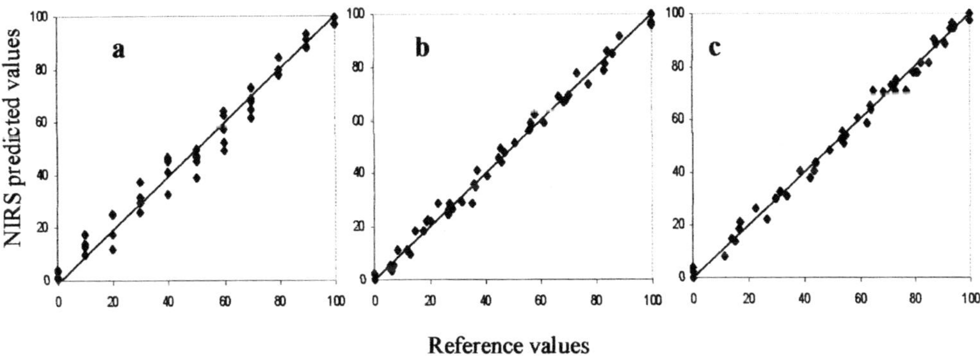

Figure 2. Actual against NIR predicted content of (a) content of clover (or ryegrass) for fresh samples (math treatment: 1–15–15–1, SNV + Detrend), (b) content of clover for dry samples (math treatment: 1–5–5–1, None) and (c) ryegrass content in dry samples (math treatment 1–5–5–1, None). See text for details.

an equation is considered to be acceptable.[3] If this criterion is assumed valid for *SECV* as well, then the equations developed can be useful for predicting botanical composition.

Best equations for fresh clover and fresh ryegrass were identical, which results from mixing complementary percentages of only two species on a fresh basis when making up the samples. In this way, reference data present increments of 10 percentage units in one species as the other decreases by the same amount, explaining identical means and s.d. An important intercorrelation can, thus, be expected. In this way the spectral "signal" for 90% clover could also be interpreted as a signal for 10% ryegrass and so on. Work is in progress to evaluate more complex and "real world" mixtures.

Not surprisingly, a better agreement is expected with dry ground samples (1-*VR* of 0.98 to 0.99) than with fresh chopped samples (1-*VR* of 0.96), as in the last case more factors could be affecting the spectra. Among them, water content, heterogeneity of plant tissues (leaves, stems, flowers), higher particle size, air pockets in the sample cup, etc.

As for math treatments, appropriate equations ranged from first to third order derivative and scatter correction (*SNV* and detrend) in general improved prediction accuracy.

Conclusions

Best equations were developed with a first or third derivative for fresh samples and with a second or third derivative for dry samples, with comparable results among math treatments for the superior equations.

In mixtures of white clover and perennial ryegrass, sward condition, in terms of botanical composition, can be successfully predicted by NIR, providing suitable calibrations are developed.

This could be accomplished in less time and with lower but still good accuracy, when using fresh chopped, instead of dry and ground samples.

Acknowledgments

This work was supported by a grant from the Chilean National Fund for Development of Science and Technology (FONDECYT). Project 1000432.

References

1. S.W. Coleman, F.E. Barton and R.D. Meyer, *Crop Sci.* **25,** 835 (1985).
2. J.A. Shaffer, G.A. Jung, J.S. Shenk and S.M. Abrams, *Agronomy J.* **82,** 669 (1990).
3. C.A. Kennedy, J.A. Shelford and P.C. Williams, in *Near Infrared Spectroscopy: The Future Waves*, Ed by A.M.C. Davies and Phil Williams. NIR Publications, Chichester, UK, pp. 524–530 (1996).

Optimising calibration to measure degradability parameters of alfalfa hays and dehydrated forages

Donato Andueza,[a] Fernando Muñoz,[a] Adela Martínez[b] and Begoña de la Roza[b*]

[a]*Animal Production Department, Servicio Regional de Investigación Agraria (SIA), Apdo 727, 50080 Zaragoza, Spain*

[b]*Animal Production Department, Servicio Regional de Investigación y Desarrollo Agroalimentario (SERIDA), Apdo 13, 33300 Villaviciosa, Asturias, Spain*

Introduction

Hay-making has been the traditional conservation method for lucerne cultivated in Spain although, in the last few years, the quantity of dehydrated forage has been higher than hay-making.[1] Among the advantages attributed to dehydration, we can mention the following: culture intensification, diversification of the forage use and independence from the weather conditions.[2] The dry matter and nitrogen ruminal degradability of these forages differ widely according to the dried process characteristics. For this reason, it is necessary to know the degradability characteristics to obtain a good and accurate application of actual feeding systems. However, the availability of *in vivo* and *in situ* degradability values for this kind of forage is limited because these methods require working with fistulated animals which is rather complicated and requires intensive and expensive labour. Even though we use the easier *in situ* methodology, the dynamics and logistics result in considerable work, due to limitations on the number of samples and number of bags that can be placed in an animal and the different time intervals required to perform kinetic studies. Therefore, a simple method is necessary to estimate the feed degradation characteristics. In this way, near infrared (NIR) reflectance spectroscopy has been widely used to predict degradation characteristics of forage.[3,4]

The aim of this study was to confirm the potential of NIR to optimise work conditions to avoid duplicated efforts in collaborative trials on animal feed evaluation between research institutions.

Materials and methods

In situ degradation

Two sets, with a total of 50 samples of lucerne preserved as hay or dehydrated forage, were evaluated using the *in situ* method: 40 from the SERIDA Research Centre located in the northwest of Spain and ten from SIA in the North of Spain. The technique of nylon bags was used to determine the kinetics of nitrogen (N) degradation. The fractions, soluble (a), insoluble potentially degradable (b), the fractional degradation rate of the slowly degradable fraction (c) and the effective degradability (ED) were estimated according to the equation obtained by Ørskov and McDonald.[5]

The trials were carried out with three four-year-old castrated rams (Fleichchaff x Rasa Aragonesa in the SIA trials and Manchega in the SERIDA trials) provided with a rumen cannula, 60–65 kg average live weight and adapted to a ration at maintenance level with alfalfa hay and concentrate [60 : 40 dry matter (DM) basis]. The forages were milled through a 2 mm screen and weighed in dacron bags with a 50 ± 10 micron pore size (Ankom Technology Corporation) at 12.5 mg cm^{-2}. The incubation times were 0, 2, 4, 8, 12, 16, 24, 48 and 72 h.

Parameters (a), (b) and (c) were estimated individually for each animal by a nonlineal regression. The adjustment of the kinetic values of rumen degradation to that equation was made by the procedure NLIN from the SAS statistical package.[6] The effective degradability (ED) of the DM and N were calculated for a rumen outflow rate (k) of 2% h^{-1},[7] by the equation $ED = a + [(b * c) * (c + k)^{-1}]$.

NIR scanning and calibration procedures

The lucerne samples were ground at 1 mm and scanned using a Foss NIRSystem 6500 monochromator with transport sampler, over a wavelength range from 400 to 2500 nm in steps of 2 nm.

Three sets of NIR spectra were created. The first set consisted of the lucerne spectra from SERIDA (40 samples) in their original form, the second included the spectra from SERIDA and five samples from SIA and the third set consisted of spectra from SERIDA and ten samples from SIA. The spectra from SIA were standardised using a single sample standardisation with one sample spectrally close to the centre of the population.[8] Every sample was measured in two replicates and the average of the replicate spectra obtained as log $1/R$ (R = Reflectance) was used in the calibration. Calibration equations were obtained with WINISI II v 1.5 sofware (Infrasoft International, Port Matilda, PA, USA), using a full wavelength range and modified partial least square as the regression method. The standard normal variate method (SNV)[9,10] was used for scatter correction and the second derivative as the mathematical treatment.

Results and discussion

The sample sets were characterised by the statistic listed in Tables 1–4: range and standard deviation (SD) for (a), (b) and ED of the dry matter and ED for nitrogen.

Generally, no significant differences were found between field-dried and dehydrated forages in terms of ED of dry matter, although the conservation in dehydrated form presented higher values than hay with respect to (b) fraction and lower values concerning the (a) fraction. This phenomenon can be explained by the larger exchange of nutrients, which can take place between fractions (b) and (a) in hays in relation to dehydration.

In terms of nitrogen ED, the lucerne preserved as dehydrated forage showed lower values of ED than hay forage (Table 4).

Tables 1–3 also summarise the calibration equation statistics that were obtained for kinetic parameters of DM. The values for ED of nitrogen are shown in Table 4. The ratio between the reference data

Table 1. Statistical results of calibrations and population characteristics for dry matter soluble fraction: a (%)

Population	*SEC*	R^2	*SECV*	r^2	*RER*	Range	*SD*
Set A (n = 40)	0.87	0.97	2.34	0.79	8.14	22.0-41.0	4.98
Set B (n = 40 + 5)	1.34	0.98	3.91	0.84	10.73	22.0-64.0	9.77
Set C (n = 40 + 10)	1.85	0.98	3.48	0.92	12.65	22.0-66.0	12.63

A: 40 alfalfa hays and dehydrated forages from SERIDA
B: 40 alfalfa hays and dehydrated forages from SERIDA + 5 alfalfa hays from SIA
C: 40 alfalfa hays and dehydrated forages from SERIDA + 0 alfalfa hays from SIA

Table 2. Statistical results of calibrations and population characteristics for dry matter degradable fraction: b (%).

Population	*SEC*	R^2	*SECV*	r^2	*RER*	Range	*SD*
Set A (*n* = 40)	1.12	0.95	3.28	0.52	7.02	36.0-59.0	4.77
Set B (*n* = 40 + 5)	1.39	0.97	4.04	0.70	9.15	22.0-59.0	7.43
Set C (*n* = 40 + 10)	1.76	0.95	3.60	0.81	10.55	21.0-59.0	8.22

A: 40 alfalfa hays and dehydrated forages from SERIDA
B: 40 alfalfa hays and dehydrated forages from SERIDA + 5 alfalfa hays from SIA
C: 40 alfalfa hays and dehydrated forages from SERIDA +10 alfalfa hays from SIA

Table 3. Statistical results of calibrations and population characteristics for dry matter effective degradability: ED (%).

Population	*SEC*	R^2	*SECV*	r^2	*RER*	Range	*SD*
Set A (*n* = 40)	0.92	0.97	2.15	0.84	10.69	50.0–73.0	5.37
Set B (*n* = 40 + 5)	0.90	0.98	2.34	0.88	12.82	50.0–80.0	6.82
Set C (*n* = 40 + 10)	1.18	0.98	2.36	0.91	13.12	50.0–81.0	7.95

A: 40 alfalfa hays and dehydrated forages from SERIDA
B: 40 alfalfa hays and dehydrated forages from SERIDA + 5 alfalfa hays from SIA
C: 40 alfalfa hays and dehydrated forages from SERIDA +10 alfalfa hays from SIA

Table 4. Statistical results of calibrations and population characteristics for nitrogen effective degradability: ED (%)

Population	*SEC*	R^2	*SECV*	r^2	*RER*	Range	*SD*
Set A (*n* =40)	2.76	0.67	3.90	0.34	4.87	65.0–84.0	4.81
Set B (*n* = 40 + 5)	2.44	0.79	3.66	0.51	5.19	65.0–84.0	5.27
Set C (*n* = 40 + 10)	2.50	0.80	3.86	0.51	5.44	63.0–84.0	5.53

A: 40 alfalfa hays and dehydrated forages from SERIDA
B: 40 alfalfa hays and dehydrated forages from SERIDA + 5 alfalfa hays from SIA
C: 40 alfalfa hays and dehydrated forages from SERIDA +10 alfalfa hays from SIA

range and standard error of cross-validation (*SECV*) was calculated (*RER*). Ideally, it should be at least ten.[11]

The relationship between the NIR data and the degradation characteristics have a tendency to improve by adding a small number of samples. The addition to samples recorded on the host instrument, which were standardised using a single sample standardisation procedure, maximise the range of composition on kinetic parameters—22.0–66.0 for (a); 21.0–59.0 for (b), 50.0–81.0 for ED of DM and from 63.0–84.0 for ED of N. The calibration statistics indicate a better correlation between NIR and reference data and good accuracy. The *RER* ratio also increased, for DM degradation parameters it was always above ten.

The relationship between the NIR data and the ED was weaker for N than DM ($r^2 = 0.51$ vs $r^2 = 0.91$) according to Atanassova *et al.*[3] and De la Roza *et al.*[4] ED of nitrogen values could be less accurate because no account was taken of possible microbial contamination of the bag residues. For this reason, the statistics obtained for (a) and (b) fractions of N also were of lower accuraty than statistics for (a) and (b) fractions of DM and are not included in this paper.

References

1. A. Ben Chabanne and I. Delgado, in *La alfalfa: Cultivo, transformación y consumo*, Ed by AIFE Lleida, Spain, p. 253 (1998).
2. M. Llorca, J. Masip, J. Fraile and I. Fraile, in *La alfalfa: Cultivo, transformación y consu*mo, Ed by AIFE Lleida, Spain, p. 253 (1998).
3. S. Atanassova, N. Todorov and D. Pavlov, in *Leaping Ahead with Near Infrared Spectroscopy,* Ed by G.D. Batten, P.C. Flinn, L.A. Welsh and A.B. Blakeney. RACI, Victoria, Australia, p. 495 (1995).
4. B. De la Roza, A. Martínez and B. Santos, *J. Near Infrared Spectrosc.* **6,** 145 (1998).
5. E.R. Ørskov and I. McDonald, *J. Agri. Sci., Camb.* **92**, 499 (1979).
6. SAS Institute Inc. *User' Guide*, Versión 6.12. SAS/stat, Cary, USA (1998).
7. AFRC, *Energy and Protein Requirements of ruminants*. CAB International, Wallinford, Oxon, UK (1993).
8. J.S. Shenk and M.O. Westerhaus, *Crop Sci.* **31,** 6 (1991).
9. J.S. Shenk and M.O. Westerhaus, NIRSystem, Inc., 12101 Tech Road, Silver Spring, MD 20904, PN IS-0119 (1996).
10. V.M. Fernández and A. Garrido, *Química Analítica* **18,** 113 (1999).
11. P. Williams and D. Sobering, in *Near Infrared Spectroscopy: The Future Waves,* Ed by A.M.C. Davies and P. Williams. NIR Publications, Chichester, UK, p. 185 (1996).

Near infrared reflectance spectroscopy as a tool to predict qualitative and quantitative meat and bone meal presence in compound feed

María Fernández, Adela Martínez, Sagrario Modroño and Begoña de la Roza*

Animal Production Department, Servicio Regional de Investigación y Desarrollo Agroalimentario (SERIDA), Apdo 13, 33300 Villaviciosa, Asturias, Spain

Introduction

Bovine spongiform encephalopathy (BSE) belongs to the group of diseases called TSE (transmissible spongiform encephalopathys), that can be transmitted between animals and humans. It is one of the most serious problems that has affected the economy of European cattles and public safety.

It is believed that the disease is caused by a natural protein which folds in the wrong way and then causes other similar proteins to change into this shape. The wrong shape form gradually builds up and spreads. These proteins are called prions. Possible sources or vehicles of infection could be imported cattle, contact with sheep, contact with wildlife and contaminated biological products, including foodstuff. In the case of the latter, the most likely vehicle for infection is meat-and-bone meal (MBM) which was included in cattle rations as a source of valuable bypass protein until it was banned.[1,2] The prohibited use of any animal tissue in animal feed means that a fast and reliable analytical methods to identify those ingredients in compound feed is needed.[3] The ban has been extended at the moment until 1st January 2002 (Regulation 1326/02001/EC of 29/06/01). Nowadays the official method for MBM detection in compound feeds is a classic microscopy technique, although other techniques such as polymerase chain reaction (PCR), ELISA, NIR and NIR microscopy are being studied.[3]

This official methodology is a subjective tool and requires exhaustive quantitative analysis and needs to differentiate between mammalian and poultry bones. In adittion, the separation of the different fractions in a sample by density before the analysis requires the use of organochlorate products such as tetracholromethane (CCl_4), which cause serious damage in the atmosphere's ozone content.

NIR methodology is a possible way to confirm and identify animal ingredients in compound feed. The capabilities for quantitative and qualitative analysis of feed by NIR has already been demonstrated many times.[3–6]

The purpose of this preliminary study was to test NIR methodology as an alternative tool to microscopy for qualitative and quantitative analysis to detect MBM in compound feed and to test the better samples presented for analysis (ground or intact samples) using NIR.

Materials and methods

Population

A total of 264 ground samples (population A) and 121 intact commercial samples, pellets, small pills and flour (population B), were used to create a spectral database.

Set A was built with 133 non-adultered compound feedstuffs, 113 with the addition of MBM as a cross-contaminant, adultered or made experimentally with different weight proportions to obtain high MBM levels and 18 raw MBM samples.

Set B was created using a similar procedure to set A which included 74, 41 and 6 samples, respectively for each group.

In the adultered compound feed, the MBM quantitative analysis was performed using microscopy technology as the reference method after making some modifications.[7]

NIR scanning and calibration procedures

The samples belonging to population A were milled at 0.75 mm. The rest (population B) were used in their intact form. Both populations were scanned using a Foss-NIRSystem 6500 monochromator with transport sampler, over a wavelength range from 400 to 2500 nm in steps of 2 nm.

Every sample was measured in two replicates and the average of the replicate spectra obtained as log $1/R$ (R = Reflectance) was used in the calibration. Population boundaries were established with a maximum standardised H distance from the average spectrum of 3.0.[8] The discriminant procedures and the calibration equations for ground and intac feedstuffs were obtained using WINISI II v 1.02 sofware (Infrasoft International, Port Matilda, PA, USA), using the full wavelength range.

For the discriminant analysis, each population was divided into three sets depending on the MBM percentage (0, 0.05–90 and 100) using partial least square (PLS) as the discriminant procedure. Modified partial least square (MPLS) was used as the regression method for quantitative analysis.

Standard normal variate and detrend[9] (SNVD) was used for scatter correction in population A and standard normal variate (SNV)[10,11] for population B. Spectral data were transformed using second derivative as the mathematical treatment.

Results and discussion

In Figures 1 and 2 we can see the absorption spectra of ground and intact compound feed samples according to their different MBM content. The major variation sources between spectra of each group are related to protein and oil absorption bands in relation to MBM percentages.

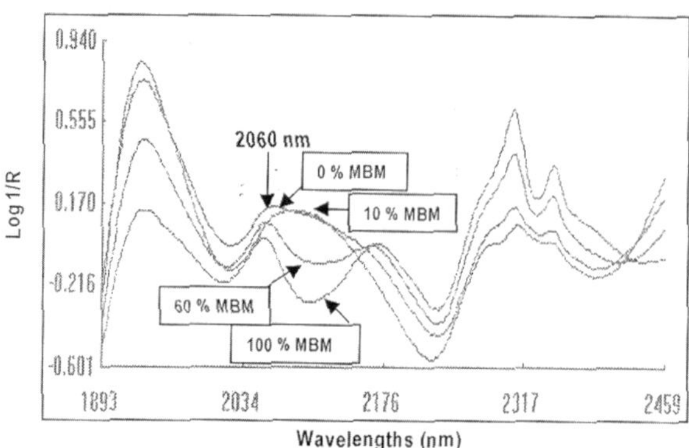

Figure 1. NIR absorption spectra of ground compound feed with different MBM percentage.

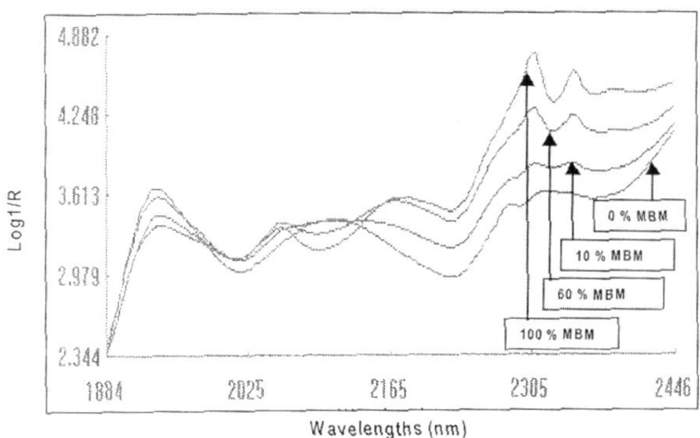

Figure 2. NIR absorption spectra of intact compound feed with different MBM percentage

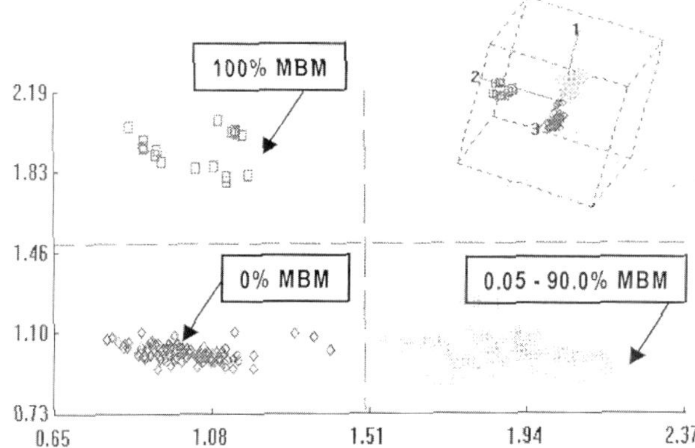

Figure 3. Spacial classification of ground compound feed according to different MBM contents after discriminant analysis.

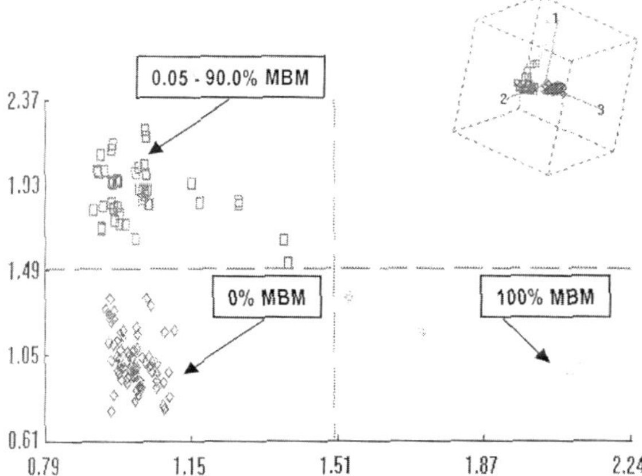

Figure 4. Spacial calssification of intact compound feed according to different MBM contents after discriminant analysis.

Table 1. Statistical results of calibrations and population characteristics for quantifying meat and bone meal in compound feed.

	SEC	R^2	*SECV*	r^2	Range	*SD*
Ground samples	0.895	0.999	0.988	0.999	0–100	31.932
Intact samples	1.008	0.999	1.328	0.998	0–100	28.571

Qualitative analysis

Discriminant PLS equations were developed to identify the presence of MBM, banned as an ingredient in compound feed composition.

The discriminant models used for ground and intact samples employed six factors for each. The general concept indicates that a predicted value of 2 is a perfect identification, 1 as not in the group and 1.5 indicates that the classification could go either way.[12]

Spacial classification for ground and intact compound feed, according to different MBM content after discriminant analysis in the calibrations sets, are show in Figures 3 and 4. The correct classification accuracy of these models was 93.25% and 91.74% for ground and intact samples, respectively. The main problem in these models is the accurate discrimination of low MBM levels, always lower than 0.2%.

A test set of six samples with known concentrations of MBM was used. These samples were a perfect match.

Quantitative analysis

Table 1 shows the characteristics of population and stastistical results for calibration and cross-validation to estimate the proportion of MBM in compound feed.

The calibrations for MBM show a good predictive capacity on ground compound feed as well as in tact compound feed, with r^2 values greater than 0.99 and low *SECV* (0.988 and 1.328, respectivily) which agree with the results reported by Garrido and Fernández.[13] Nevertheless, these error values are elevated to predict cross-contamination or low level adulteration by MBM in compound feed.

Conclusions

The preliminary results obtained showed that it is possible to achieve similar accuracy in identifying the presence of meat and bone meal using either ground or intact compound feed. To get the best results using NIR it is recommended that intact samples are used for analysis. However, it is necessary to add more samples to achieve the best possible representation of low levels of MBM, especially between 0.05 and 1% of the population.

Acknowledgments

The authors thank INIA for the financial support for this project (OT00-037-C17). Also we acknowledge the assistance of Alfonso Carballal for his help in supplying information.

References

1. J.W. Wilesmith, G.A.H. Wells, M.P. Cranwell, M.P. Ryan and J.B.M. Ryan, *Vet. Rec.* **123,** 638 (1988).
2. N. Healy, S. Ward, F. Butler and K. McDonnell, in *Proceedings of Agricultural Research Forum*, Ed by P. O'Kiely, T. Storey and J.F. Collins. Dublin. 7 (1998).

3. F. Piraux and P. Dardenne, in *Near Infrared Spectroscopy: Proceedings of the 9th International Conference*, Ed by A.M.C. Davies and R. Giangiacomo. NIR Publications. 535 (2000).
4. G. Xicatto, A. Trocino, A. Carazzolo, M. Meurens, L. Maertens and R. Carabaño, *Anim. Feed Sci. Techon.* **77,** 201 (1999).
5. B. de la Roza, L. Sánchez, S. Modroño and A. Martínez, in *Actas III Reunión Ibérica de Pastos y Forrajes*, 461 (2000).
6. A. Garrido, M.D. Pérez, A. Gómez, J.E. Guerrero, F. de Paz and N. Delgado, *ITEA*. **22(2),** 592 (2001).
7. B.O.E. 2/07/1999, ORDEN 24/06/1999, 25096 (1999).
8. J.S. Shenk and M.O. Westerhaus, *Crop Science* **31,** 1148 (1991).
9. R.J. Barners, M.S. Dhanoa and S.J. Lister, *Appl. Spectrosc.* **43(5),** 772 (1989).
10. J.S. Shenk and M.O. Westerhaus, NIRSystem, Inc., 12101 Tech Road, Silver Spring, MD 20904, USA, PN IS-0119 (1996).
11. V.M. Fernández and A. Garrido, *Química Analítica.* **18,** 113 (1999).
12. ISI-WINDOWS Near Infrared Software. Infrasoft International, LLC (1998).
13. A. Garrido and V. Fernández, in *Report of the Worshop on Identification of animal ingredients in compound feed focusing on the microscopy method for identification.* CEMA Group (1998).

Near infrared reflectance analysis of molasses and fats used in compound feed

A. Garrido-Varo,[a] M.D. Pérez-Marín,[a] A. Gómez-Cabrera,[a] J.E. Guerrero,[a] F. De Paz[b] and N. Delgado[b]

[a]*Escuela Técnica Superior de Ingenieros Agrónomos y Montes, University of Córdoba, Avda Menéndez Pidal s/n, E-14080 Córdoba, Spain*

[b]*SAPROGAL SA*

Introduction

Molasses and fats are today widely used in feedstuff manufacture, chiefly as energy sources but also because they share certain technological and nutritive advantages as feed components. Since these simple ingredients are derived by agroindustrial processing from a basic commodity, their composition varies depending, *inter alia,* on agroclimatic and technology-related factors. In view of this variability, it is recommended that analysis includes measurement of moisture and sugars in molasses and MIU (moisture + impurities + unsaponifiable), free fatty acids and both saturated and unsaturated fatty acids in fat.[1–3] Liquid samples are generally more awkward to analyse, in terms of both laboratory handling and the need for more expensive, more sophisticated but also slower equipment than that traditionally used for the analysis of solid ingredients. For that reason, quality controls of these key ingredients may often be inadequate and, in some cases, certain parameters of nutritional interest may not be determined (for example fatty acids).

The purpose of the present study was to demonstrate the viability of NIR for quality analysis of fat and molasses used in the manufacture of feedstuffs.

Material and methods

Experimental material

A total of 42 samples of molasses (cane and beet) and 61 samples of fats (animal and oleins) were supplied by a Spanish compound feed manufacturer (SAPROGAL S.A).

Collection of NIR spectra

Samples of molasses and of melted fat were analysed using a Foss NIRystems 6500 SY-I scanning monochromator fitted with a spinning cup, working in reflectance mode in the spectral range 400–2500 nm. Measurements were made in folded-transmission gold reflector cam-lock cups, with a pathlength of 0.1 mm. Two spectra were measured per sample, the mean spectrum being used for subsequent analysis.

Reference analyses

Molasses

The following parameters were determined: moisture [(1) infrared balance and (2) oven-dried with marine sand), crude protein (Kjeldahl), ash (heating furnace oven at 550°C), sodium chloride (potentiometric method) and total sugars (Luff–Schoorl).

Fats

The following parameters were determined: moisture (oven drying), acidity (volumetric method expressed as % oleic), insoluble impurities (petroleum ether filtration), unsaponifiable (ethyl ether + KOH method) and fatty acids profile (gas chromatography).

Chemometric treatment of data

Spectroscopic and chemical data were subjected to chemometric treatment using WinISI ver. 1.04 software.[4] NIR calibration equations were obtained and evaluated following Shenk and Westerhaus[5] and Williams and Sobering.[6] The following statistical parameters were used to select the best calibration equations: standard error for the calibration set (*SEC*), standard error of cross validation (*SECV*), coefficient of determination for the calibration process (R^2) or the cross-validation process (r^2) and ratios *RPD*[6] (*DT*/*SECV*) and *RER* (Range/*SECV*).

Results and discussion

Table 1 shows statistics for the sample population and results for NIR calibration equations for predicting the chemical composition of molasses.

Calibrations obtained for crude protein ($SECV = 0.14\%$; $r^2 = 0.99$) and NaCl ($SECV = 0.05\%$; $r^2=0.99$) afforded a high degree of precision and accuracy. The equations obtained for sugars and ash accounted for 86% and 84%, respectively, of variability within the sample set used; and although in both cases the *RPD* ratio fell short of the value of three recommended by Williams and Sobering[6] for routine analyses, it is likely that a larger sample set would yield better statistics. In any case, the equation clearly enables swift and precise analysis of a parameter which is of great value in characterising molasses and which, at present, prompts considerable inter-laboratory error.[7] The NIR equation obtained for measuring moisture content, using the infrared balance as reference method, displayed poor predictive capacity ($SECV = 1.69\%$, $r^2 = 0.42$). This may be due to the use of an indirect reference method, leading to a non-error-free prior calibration. To confirm this hypothesis, a new calibration was

Table 1. Characterisation of molasses from the calibration set and resulting calibration statistics.

	Mean	Range	*SD*	*SECV*	r^2	*RPD*	*RER*
Moisture[a]	23.06	19.3–28.6	2.10	1.69	0.42	1.2	5.5
Moisture[b]	22.40	18.5–28.1	1.81	0.34	0.96	5.3	28.2
Protein	8.37	3.9–10.3	2.64	0.14	0.99	19.1	45.7
Ashes	10.61	8.0–13.8	1.47	0.60	0.84	2.5	9.7
NaCl	2.17	1.16–4.03	1.11	0.05	0.99	20.9	57.4
Total sugars	48.48	43.9–53.9	2.83	1.04	0.86	2.7	10.0

[a]Reference method: infrared balance
[b]Reference method: heating oven with marine sand

obtained for moisture content using oven heating with marine sand as the reference method; this afforded an excellent predictive capacity (*SECV* = 0.34%; r^2 = 0.96).

Gillespie[7] obtained *SECV* values of 0.42%, 0.63%, 0.78% and 0.79% for crude protein, moisture, total sugars and ash, respectively, using a molasses sample set whose mean values for these parameters were very similar to those recorded here.

Chemical and NIR calibration statistics for the fats studied are shown in Table 2. In overall terms, NIR technology predicts a large number of quality parameters in fats destined for use in animal feeds. Of all the parameters studied here, only insoluble impurities correlated poorly (r^2 = 0.31) with spectroscopic data. This is not surprising, given that we are not dealing with a chemically-defined entity[8]. For practical commercial analysis of fat, the parameter generally used is MIU, which jointly covers values for insoluble impurities, moisture and the unsaponifiable fraction. This analytical parameter provides considerable information on the non-nutritional fraction of a fat,[2] and is one of the criteria to be used in energy evaluation.[3] As Table 2 shows, the equation for predicting MIU affords a high degree of precision and accuracy (*SECV* = 0.38%, r^2 = 0.94). The NIR equation obtained for predicting moisture content ensures precise evaluation of this parameter (*SECV* = 0.14%, r^2 = 0.88). Moisture assessment is of major importance, since moisture accelerates oxidation by reacting with the equipment (tanks) and releasing oxidation–catalysing metals.

There are no studies in the NIR literature addressing the use of this technology for the analysis of fats used in the feed industry, so some comment is required on the precision and accuracy of the equations obtained in fats and oils for human consumption, an area where NIR applications are becoming

Table 2. Characterisation of animal fats from calibration set and resulting calibration statistics.

	Mean	Range	*SD*	*SECV*	r^2	*RPD*	*RER*
Moisture	0.44	0.07–2.10	0.42	0.14	0.88	3.0	14.0
Free fatty acids	7.22	0.65–11.40	1.90	0.83	0.82	2.3	13.0
Insol. impurities	0.32	0.01–0.69	0.16	0.13	0.31	1.2	5.1
Unsaponifiable	1.45	0.18–5.80	1.23	0.45	0.87	2.7	12.5
MIU[a]	2.26	1.04–7.59	1.60	0.38	0.94	4.2	17.33
Lauric acid	0.11	0.06–0.13	0.02	0.009	0.67	2.2	7.8
Miristic acid	1.95	0.11–3.01	0.75	0.13	0.97	5.8	22.1
Palmitic acid	21.72	7.16–25.89	5.47	0.76	0.98	7.2	24.5
Palmitoleic acid	2.27	0.09–3.34	0.86	0.33	0.86	2.6	10.0
Stearic acid	13.65	4.05–20.38	4.60	0.74	0.98	6.2	22.2
Oleic acid	37.73	24.59–45.55	3.43	0.97	0.92	3.5	21.7
Linoleic acid	13.55	3.82–55.95	14.56	0.65	0.99	22.4	80.4
Margaric acid	0.67	0.05–1.13	0.28	0.06	0.96	4.7	18.6
Margaroleic acid	0.42	0.19–0.58	0.08	0.039	0.77	2.1	10.0
Araquidic acid	0.19	0.14–0.43	0.06	0.024	0.86	2.5	12.1
Behenic acid	0.21	0.01–0.72	0.21	0.035	0.97	6.0	20.3

[a]MIU: moisture + impurities + unsaponifiable

increasingly important. Although the acidity equation displayed satisfactory predictive capacity ($SECV = 0.83\%$, $r^2 = 0.82$), greater precision and accuracy have been obtained in oils for human consumption.[9] Precise knowledge of the proportion and/or the ratio of saturated to unsaturated fatty acids is among the most crucial criteria when assessing the energy and nutritional value of fats.[8] In practice, the high cost of fatty acid analysis has led analysts to prefer the so-called iodine index as a measure of the degree of unsaturation. This index enables assessment of the fatty acid profile and of the saturated/unsaturated ratio. The results obtained here (Table 2) show that NIR technology affords a high degree of precision and accuracy for predicting the major fatty acids, such as palmitic, stearic, oleic and linoleic acids, although even greater precision has been obtained for rendered Iberian pig fat using a similar analytical method.[10,11] Differences in precision and accuracy may be accounted for by the greater presence of impurities in fats used in the feed industry and the small number of samples displaying the calibrations shown in Table 2. NIR technology also enables excellent, or at least acceptable, prediction of other fatty acids present in smaller amounts.

Conclusions

NIR technology provides a viable method of chemical and, therefore, nutritional characterisation of two major liquid ingredients in animal feed. Given the current crisis of consumer confidence in ingredients of animal origin, characterisation of animal fats could usefully be extended; NIR technology offers a perfectly feasible means of doing this.

Acknowledgements

This study formed part of Project CICYT-Feder IFD1997-0990, and was performed using equipment and infrastructure belonging to SCAI (Unidad NIR/MIR), University of Córdoba (Spain). The authors thank Mr. Antonio López, Mr. Alberto Sánchez de Puerta and Ms. Isabel Leiva of the Animal Production Department (ETSIAM-UCO) for technical assistance.

References

1. C.V. Boucqué and L.O. Fiems, in *Livestock Feed Resources and Feed Evaluation in Europe. Present situation and Future Prospects,* Ed by F. De Boer and H. Bickel. EAAP Publication No. 37. Elsevier, Amsterdam, The Netherlands, p. 97 (1988).
2. B.K. Edmunds in *Feedstuffs Evaluation, Ed by J. Wiseman and D.J.A. Cole. Butterworths, London, UK, p. 197 (1990).*
3. G.G. Mateos, P.G. Rebollar and P. Medel, in *Avances en Nutrición y Alimentación Animal. XII Curso de Especialización FEDNA,* Ed by P.G. Rebollar, G.G. Mateos and C. de Blas. FEDNA, Madrid, Spain, p. 3 (1996).
4. ISI. The complete software solution for routine analysis, robust calibrations, and networking manual. Foss NIRystems/Tecator. Infrasoft International, Silver Spring MD, USA (1998).
5. J.S. Shenk and M.O. Westerhaus, in *Monograph.* NIRystems, Infrasoft International, Silver Spring MD, USA (1995).
6. P.C. Williams and D. Sobering, in *Near Infrared Spectroscopy: The Future Waves*, Ed by A.M.C. Davies and P.C. Williams. NIR Publications, Chichester, UK, pp. 185–188 (1996).
7. F. Gillespie. *In Focus*, **24(1),** 4 (2000).
8. F.D. Bisplinghoff, in *Collection Feeding and Nutrition.* Texas A & M University, Texas, USA, p. 11 (1993).
9. A. Garrido, C. Cobo, J. García, M.T. Sánchez, R. Alcalá, J.M. Horcas and A. Jiménez, in *Near Infrared Spectroscopy: Proceedings of the 9th International Conference,* Ed by A.M.C. Davies and R. Giangiacomo. NIR Publications, Chichester, UK, p. 867 (2000).

10. J. García, A. Garrido and E. De Pedro, in *Near Infrared Spectroscopy: Proceedings of the 9th International Conference,* Ed by A.M.C. Davies and R. Giangiacomo. NIR Publications, Chichester, UK, p. 253 (2001).
11. D. Pérez, A. Garrido, E. De Pedro and J. García. *ITEA* **22(II),** 610 (2001).

Development of robust calibration for determining apple sweetness by near infrared spectroscopy

Mi-Ryeong Sohn, Young-Kil Kwon and Rae-Kwang Cho

Department of Agricultural Chemistry, Kyungpook National University, Taegu 702-701, Korea

Introduction

Sweetness is a main quality factor contributing to the fruit taste. Most consumers require tasty and fresh fruit at a reasonable price. Near infrared (NIR) spectroscopy has been used successfully to estimate sweetness of various fruits[1–3] non-destructively, allowing the sweetness grading of individual fruit. In previous studies we expressed the sweetness as a Brix (by refractometer), total or individual sugar content (by HPLC analysis) and sweetness score (by calculating the sweetness index), comparing the calibration accuracy of each expression method.[4,5] Calibration samples for these models involved only one cultivar, one growing district and one harvest year. According to previous reports, some factors, such as harvest year, variety and growing season, affect the calibration model for sweetness of fruit. Peiris *et al.*[6] reported on the calibration of NIR spectra with the soluble solid of peaches which were collected over three years. Guthrie *et al.*[7] reported on the robust calibration for pineapple Brix across growing seasons and Miyamoto *et al.*[8] investigated the influence of various factors on a calibration model for mandarin Brix.

In this study, we investigated the influence of two variables, growing district and harvest year, for calibration and developed a robust model for the determination of sweetness of *Fuji* apple fruit.

Materials and methods

Apple fruits

About 2000–3000 apple fruits (*Fuji*) from 1995 to 1999 were collected from Andong, Youngchun and Chungsong in Korea and used in calibration and prediction.

Chemical analysis

After the NIR spectra were collected, part of the sample was squeezed and the Brix value was measured with an Atago (Japan) digital refractometer.

NIR spectra

NIR spectroscopic analyses were performed using an InfraAlyzer 500C (Bran+Luebbe, Germany) and the instrument was operated by the software package IDAS. For spectra collection a sample holder[9] was used. The reference spectral data were scanned over the range of 1100 to 2500 nm at 2 nm intervals to give701 data points and stored as log (1/*R*). A polystylene was used as background material. The temperature of the fruit was maintained at 15°C.

Table 1. Modelling of sample sets and range of reference Brix values.

Model	Growing district	Harvest year	Brix		
			Minimum	Maximum	Mean
Model 1	Andong	1995	12.5	17.6	14.8
Model 2	Youngchun	1995	10.9	16.9	14.3
Model 3	Chungsong	1995	11.4	15.9	13.8
Model 4	Andong	1996	13.5	16.6	15.0
Model 5	Andong	1997	12.5	15.7	14.6

Data analysis

Spectral data were imported into Sesame software, version 3.0 and reference values inputted. Samples were assigned to calibraiton and prediction sets according to Brix value. Two samples with maximum and minimum value were located to the calibration set and the other samples were allocated to the calibration set and prediction set in the ratio of 3 to 2, respectively. The data analysis was performed by stepwise multiple linear regression (MLR) using raw spectra data with no pre-processing. Calibration statistics included the correlation of coefficient (*R*), standard error of prediction (*SEP*) and bias.

Results and discussion

Characteristics of apple Brix

Table 1 shows modelling for the calibration set and Brix value of each sample set. Brix value varies from 10.9 to 17.6 and each sample set has a different range of Brix value. The Chungsong sample set has a relatively narrow range, (11.4 to 15.9) and low mean value of Brix compared with the Andong and Youngchun sample sets. The Andong samples showed higher Brix value than the other district's samples. Each model has different variables such as growing districts and harvest years, respectively.

Influence of growing districts

Table 2 shows the calibration and prediction results of models 1, 2 and 3. Correlation coefficients (*R*) of the models were over 0.95 with the exception of model 2. Each calibration model was successfully predicted to its own sample set. *SEP* were 0.47 for Andong, 0.85 for Youngchun and 0.80 for

Table 2. The results of MLR calibration and prediction for models 1, 2 and 3 of Table 1.

Calibration		Prediction					
Model	*R*	Andong		Youngchun		Chungsong	
		SEP	Bias	*SEP*	Bias	*SEP*	Bias
Model 1	0.96	0.47	0.044	1.32	–0.304	1.05	–0.487
Model 2	0.85	1.32	–0.447	0.85	0.280	1.83	1.303
Model 3	0.95	1.64	–1.430	1.16	–0.240	0.80	0.448

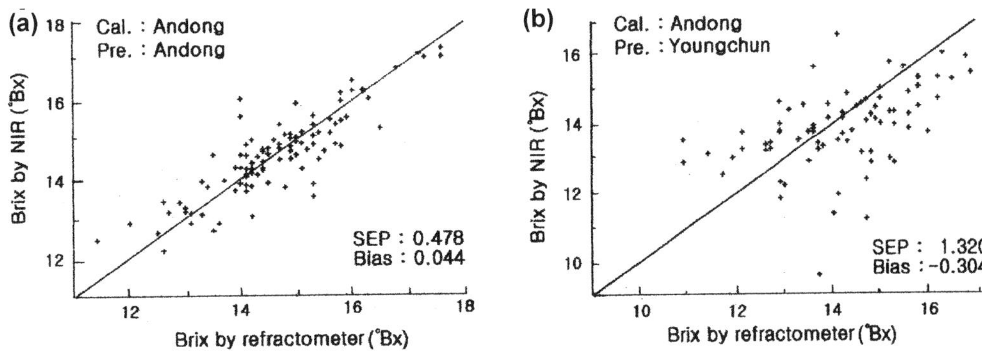

Figure 1. NIR *vs* actual Brix value of the apple fruit, representing the prediction results of (a) the Andong sample set and (b) the Youngchun sample set with the Andong calibration model.

Table 3. Prediction result of each sample set with the combined calibration model for growing districts.

Calibration		Prediction							
Model	*R*	Andong		Youngchun		Chungsong		Combined	
		SEP	Bias	*SEP*	Bias	*SEP*	Bias	*SEP*	Bias
Combined	0.94	0.78	–0.290	0.81	0.060	0.58	0.121	0.69	–0.075

Chungsong, respectively. However, each model was poorly predicted to the other sample sets, indicating a higher *SEP* and a higher bias. For example, the Youngchun and Chungsong sample sets were predicted poorly to the Andong calibration (model 1) with higher *SEP* and bias. As shown in Figure 1(b), there is an unaccesptable scatter relationship between NIR analysis and the reference method.The same results were obtained when the Younchun samples were predicted to models 1 and 3 and the Chungsong samples were predicted to models 2 and 3. These results indicate that the calibration model developed using one district's samples could not be used with the samples grown in other districts.

Table 3 shows the calibration statistics of the combination sample set which combined three sample sets, models 1, 2 and 3. This model resulted in an acceptable error level. *SEP* were 0.78 for Andong, 0.58 for Younchun and 0.69 for Chungsong samples, respectively.

Figure 2 shows the scatter plot of NIR predicted vs reference Brix of apple using the combined calibration model.

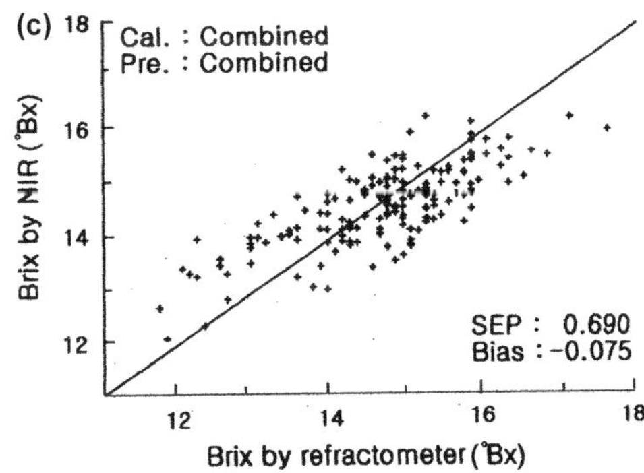

Figure 2. NIR *vs* actual Brix of apple fruit, showing the prediction result of thecombined prediction sets with the combined calibration models for growing the districts.

Influence of harvest year

Table 4 shows the calibration and prediction results of models 1, 4 and 5. The correlation coef-

Table 4. Calibration and prediction results of MLR models 1, 4 and 5 of Table 1.

Calibration	Prediction						
Model	*R*	1995		1996		1997	
		SEP	Bias	*SEP*	Bias	*SEP*	Bias
Model 1	0.96	0.47	0.044	4.75	–4.640	2.40	–2.250
Model 4	0.94	1.13	0.945	0.51	0.075	1.67	1.546
Model 5	0.91	0.79	–0.023	3.90	–3.827	0.41	0.001

ficients (*R*) of the three models were all over 0.9. Each calibration model predicted its own sample set successfully, as shown in Figure 3(a). The *SEP* was 0.47 for 1995, 0.51 for 1996 and 0.41 for 1997, respectively. But each model could not predict to the other sample sets successfully. When the 1997 samples predicted to the 1995 model, high *SEP* and bias were obtained (2.40 and –2.25, respectively), as shown in Figure 3(b). The scatter plots show a big bias effect. This indicates that the calibration model developed using samples having a limited growing year could be predicted to its own sample set well but could not be predicted to the samples grown in other years.

Table 5 shows the result of each sample set predicted with the combined model for the harvest year. The combined model was developed using a combination sample set of 1995, 1996 and 1997. Prediction resulted in SEP of 0.53 and 0.58 for 1995 and 1997 samples, respectively. This is less accurate than those by their own calibration models. However, the 1996 set showed better results. The *SEP* decreased to 0.47. Figure 4 shows the scatter plot for the reference sweetness vs predicted values from the NIR spectra using the above comination model. The *SEP* was 0.53 and bias was 0.004. These results indicate that the combined model could be used for determining sweetness of apples harvested for each year with acceptable accuracy.

Consequently, when the calibration sample has a sufficient variable range for different origins such as growing district and harvest year, it is possible to develop a robust calibration equation.

Climate and soil conditions could affect fruit sweetness. The best prediction results can be obtained when the calibration set and prediction set have the same conditions. However, it is impossible

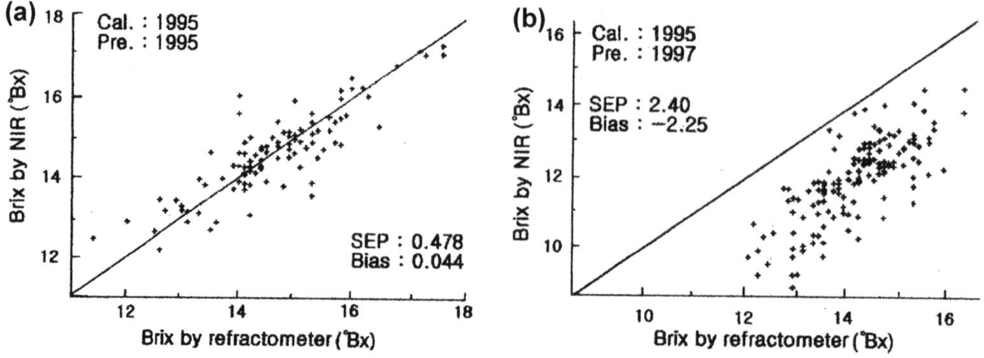

Figure 3. NIR *vs* actual Brix of apple fruit, representing the prediction results of the (a) 1995 sample set and (b) 1997 sample set with the 1995 calibration model.

Table 5. Prediction result of each sample set with the combined calibration model for harvest years.

Calibration		Prediction							
Model	*R*	1995		1996		1997		Combined	
		SEP	Bias	*SEP*	Bias	*SEP*	Bias	*SEP*	Bias
Combined	0.94	0.53	0.005	0.47	–0.117	0.58	0.090	0.53	0.004

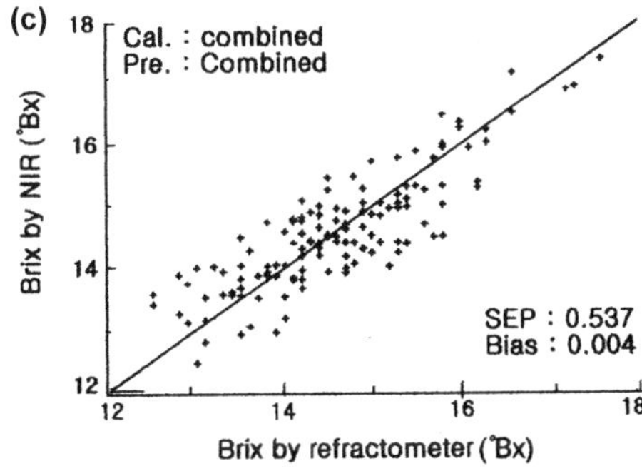

Figure 4. NIR *vs* actual Brix of apple fruit, showing the prediction results of the combined prediction sets with the combined calibration models for the harvest year.

to make a new calibration model every year for each district. A great deal of effort should be made to establish a robust calibration.

Conclusion

A robust calibration model for determining sweetness of apple fruit was developed using near infrared spectroscopy and the influences of growing district and harvest year were investigated. The calibration model for each growing district predicted was well predicted to their own sample set but poorly predicted to the other sample sets. The combined calibration model for three growing districts predicted samples reasonably well with an *SEP* of 0.69% and a bias of –0.075%. The calibration model for each harvest year was not transferable across harvest years but the combined calibration model for three harvest years was sufficiently robust to predict each sample set (*SEP* = 0.53%, bias = 0.004).

Reference

1. T. Temma, M. Chiba, K. Hanamatsu, F. Shinoki and T. Tsushima, *Proceedings of 7th International NIRS Confererence*, pp. 638–643 (1995).
2. D.C. Slaughter, *Near infrared analysis of soluble solids in peaches*, ASAE summer meeting, Charlotte, USA, (1992).
3. G.G. Dull, R.G. Leffler, G.S. Birth and D.A. Smittle, *Trans. of the ASAE* **35,** 735 (1992).
4. R.K. Cho and M.R. Sohn, *J. Korean Soc. Hort. Sci.* **39,** 745 (1998).
5. R.K. Cho, M.R. Sohn and Y.K. Kwon, *J. Near Infrared Spectrosc.* **6,** pp. A75–A78 (1998).
6. K.H.S. Peiris, G.G. Dull, R.G. Leffler and S.J. Kays, *Proc. of the Sensors for Nondestructive Testing,* p. 17 (1997).
7. J. Guthrie and B. Wedding, *J. Near Infrared Spectrosc.* **6,** 259 (1998).
8. K. Miyamoto and Y. Kitano, *J. Near Infrared Spectrosc.* **3,** 227 (1995).
9. M.R. Sohn, W.C. Park and R.K. Cho, *J. Korean Near Infrared Analysis* **1,** 27 (2000).

Estimating soil properties using near infrared spectroscopy to assess amendments in intensive horticultural production

Francisco Peña,[ac] Natalia Gallardo,[a] Carmen del Campillo,[a] Ana Garrido-Varo,[a] Víctor Fernández-Cabanás[b] and Antonio Delgado[b]

[a]*Escuela Técnica Superior de Ingenieros Agrónomos y de Montes, University of Córdoba, Apdo 3048, 14080 Córdoba, Spain*

[b]*Escuela Universitaria de Ingeniería Técnica Agrícola, University of Seville, Ctra Utrera Km. 1, 41013 Sevilla, Spain*

[c]*Centro de Investigación y Formación Agraria "Alameda del Obispo", DGIFAP, Apdo 3092, 14080 Córdoba, Spain*

Introduction

Over the past ten years, near infrared (NIR) spectroscopy has been successfully applied to the analysis of a wide range of agriculture products. Previous studies[1,2] have shown the potential of this technology for soil analysis, estimating a number of parameters with a single scan. The main advantages of NIR applications in soils are the speed of response, which enables an increase in the number of samples analysed for defining a particular soil and the instantaneous production of recommendations for fertilisation and soil amendment. A further advantage of this technique is that it obviates the need for chemical reagents and is, therefore, environmentally safe. The measurement of soil organic carbon (C_{org}) and organic nitrogen (N_{org}) is essential to proper management of the organic amendments which are of major importance in intensive horticultural crops in Mediterranean regions. In a wide area of the Mediterranean, low organic carbon content poses a considerable limitation for horticultural use, since it may be associated with soil structure degradation. Fast routine analysis for C_{org} and N_{org} may be valuable in assessing the organic amendment rate. Traditional methods of measuring C_{org} and N_{org} are time-consuming and involve the use of dangerous reagents such as dichromate, Hg or Se. These problems could be minimised with the use of NIR spectroscopy.

Material and methods

Samples

This study analysed a set of 125 samples of representative greenhouse soils in southern Spain. Samples were selected in such a way as to encompass a wide range of agricultural and soil uses. Samples were taken from the surface horizon, processed (dried, milled and filtered through a 2 mm sieve) and analysed for organic carbon (C_{org}, by dichromate oxidation), total nitrogen (N_t, by salicylic thiosulphate modification of the Kjeldahl method according to Bremner[3]), inorganic nitrogen (nitrate

Table 1. Soil composition of analysed samples.

	N	Mean	Minimum	Maximum	*SD*
C : N ratio	108	4.94	0.65	16.86	2.856
C_{org}	117	0.371	0.05	1.37	0.260
EC	117	279	97.5	869	118.317
f_{HH}	125	0.982	0.947	0.993	0.008
HH	125	1.84	0.70	5.51	0.798
NH_4^+	115	27	13.5	38	4.320
$NO_3^- + NO_2^-$	117	31	0.57	141	27.376
N_{org}	105	0.076	0.013	0.227	0.038
N_t	108	0.081	0.01	0.23	0.038
pH	117	8.48	7.69	9.07	0.229

f_{HH} is the factor of HH; organic nitrogen (N_{org}) was estimated as the difference between total (N_t) and inorganic ($NO_3NO_2 + NH_4$)
In % data for HH, C_{org}, N_t and N_{org}; in mg kg^{-1} for NO_3^-, NO_2^- and NH_4^+; in µS cm^{-1} for EC

and ammonium) following Mulvaney,[4] hygroscopic humidity (HH), humidity factor (f_{HH} = 100-HH / 100) pH and electrolytic conductivity (EC) in a 1 : 1 extract. Soil characterisation was performed using ten analytical parameters, as shown in Table1.

NIR hardware

A Foss-NIRSystems model 6500 scanning monochromator (NIRSystems Inc., Silver Spring, MD, USA) was used to measure reflectance spectra from 400 to 2498 nm at 2 nm intervals. Analysis was performed using a spinning module, which employs standard ring cells.

NIR software

All spectra were manipulated and processed and all calibrations were obtained using ISI software NIRS3 ver. 4.0 and WINISI ver. 1.50 (Infrasoft International, Port Matilda, PA, USA).

Results and discussion

Table 2 shows the calibration statistics for the different parameters used in the study. In general, there is a significant correlation between the value obtained by traditional methods and that obtained using NIR spectra for several properties relating to organic matter and moisture content, i.e. C_{org} and f_{HH}, respectively.

The high correlation between C_{org}, N_{org} and estimated values using NIR (1-*VR* values were 0.862 and 0.886, respectively) can be accounted for by modifications in the NIR spectra associated with absorbance of organic compounds in the NIR range. This would also account for N_t values, since most of the total N corresponds to organic N (N_{org}): in most cases, inorganic N ($NO_3NO_2 + NH_4$) accounted for less than 5% of N_t. These properties are important in evaluating the potential of Mediterranean agricultural soils, since organic matter is related to good physical properties and N_{org} represents forms not easily available for leaching, such as NO_3^-, which is normally the main inorganic compound among N forms. Thus, the organic N pool represents the soil N reserve available to plants in the medium to long

Table 2. Calibration statistics for selected equations.

	N	Mean	*SD*	*SECV*	1-*VR*	Derivative	Region
C : N ratio	104	4.648	2.422	1.421	0.658	1,4,4,1	vis + NIR
C_{org}	113	0.359	0.240	0.089	0.862	1,10,10,1	vis + NIR
EC	113	268.826	100.618	67.837	0.548	1,4,4,1	NIR
f_{HH}	120	0.983	0.006	0.003	0.690	1,10,10,1	NIR
HH	120	1.791	0.657	0.365	0.690	2,5,5,1	vis + NIR
NH_4^+	108	26.824	3.532	3.245	0.169	2,10,10,1	vis + NIR
$NO_3^- + NO_2^-$	113	28.857	23.696	18.986	0.358	1,4,4,1	NIR
N_{org}	99	0.073	0.035	0.012	0.886	1,4,4,1	vis + NIR
N_t	102	0.079	0.036	0.012	0.897	2,10,10,1	vis + NIR
PH	115	8.492	0.218	0.161	0.451	0,0,1,1	vis + NIR

term. The information provided concerning the C : N ratio may also be of interest. Although the calibration statistics are not as good as those obtained for other properties relating to organic matter, the technique may be reasonably useful in providing an approximate value that enables us to establish the degree of decomposition of organic matter in soils (Table 2).

The good statistics obtained for f_{HH} calibrations may be related to the high absorbance of water in the NIR range and might even be improved by reducing the gap between wet chemistry and NIR analysis of soil samples. HH or f_{HH} calibrations may be of great value, since these parameters are highly correlated with soil clay content and could provide an estimation of clay content when used in a group of soils with uniform clay-fraction mineralogy. This might account for the good correlation between clay content and NIR estimated data previously reported by other authors.[4]

Salinizaqtion is a major agronomic problem in arid and semi-arid areas. Wide horticultural areas of southern Europe (for example, south-eastern Spain) are saline, due to unsuitable use of irrigation water and to the intrusion of seawater in underground water deposits. A fast method for determining soil salt content might also be of value in establishing reclamation procedures. Although there is no chemical reason for any correlation between salt content (measured as EC) and NIR spectra, a significant correlation was recorded between observed and estimated values. Cross-validation determination coefficients were not high (1-*VR* = 0.55), but this calibration would at least enable the division of samples into different classes, i.e. high, medium and low saline soils.[5] Although the standard error of cross-validation was relatively high, there are reasons to consider the technique as viable not only for the implementation of a classification model but also for the development of a preliminary method to assess the EC of soils in the 1 : 1 extract. This can be explained in terms of the effect of texture type on the fraction of applied water that is lost by drainage. A sandy soil has less water retention capacity and thus a higher drainage fraction than a clay soil for the same irrigation rate and the same crop water consumption, giving rise to lower salt accumulation in the soil matrix.

Conclusions

NIR affords a valuable method for estimating organic carbon, organic nitrogen, total nitrogen and hygroscopic humidity in soils. Organic carbon and nitrogen data are of great value when deciding on organic amendments. Hygroscopic humidity is related to the textural class of the soil (clay content) and indicates the utility of the method for determining soil texture. Saline classes might also be established on the basis of NIR data.

Acknowledgments

This work was carried out using NIR hardware and software at the Centralised NIR and MIR Unit (SCAI), University of Córdoba (Spain).

References

1. M.J. Morra, M.H. Hall and L.L. Freeborn, *Soil Sci. Soc. Am. J.* **55,** 288 (1991).
2. A. Salgó, J. Nagy and J. Tarnóy. *J. Near Infrared Spectrosc.* **6,** 199 (1998).
3. J.M. Bremmner, in *Methods of soil analysis: Part 3, Chemical Methods,* Ed by J. Bigham. SSSA, Madison, WI, USA (1996).
4. R.L Mulvaney, in *Methods of soil analysis: Part 3, Chemical Methods,* Ed by J. Bigham. SSSA, Madison, WI, USA, p. 869 (1996).
5. J.S. Shenk and M.O. Westerhaus, in: *Near Infrared Spectroscopy. The Future Waves*, Ed by A.M.C. Davies and P.C. Williams. NIR Publication, Chichester, UK, p. 198. (1996).

Development of a continuous, high-speed, single-kernel brown rice sorting machine based on rice protein content

Motoyasu Natsuga,[a] Akitoshi Nakamura[b] and Sumio Kawano[c]

[a]*Faculty of Agriculture, Yamagata University, 1-23 Wakaba-machi, Tsuruoka 997-8555, Japan*

[b]*Shizuoka Seiki Co. Ltd, 2753-116 Takao, Fukuroi 437-0023, Japan*

[c]*National Food Research Institute, 2-1-2 Kannondai, Tsukuba 305-8642, Japan*

Introduction

Recently, consumers have been purchasing rice with an increasing interest for its taste. Consequently, every agricultural research station in Japan has been conducting its breeding programme by systematically selecting rice based on taste-related constituents. To improve this process, it is desirable to be able to estimate the constituent content of rice kernel by kernel and many researchers have been conducting single-kernel estimation using commercial bench-type NIR instruments.[1] However, those estimations usually take a long time—approximately 30 seconds to 1 minute—so they are not suitable for sorting the large number of samples that have either natural distribution or have been artificially mutated in a short time. To improve sorting speed, Shizuoka Seiki (Fukuroi City, Japan) and NFRI (National Food Research Institute) of MAFF (Ministry of Agriculture, Forestry and Fisheries of Japan) have jointly developed a continuous, high-speed, single-kernel brown rice sorting machine based on rice protein content.

Instrumentation

Figures 1 and 2 show the schematic diagram of the developed instrument and sample transport/sorting mechanism, respectively. The instrument consists of several sections, which include a feeding mechanism, measuring unit, sorting mechanism and controlling PC. As shown in Figure 2, the feeding mechanism picks up single-kernel brown rice from the hopper (maximum 5 kg storage capacity) and sends it to the measuring unit. A spectrum of the brown rice is obtained in the measuring unit, which consists of a near infrared array sensor. The brown rice is then sorted in the sorting mechanism based on its protein content estimated by the controlling PC.

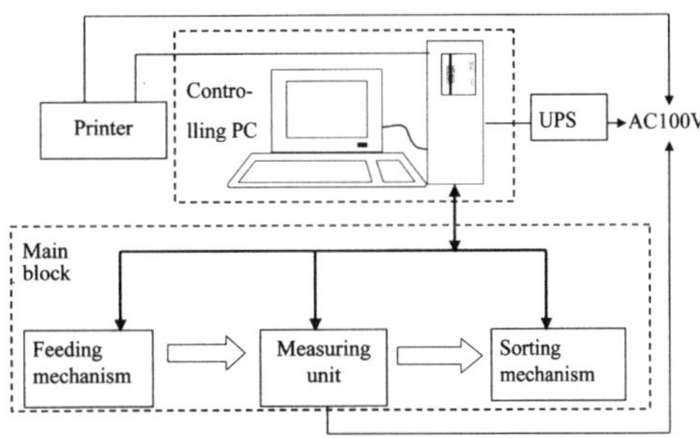

Figure 1. Schematic diagram of a continuous high-speed single-kernel brown rice sorting machine.

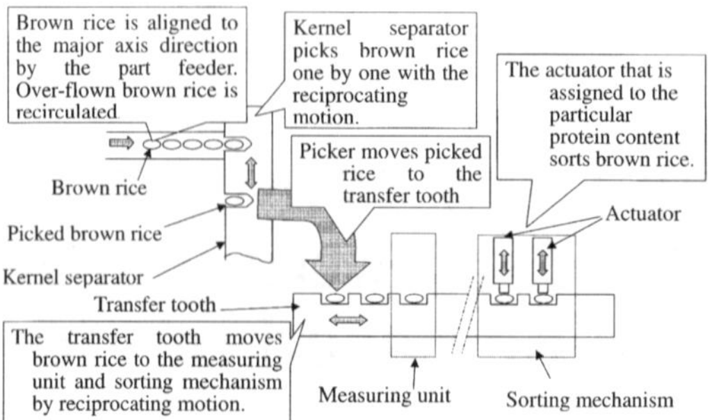

Figure 2. Schematic diagram of sample transport/sorting mechanism.

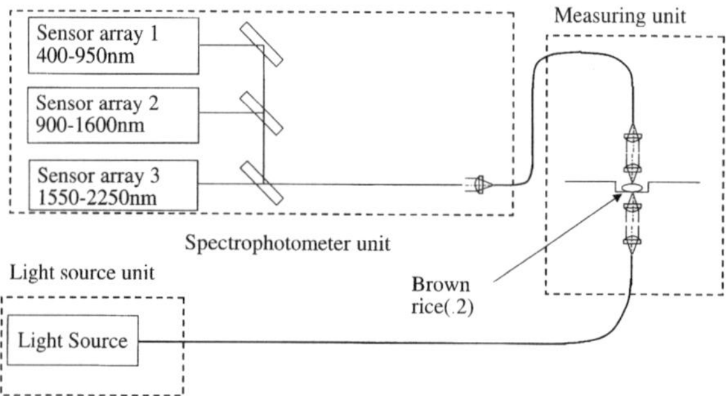

Figure 3. Optics layout (transmission).

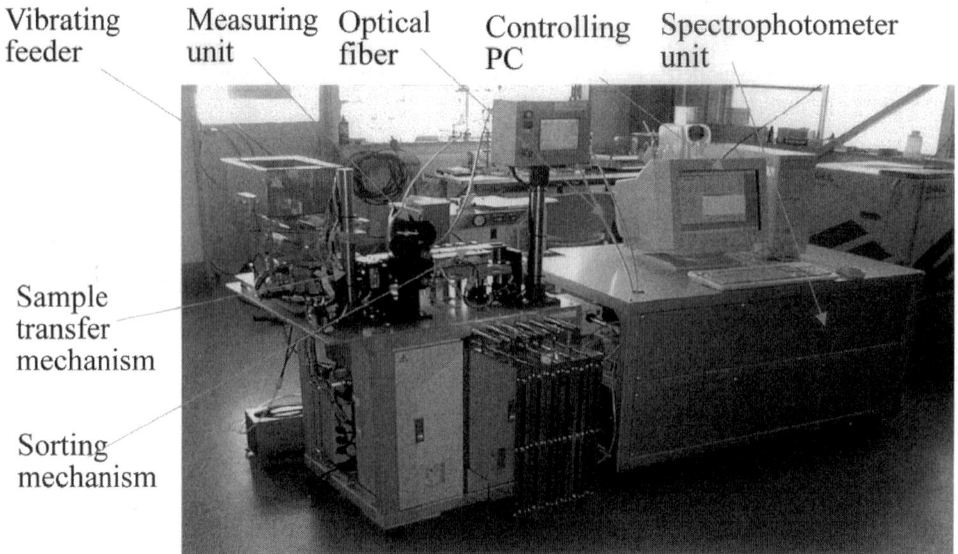

Photo 1. The developed instrument.

Figure 3 shows the optics layout. The light induced by input from the optical fibre to the measuring unit is focused approximately 2 mm diameter on the surface of brown rice. Light transmitted through the rice is then focused on an output optical fibre and is guided to a detector consisting of three array sensors, (400–950 nm Si, 900–1600 nm InGaAs and 1550–2250 nm InGaAs) after being monochromised by gratings. Measuring speed was approximately 500 ms for the full spectrum range of 400–2250 nm, 1 nm interval and overall sorting speed was approximately 2.8 s for one kernel.

Photo 1 shows the appearance and Table 1 shows the major specifications of the instrument.

Materials and methods

Samples

We used 363 brown rice samples (cultivar: Hitomebore) harvested in 1997, 1998 and 1999 provided by the Shizuoka Agricultural Experiment Station.

Spectrum measurement

Single-kernel brown rice spectra were obtained for both transmission and reflectance mode, configured to single-kernel measurement mode in the developed instrument set-up to obtain optimum S/N ratio when measurement was carried out at 500 ms. The total measuring time, including sample preparation and removal, was approximately 30 to 35 seconds.

Table 1. Major specifications of the developed instrument.

Model	CTC-3	
Feeding mechanism	Sample hopper Feeding method Transfer method	5 kg brown rice Automatic feeding using vibrating feeder Single kernel transfer mechanism
Measuring unit	Method Wavelength range Detector Measuring constituent	Transmission / Reflectance 400–2500 nm spectrum (1 nm interval) Si and InGaAs array sensors0 Protein content(DM)
Control unit	Controlling PC OS I/O	DOS/V PC Windows98 High-speed data transfer using SCSI
Sorting mechanism	Sorting Sorting method	Any given 5 classes of 0.1% pitch + NG Shutter
Power requirement	Instrument Air compressor	AC100V 10A AC100V 10A
Dimension		App. 1500 W × 700 L × 1500 H
Weight		App. 200 kg (air compressor excluded)
Spectrophotometric measuring time		App. 500 ms
Feeding speed		1 Kernel / 1.8 s
Accuracy	*SEP*	0.5% Protein content (DM)
Overall sorting speed		1 Kernel / 2.8 s

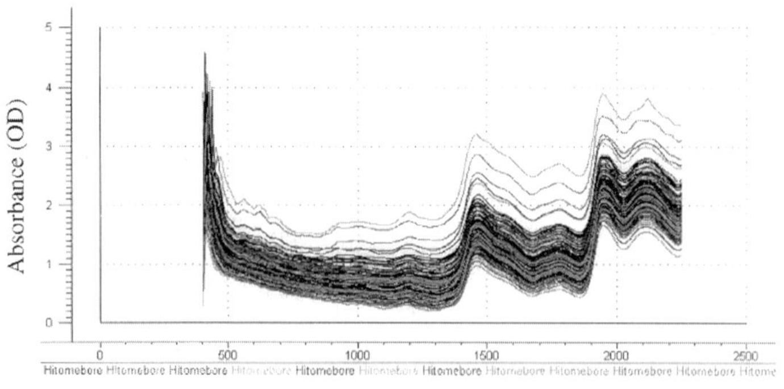

Figure 4. Brown rice spectra (transmission: 400–2500 nm.)

Chemical analysis

The spectrum-measured brown rice kernel first was dried completely by the oven method for 135°C for 24 h. Its nitrogen content was then measured using an NC-1000 (Sumika Chemical Analysis Service, Osaka, Japan) combustion nitrogen analyser and the protein content was calculated by multiplying 5.95 times the nitrogen–protein conversion factor. SDD (standard deviation of the difference) was 0.04%.

Developing calibrations

PLS regression analysis was carried out to developing calibrations using Unscrambler Version 7.5 (CAMO, Norway). Samples were randomly divided into three parts; two were used for developing calibrations and the other was used for validation.

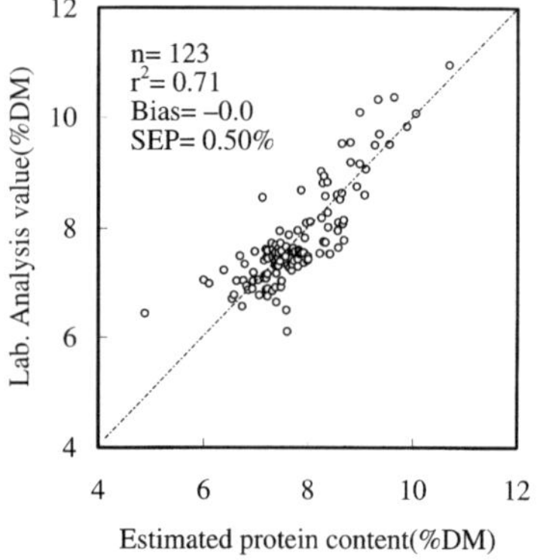

Figure 5. Scatter plot of estimated protein content *vs* laboratory analysis value (transmission 1000–2000 nm).

Results and discussion

As a consequence of examining both measuring mode and wavelength range, the best calibration was obtained when using 1000–2000 nm as the wavelength range in the transmission mode with a *SEP* value of 0.50%. Figure 4 shows the spectra of brown rice and Figure 5 shows the scatter plot. No good calibrations were obtained in the reflectance mode.

Concluding remarks

The new instrument reported here is now being used at the Shizuoka Agriculture Experiment Station. While good *SEP* results were obtained in the present study, over-all sorting speed and the consistency of measurements should be improved. These improvements and further improvement for *SEP* are currently underway.

Acknowledgement

We would like to express our gratitude to Yuji Miyata of the Shizuoka Agriculture Experiment Station for providing the brown rice samples.

References

1. S. Kawano, C. Iyo, H. Abe and M. Iwamoto in *Proceedings of 8th International Conference on Near-Infrared Spectroscopy*, p. 15 Food (1997).

Authentication and classification of strawberry varieties by near infrared spectral analysis of their leaves

Mercedes G. López

Unidad de Biotecnología e Ingeniería Genética de Plantas, Centro de Investigación y de Estudios Avanzados (CINVESTAV), Unidad Irapuato, Apartado Postal 629, Irapuato, Gto, 36500, México

Introduction

In Mexico, Irapuato is known as the Capital of Strawberries. *Fragaria x ananassa* was introduced by the Spaniards in the XI century. Since then it has become an icon for Irapuato, Guanajuato in Mexico. Most of the strawberry production is exported, principally to Asian countries like Japan where they demand a very high quality product. Most techniques to check crop status such as compositions are time-consuming and destructive. Therefore, a rapid and cheap tool is needed to verify the quality of many food items. Near infrared (NIR) spectroscopy[1] is known as one of the widely used non-destructive techniques. The NIR instrument is a fast and inexpensive analytical tool that could be used to classify, identify and authenticate a wide range of foods and food items. Therefore, the major objectives of this study were to provide a new insight into the authentication of two wildtype species and three varieties and to correlate the results with geographical origin and the propagating method used.

Materials and methods

Three weeks-old plants of five different strawberry varieties (*Fragaria* x *ananassa* Duch. cv Camarosa-California, Camarosa-Cinvestav, Seascape, *F. chiloensis* and *F. virginiana*) were cultivated *in vitro* first, then transferred to pots with special soil and grown in a greenhouse at Cinvestav. All varieties were acquired from California (USA). After eight months, ten leaves from each variety were collected. Transmittance spectra of each leave were recorded over a range of 10,000–4,000 cm^{-1}, 32 scans of each leave were collected at the resolution of 4 cm^{-1} with a Paragon IdentiCheck FT-NIR System Spectrometer. Spectra of both sides of the leaves and of the vascular system (N) were recorded. The number of replicates varied from 40 to 47 for each leave. All spectra were analysed using principal component analysis (PCA) and soft independent modelling class analogy (SIMCA). The optimum number of components to be used in the regression was automatically determined by the software (13–17).

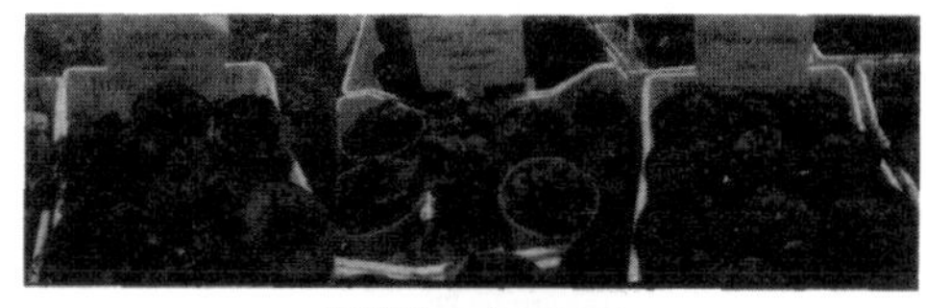

Figure 1. Some of the biological materials used in this study.

Results and discussion

All strawberry varieties displayed a very similar spectrum in the NIR range (10,000–4,000 cm^{-1}). However, the small differences among them allowed us to establish inter-class distances

Table 1. Inter-class distances among all strawberry varieties.

Samples	Seascape	Cam-CA	Cam-Cin	*Virginiana*	*Chiloensis*
Seascape	—	4.24	8.62	9.98	15.91
Cam-CA	—	—	5.74	14.48	19.68
Cam-Cin	—	—	—	19.18	29.32
virginiana	—	—	—	—	28.99

Table 2. Percentages of recognition and rejection among strawberry varieties.

Strawberry Variety	Recognition	Rejection
Seascape	100 (44 / 44)	93 (210 / 226)
Cam-CA	95 (40 / 42)	97 (222 / 228)
Cam-Cin	95 (41 / 43)	97 (221 / 227)
virginiana	100 (47 / 47)	100 (223 / 223)
chiloensis	96 (46 / 48)	100 (222 / 222)

and, therefore, the identification of all materials becomes possible. Camarosa was the only variety grown from the same shoot but propagated by a different method (direct or *in vitro*). Seascape from California presented the shortest inter-class distance (Table 1) to Camarosa California (this distance is even shorter than the inter-class distance between Camarosa samples). On the other hand, the Seascape variety presented the smallest rejection percentage among all varieties (more similarities with the rest of the samples) and it had a short inter-class distance to *F. chiloensis*. This means that it conserves some of its ancestral characteristics. Five different clusters can be observed in Figure 2. It can also be observed that the larger differences are displayed among wildtype samples, *F. chiloensis* (A) and *F. virginiana* (B). Camarosa-California (C) and Camarosa-Cinvestav (D) (*in vitro*) displayed a small overlapping region between them. Seascape samples are localised in (E). Figures 3 shows the validation methods for *F. chiloensis* and *F. virginiana*, in which samples with a N in their nomenclature

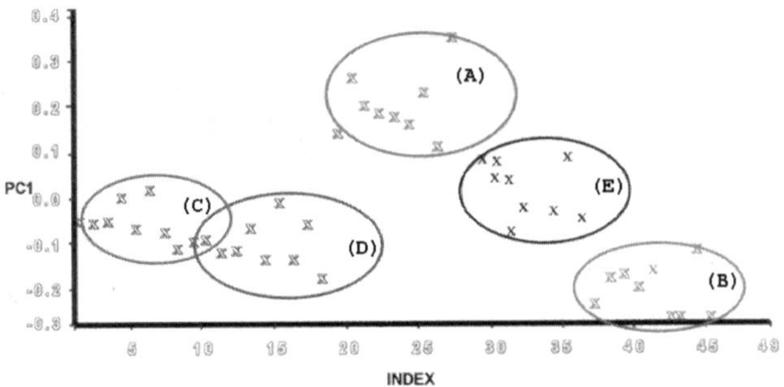

Figure 2. Clusters formed of the five strawberry varieties.

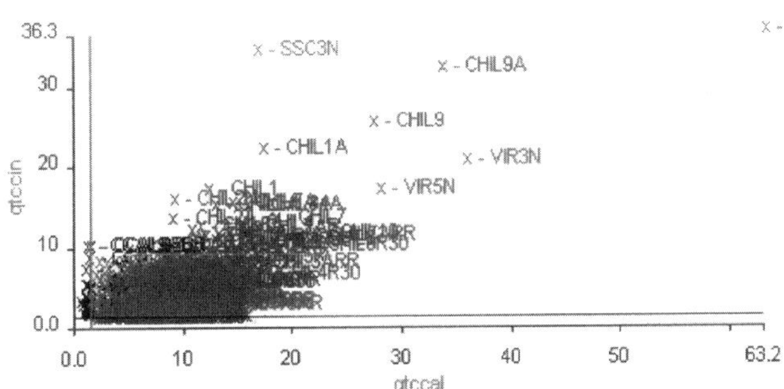

Figure 3. Validation method using Camarosa-California and Camarosa-Cinvestav.

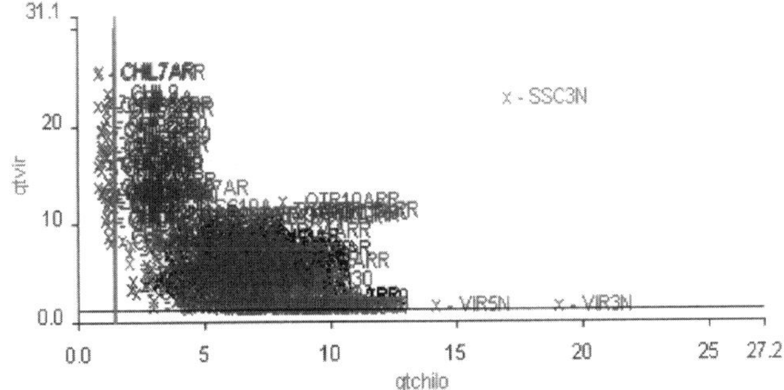

Figure 4. Validation method using *F. virginiana* and *F. chiloensis.*

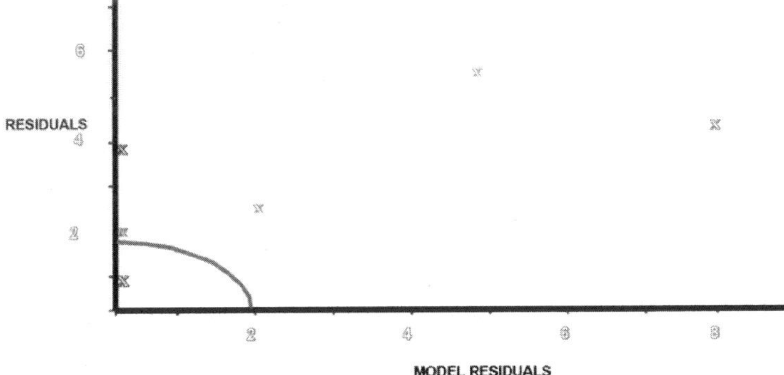

Figure 5. Classification of an unknown.

presented different behaviour. As for Seascape-CA and Camarosa-CA in Figure 4, it can be observed that only SSN (outlier) is out from the minimum distance allowed and the reason is that the spectrum of this leaf includes part of the nerve. Finally, Figure 5 show an example of a classification, where a leaf from an unknown strawberry variety was analysed using a blind number and it was positively classified (corresponding to the X inside the angle in Figure 5). Therefore, it can be concluded that NIR enables the authentication of all strawberry varieties and their geographical origin as well. It was also possible to form subclasses of the same materials. The results presented showed that NIR spectroscopy is a very powerful and promising analytical tool.

Conclusions

It was possible to create methods to classify and authenticate strawberry varieties by near infrared spectroscopy. Seascape variety was the sample that presented the largest similarities to other varieties, especially to Camarosa-California and *F. chiloensis*. On the other hand, the largest inter-class distance was observed for the two wildtype strawberry samples. Only *F. virginiana* displayed 100% recognition and 100% rejection and it is highly probable that Seascape is a cultivated variety with an *F. virginiana* ancestor. This study presented an overall of 14.1% of misclassified samples.

References

1. T. Gale, H. Yto and C. Miller, *J. Near Infrared Spectrosc.* **1,** 1 (1999).
2. M.G. López, *Comparison of beers by near infrared spectroscopy (NIRS).* Master Brewing Beer Association 114th Anniversary, (2001).
3. C.I. Noriega-Ayala and M.G. López, in *Bebidas Mexicanas.* Alfa Ed, Mexico City, Mexico, pp. 11–21 (1999).
4. M.J. Adams, *Chemometrics in Analytical Spectroscopy.* The Royal Society of Chemistry (1995).

Predicting malting quality in whole grain malt compared with whole grain barley by near infrared spectroscopy

Cassandra K. Black and Joseph F. Panozzo

Agriculture Victoria - Horsham, Victorian Institute for Dryland Agriculture, Private Bag 260, Horsham, 3401 Australia

Introduction

Near infrared (NIR) reflectance technology is a non-destructive, cost effective and rapid tool for simultaneous prediction of multiple constituents in agricultural products. In southern Australia, NIR technology is currently implemented to streamline the breeding of new, improved cultivars of malting barley.[4] Constituents of particular interest to the malting and brewing industry include hot water extract, diastatic power, free α-amino nitrogen (FAAN), soluble protein, wort β-glucan and β-glucanase. After harvest, the barley requires approximately two months of post-harvest maturation to break dormancy and allow the assessment of brewing attributes. Post-harvest dormancy in barley is the main contributor to delays in providing quality data to the breeder for the selection of superior malting quality breeding lines before the next sowing season (a four-month period). NIR calibrations developed on whole grain barley, before the grain was malted, would allow breeding lines to be evaluated for malting quality immediately after harvest.

The aim of this study was to determine the applicability of NIR spectroscopy in predicting malting quality of whole grain barley as compared with whole grain malt.

Materials and methods

Barley samples

The barley-breeding programme at the Victorian Institute for Dryland Agriculture generously supplied the barley samples. Thesamples represented an array of breeding generations grown at various sites throughout Victoria in 1999. The barley samples were tested by NIR before and after they were malted and then analysed for the various quality constituents.

Laboratory reference methods

The barley samples were malted[3] and analysed for hot water extract according to a small-scale version of the European Brewing Convention (EBC) fine grind method (grist : liquor ratio of 10 g : 70 mL).[2] Malt samples were ground on an FN3100 Mill (Falling Number, Stockholm, Sweden) to pass a 0.8 mm sieve and analysed for diastatic power[1] and β-glucanase according to the Megazyme method (Megazyme, Australia). Wort from the hot water extract was analysed for FAAN using the ninhydrin method;[3] soluble protein using the Dumas combustion method on a CNS-2000 (Leco Corporation, St Joseph, MI, USA) and wort β-glucan using the Megazyme method (Megazyme Australia).

NIR analysis

Instrumentation

Reflectance spectra, log (1/*R*), were collected on a Model 6500 monochromator (NIRSystems, Silver Springs, MD, USA) equipped with a transport module and a standard coarse NIR sample cell. Spectra were recorded across a range of 400–2498 nm with 2 nm wavelength increment. Diffuse reflectance readings off a ceramic tile were referenced before and after each sample scan. Both barley and malt samples were equilibrated to 21°C for 24 hours prior to analysis.

Math treatment

The spectra were first corrected for scatter with standard normal variate (SNV) and detrending and then a mathematical treatment of second order derivative with gap and smooth sizes of five and five data points, respectively, was applied.

Calibration development

The spectral population was structured to lie within three Global H units. The calibration set was optimised with a neighbourhood *H* value of 0.3. Calibrations were developed using a modified partial

Table 1. Statistical data of near infrared calibration and validation samples for malt quality constituents of whole grain barley and malt.

	Calibration Set					Validation Set			
Quality Parameter	N[a]	Range	No. of Terms	R^2[b]	SEC[c]	N[a]	Range	R^2[b]	SEV[d]
Whole grain barley set									
Hot water extract (%)	277	77–87	7	0.87	0.6	131	76–81	0.78	1.1
Diastatic power (WKE)	279	179–549	5	0.57	45	131	225–545	0.39	57
FAAN (mg L^{-1})	263	92–228	5	0.54	16	131	116–239	0.10	31
Soluble protein (%)	201	4–6	5	0.60	0.2	131	4–6	0.01	0.5
Wort β-glucan (mg L^{-1})	267	0–1089	9	0.77	104	131	0–760	0.25	240
β-Glucanase (units kg^{-1})	276	297–868	7	0.60	66	131	322–788	0.02	135
Whole grain malt set									
Hot water extract (%)	276	77–87	7	0.89	0.6	131	76–81	0.76	1.0
Diastatic power (WKE)	279	179–549	8	0.75	35	131	225–545	0.54	54
FAAN (mg L^{-1})	268	92–228	9	0.89	8	131	116–239	0.63	17
Soluble protein (%)	203	4–6	9	0.79	0.2	131	4–6	0.53	0.3
Wort β-glucan (mg L^{-1})	268	0–1089	9	0.83	93	131	0–760	0.51	165
β-Glucanase (units kg^{-1})	271	297–868	8	0.70	56	131	322–788	0.47	97

[a] Number of samples
[b] Coefficient of determination
[c] Standard error of calibration
[d] Standard error of validation

least squares (MPLS) algorithm using a cross-validation technique. ISI software V 4.01 (Infrasoft International, Silver Springs, MD, USA) was used for all data processing.

A population of 131 barley samples from separate trials of the 1999 harvest was used for validating the performance of calibrations.

Results and discussion

The NIR calibration and validation statistics for each constituent are presented in Table 1. The hot water extract value indicates the concentration of fermentable sugars in the malt. Near infrared (NIR)

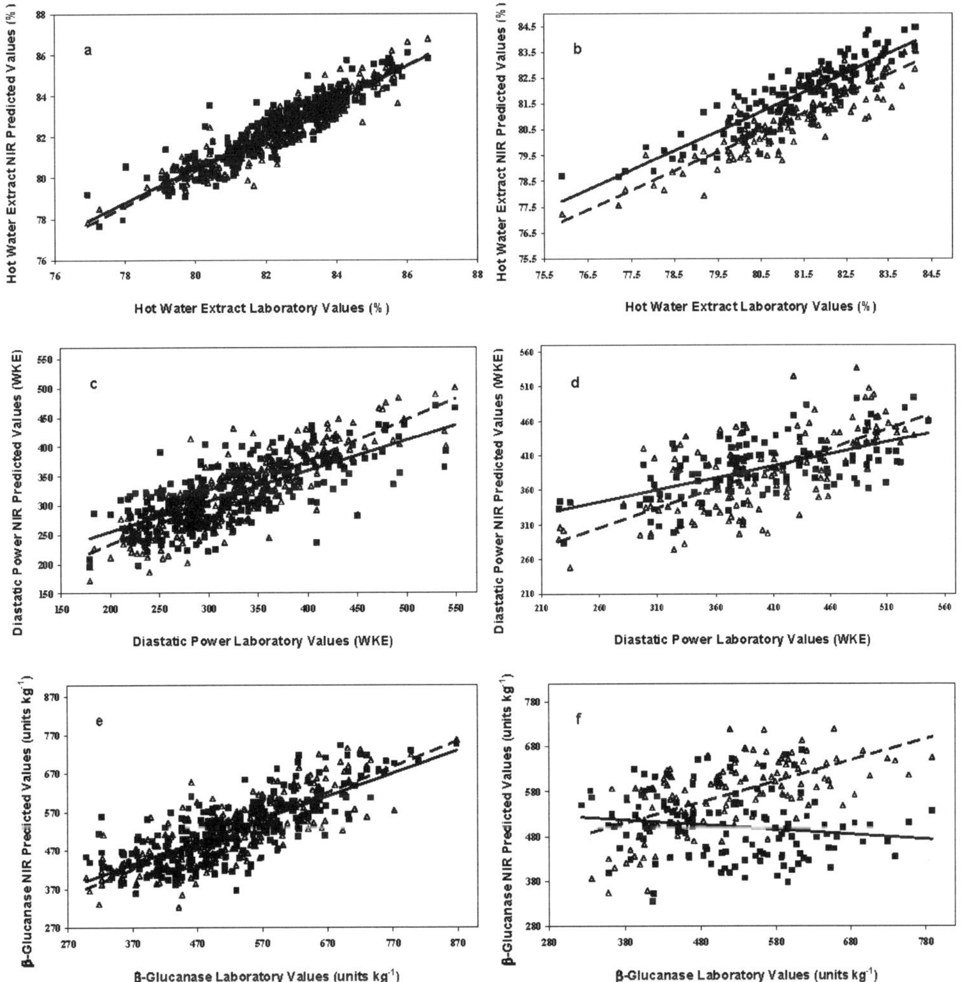

Figure 1. Scatter plots of the laboratory reference data and NIR predicted values for three constituents for whole barley and whole malt samples. (a) Hot water extract, calibration set; (b) hot water extract, validation set; (c) Diastatic Power, calibration set; (d) Diastatic Power, validation set; (e) β-Glucanase, calibration set; and (f) β-Glucanase, validation set (barley samples, △ malt samples).

predicted hot water extract values of the calibration samples were highly correlated with the laboratory reference data for both whole grain barley ($R^2 = 0.87$) and whole grain malt ($R^2 = 0.89$), and were associated with low standard errors of calibration (0.6 and 0.6, respectively). The validation data for hot water extract reflected those of the calibration samples, displaying good correlations and low standard errors (Table 1). Scatter plots of the laboratory reference data and the NIR predicted values of calibration and validation barley samples are presented in Figure 1(a) and 1(b). The data in this study are in agreement with findings of other researchers and this constituent has been routinely analysed by NIR for the past three years on whole grain barley to aid in the screening of early generation lines in southern Australian barley breeding programmes.[3,4]

Diastatic power is the ability of the enzymes present in the malt to convert starch to fermentable sugars.[1] NIR predicted values for diastatic power were moderately correlated to the laboratory reference values ($R^2 = 0.57$) for calibrations developed for whole grain barley [Table 1 and Figure 1(c)]. Validation statistics for diastatic power indicate a weak relationship between the NIR predicted values and the reference data [Table 1 and Figure 1(d)]. However, the size of the standard error of validation may suggest that NIR analysis could be useful for classification of early generation barley cultivars into high or low groups. In contrast, high correlation was observed between NIR predicted values and laboratory method for whole grain malt ($R^2 = 0.75$). Calibrations developed on whole grain malt are currently used for the evaluation of diastatic power in lines pre-selected with the aid of whole barley calibrations.

Soluble protein, FAAN, wort β-glucan and β-glucanase are complex malting constituents that influence the production of beer during the brewing process. Calibrations for these constituents, developed on whole grain barley, displayed a reasonably good relationship between the NIR predicted and laboratory reference values [Table 1 and Figure 1(e)]. Unexpectedly, this relationship was not confirmed by statistics observed for the validation samples (Table 1). The scatter plot of the laboratory reference data and the NIR predicted data for the constituent β-glucanase is depicted in Figure 1(f).

Calibrations developed on whole grain malt for soluble protein, FAAN, wort β-glucan and β-glucanase had high values for the coefficient of determination ($R^2 > 0.70$) and were associated with low standard errors of calibration (Table 1). The statistics observed for the validation samples were in agreement, displaying acceptable correlations with low standard errors (Table 1). Calibrations developed for these complex constituents on whole grain malt are suitable for the evaluation of early to mid-generation barley lines from a breeding programme.

Conclusion

NIR calibrations developed on whole grain barley for hot water extract display similar precision and accuracy as calibrations developed on malted grain. The validation data for diastatic power indicated a weak relationship between NIR predicted values and reference values, therefore a two step operation was employed. The calibration developed on whole grain barley was used to categorise early generation barley samples into low–high groups. The selected samples of interest were malted and the calibration developed on malted grain applied. The gain is in the reduced number of samples going through the lengthy process of micro-malting. The validation data for soluble protein, FAAN, wort β-glucan and β-glucanase demonstrated that the calibrations developed on whole grain barley could not be applied to the analysis of these constituents. This can be explained by the complex nature of these constituents. The unmalted barley has proteins and starches in the storage form. The composition of barley is modified by the action of enzymes throughout the steeping and germination stages and by heating during the kilning stage of the malting process.

Calibrations developed for hot water extract, diastatic power, soluble protein, FAAN, wort β-glucan and β-glucanase on malted barley are suitable for the evaluation of early to mid-generation barley lines from a breeding programme, although there is the added expense of micro-malting the samples.

The ability to apply barley and malt calibrations to different generations is an advantage to a barley breeding programme that requires thousands of samples to be assessed each year.

Acknowledgments

Thanks to staff of the Grains Chemistry Laboratory at VIDA – Horsham, David Moody and Dr Tadeusz Golebiowski at Ag-Seed Research Pty Ltd for their contribution. We acknowledge the financial support of the Malting Barley Quality Improvement Program (MBQIP), the Victorian Department of Natural Resources and Environment (NRE) and the Grains Research and Development Corporation (GRDC).

References

1. G. Fox, S. Logue, S. Harasymow, H. Taylor, M. Ratcliffe, S. Roumeliotis, K. Onley, P. Tansing, R. Ferguson, M. Glennie-Holmes, A. Inkerman, A. Tarr, B. Evans, J. Panozzo, A. Osman and A. Smith, in *The Proc. 9th Barley Tech. Sym.* Melbourne, Australia, pp. 2.35.1–2.35.5 (1999).
2. L.C. Macleod, M.A. Dowling, D.H.B. Sparrow and R.C.M. Lance, in *The Proc. 5th Barley Tech. Sym.* Australia, pp. 125–128 (1991).
3. M. Nilsen and J. Panozzo, in *Leaping ahead with Near Infrared Spectroscopy*, Ed by G.D. Batten, P.C. Flinn, L.A. Welsh and A.B. Blakeney. NIR Spectroscopy Group, Melbourne, Australia, pp. 174–177 (1995).
4. S. Roumeliotis, S.J. Logue, S.P. Jefferies and A.R. Barr, in *Near Infrared Spectroscopy: Proceedings of the 9th International Conference*, Ed by A.M.C. Davies and R. Giangiacomo. NIR Publications, Chichester, UK, pp. 673–678 (2000).

Compositional analysis by near infrared diode array instrumentation on forage harvesters

Andreas Haeusler,[a] Michael Rode[b] and Christian Paul[a]

[a]*Institute of Crop and Grassland Science, Federal Agricultural Research Centre (FAL), Braunschweig, Germany, christian.paul@fal.de*

[b]*Carl Zeiss Jena GmbH, Jena, Germany, rode@zeiss.de*

Introduction

Near infrared (NIR) spectroscopy, as a non-invasive optical method, offers the opportunity to measure plant constituents very rapidly. As robust NIR diode array spectrometers have been shown to be ideally suited for on-line measurements in several different applications in industry, it is an obvious next step to examine these instruments for agricultural purposes.[1] It has already been shown that, for assessing a multitude of quality parameters in dried forage samples, NIR diode array instruments can perform analytically as satisfactorily as NIR scanning monochromator instruments.[2] Our work is an elaboration of previous attempts to utilise the potential of diode array instrumentation for the on-line compositional analysis on agricultural plot harvesters.[2,3] It first aims to assess the dry matter content in forages without having to handle samples or transport them to a laboratory and dry them.

Materials and methods

This work was performed with the world's first commercially available experimental forage plot harvester equipped with an NIR module for the collection, compression and scanning of forage samples during harvesting (Haldrup NIRS Harvest Line; see Figure 1). The NIR-module includes an InGaAs diode array spectrometer (Zeiss Corona 45 NIR 1.7) and installations for spectrometer referencing, sample preparation and presentation. A representative subsample of each individual plot is automatically chopped and filled into the measuring chamber where it is compressed. Subsequently, the diffuse reflectance of the sample surface is measured by the diode array spectrometer while it moves downwards in front of the measuring chamber. This whole procedure ensures that exactly the same sample that was scanned is available for conventional analysis in order to build up a robust calibration database.

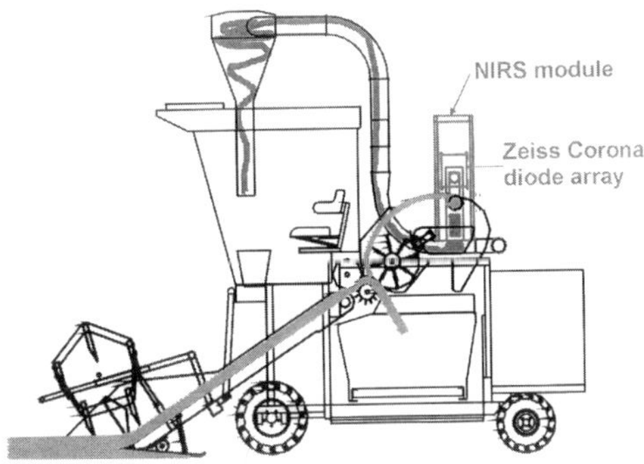

Figure 1. Cross-section of the Haldrup forage plot harvester with NIR module.

In the year 2000, the plot harvester was used for harvesting and analysing typical forage plots (1.4 m × 9 m) consisting of several grass species

(*Lolium perenne*, *Lolium multiflorum*, *Festuca pratensis*), a legume species (*Trifolium repens*) and mixtures of both. After NIR measurements were taken in the field, each sample was also measured in the laboratory using a conventional grating spectrometer equipped with both an Si- and a PbS-detector (NIRSystems 6500). Dry matter content (DM) was then determined conventionally by means of oven drying at 105°C for 36 h. Routine chemometric procedures were employed to assess the comparative accuracy and precision of the DM assessment in the spectral range between 960 and 1690 nm by the NIR diode array, as well as by the conventional NIR grating instrument. On-line data acquisition and control of the NIR module was done by means of the Zeiss software CORA. The WINISI II software from Infrasoft International was used to operate the laboratory spectrometer and for chemometric procedures on both types of spectrometers. For calibration, modified partial least squares (MPLS) regression analysis was performed on transformed NIR absorbance values vs DM contents. The accuracy of the DM assessment was estimated through cross-validation by means of the statistical parameters *SECV* (standard error of cross validation) and $1/VR$ (1 minus variance ratio which corresponds to the r^2 value).

Results and discussion

Because of the high absorptivity of water in the near infrared, an NIR method for assessing moisture content (or alternatively dry matter) has become the obvious first target of our on-line forage measurements under field conditions. The NIR spectra obtained during harvesting, using the Zeiss diode array spectrometer, are typified by broad, rolling absorption bands dominated by the strong, first OH stretch at 1450 nm attributable to water in forage. Spectra with minimal and maximal absorbance levels, as well as the average spectrum, are shown to indicate the overall spectral variability of the forages under investigation (Figure 2).

Under these circumstances, calibration for dry matter content was performed on forages in two separate concentration ranges: (a) the range typically observed in forage trials, i.e. from 12% to 30% DM and (b) in forages with excessive maturity ranging from 30% to 44% DM.

Calibration of the diode array spectrometer, in the typical range of forage maturity, resulted in an *SECV* value of 1.19%, while in forages with excessive maturity the *SECV* amounted to 1.64% DM (Figure 3). The corresponding error levels obtained for the same subsets of samples in post-harvest laboratory measurements on the conventional grating NIR spectrometer were 1.04% DM and 1.42% DM for forages with typical and excessive maturity, respectively (Figure 4). The $1/VR$ values also sup-

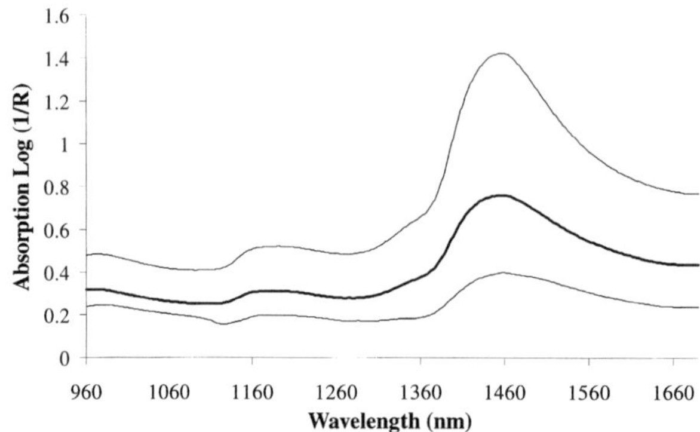

Figure 2. NIR spectra of fresh forage samples measured by Zeiss Corona NIR 1.7 diode array spectrometer during harvesting [maximum, minimum and average spectrum (thick line)].

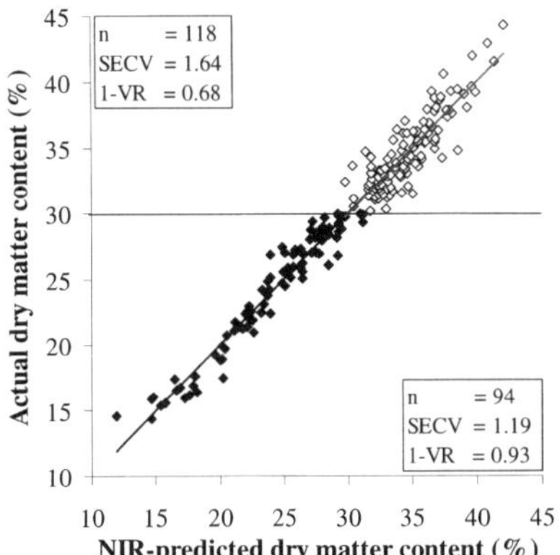

Figure 3. On-line assessment of dry matter content (DM%) in fresh forages by means of the diode array spectrometer Zeiss Corona 45 NIR 1.7 (samples with DM ≤ 30%: black symbols, samples with DM > 30%: white symbols).

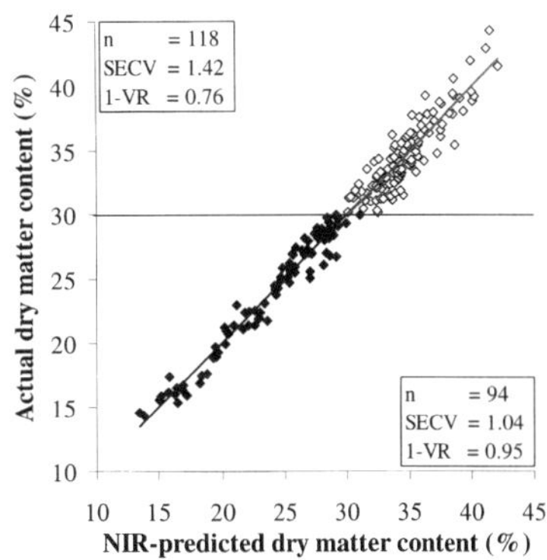

Figure 4. Assessment of dry matter content (DM %) in fresh forages by means of conventional grating spectrometer NIRSystems 6500 in the laboratory (samples with DM ≤ 30 %: black symbols, samples with DM > 30%: white symbols).

port the conclusion that the degree of fit between observed and predicted DM values differs strongly between forages of typical and excessive maturity. It is assumed that—because of increasing morphological differentiation and heterogeneity within plants—the sampling error in forages increases with maturity. Consequently, in forages of excessive maturity, the increase in *SECV* is seen as being caused by the increased sampling error. As a matter of fact, this effect is more noteworthy than the small difference in analytical precision between on-line diode array measurements in the field and post-harvest measurements using the laboratory spectrometer.

The results of this study show that the forage plot harvester with the integrated NIR diode array instrument employed here functions well for the assessment of dry matter even under rugged field conditions. Further refinements appear to be necessary for optimising the process of automated sampling, chopping and filling of the sample compartment to minimise sampling error in highly heterogeneous forage materials. In any case, it is expected that, in future, fundamental savings of energy and labour will be realised by the use of NIR diode array instruments on agricultural plot harvesters, thus replacing the costly and labour-intensive conventional oven drying method for dry matter determination.

Acknowledgments

Financial support by the German Federal Ministry of Consumer Protection, Food and Agriculture under R & D Project BLE 514–33.18/UM 137 is acknowledged. The splendid cooperation with Jens Haldrup a/s has contributed enormously.

References

1. C. Paul, Landbauforshung Volkenrode, *Sonderheft* **206,** 77 (1999).
2. M. Rode and C. Paul, in Book of Abstracts: *Proc. of 9th Int. Conf. near infrared Spectroscopy*, Verona/Italy, pp. 4–50 (1999).
3. P. Dardenne and N. Femenias, in Book of Abstracts: *Proc. of 9th Int. Conf. near infrared Spectroscopy*, Verona/Italy, p. 8 (1999).

Near infrared spectroscopic quality analysis of pre- and post-harvest sugarcane

Sarah E. Johnson and Nils Berding

BSES, PO Box 122, Gordonvale, Queensland 4865, Australia

Introduction

The quality of sugarcane grown on Australia's northeast tropical coast (16° 15′–18° 15′ S Lat.) has declined markedly in the past seven years. This reduction has been linked to dilution of mill-supply cane with increasing levels of non mature-stalk material, or extraneous matter, consisting of leaves and sucker culms. Extraneous matter is undesirable in mill-supply cane because of the high fibre levels and the associated increases in processing and production costs. The reasons for increased levels of extraneous matter in mill-supply cane are higher than average harvest-season rainfall and its effect on crop condition, record or near record crop yields and cyclones. Heavy sugarcane crops are predisposed to lodging (falling over), resulting in an open canopy and increased light levels, triggering sucker, or water shoot, development. Current harvesting technology is inefficient, particularly in lodged crops, resulting in sub-standard cane being delivered for milling. This paper reports results of research that primarily aimed to examine the transition from the pre-harvest, in-field crop to the harvested material sent for processing, in terms of quality and crop composition.

Materials and methods

Fifty-four random crop sites were sampled (17 in 1999 and 37 in 2000) and encompassed a wide range of variables including crop (cultivar and crop class) and environmental factors (edaphic, topographic, climatic and temporal). Ten quadrat samples (2 m row length) were taken from each site immediately prior to harvest. These were partitioned into four crop fractions—sound and unsound mature stalks (culms), sucker culms (water shoots) and extraneous matter (leaves). Ten pail samples (45 L) were taken from the mill-consigned material, immediately after harvest at each site. These also were partitioned into four fractions—sound and unsound billets (culm pieces), culm spindle pieces (cabbage) and leaf. In 2000, before the harvest season, 14 additional sites were sampled monthly, on three occasions, between March and June. Erect stalks, (standing between 0° and 45° from vertical), and non-erect (lodged) stalks, (lying between 60° and 90° from vertical) were divided further into sound and unsound classes.

All samples were processed immediately, or, if necessary, stored at 4°C until required. Samples were weighed, disintegrated using a Brazilian sugarcane disintegrator (Dedini Codistil, Piracicaba, SP) and mixed in a rotary stainless steel drum mixer for 90 s to ensure homogeneity. Prepared samples (≈ 3 kg) were presented to an NIRSystems (Silver Spring, MD, USA) remote reflectance module using the BSES large cassette module[1,2] in which a one metre long, bottomless cassette moved over a quartz plate window. The remote reflectance module was linked fibre-optically to an NIRSystems Model 6500 spectrophotometer fitted with a fast motherboard. Ninety-five sample scans were taken along the entire length of the cassette, with 32 reference scans being taken after the sample scans. The average spectra of the scans was stored and, by using the *H* and *t* statistics, was classified as normal, or either a spectral, (*H*), or component, (*t*), outlier.

Near infrared (NIR) spectroscopic analyses were developed for the rapid determination of the sugarcane quality components:- Brix (soluble solids, g kg^{-1}), commercial cane sugar (CCS; g kg^{-1}), fibre (insoluble carbohydrates, g kg^{-1}), moisture (g kg^{-1}) and polariscope reading (optical rotation, °Z). The material was classed into three broad groups for calibration: culm (mature-stalk and sucker culms; $n = 639$), non-culm (leaf material and culm spindle pieces; $n = 496$) and combined (all crop fractions) (Table 1). Calibrations were developed on the 1999 harvest season population ($n = 1{,}135$) using these three material groups. Two random sub-sets ($n = 178$ and 190), consisting of approximately 10% of the pre-harvest-season and harvest-season populations analysed in 2000 also were subjected to full routine laboratory analyses and later incorporated into these calibrations.

Calibrations were developed using two elimination passes for the removal of *H* and *t* statistic outliers. Samples were scanned using the wavelength range 800 to 2200 nm, with 2 nm increments. All calibrations were developed using SNV and Detrend scatter and various math treatments up to the second derivative were applied. Cross-validation was performed over four groups. The culm and combined calibrations were developed further to include additional samples collected in 2000.

Results and discussion

Calibration development was based on the three material classes described above (Table 1). Samples collected during the 2000 pre-harvest season consisted of culm material only and samples collected during the harvest season for both years consisted of all crop fractions. The 1999 combined calibrations were excellent. However, the culm calibrations produced consistently lower standard errors. The non-culm calibrations were marginally better than the combined calibrations for only CCS and pol. reading (Table 2). The performance of the combined calibration developed in 1999 was acceptable when used to predict, at-line, quality components of the captured spectral data from the 2000 harvest season. The standard errors of prediction ranged from 6.63 for pol. reading to 16.04 for CCS and correlation coefficients were good, ranging from 0.948 to 0.972. Recalibration, using additional data from 2000 ($n = 368$), improved the results considerably (Table 3) as evidenced by reduced standard errors, stronger correlations for all quality components, reduced bias and lower differentials between actual and predicted means.

Analysis of the 2000 culm data, using calibrations developed with 1999 and 2000 culm data, resulted in better predictions relative to the 1999 culm calibrations. Results of the 1999 culm calibrations, applied to the 2000 pre-harvest-season spectral population, showed (Table 4) standard errors of prediction were reasonably high and correlations between predicted and actual values were weaker for fibre and moisture. Re-calibration, using 298 additional samples from the 2000 culm population, produced consistently lower standard errors and stronger correlations for all components, except CCS (Table 4).

Table 1. Three approaches to calibration development, based on 1999 data.

Group	*n*	In-field	Post-harvest
Combined	1,135	stalks, tops, trash, suckers	billets, cabbage, leaf
Culm	639	stalks, suckers	billets
Non-culm	496	tops, trash	cabbage, leaf

Table 2. Development of the combined, culm and non-culm calibrations, from 1999.

Group	Component	Treat.	# terms	*SEC*	R^2	Mean	*CV%*
Comb.	Brix	1,4,4,1	11	4.17	0.995	156.29	2.67
	CCS	2,8,8,1	14	4.09	0.997	71.76	5.70
	Fibre	1,4,4,1	10	6.87	0.995	199.07	3.45
	Moisture	1,8,8,1	12	5.45	0.996	670.46	0.81
	Pol. reading	2,8,8,1	15	1.71	0.998	46.73	3.66
Culm	Brix	2,8,8,1	9	3.07	0.996	189.40	1.62
	CCS	1,8,8,1	13	3.65	0.996	123.74	2.95
	Fibre	1,8,8,1	12	5.25	0.913	132.41	3.96
	Moisture	2,8,8,1	11	3.63	0.995	705.34	0.51
	Pol. reading	1,4,4,1	10	1.46	0.997	69.09	2.11
Non-culm	Brix	1,8,8,1	9	4.38	0.971	107.90	4.06
	CCS	1,8,8,1	16	3.73	0.922	–3.30	–112.84
	Fibre	1,8,8,1	13	7.82	0.990	310.15	2.52
	Moisture	1,8,8,1	14	5.14	0.996	609.71	0.84
	Pol. reading	1,4,4,1	15	1.64	0.947	13.18	12.47

Table 3. Combined calibrations applied to 1999 and 2000 harvest season spectral populations.

Calibration	Component	*SEP*(*C*)	Bias	r^2	Means	
					RLA	NIR
Comb. 1999 (n – 1,135)	Brix	10.50	–1.61	0.972	154.48	156.08
	CCS	16.04	–14.61	0.918	86.44	101.05
	Fibre	11.35	–6.54	0.964	152.51	159.04
	Moisture	12.10	6.16	0.955	715.68	709.52
	Pol. reading	6.63	–6.83	0.960	50.75	57.58
Comb. 99-00 (n = 1,503)	Brix	8.85	0.98	0.979	154.48	153.49
	CCS	12.20	–0.05	0.969	86.44	86.50
	Fibre	9.67	0.36	0.972	152.51	152.14
	Moisture	9.18	–1.83	0.974	715.68	717.50
	Pol. reading	4.46	0.07	0.981	50.75	50.68

Table 4. Culm calibrations applied to 1999 harvest season, and 2000 pre-harvest season culm populations.

Calibration	Population	n	Comp.	SEP(C)	$\bar{x}$	r^2
1999 (n = 639)	2000	298	Brix	8.67	163.03	0.962
			CCS	22.09	98.72	0.903
			Fibre	15.09	117.48	0.620
			Moisture	21.76	744.02	0.846
			Pol. reading	4.32	56.80	0.965
	1999 + 2000	937	Brix	8.72	179.80	0.968
			CCS	16.17	114.84	0.925
			Fibre	10.37	129.27	0.748
			Moisture	13.89	118.62	0.938
			Pol. reading	4.65	64.67	0.968
1999 + 2000 (n = 937)	2000	298	Brix	6.60	163.03	0.978
			CCS	23.56	98.72	0.863
			Fibre	9.09	117.48	0.831
			Moisture	11.40	744.02	0.949
			Pol. reading	3.56	56.80	0.976
	1999 + 2000	937	Brix	8.15	179.80	0.972
			CCS	16.74	114.84	0.917
			Fibre	8.13	129.27	0.841
			Moisture	9.10	718.62	0.973
			Pol. reading	4.34	64.67	0.972

The results obtained from the 2000 pre-harvest season sampling (Table 5) showed that CCS of lodged cane was 7.6% lower than CCS of erect, sound cane. Similarly, a 12.2% reduction in CCS was apparent when unsound cane was compared. In combination, these two undesirable traits resulted in a CCS reduction of 24.1% (Table 5). From results obtained during the 2000 harvest-season (Table 6), potential CCS of mature culms and billets is high, at 147.2 and 144.2 g kg^{-1}, respectively. These concentrations are being heavily discounted by dilution with extraneous matter, resulting in an 18.5 g kg^{-1} and 16.9 g kg^{-1} reduction, respectively, in in-field and post-harvest CCS. The CCS of the harvested material was fractionally lower than the CCS of the entire in-field crop, with a reduction of 1.4 g kg^{-1}. In-field extraneous matter, not including dead leaf clinging to the stalks (trash), constituted 13.3% of the

Table 5. CCS values of combinations of crop condition and crop habit.

Condition	Erect	Lodged
Sound	109.94	101.63 (92.44)
Unsound	96.52 (87.79)	83.42 (75.88)

Table 6. Summation of crop fractions and commercial cane sugar (CCS) distribution based on 2000 harvest season data.

Sample origin	Groups	Individual components	%	CCS	Weighted CCS	
					Groups	Overall
In-field	Mature stalks	Sound stalks	53.5	164.5	154.6	110.6
		Unsound stalks	16.7	122.8		
	Extraneous matter	Sucker culms	10.8	19.8	1.8	
		Extraneous matter	18.9	–8.1		
Post-harvest	Mature stalks	Sound billets	72.2	148.5	141.7	117.6
		Unsound billets	8.9	85.6		
	Extraneous matter	Cabbage	10.5	15.2	14.3	
		Leaf	8.4	10.5		

sample weight, which, after harvesting, was reduced marginally to 13.1%. Of the total in-field sample weight, trash constituted 7.0, increasing the actual extraneous matter quantity from what is reported here.

Conclusions

Assessment of quality components in pre- and post-harvest sugarcane using NIR (combined calibration) was more cost effective than routine laboratory methods. Outcomes from this NIR-facilitated research will have important economic consequences for the Australian sugarcane industry. Potential CCS present in mature culms is being discounted by dilution with leaves and sucker culms, threatening farm viability. The CCS of harvested cane was improved only marginally over that of the entire in-field crop. The results question the efficacy of current harvesting technology, highlighting a need for either supplementary, innovative pre-mill processing or a design revolution to improve mill- supply cane quality and, therefore, whole-of-industry economics. NIR facilitated analyses during the growth or pre-harvest-season, quantified the benefits of growing erect, sound crops. Loss of CCS, therefore, can be minimised only by a combination of crop improvement and agronomic solutions applied as part of sound, on-farm management regimes.

Acknowledgements

The Australian Sugar Research and Development Corporation is gratefully thanked for the funding of project BSS220, under which this research was conducted. The sampling team is thanked for their commitment and enthusiasm.

References

1. N. Berding and G.A. Brotherton, in *Sugarcane: Research towards Efficient and Sustainable Production,* Ed by J.R. Wilson, D.M. Hogarth, J.A. Campbell and A.L. Garside. CSIRO Division of Tropical Crops and Pastures, Brisbane, Australia, p. 57 (1996).
2. N. Berding and G.A. Brotherton. *NIR news* **7(6),** 14 (1996).

Selection of wavelength region for partial least squares Brix calibration of mango using the multiple linear regression method

Sirinnapa Saranwong,[a] Jinda Sornsrivichai[a] and Sumio Kawano[b]

[a]*Department of Biology, Faculty of Science, Chiang Mai University, Chiangmai 50002, Thailand*

[b]*National Food Research Institute, 2-1-12 Kannondai, Tsukuba 305-8642, Japan*

Introduction

In order to establish the calibration equation of near infrared (NIR) spectroscopy, multiple linear regression (MLR) was used during the early period[1–11] while partial least squares (PLS) regression has been widely applied recently because no selection of wavelength is needed.[9–13] However, even in a PLS calibration, the selection of a wavelength region is needed to make a good calibration, which is more complicated and time-consuming.[14,15] Therefore, in this work, the performance of MLR and PLS calibration equations were compared and the relationship between them was also investigated by using the Brix value of mango as a model.

Materials and methods

Materials

A total of 138 Philippine mango fruits (*Mangifera indica* cv. Caraboa), which were commercially available, were used as samples for this experiment. The fruits packed with corrugated fibreboard boxes were purchased at the wholesale market, transported to our laboratory at the National Food Research Institute (NFRI) and then kept in the cold storage room at a temperature of 5°C for the experiment the following day. Six hours before NIR spectra acquisition, the fruits were moved from the cold storage room to a room with a temperature of 25°C so that the sample temperature reached room temperature.

Spectral acquisition

NIR spectra of mango fruits were measured in the short wavelength region from 700 nm to 1100 nm. A commercially available NIR instrument (NIRSystems 6500) with a fibre-optic "Interactance Probe" was used to measure the NIR spectra of intact mango fruits in a similar way to our previous study.[4] The NIR spectra were obtained at the fruit shoulder by averaging 50 scans. A standard measurement of a teflon sphere (8 cm diameter) was performed on every six fruits.

Prior to NIR measurement of samples, control of sample temperature was performed by dipping each sample into a water bath controlled at 25°C. In order to prevent samples from getting wet, the surface of the water bath was covered with a polyethylene film and then the samples were dipped into the water, together with the film, for ten minutes.

Table 1. Characteristics of calibration and validation sample sets of mango used.

Item	Calibration set	Validation set
Number of sample	74	64
Range	10.80 – 18.25	11.15 – 16.65
Mean	14.44	14.22
Standard deviation	1.72	1.38
LE[a]	0.23	0.25
Unit	°Brix	°Brix

[a]LE is laboratory error calculated from standard deviation of error between duplicate measurements of each reference analysis

Chemical analysis

A portion of flesh, about 10 mm deep from the peel, which was illuminated by NIR, was sampled and analysed for Brix value with a digital refractometer (ATAGO, Model PR101).

Each constituent value used for the following calculation was given by averaging of duplicate measurements. Statistical characteristics of calibration and validation sets in both wavelength regions are shown in Table 1.

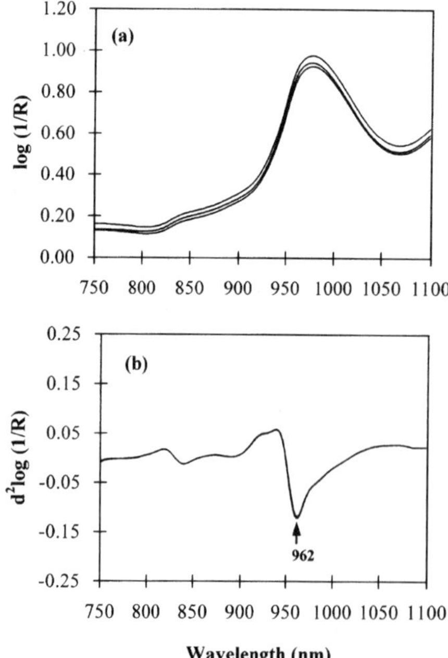

Figure 1. (a) Original spectra and (b) second derivative spectra of typical mango fruits having low, medium and high Brix values.

Data analysis

The Near Infrared Spectral Analysis Software (NSAS) program and the Unscrambler program were used for MLR and PLS regression, respectively.

Results and discussion

NIR spectra of intact mango

The original spectra and second derivative spectra of typical intact mangoes having low, medium and high Brix values are shown in Figure 1(a) and 1(b), respectively. Strong absorption bands due to water were observed at the wavelength of 962 nm.

Calibration and validation by MLR

The correlation plot which were correlation coefficients between $d^2\log(1/R)$ and Brix value plotted against wavelength is shown in Figure 2. Large negative peaks were observed at 848, 906 and 990 nm. The wavelength of 906 nm was selected manually as the first wavelength in a similar way to our previous study.[4] The calibration

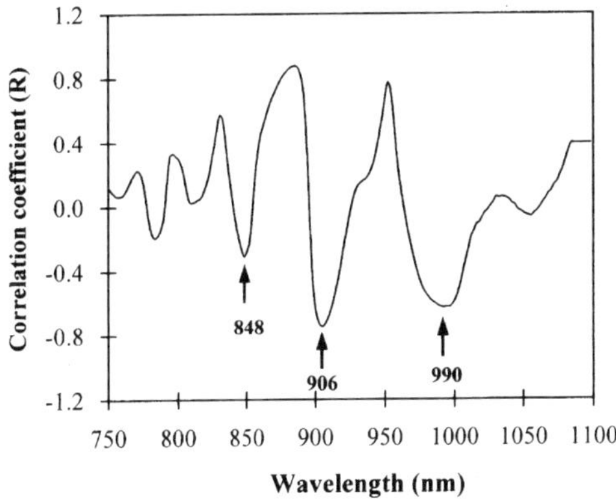

Figure 2. Correlation plots for selecting the first calibration wavelength of Brix value.

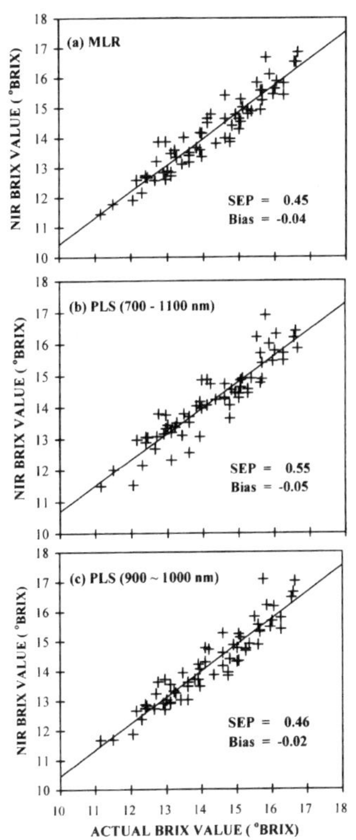

Figure 3. Scatter plot of Brix value of mango fruits. The Brix values are predicted by (a) MLR, (b) PLS (700–1100nm), and (c) PLS (900–1000 nm).

and validation results are shown in Table 2. The scatter plot of Brix values of mango fruits predicted by MLR developed calibration equation are shown in Figure 3(a).

Calibration and validation by PLS

The PLS calibration and validation results using different wavelength regions are shown in Table 3. The *SEP* was smallest when the wavelength region from 900 nm to 1000 nm was se-

Table 2. Calibration and validation results obtained from MLR regression regions for determining Brix values of mango fruits.

Selected Wavelengths (nm)				*R*	*SEC*	*SEP*	Bias
λ_1	λ_2	λ_3	λ_4		(°Brix)	(°Brix)	(°Brix)
906[M]	—	—	—	–0.74	1.17	1.20	0.27
906[M]	996[M]	—	—	0.91	0.73	0.81	0.13
906[M]	996[M]	978[M]	—	0.91	0.73	0.77	0.15
906[M]	996[M]	978[M]	942[C]	0.96	0.49	0.45	–0.04

[M] Selected manually
[C]: Selected by computer
R: Multiple correlation coefficients
SEC: Standard error of calibration,
SEP: Bias-corrected standard error of prediction
Bias: The average of difference between actual value and NIR value

Table 3. Calibration and validation results of PLS calibration on various wavelength regions for determining Brix values of mango fruits.

Wavelength region (nm)	Factors	*R*	*SEC* (°Brix)	*SEP* (°Brix)	Bias (°Brix)
700 – 1100	9	0.97	0.43	0.55	–0.05
800 – 1100	4	0.95	0.52	0.53	–0.02
900 – 1100	5	0.96	0.48	0.49	0.01
1000 – 1100	3	0.92	0.68	0.77	0.23
900 – 1000	5	0.96	0.47	0.46	–0.02
900 – 950	3	0.93	0.62	0.72	0.08

R: multiple correlation coefficients
SEC: standard error of calibration
SEP: bias-corrected standard error of prediction
Bias: the average of difference between actual value and NIR value

lected. The scatter plots of Brix values of mango fruits predicted by PLS developed calibration equation are shown in Figure 3.

Relationship between MLR and PLS calibrations

It was found that the best wavelength region for PLS calibration from 900 nm to 1000 nm was identical to the wavelength region which involved the wavelengths selected by MLR.

The regression coefficient (*K*) plot of the best PLS calibration is shown in Figure 4. Four negative peaks were observed at 908, 942, 978 and 1000 nm, which corresponded to the wavelengths selected for the MLR calibration equation as shown in Table 2. The MLR calibration equation can be written as follows:

$$\text{Brix value} = 13.61 - 1245.02\,d^2\log(1/R_{906}) - 329.01d^2\log(1/R_{996}) - 1052.89d^2\log(1/R_{978}) - 824.89\,d^2\log(1/R_{942}) \quad (1)$$

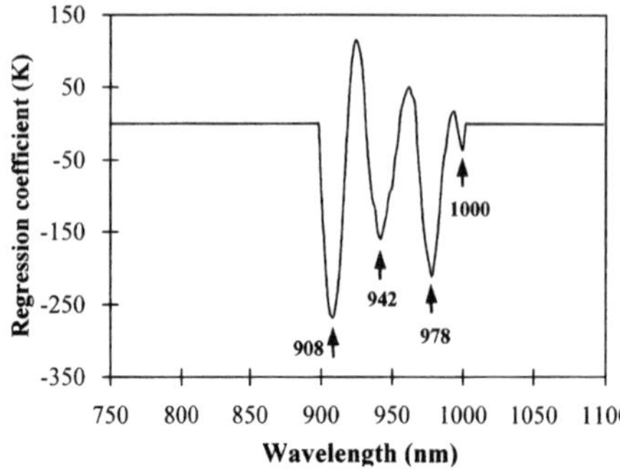

Figure 4. Regression coefficient plot for PLS calibrations using the wavelength region from 900 nm to 1000 nm.

In addition, it was found that there is also high similarity in the order of the regression coefficients. The order for MLR and PLS calibration equations is:

$$K_{906} < K_{978} < K_{942} < K_{996} \text{ for MLR}$$
$$\text{and } K_{908} < K_{978} < K_{942} < K_{1000} \text{ for PLS}$$

where, $K_{\lambda i}$ is the regression coefficient at the wavelength λ_i.

These results indicated that even if the procedures for getting the calibration equation are different, the MLR and PLS calibration equations to predict Brix value were based on the same important wavelengths.

Conclusion

It was concluded that NIR spectroscopy in the short wavelength region is sufficiently accurate to determine Brix value of mango fruit nondestructively. Both MLR and PLS could be used in making calibration equations with similar predictive efficiency. It was found that the selection of wavelength region is needed in PLS calibration in the same way as in MLR calibration. The wavelength region selected by MLR could be used as a good navigator for the selection of the wavelength region in the PLS calibration.

Acknowledgement

This research was supported by the Royal Golden Jubilee PhD Programme under the Thailand Research Fund (TRF).

References

1. G.S. Birth, G.G. Dull, W.T. Renfroe and S.J. Kays, *J. Amer. Soc. Hort. Sci.* **110,** 297 (1985).
2. G.G. Dull, G.S. Birth, D.A. Smittle and R.G. Leffler, *J. Food Sci.* **54,** 393 (1989).
3. S. Kawano, H. Watanabe and M. Iwamoto, in *The Proceedings of the 2nd International Near Infrared Spectroscopy Conference,* Ed by M. Iwamoto and S. Kawano. Korin Publishing, Japan, p. 343 (1990).
4. S. Kawano, H. Watanabe and M. Iwamoto, *J. Japan. Soc. Hort. Sci.* **61(2),** 445 (1992).
5. S. Kawano, H. Abe and M. Iwamoto, *J. Near Infrared Spectrosc.* **3,** 211 (1995).
6. S. Kawano, T. Fujiwara and M. Iwamoto, *J. Japan. Soc. Hort. Sci.* **62(2),** 465 (1993).
7. T. Temma, M. Chiba, K. Hanamatsu and F. Shinoki, in *Near Infrared Spectroscopy: The Future Waves*, Ed by A.M.C. Davies and P. Williams. NIR Publications, Chichester, UK, p. 638 (1996).
8. R.K. Cho, M.R. Sohn and Y.K. Kwon, *J. Near Infrared Spectrosc.* **6,** A75 (1998).
9. J. Guthrie and K. Walsh, *Aust. J. Exp. Agr.* **37,** 253 (1997).
10. Z. Schmilovitch, A. Mizrach, A. Hoffman, H. Egozi and Y. Fuchs, *Postharv. Biol. Technol.* **19,** 245 (2000).
11. P. Dardenne, G. Sinnaeve and V. Baeten, *J. Near Infrared Spectrosc.* **8,** 229 (2000).
12. V.A. McGlone and S. Kawano, *Postharv. Biol. Technol.* **13,** 131 (1998).
13. K. Miyamoto, M. Kawauchi and T. Fukuda, *J. Near Infrared Spectrosc.* **6,** 267 (1998).
14. S.D. Osborne, R. Kunnemeyer and R.B. Jordan, *J. Near Infrared Spectrosc.* **7,** 9 (1999).
15. P. Carlini, R. Massantini and F. Mencarelli, *J. Agric. Food Chem.* **48,** 5236 (2000).

Application of time-of-flight near infrared spectroscopy to Satsuma mandarin

Satoru Tsuchikawa,[a] Satomi Ito,[a] Kinuyo Inoue[a] and Kumi Miyamoto[b]

[a]*Graduate School of Bioagricultural Sciences, Nagoya University, Nagoya 464-8601, Japan*

[b]*Wakayama Fruit Tree Experiment Station, Kibi, Wakayama 643-0022, Japan*

Introduction

The detection of near infrared (NIR) light from the sample is through either transmittance or reflectance.[1] The transmittance NIR method is very desirable for detecting internal information, whereas the optical information from diffuse reflectance spectra is confined to the subsurface layer of samples. Therefore, it is especially important that progress be made in developing an NIR transmission device for detection of internal characteristics of high-moisture fruit. For example, the nondestructive detection of sugar content in Satsuma mandarin would be of considerable value to the fresh fruit industry. Some techniques and devices for the transmittance method, which could detect inner information of Satsuma mandarin have been proposed previously.[2,3] However, the behaviour of transmitted light from an agricultural product is directly affected by physical and chemical properties of tissues, so that it is preferable to examine the optical characteristics of the tissue and its origin in detail from a new concept.

In this study, an optical measurement system, which was mainly composed of a parametric tunable laser and a near infrared photoelectric multiplier, was introduced to accomplish this purpose from the viewpoint of time-of-flight near infrared spectroscopy (ToF-NIR).[4,5] This system combines the best features of the spectrophotometer and the laser beam, and more advantageously, the time-resolved profile of transmitted output power can be measured sensitively in nanoseconds. The combined effects on the time-resolved profiles of sample diameter, sugar content, the wavelength of laser beam and the detection position of transmitted light were investigated in detail.

Material and methods

The samples used were Satsuma mandarin (*Citrus unshu* $M_{ARC.}$) (location: Wakayama, Japan) having the diameters of 50–84 mm. The sugar content measured by a refractometer varied from 9.9 to 16.3 Brix %.

The wavelength of the pulsed laser λ was tuned from 600 nm to 1100 nm by the optical parametric oscillation of a BBO (β-BaB_2O_4) crystal.[6,7] The transmitted output power was measured by an NIR photoelectric multiplier with a spectral response ranging between 300 nm and 1700 nm through an optical fibre cable having a diameter of 7 mm. The equator of the sample was irradiated vertically with the pulsed laser and the detection position was varied at the equator of the sample. The sampling time and the number of averaging of the transmitted output power were 100 ns and 300 times, respectively.

The time-resolved profile refers to the variation of the intensity of the detected light beam with time. In this study, the time-resolved profile of the cuticle was employed as the reference. The measure of attenuance A_t is defined as follows:

$$A_t = \log\left(\frac{Pw_0}{Pw}\right) \tag{1}$$

where Pw_0 and Pw indicate the peak maxima of the reference and the object, respectively. The measure of time delay of peak maxima Δt is expressed as follows:

$$\Delta t = t - t_0 \tag{2}$$

where t_0 and t indicate the time at peak maxima of the reference and the object, respectively. The variation of the full width at half maximum value of the profile Δw is also expressed as follows:

$$\Delta w = w - w_0 \tag{3}$$

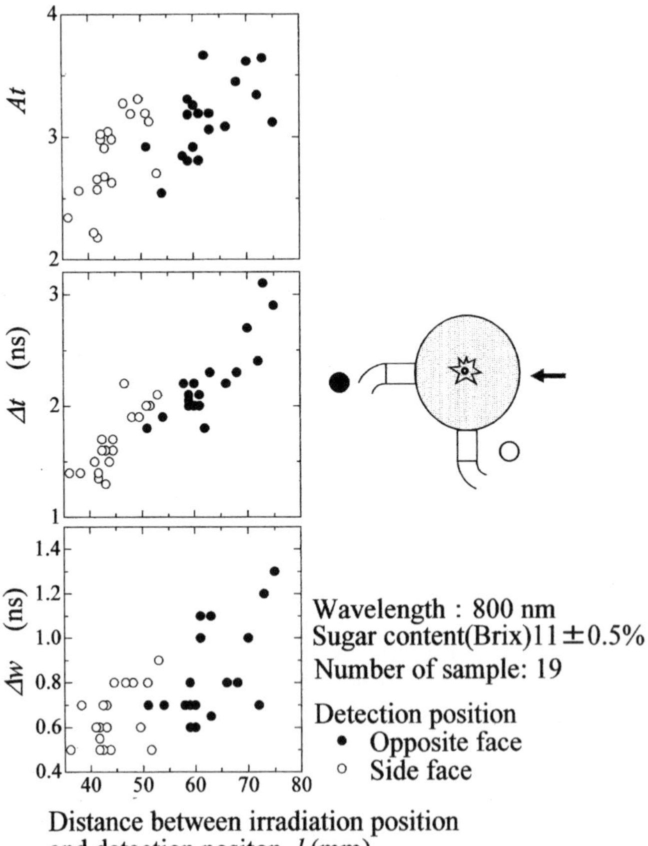

Figure 1. Variation of attenuance A_t, time delay of peak maxima Δt and variation of full width at half maximum Δw with detection position.

where w_0 and w indicate the full width at half maximum value of the profile for the reference and the object, respectively.

Results and discussion

Figure 1 indicates the variation of A_t, Δt and Δw with the distance between irradiation position and detection position l, respectively. The detection position is either opposite face or sider face (see Figure 1). The diameter of the sample varied from 50 mm to 84 mm. The sugar content in the population is nearly constant at 11 ± 0.5%. These optical parameters increased gradually as l increased to be greatly absorbed and vigorously scattered. In particular, Δt is positively correlated with l.

Figure 2 indicates the variation of A_t, Δt and Δw with the sugar content, respectively. The diameter in the population is nearly constant at 60 ± 1 mm. Each optical parameter had a tendency to increase as the sugar content increased. Such behaviour was remarkable when the transmitted light was detected at the side face of a sample. However, there is little correlation of sugar content with Δt and Δw when the transmitted light is detected at the opposite face.

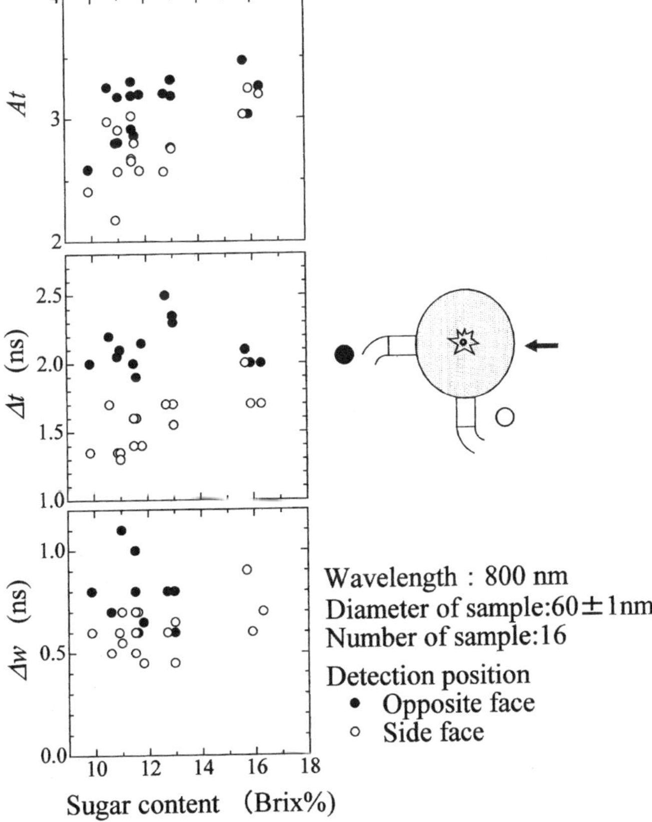

Figure 2. Variation of attenuance A_t, time delay of peak maxima Δt and variation of full width at half maximum Δw with sugar content.

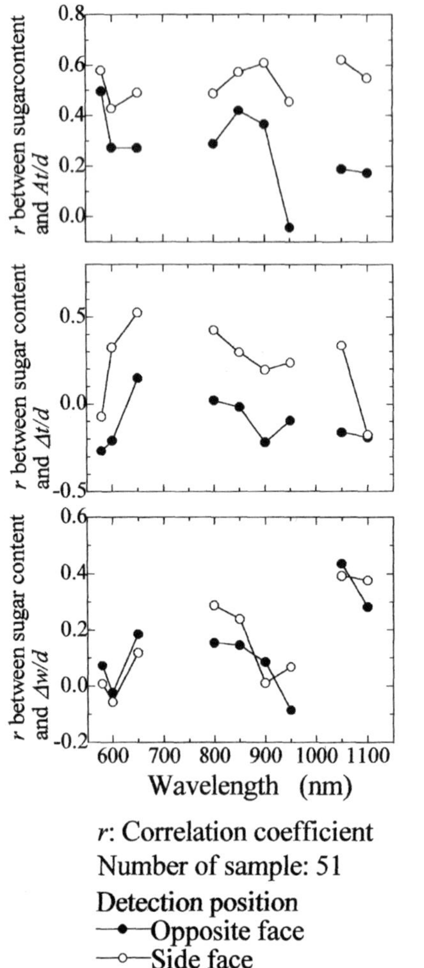

Figure 3. Correlation coefficient between sugar content and each normalised parameter.

Next, we examined the relationship between each optical parameter and sugar content of the Satsuma mandarin having various sizes. To compensate the effect of sample size on the time- resolved profile, each parameter was normalised by *l*. Figure 3 shows the correlation coefficient between sugar content and each normalised parameter. The correlation coefficient at the side face shows a larger value than that at the opposite face, independent of wavelength. Furthermore, it is known that the relationship between sugar content and A_t is not necessarily coupled with the relationship between sugar content and Δt. To correctly interpret these results, however, we may consider that the characteristics of the transmitted light differ with the detection position. The output detected at directly opposite the irradiation position may include the straight-through or near-straight beams. The output at other detection positions could be defined as diffusely scattered light. The difference in the optical characteristics of transmitted beams directly reflects the substantial optical path length.

When we apply ToF-NIR to detection of the information for the inside of fruit with high moisture content such as Satsuma mandarin, it is very important to give attention to the difference in the scattered light within tissues and the semi-straight propagated light. Furthermore, we tried to express the resulting phenomena by using model samples composed of water, sucrose and milk. The variation of the time-resolved profile is strongly governed by the combination of the light absorption component, scattering medium and refractive index.

Conclusions

In this study, a newly constructed optical measurement system, whose main components were a parametric tunable laser and a near infrared photoelectric multiplier, was applied to detection of the sugar content of Satsuma mandarin using ToF-NIR spectroscopy. The combined effects on the time-resolved profile of sample diameter, sugar content, the wavelength of the laser beam and the detection position of transmitted light were investigated in detail.

The variation of the attenuance of peak maxima *At*, the time delay of peak maxima Δt and the variation of full width at half maximum Δw were strongly dependent on the detection position and the wavelength of the laser beam. *At*, Δt and D*w* increased gradually as the sample diameter increased to be greatly absorbed and vigorously scattered. On the other hand, each optical parameter had a tendency to increase as the sugar content increased. Such behaviour was remarkable when the transmitted light was detected at the side face of a sample. To find the resulting phenomena of the time-resolved profile, it is very important to give attention to the difference in the scattered light within tissues and the semi-straightly propagated light.

References

1. B.G. Osborne, T. Fearn and P.H. Hindle, *Practical NIR Spectroscopy with Applications in Food and Beverage Analysis*. Longman Scientific & Technical, Harlow, UK (1993).
2. K. Miyamoto, M. Kawauchi and T. Fukuda, *J. Near Infrared Spectrosc.* **6,** 267 (1998).
3. S. Kawano, T. Fujiwara and M. Iwamoto, *J. Japan Soc. Hort. Sci.* **62,** 465 (1993).
4. L. Leonardi and D.H. Burns, *Appl. Spectrosc.* **53,** 628 (1999).
5. L. Leonardi and D.H. Burns, *Appl. Spectrosc.* **53,** 637 (1999).
6. A. Yariv, *Quantum Electronics,* 2nd Edn. John Wiley & Sons, New York, USA (1975).
7. P.G. Harper and B.S. Wherrett (Eds), *Nonlinear Optics*. Academic Press, New York, USA (1977).

Application of time-of-flight near infrared spectroscopy to apples

Satoru Tsuchikawa,[a] Takahiro Hamada,[a] Kinuyo Inoue[a] and Rae-Kwang Cho[b]

[a]*Graduate School of Bioagricultural Sciences, Nagoya University, Nagoya 464-8601, Japan*

[b]*Department of Agricultural Chemistry, Kyungpook National University, Taegu 702-701, Korea*

Introduction

Consumers' demands, with respect to agricultural products, are becoming increasingly diverse. The producer must not only supply them safely while maintaining freshness, but also assure the taste and nutritional values. Therefore, it is desirable for the horticultural industry to evaluate these indices correctly without waste of time and energy. For example, the nondestructive detection of water core and sugar content in apples would be of considerable value to the fresh-apple industry.

A number of techniques and devices for either the transmittance or reflectance method, which could detect inner information of apples (for example, sugar content or water core), were proposed previously.[1–4] However, the behaviour of transmitted light from an agricultural product is directly affected by the physical and chemical properties of tissues, so that it is very important to examine the optical characteristics of the tissue and its origin in detail from a new viewpoint. In this study, an optical measurement system, which was mainly composed of a parametric tunable laser and a near infrared (NIR) photoelectric multiplier, was introduced to achieve this purpose by using time-of-flight near infrared spectrometer (ToF-NIR).[5,6] This system combines the best features of the spectrophotometer and the laser beam and, more advantageously, that the time-resolved profile of transmitted output power can be measured sensitively in nanoseconds. The combined effects of the condition of water core, sugar content and sample diameter on the time-resolved profiles were investigated in detail.

Material and methods

The samples used were "Fuji" apple (*Malus domestica* Borkh.cv.Fuji) (location:Aomori, Japan) which were between 80–90 mm in diameter. The condition of water core was defined by visual evaluation. The sugar content, measured by a refractometer, varied from 11.6 to 14.8 °Brix. The Q-switched Nd:YAG laser was employed as the exciter laser. The wavelength of the pulsed laser could be turned from 500 to 1100 nm by an optical parametric oscillation of a BBO crystal.[7,8]

The transmitted output power from the sample was measured by an NIR photoelectric multiplier, having a spectral response ranging from 300 nm to 1700 nm, which was cooled to –80°, through an fibre-opic cable, having a diameter of 7 mm. A Si pin-type photodiode was placed near the optical parametric oscillator to generate a trigger signal. The fibre-opic cable was directly in contact with the apple, which was enclosed in aluminum foil to keep out stray light. Furthermore, the sample, the cooling box containing the NIR photoelectric multiplier and the fibre-opic cable were also covered with black cloth. The sampling time and the average number of the transmitted output power were 100 ns and 100 times, respectively. The equator of the apple was irradiated vertically with the pulsed laser and the transmitted output power was measured on the restricted position of the equator using the fibre-opic cable.

Outline of time-resolved profile

The time-resolved profile refers to the variation of the intensity of the detected light beam with time. We focused on some typical parameters representing the variation of the time-resolved profile. The normalised time-resolved profiles of a sound apple (89 mm in diameter) and its cuticle (thickness = 1.0 mm) are shown in Figure 1. In this study, the time-resolved profile of the cuticle was employed as the reference.

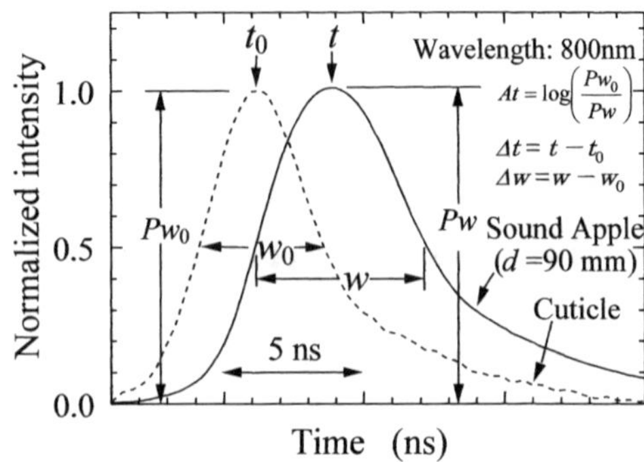

Figure 1. Normalised time-resolved profiles.

The variations of the peak maxima A_t, the time delay of peak maxima Δt and the variation of the full width at half maximum of the profile Δw were examined under the experimental measuring conditions, respectively.

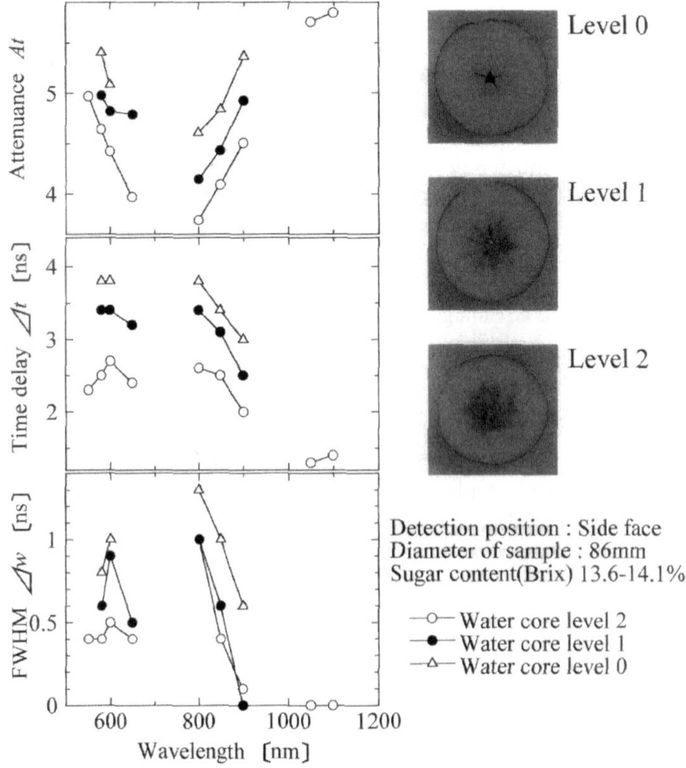

Figure 2. Spectral variation of the attenuance *At*, the time delay of peak maxima Δt and the variation of full width at half maximum Δw.

Results and discussion

Variation of time-resolved profile with water core

Figure 2 indicates the spectral variation of A_t, Δt, and Δw with the presence and level of the water core, respectively. In visual evaluation, we used a scale of "level 0" for a sound apple to "level 2" for apples having solid water core.

It is known that, independently, A_t decreases gradually as the size of the water core increases. In the presence of water-cored tissue (level 2), we could find an output signal at around 1000 nm which may be assigned to water absorption; however, other samples had no data because of a strongly attenuated output signal. This means that the water-cored tissue would transmit much more energy.

Δt and Δw also decrease gradually as the size of the water core increases; however, its wavelength dependency is inversely related to A_t. When the size of the water core increases, the intercellular spaces are filled with liquid, or the cells have become swelled, eliminating air spaces which results in a translucent or water-soaked appearance. This results in less light scattering, so that the light path time through a sample decreases.

Variation of A_t, Δt, and Δw with sugar content

We examined the variation of A_t, Δt, and Δw with sugar content. Figure 3 shows the relationship between the sugar content and each optical parameter normalised by diameter. The wavelength of the laser beam was $\lambda = 850$ nm and the output detected at directly opposite the irradiation position.

A_t/d, $\Delta t/d$ and $\Delta w/d$ decreased gradually as the water core increased. Such a tendency was noticeable in water-cored apples. As shown in Figure 3(c), sugar content is correlated with the normalised Δw independently of water-cored condition. To correctly interpret these results, we may consider that the difference in the refractive index between the cellular structure and the intercellular spaces of an apple differ with the condition of water core. It is known that the refractive index of the cellular structure may approach those of the intercellular spaces as water core increases. So, we know that the variations of the refractive index with the water core or sugar content relates directly to the time-resolved profiles. In any optical parameters, apples with water core have a higher correlation coefficient than those without water core.

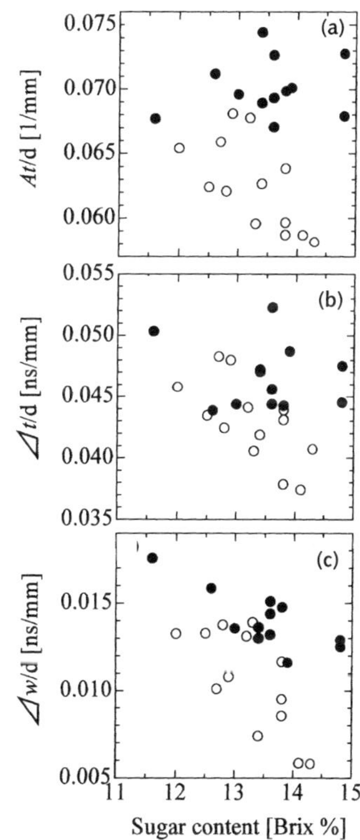

Figure 3. Relationship between sugar content and each normalised parameter.

Conclusions

An optical measurement system, which was mainly composed of a parametric tunable laser and a near infrared photoelectric multiplier, was introduced to detect inner information of apples using time-of-flight near infrared spectroscopy (ToF-NIR). The combined effects of the condition of water core, the sugar content and the sample diameter on the time-resolved profiles were investigated in detail. The attenuance of peak maxima At, the time delay of peak maxima Δt and the variation of full width at half maximum Δw decreased gradually as the water core increased. The water-cored tissue would transmit much more energy because of the filling of the intercellular spaces of an apple with liquid, so that the light path time through a sample decreased. These optical parameters were also governed by the wavelength dependency of light scattering and absorption characteristics. It became clear that At, Δt and Δw decreased as the sugar content increased. Such a tendency was noticeable in water-cored apples. The variations of the refractive index with the water core or sugar content relates directly to the time-resolved profiles. Thus, we can know the sugar content by measuring the optical parameters.

Acknowledgment

The authors would like to thank J.A., Kuroishi City, for providing apples.

References

1. G.S. Birth and K.L. Olsen, *Proceedings of the American Society for Horticultural Science* **85,**74 (1964).
2. R. Cho, M. Sohn and Y. Kwon, *J. Near Infrared Spectrosc.* **6,** A75 (1998).
3. I. Hwang and S. Noh, in *Near Infrared Spectroscopy: Proceedings of the 9th International Conference,* Ed by A.M.C. Davies and R. Giangiacomo. NIR Publications, Chichester, UK, p. 379 (2000).
4. M. Meurens and E. Moons, *Near Infrared Spectroscopy: Proceedings of the 9th International Conference,* Ed by A.M.C. Davies and R. Giangiacomo. NIR Publications, Chichester, UK, p. 125 (2000).
5. L. Leonardi and D.H. Burns, *Appl. Spectrosc.* **53,** 628 (1999).
6. L. Leonardi and D.H. Burns, *Appl. Spectrosc.* **53,** 637 (1999).
7. A. Yariv, *Quantum Electronics, 2nd* Edn. John Wiley & Sons, New York, USA (1975).
8. P.G. Harper and B.S. Wherrett (Eds), *Nonlinear Optics.* Academic Press, New York, USA (1977).

Measurement of metabolic parameters in lactating dairy cows by near infrared reflectance spectroscopy analysis using cattle faecal samples

Begoña de la Roza,* **Adela Martínez, Sagrario Modroño and Alejandro Argamentería**

Animal Production Department. Servicio Regional de Investigación y Desarrollo Agroalimentario (SERIDA), Apdo 13, 33300 Villaviciosa, Asturias, Spain

Introduction

Nutritional information about animal production is difficult to assess due to the difficulty in determining the amount and quality of food intake and, in particular, if forages are the most important compound in the diet for milk production. One of the most common procedures for the evaluation of nutritional value in foods is *in vivo* measurement. These metabolic studies with cows requires at least a knowledge of the amount of all food consumed and the excretion of milk faeces and urine. Taking these criteria as a basis, forage and total intake, dietary digestibility and balances of nitrogen and energy can be calculated. However, these food evaluation experiments with animals involve a lot of time and are very expensive.

The quality of individual forage varies widely. This fact makes it necessary to find alternative evaluation methods which are more rapid and cheaper to operate. An effective method for predicting grass and silages intake is an essential prerequisite to the accurate rationing of dairy cows. However, most authors have used simple or multiple regression analysis to examine the relationships between chemical composition and individual animal parameters.[1] In these situations near infrared (NIR) reflectance spectroscopy has produced promising results for predicting intake and biological parameters.[2–4]

The faeces excreted by animals contain undigested residues of the diet consumed. For this reason, their analysis can be a successful tool to determine the amount and quality of food intake and other important biological parameters.

The aim of this work has been to find out if faecal analysis by NIR could be used to determine, with enough accuracy, some attributes of the diets of different lactating dairy cows.

Materials and methods

Diet

The basic diets were eight grass, three grass silages and two maize silages, *ad libitum*, twice a day, supplemented with 0–5 kg concentrate cow^{-1} day^{-1}.

Experimental design and animals

Six Friesian dairy cows, in the 29th week of lactation, were housed and milked in individual stalls. The animals were divided into three groups of two and allocated, at random, to three treatments according to design,[5] were used in a latin square design experiment with three periods each of twenty one days duration, fifteen days for adaptation period and seven days for collection period.

During the collection period of the trial, daily samples of food offered and refused, milk, urine and faeces were colleted. All subsequently were analysed. A total of 79 faecal samples were used for the initial analysis with the aim of obtaining information on a wide range of forage and total intake, organic matter digestibility, gross energy digestibility and digestible and metabolisable energy.

NIR scanning and calibration procedures

The samples of faeces collected were freeze-dried and ground at 0.75 mm. The samples were then scanned using a Foss-NIRSystem 6500 scanning monochromator over a wavelength from 1100 to 2500 nm in steps of 2 nm. Due of the heterogeneous nature of the material, each sample was measured in two replicates and the mean of the replicate spectra obtained (log $1/R$) was used in the calibration.

Population boundaries were established with a maximum standardised H distance from the average spectrum of 3.0.[6] Calibration equations were obtained by WINISI II v. 1.5 sofware (Infrasoft International, Port Matilda, PA, USA), using a full wavelength range and modified partial least square as the regression method. Standard normal variate[7,8] (SNV) was used for scatter correction and second derivative as the mathematical treatment. Cross-validation, to minimise overfitting of the equations, was used to test the calibration equations. Those were selected according to the lowest standard error of cross-validation (*SECV*), the higest coefficient of determination (r^2) and the *RER*[9] (ratio range $SECV^{-1}$).

Results and discussion

The mean, range and standard deviation for forage and total intake (FI and TI, respectively), organic matter digestibility (OMD), gross energy digestibility (GED) and digestible and metabolisable energy (DE and ME, respectibily) values for the 79 faecal samples are given in Table 1. These values are typical of metabolic trials on lactating dairy cows with diets based on green and preserved forages and confirm that the samples had a wide range for all parameters. The amount of total and forage intake decreased with progressing maturity. As a result OMD, GED, DE and ME obviously also decreased.

The untreated NIR average spectrum (log $1/R$) of global population and symmetry plot of PCA scores for faecal samples in the wavelength range from 1100 to 2500 nm, are shown in Figure 1. The three-dimensional display of the sample scores provides a good sample distribution in the population.

In general, equation statistics were acceptable for all parameters of interest. The best NIR calibrations were obtained when the spectral data were converted into second derivative (Table 2). *SECV*

Table 1. Mean, range and standard deviation of reference values for calibration sets of faeces samples.

Parameters	Range	Mean	SD
Total intake (kg DM $cow^{-1}day^{-1}$)	15.4	10.3–19.8	2.216
Forage intake (kg DM $cow^{-1}day^{-1}$)	10.9	5.3–15.8	2.654
Organic matter digestibility (%)	72.0	54.2–80.8	5.271
Gross energy digestilbility (%)	66.6	52.2–74.4	4.974
Digestible energy (MJ kg DM^{-1})	12.2	10.1–13.8	0.864
Metabolisable energy (MJ kg DM^{-1})	10.1	8.2–11.7	1.012

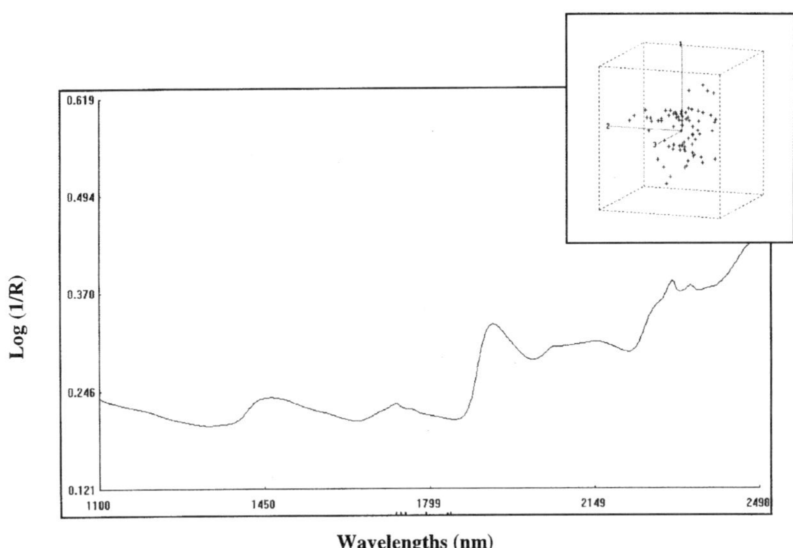

Figure 1. Untreated NIR average spectrum of global population and symmetry plot of PCA scores for faeces samples (n = 79).

ranged from 0.434 for ME to 2.626 for OMD. Coefficients of determination to cross-validation ranged from 0.560 for TI to 0.819 for EM. Values for the *RER* ranged from 6.50 for TI to 10.13 for OMD.

The equation statistics for a direct relationships between faecal NIR spectra and intake, OMD, GED, DE and ME clearly indicate the superiority of the relationship between the composition of green and preserved forages determined by conventional laboratory methods and these parameters. Steen *et al.*[1]cited lower coefficients when intakes were predicted using relationships based on regressions involving a range of chemical parameters.

In this study, OMD can be predicted with a high degree of accuracy. Although several reports[2] have noted the limitation in the usefulness of faecal samples for predicting OMD over a wide range of for-

Table 2. Coefficients of determination for calibration (R^2) and cross validation (r^2), standard error of calibration (*SEC*) and cross-validation (*SECV*), with different sample sets in NIR equations.

Parameters	*SEC*	R^2	*SECV*	r^2	*RER*
Total intake (kg DM cow^{-1}day^{-1})	1.349	0.629	1.467	0.560	6.50
Forage intake (kg DM cow^{-1}day^{-1}	0.967	0.867	1.309	0.757	8.02
Organic matter digestibility (%)	1.994	0.857	2.626	0.750	10.13
Gross energy digestilbility (%)	1.874	0.856	2.622	0.715	8.49
Digestible energy (MJ kg DM^{-1})	0.344	0.841	0.475	0.695	7.81
Metabolisable energy (MJ kg DM^{-1})	0.262	0.933	0.434	0.819	8.16

ages, these results agree with those reported by other authors[10–12] based on direct OMD prediction results using regression equations developed directly on forage samples.

The accuracy for predicting GED, DE and ME are acceptable. The data reported here contains an inherent variability on the reference values.

In order to know if a calibration equation is acceptable for prediction, the *RER* were calculated for each parameter. Values for the ratio range $SECV^{-1}$ ideally should be at least ten.[9] Only for OMD has the *RER* value been higher than ten. Nevertheless, *SECV* have not been too different from the required parameters, which are associated with a relatively high standard error.

The variation coefficients in the present study were 9.5, 12.0, 3.6, 3.9, 3.9 and 4.3 % for TI, FI, OMD, GE, DE and ME, respectively, in relation to the population mean.

Conclusion

In conclusion, faecal analysis by NIR can be used to determine diet quality. The results indicate that the forage and total intake, organic matter digestibility, gross energy digestibility and digestible and metabolisable energy can be directly predicted with enought accuracy from the NIR spectra of freeze-dried faecal samples.

Acknowledgements

The authors would like to thank to the staff of the SERIDA Animal Nutrition Laboratory for undertaking laboratory determinations. We also thank the farm staff for cattle handling and care.

References

1. R.W.J. Steen, F.J. Gordon, L.E.R. Dawson, R.S. Park, C.S. Mayne, R.E. Agnew, D.J. Kilpatrick and G. Porter, *Animal Science* **66,** 115 (1998).
2. N.P. Kjos, *Norwegian Journal of Agricultural Sciences* **4,** 305 (1990).
3. S.W. Coleman, J.W. Stuth and J.W. Holloway, in *Proceedings of the XVI International Grassland Congress*. Niza, 881.
4. E.R. Leite and J.W. Stuth, *Small Ruminant Research* **15,** 223 (1995).
5. B. De la Roza, A. Martínez and A. Argamentería, *ITEA* **20(2),** 526 (1999).
6. J.S. Shenk and M.O. Westerhaus, *Crop Science* **31,** 1148 (1991).
7. J.S. Shenk and M.O. Westerhaus, NIRSystem Inc., 12101 Tech Road, Silver Spring, MD 20904, USA, PN IS-0119 (1996).
8. V.M. Fernández and A. Garrido, *Química Analítica* **18,** 113 (1999).
9. P. Williams and D. Sobering, in *Near Infrared Spectroscopy: The Future Waves,* Ed by A.M.C. Davies and P. Williams. NIR Publications, Chichester, UK, p. 185 (1996).
10. D.I. Givens, J.M. Everington and A.H. Adamson, *An. Feed. Sci. Technol.* **24,** 27 (1989).
11. B.C. Grabrielsen, K.P. Vogel and D. Knudsen, *Crop Sci.* **28,** 44 (1988).
12. B. de la Roza, A. Martínez, S. Modroño, G. Flores and A. Argamentería, in *Near Infrared Spectroscopy: Proceedings of 9th International Conference*, Ed by A.M.C. Davies and R. Giangiacomo. NIR Publications, Chichester, UK, p. 661 (2000).

Nondestructive germinability assessment of radish seeds by near infrared spectroscopy

T.G. Min, W.S. Kang and K.S. Ryu

College of Natural Resources, Taegu University, Kyungsan, Kyungbuk 712-714, Korea

Introduction

Near infrared (NIR) spectroscopy is widely used today as a quantitative technique for predicting the chemical composition of various agricultural products. Nowadays NIR spectroscopy has been used to measure protein, fat, sugar and fibre content of many crops, such as wheat, oats, rice, rye, corn, millet and so on on a single grain or whole grain basis. However, few applications exist for seed quality assessment, especially for seed germinability. Previous works in germinability testing have usually focused on biochemical tests such as tetrazolium(TZ) or conductivity tests. The terazolium test was based on the relative respiration rate, at the hydrate state, for the viable and dead tissues of the embryo. This test showed the activity of dehydrogenase enzymes as an index to the respiration rate and seed viability. The conductivity test was based on the premise that the cell membranes become less rigid and more water-permeable, allowing the cell contents to escape into water solution and increasing its electrical conductivity as seed deterioration progressed. However, these tests have experienced some difficulty and necessary experience to interpret and need bulk testing. The sinapine or amino acid leakage method recently developed was used to test seed viability as a single seed base from *Brassica* seeds, but the seeds should be pretreated before evaluation. Therefore, the purpose of this study is to show the possible application of NIR spectroscopy as a nondestructive germinability test on radish (*Raphanus sativus* L) seeds in a single seed base without any pretreatment.

Materials and methods

The radish seeds (cultivar, Chung Su Gung Jung, Nong woo Bio Co. Ltd, Korea) harvested in 1993 was used in this experiment. NIR spectral measurements were carried out on the flat side of the seed surface of a single grain kernel using an NIRSystems 5000 (Foss NIRSystems, Silver Springs, Maryland, USA). Each seed was scanned, in the NIR reflectance mode, from 1100 to 2500 nm at 2 mm increments in a sample hole (4 mm diameter) of seed holder plate. The WinISI II program (Foss NIRSystems, Infrasoft, International, LLC.) was used to process the data. The seeds, after spectral measurement, were planted on a blotter individually and germination was observed. The seeds were characterised into non-germination and germination, then grouped again into normal and abnormal germination by the rules of the Association of Official Seed Analysts (AOSA) and then compared with that of NIR spectra. Figure 1 illustrates the normal, abnormal and non-germination of radish seeds according to the rules of AOSA.

Results and discussions

A total of 571 seeds were planted after NIR scanning. 300 germinated with 25 outliers and 226 non-germinated with 20 outlier seeds were observed and outliers were excluded in interpreting the

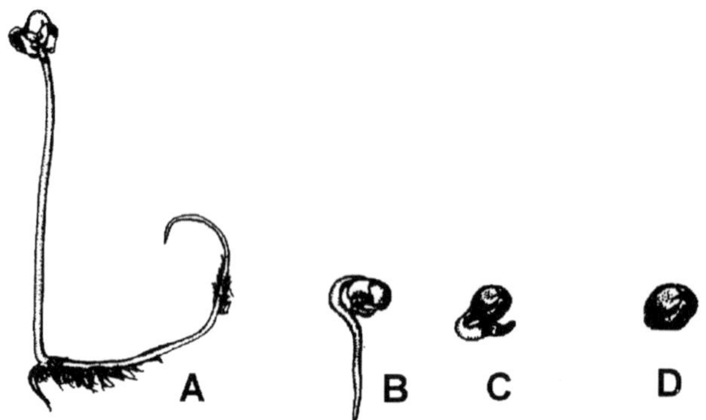

Figure 1. Normal (a), abnormal (b, c) and dead radish seeds (d) evaluated after germination.

NIR spectra. 261 out of 300 germinated seeds were classified as normal and 39 seeds were classified as abnormal. Figure 2 shows a PC scores plot for radish seeds which has been classified either as 'germinated seeds' or 'non-germinated seeds'. The plot represents the three PCs that explain the large variation in the spectral data with respect to the 'germinated' and 'non-germinated' characteristics of radish seeds.

Figure 3, using PC score plots, demonstrates how the characteristics of normal germinated radish seeds differ from those of abnormal germinated seeds. It shows a clear difference in pattern of the distribution of the spectra of normal from abnormal germination.

The 3D plots from Figures 2 and 3 showed that the application of principle component analysis (PCA) for the spectral data could be successfully performed to characterise germination to non- germination and normal to abnormal germination of radish seeds. Therefore, the PCA of NIR reflectance spectra for the radish seeds of germination and non-germination and normal and abnormal germina-

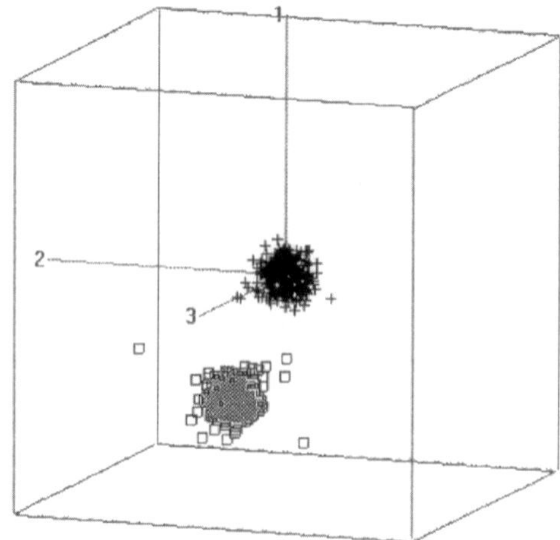

Figure 2. 3D PC scores plot for 'germinated' and 'non-germinated' radish seed characteristics. The symbol + denotes spectra from 'germinated seeds' and □ denotes the spectra from 'non-germinated seeds'.

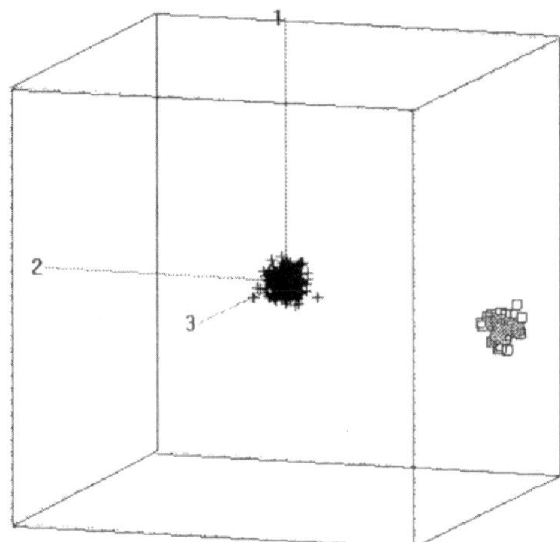

Figure 3. PC scores plot showing 'normal' and 'abnormal' germinated radish see characteristics. The symbol + denotes spectra from 'normal' and □ denotes the spectra from 'abnormal' germinated seeds.

tion showed quite different patterns for each group and the results suggest that NIR spectra could be applicable to separate radish seeds on the germinability.

Conclusion

Analysis of NIR spectra on radish seeds was clearly characterised into non-germination (non- viable) and germination (viable) groups, which were, in turn, grouped into normal and abnormal germination. The results indicated that the NIR technique could be used as a means of discriminating physiological seed quality.

Acknowledgements

The authors are grateful to Kyu-Chae Cho (Doo Ree Tech. Inc., Korea) for assistance in processing the data.

Reference

1. S. Maxon, *Rules for Testing Seeds*. Association of Official Seed Analysts. (1993).
2. The Seed Vigor Test Committee of the AOSA, *Seed Vigor Testing Handbook*. Association of Official Seed Analysts (1983).
3. C.N.G. Scotter, in *Advances in Near Infrared Spectroscopy*, Ed by I. Murray and I.A. Cowe. VCH, Weinheim, Germany, pp. 849–854 (1992).
4. H. Schulz, B. Steuer and H. Kruger, in *Advances in Near Infrared Spectroscopy*, Ed by I. Murray and I.A. Cowe. VCH, Weinheim, Germany, pp. 447–453 (1992).
5. P.C. Flinn, N.J. Edward, C.M. Oldham and D.M. McNeill, in *Near Infrared Spectroscopy: The Future Waves*, Ed by A.M.C. Davis and Phil Williams. NIR Publications, Chichester, UK, pp. 576–580 (1996).
6. A.G. Taylor, T.G. Min and C.A. Mallaber, *Seed Science & Technology* **19,** 423 (1991).
7. T.G. Min, *J. of the Korean Society for Horticultural Science* **41,** 576 (2000).

Mathematical pretreatments of spectra—useful tools for reducing instrument differences

Henryk W. Czarnik-Matusewicz,[a] Ana Garrido-Varo,[*] Juan Garcia-Olmo and Maria Dolores Perez-Marin

Animal Production Department, School of Agriculture and Forestry Engineering, University of Cordoba, PO Box 3048, E-14080 Cordoba, Spain

Introduction

Near infrared (NIR) spectra are composed of a mixture of diffuse and specular components that are strongly influenced by the scattering and absorption characteristics of the sample. Diversity of the particle size of the material will affect the scattering and is a major source of variation in NIR spectra. These effects vary from sample to sample and are both additive and multiplicative in nature. Additive effects cause vertical displacement (or shift of the spectra upward or downward) whereas multiplicative effects appear as non-linear slope changes when compared to an "ideal" or reference spectrum. In practice, "ideal" spectra are impossible to obtain, therefore the mean spectrum of a set is assumed to be good representation for the reference base.

Many mathematical manipulations (or treatments) which have been proposed to correct for particle size anomalies are used in NIR spectroscopy. These include calculating first and second derivatives, multiplicative scatter correction (MSC), standard normal variate (SNV) and detrending (DT) transformation. All transformations are applied before entering the calibration process.

The objective of the study reported in this paper was to determine the potential of math pretreatments for reducing differences between two different types of NIR spectrophotometers—a Foss NIRSystems Model 6500 (Silver Spring, MD, USA) and an InfraAlyzer 500 (Braun+Luebbe, Nordersted, Germany). Only four math treatments (and their combination) were considered: (1) First derivative, (2) second derivative, (3) standard normal variate (SNV) and (4) detrending (DT). It would be advantageous if a calibration developed on one NIR instrument could be successfully transferred to and utilised on another NIR instrument.

Material and methods

Mathematical treatments of all the spectral raw Bran+Luebbe InfraAlyzer 500 data of Polish pig fat samples with chemically determined iodine values[1] were performed using the ISI NIR 3 software ver. 3.11 (Infrasoft International, Port Matilda, PA, USA). Calibration equations for the prediction of four main fatty acids of Iberian pig fat: palmitic acid (C16 : 0), stearic acid (C18 : 0), oleic acid

[a]Collaborator, via a fellowship under the OECD Co-operative Research Programme, Biological Resource Management for Sustainable Agriculture Systems

Present address: Wroclaw Medical University, Department of Clinical Pharmacology, Bujwida 44, PL-50345 Wroclaw, Poland.

(C18 : 1) and linoleic acid (C18 : 2) were also developed using the same software.[2] The NIR spectra of fatty acids contain bands at 1180, 2143 and 2190 nm which may be attributed to *cis* unsaturation.[3]

The modified partial least squares (MPLS) regression technique[4] was used to develop all NIR calibrations. In MPLS regressions, the spectral data are reduced to a few independent factors that retain most of the spectral information—factors correlatable to chemical (reference data) of the samples. The standard normal variate (SNV) and detrend (DT) scatter correction procedure were applied to the spectral data in the following manner.[5] The SNV pretreatment required that the standard deviation of each spectrum be 1.0. DT removed the linear and quadratic components of each spectrum. Then a first order mathematical calculation was made, using ISI default settings of 1, 4, 4, 1 where the first digit is the number of the derivative, the second is the gap over which the derivative is calculated and the third is the number of data points in the convolution interval (segment or running average).[6]

Figure 1 shows the mean absorbance, SNV, DT, SNV + DT, SNV + DT + 1st derivative and SNV + DT + 2nd derivative spectra of Iberian pig fat recorded on an NIRS 6500 and Polish pig fat (recorded on an IA 500). The spectral wavelength range for all analyses in this paper was restricted to 1100–2500 nm.

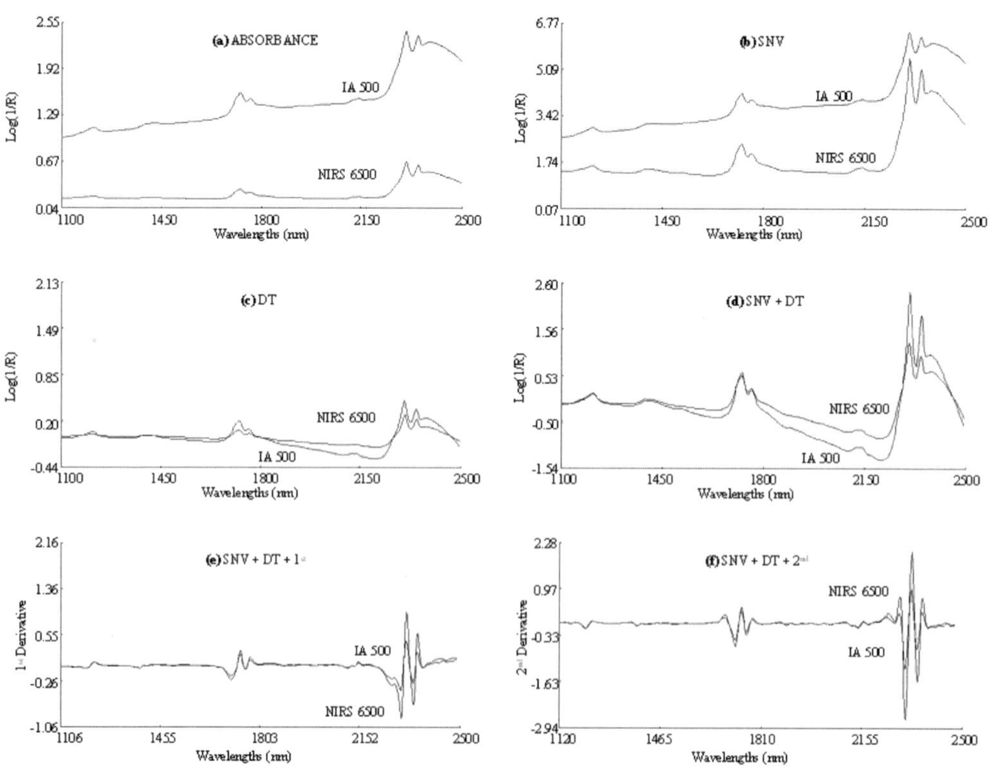

Figure 1. NIR mean spectra of pig fat sample sets: (a) absorbance, (b) SNV, (c) DT, (d) SNV + DT, (e) SNV + DT + 1st derivative and (f) SNV + DT + 2nd derivative.

Results and discussion

Correlations between the main fatty acids content of Iberian-pig-fat samples and iodine value in Polish-pig-fat samples are given in Table 1.

The results of calculation using MPLS regression equations[2] show (Table 1) that linoleic acid (C 18:2) contents (%) of Iberian pig fat samples were highly correlated ($r = 0.897$) with the iodine values of Polish pig fat samples. The degree of unsaturation of fats is proportional to the iodine value, defined as the number of grams of iodine required to react with 100 g of fat under test conditions.

The graphic presentation of these dependencies is illustrated in Figure 2. To validate the calibrations, a new independent sample set of Polish pig fat was collected in from meat grown in an experiment to determine the efficiency of mineral–fat supplement in boars and gilts nutrition.[7] The vegetable dietary fat supplement had a significant effect on increasing the iodine values of pig back-fat from 57.9 (control group) to 62.3, 63.0 and 67.0 (experimental groups). The effect of vegetable oil on animal fat quality was reflected in the composition of the fatty acids determined by gas chromatography (GC).[8] The

Table 1. Correlation between main fatty acids contents of Iberian pig fat samples and iodine value in Polish pig fat samples.

Fatty acid	Correlation coefficient (r)
C 16 : 0 (palmitic)	0.272
C 18 : 0 (stearic)	0.071
C 18 : 1 (oleic)	0.158
C 18 : 2 (linoleic)	0.897

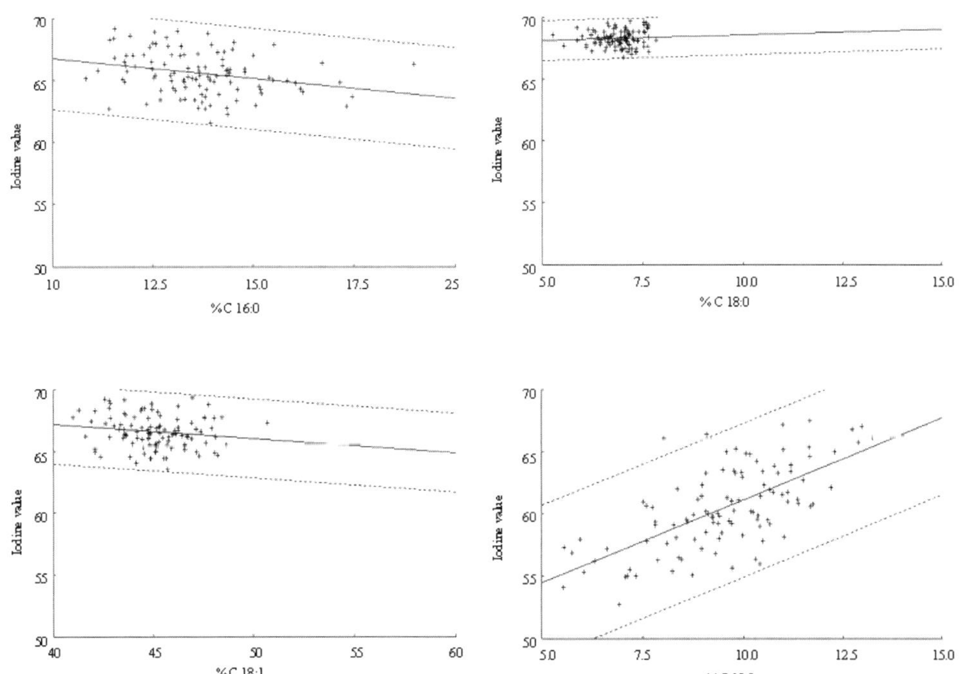

Figure 2. Graphic relationship of correlation between Iberian pig fat samples fatty acids contents and iodine values of Polish pig fat samples.

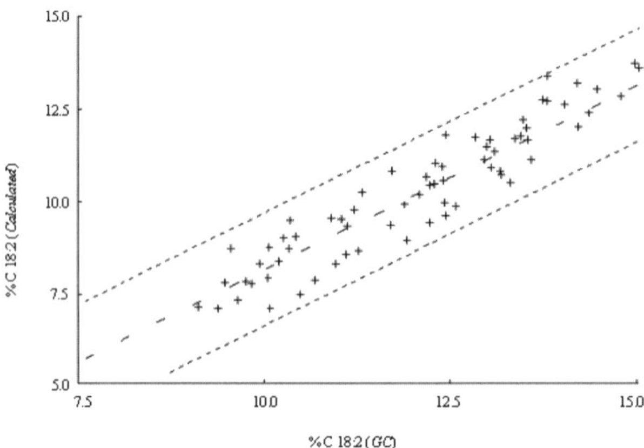

Figure 3. Relationship between GC-determined and calculated lineolic acid contents for independent prediction set of Polish pig fat samples.

contents of unsaturated (oleic, linoleic) fatty acids in the experimental groups were found to increase significantly (linoleic acid—from 9.6 to 12.8%).

The prediction ability of the calibrations for the determination of linoleic acid content is plotted in Figure 3, comparing the calculated contents to their corresponding linoleic acid concentration determined by GC method. Although the regression statistics are good (Figure 3), the deviations of predicted contents of linoleic acid are high. The relatively high correlation coefficient ($r = 0.878$) with the worse *SEP* (*SEP* = 0.37%) is probably due to fact that *r* only measures the degree to which the calibrations fits the data. It does not give an indication of how reliable these predicted values are.

Data presented in Figure 3 suggest the need for a bias correction. The different spectral response of spectrophotometers and/or the different GC methods[8,9] of the fatty acids determination can be the sources of bias (bias = 2.4%).

These results demonstrate that the linoleic acid concentration in pig fat can be estimated from the "calibration" curve calculated from the NIR data transferred from other type of NIR monochromator spectrophotometer. However, the high error and low precision, at this stage of our investigations, are less than satisfactory.

Acknowledgements

The first author (HWCM) would like to thank the OECD Co-operative Research Programme for providing a fellowship that allowed him to study NIR spectroscopy at ETSIAM, Cordoba, Spain. Also, gratitude is expressed to Professor Ana Garrido-Varo and all members of ETSIAM Animal Production Department for their kindness, generosity and support throughout this research experience.

References

1. H.W. Czarnik-Matusewicz and A. Korniewicz, in *Leaping Ahead with Near Infrared Spectroscopy*, Ed by G.D. Batten, P.C. Flinn, L.A. Welsh and A.B. Blakeney. NIR Spectroscopy Group, Royal Australian Chemical Institute, North Melbourne, Victoria, Australia, pp. 300–302 (1995).
2. J. Garcia-Olmo, A. Garrido-Varo and E. de Pedro, in *Near Infrared Spectroscopy: Proceedings of the 9th International Conference*, Ed by A.M.C. Davies and R. Giangiacomo. NIR Publications, Chichester, UK, pp. 253–258 (2000).

3. R.T. Holman and P.R. Edmondson, *Anal. Chem.* **28,** 1533 (1956).
4. H. Martens and T. Naes, *Multivariate Calibration*. John Wiley & Sons, Chichester, UK (1989).
5. R.J. Barnes, M.S. Dhanoa and S.J. Lister, *Appl. Spectrosc.* **43,** 772 (1989).
6. J.S. Shenk and M.O. Westerhaus, Routine operation, calibration, development and network system management manual. NIRSystems, Silver Spring, MD, USA (1995).
7. D. Korniewicz, A. Korniewicz, Z. Dobrzanski, R. Kolacz, H. Czarnik-Matusewicz and B. Paleczek, *Ann. Anim. Sci. – Rocz. Nauk. Zoot.* **26(4),** 275 (1999).
8. J. Folch, M. Lees and G.H.S. Stanley, *J. Biol. Chem.* **226,** 497 (1957).
9. Y.B.C. Man and M.H. Moh, *J. Am. Oil Chem. Soc.* **75(5),** 557 (1998).

2- and 3-way analysis of near infrared scans from seed crossings

Torbjörn A. Lestander,[a] Paul Geladi[b] and Per Christer Odén[a]

[a]*SLU Forest Seed Science Centre, Department of Silviculture, Swedish University of Agricultural Sciences, SE – 901 83 Umeå, Sweden*

[b]*Department of Chemistry, Umeå University, SE – 901 87 Umeå, Sweden*

Introduction

Seed quality from a physiological and genetic viewpoint is of major interest in plant breeding and reforestation programmes. Fast and automatic selection of good tree seeds is a vision within forest seed management programmes.[1] One aspect that needs studying within this vision is the possibility of collecting genetic information about the seeds using non-destructive methods such as near infrared (NIR) spectroscopy. In this paper we have used Scots pine (*Pinus sylvestris* L.) full sib seeds measured by NIR reflectance spectroscopy. This study is, to our knowledge, one of the first of its kind and only a few studies have explored the field to collect genetic information directly from NIR spectra,[2] whereas there are many studies that utilise the calibrations between NIR spectra and defined traits [3–6] of interest in plant [7–12] and animal[13] breeding.

In a Scots pine seed, the embryo, which is of both maternal and paternal origin, is embedded in a megagametophyte surrounded by a seed coat. The two latter components are in gymnosperms of only maternal origin. These components have been analysed[14] in Scots pine giving the ratio 25 : 66 : 9 between the mass of seed coat, megagametophyte and embryo, respectively. Thus, seed tissue of maternal origin is expected to have a high influence on NIR spectra, whereas there can be problems in collecting information about tissues of paternal origin.

The aim of the study was to use 2- and 3-way analysis to determine maternal, paternal and interaction effects in NIR spectra from controlled seed crossings.

Material and methods

Genetic material

The crossings emanated from a factorial mating design of Scots pine (*Pinus sylvestris* L.) using six maternal clones and ten paternal clones growing in clonal tree archives in northern Sweden. Three samples (one sample and two re-samples) containing *ca* 30 seeds from each of the 60 controlled crossings (full sib families) were measured by spinning cup NIR reflectance spectroscopy.

Spectra

An NIR spectrophotometer (Foss NIRSystems model 6500) was used to collect reflectance (R) from each seed sample as the mean spectrum of 32 scans at every 2nd wavelength between 400 and 2498 nm, *i.e.* 1050 wavelengths. Absorbance (A) was calculated as $A = \log(1/R)$ for each wavelength.

Data treatment

The absorbances at 1050 wavelengths were replaced by seven significant ($p < 0.05$) latent variables. The latent variables were principal components of the 180 × 1050 data set and calculated by the software SIMCA[15] after mean-centring.

The used models for univariate (ANOVA, random effects) and multivariate (MANOVA, fixed effects) analysis of variance were:

$$z_{ijk} = b_{0k} + M_{ik} + F_{jk} + M_{ik} \times F_{jk} + E_{ijk}$$

where at the *k*th principal component z_{ij} is the score value of the *ij*th seed crossing, b_0 is the intercept, M_i is the contribution from the *i*th mother, F_j is the contribution from the *j*th father, $M_i \times F_j$ is the interaction between mother *i* and father *j* and E_{ij} is the residual. The model was calculated in SPSS[16] and was one of the possible interpretations of the genetic background information.

For the parallel factor analysis [17] (PARAFAC) the replicates were replaced by their mean spectra. The Savitzky–Golay (window size 31, order 4, 1st derivative) derivative smoothing and the PARAFAC analysis were carried out in MATLAB[18] using the PLS-Toolbox.[19] Reduction of wavelengths was not used. Another PARAFAC model was also made using mean-centred raw data.

A three-way data array has three ways or modes. The array used in this paper has the modes; mother, father and wavelength. PARAFAC allows the decomposition of the 6 × 10 × 1050 array in factors (called loadings) representing the mothers, the fathers and the wavelengths and gives a residual three-way array. The number of factors used is the three-way pseudorank.

Results

The NIR spectra all look the same by visual inspection. Any interpretation of the raw spectra only relates to common properties of the seeds and not to genetic differences.

The seven PCA components used in the analysis of variance explained 99.6% of the sum of squares of the NIR data. The random effect ANOVA-model on each of the seven principal components showed that the maternal effect was highly ($p > 0.001$) to moderately ($p > 0.01$) significant for the first six principal components. The interaction effects were highly significant ($p > 0.001$) for all components and the paternal effect was only significant ($p > 0.05$) for the second component. For the fixed effect MANOVA model on the seven components together, highly to moderately significant maternal, paternal and interaction effects were found for all components.

The PARAFAC analysis on the first-order smoothing derivative gave a model explaining almost 100% of the total sum of squares of the data. A pseudorank of three was considered meaningful. The first major component explained the average spectral content of all the seed samples. The second and third components were the main ones used for showing genetic differences. The results of the three-way analysis are best seen in loading plots (Figures 1 and 2).

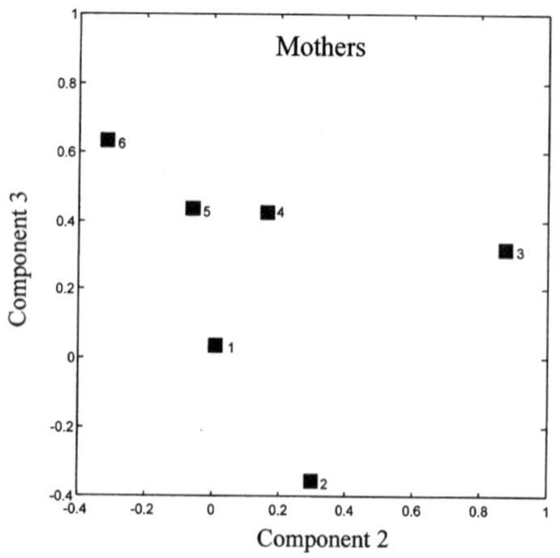

Figure 1. Second and third PARAFAC loading plot for maternal influence (mothers) on NIR spectra from controlled Scots pine seed crossings.

Because of the huge loading for the first PARAFAC component, a new model was made

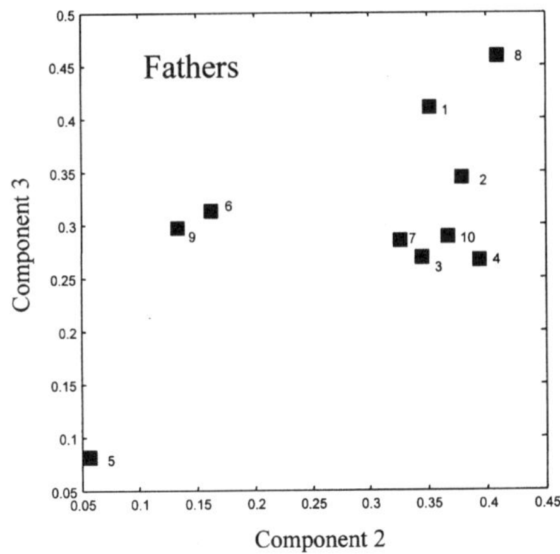

Figure 2. Second and third PARAFAC loading plot for paternal influence (fathers) on NIR spectra from controlled Scots pine seed crossings.

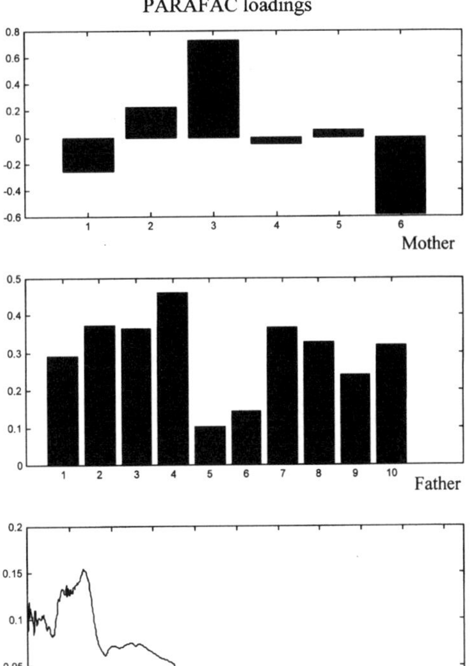

Figure 3. The PARAFAC loadings for the model on mean-centred NIR data from controlled Scots pine seed crossings.

using mean-centred raw data. This model gave one component explaining 70% of the sum of squares (Figure 3).

Discussion

The ANOVA and MANOVA on the seven significant principal components showed the importance of the maternal contribution to the seeds, which was expected, but there were also significant paternal and interaction contributions. This is in accordance with the genetic background knowledge.

Both PARAFAC models showed similar results. There was a huge influence of the mean spectrum and this had nothing to do with genetic influences. The PARAFAC loadings allowed visualisation of mainly maternal, but also of paternal influences. Because of the nature of the model, these remain mixed with the interactions.

Three-way analysis gives an advantage compared to doing a principal component analysis on a 60 × 1050 matrix. The PARAFAC decomposition is more parsimonious and the maternal and paternal effects are separated in the loadings. PARAFAC was used on NIR data earlier.[20,21] It has been shown [21] that raw NIR data are not ideal for PARAFAC analysis and that Savitzky–Golay smoothing derivatives are much better.

The loading plot of wavelengths showed a high contribution from the visual wavelength range, which is mainly an effect by the maternal seed coat. The region where NIR-radiation penetrates deepest into organic matter seems to have a high contribution to the PARAFAC model.

We have shown that it is possible to determine maternal and paternal influence as well as interactions of maternal and paternal origin in NIR spectra on seeds from controlled seed crossings. This finding opens up new possibilities in plant breeding and seed separation. The bulk of the seed has high maternal influences as expected, but only the embryo inside the seed has the detailed genetic information. Therefore, NIR imaging or confocal microscopy of single seeds is expected to be useful for extracting that information in more detail.

Acknowledgement

The authors acknowledge financial support from the Swedish Research Council for Environment, Agricultural Sciences and Spatial Planning and the Unizon-project NIRCE within the EU Interreg Kvarken-MidScandia cooperation. We also thank associate professors Bengt Andersson and Jan-Erik Nilsson for lending us the seed crossings for non-destructive NIR measurements and finally Margareta Söderström and Josefina Nyström for carrying out the NIR measurements.

References

1. T.A. Lestander and P.C. Odén, *Seed Science and Technology* **30,** 383 (2002).
2. B.L. Greaves, L.R. Schimleck, N.M.G. Borralho, A.J. Michell and C.A. Raymond, *Appita Journal* **49(6),** 423 (1996).
3. B.G. Osborne, T. Fearn and P.H. Hindle, *Practical NIR Spectroscopy with Applications in Food and Beverage Analysis*, 2nd Edition. Longman Scientific & Technical, Harlow, UK (1993).
4. S.J. Lister, M.S. Dhanoa, J.L. Stewart and M. Gill, *Animal Feed Science and Technology* **86,** 231 (2000).
5. A. Reverter, D.J. Johnston, H.-U. Graser, M.L. Wolcott and W.H. Upton, *J. Animal Science* **78,** 1786 (2000).
6. L. Velasco and H.C. Becker, *Genetic Resources and Crop Evolution* **47(3),** 231 (2000).
7. B.E. Zher, S.R. Eckhoff, W.E. Nyquist and P.L Keeling, *Crop Science* **36,** 1159 (1996).
8. S.R. Delwiche, R.A. Graybosch and C.J. Peterson, *Cereal Chem.* **76(2),** 255 (1999).
9. K.G. Campbell, C.J. Bergman, D.G. Gualberto, J.A. Anderson, M.J. Giroux, G. Hareland, R.G. Fulcher, M.E. Sorrells and P.L. Finney, *Crop Sci.* **39,** 1184 (1999).
10. M.Q. Lu, L. O'Brien and I.M. Stuart, *Australian J. Agric. Res.* **50,** 1425 (1999).
11. M.Q. Lu, L. O'Brien and I.M. Stuart, *Australian J. Agric. Res.* **51,** 247 (2000).
12. M.Q. Lu, L. O'Brien and I.M. Stuart, *Australian J. Agric. Res.* **52,** 85 (2001).
13. G.D. Snowder, J.W. Walker, K.L. Launchbaugh and L.D. van Vleck, *J. Animal Science* **79,** 486 (2001).
14. P.B. Reich, J. Oleksyn and M.G. Tjoelker, *Canadian Journal of Forest Research* **24,** 306 (1994).
15. Umetrics AB, *Simca-P8.0*, Umetrics AB, Umeå, Sweden.
16. SPSS Inc., *SPSS for Windows 10.0.5*, SPSS Inc. Headquarters, Chicago, Illinois, USA (1999).
17. R. Coppi and S. Bolasco (Eds), *Multiway Data Analysis*. North Holland, Amsterdam, The Netherlands (1989).
18. The MathWorks Inc., *Matlab* 5.3.0, The MathWorks, Inc., Natick, MA, USA (1999).
19. B. Wise and N. Gallagher, *PLS Toolbox 2.0 for use with MATLAB*, Eigenvector Research, Manson, WA, USA (1998).
20. N. Allosio, P. Boivin, D. Bertrand and P. Courcoux, *J. Near Infrared Spectrosc.* **5,** 157 (1997).
21. P. Geladi and P. Åberg, *J. Near Infrared Spectrosc.* **9,** 1 (2001).

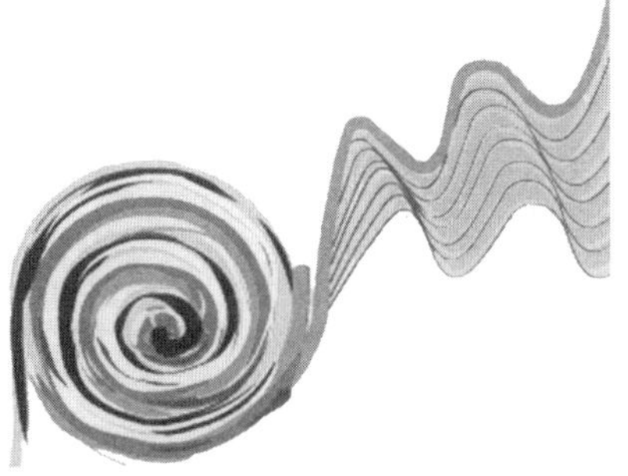

Environment

Emerging possibilities for near infrared spectroscopy to contribute to environmental analysis

D.F. Malley

PDK Projects, Inc., 365 Wildwood Park, Winnipeg, MB R3T 0E7, Canada

Introduction

Near infrared (NIR) spectroscopy is potentially a powerful and revolutionary technology for environmental analysis. Near infrared technology is supported by a large body of theoretical knowledge and results from applications (Council for NIR Spectroscopy Bibliography of over 15,000 citations). The technique is well developed with easy-to-use, highly dependable instruments, but at the same time it is evolving, particularly with the production of more portable and rapid instruments and more powerful and flexible software. NIR is used globally in numerous industries for commodity analysis. Its desirable features include accuracy, precision, discrimination capacity, speed and adaptability to various operating circumstances, including harsh or hazardous conditions.

This paper (a) examines the unique capabilities of NIR that make it a candidate method for environmental analysis, (b) identifies barriers to the use of NIR and (c) explores several key environmental problems to which NIR can contribute unique analytical capability.

Capabilities of NIR for environmental analysis

Near infrared spectroscopy is largely unknown in the field of environmental chemistry and monitoring, despite the fact that numerous other spectroscopic techniques are widely used (Table 1). Even in soil analysis, where the research literature on NIR extends over four decades, NIR is not yet routinely used.

A number of field instruments exist that are used to continuously monitor various parameters including temperature, pH, conductivity, turbidity, suspended solids, salinity, dissolved O_2, redox potential and humidity. Near infrared spectroscopy shares many of the features of other portable technologies that operate in the field and, in addition, brings unique capability to analyse the compoposition of largely organic substances in and functional properties of samples (Table 2).

Table 1. Some laboratory instrumentation used commonly for environmental analytical chemistry.

Acid-base titrations	Mass spectroscopy
Oxidation–reduction	Atomic absorption spectroscopy
UV-vis spectrophotometry	Gas chromatography
Infrared spectrometry	High performance liquid chromatography
Nuclear magnetic resonance spectroscopy	Electroanalytical methods

Table 2. Features of NIR spectroscopy relevant to environmental analysis.

Portability	Non-destruction of the sample; availability for future analysis; suitability for live material
Speed	Analysis of intact sample matrices
Real-time analysis	Weather resistance
Simultaneous prediction of multiple constituents	Simplicity of the method
Qualitative and quantitative parameter	Constancy of the method
Composition and function	Stability of instrument performance
	Cost-effectiveness for large sample sets

Field-portable NIR instruments have been newly developed or adapted from laboratory instruments. A moving field sensor was developed to operate in-furrow in agricultural fields for measuring soil organic C.[1] A Bran+Luebbe InfraProver System was operated on board a vessel in the river Nieuwe Merwede, The Netherlands. Cores were taken from the sediment, sectioned and scanned on board for the prediction of dry solids content, organic matter content in various particle size fractions and density.[2] Subsequently, the instrument was placed in a water-tight housing and deployed in the water column with a fibre-optic probe inserted into the sediment for *in situ* determinations (W.R. Doorenweerd and G.L.J. Haaij, Delta Consult bv, The Netherlands, personal communication). Two instruments developed for on-the-go analysis of agricultural crops during harvest, the Textron/Case NH ProSpectra [3,4] and the Zeiss Corona spectrometers,[5] have been mounted on a pick-up truck to create a mobile NIR laboratory for on-site analysis of hog manure.[6]

Barriers to the use of NIR in environmental analysis

Although there are a number of challenges to the use of NIR for environmental analysis (Table 3), it is probable that the lack of familiarity of environmental chemists and regulators with NIR is the ma-

Table 3. Barriers to the use of NIR spectroscopy in environmental analysis.

Not taught in university Chemistry programs, where most environmental chemists and environmental technicians are trained.	Need to calibrate for samples of each type
Virtually unknown to government scientists, regulators, consultants, and academics	Need to calibrate for each make and model of instrument
Indirect nature of the predictions raises scepticism; acceptance requires a paradigm shift in thinking	Difficulty of identifying the major variables affecting extrapolability of calibrations. What is an appropriate "training set"?
If predictions are poor, NIR is blamed, not reference analyses	The high cost of calibration for small sample sets makes it not useful for small sample sets
Need to calibrate for each constituent	Capital outlay for instrumentation in addition to maintaining conventional analytical capability

jor explanation for the present low usage of NIR. This stems from the absence of NIR in the chemistry curriculum in most universities based on a prevalent opinion held in academia that the NIR region is difficult to explain from a spectroscopic viewpoint and not of much use for analysis.

In practice, it is likely that the largest future challenge to the widespread use of NIR for environmental analysis will be the calibration process. Most NIR determinations in environmental samples will be for naturally-occurring organic matter (NOM) or for properties influenced by or correlated with NOM. The composition of NOM and the factors affecting its variability over time and space are poorly known. Therefore, the identification of an appropriate "training set" for environmental applications will require care.

Global environmental issues

Many crucial environmental issues today have organic substances at their core. These include:
(a) nutrient management

- manure and compost
- biosolids
- protection of aquatic ecosystems

(b) climate change

- carbon inventories

(c) oil and gas industry environmental impacts

- total petroleum hydrocarbons

Manure and compost

Degradation of land and surface and ground water by an overabundance of nutrients is one of the oldest and most widespread of pollution problems. Natural processes are estimated to incorporate about 140 million tons of nitrogen into the terrestrial nitrogen cycle every year. Human activity has doubled that amount.[7] The management of N and P in manures and biosolids can potentially be greatly assisted by NIR analysis.

Animal manure and compost, i.e. animal and plant wastes in the process of biodegradation, are by-products of agriculture that contain nutrients useful to crops. If managed properly, their application to agricultural land potentially solves a waste management problem and replaces some of the inorganic fertilizer that would otherwise be used. Nevertheless, the high variability in nutrient composition of manure and compost from one operation to another and with time makes reliable loading of nutrients to land difficult. Hog manure can be particularly variable in composition even during the pump-out of a manure store, largely because the particulate material containing most of the P and metals is heavy and difficult to keep in suspension. Over the 8-day pump-out of a two-celled earthen manure store in Manitoba, Canada, total N ranged over 3.5-fold and total P over 20-fold.[8]

NIR is potentially applicable to on-site determination and control of nutrient loading during manure and compost application. Progress is being made using laboratory instruments such as the Foss NIRystems Inc. model 6500 visible/near infrared scanning spectrophotometer towards documenting capabilities of NIR analysis of hog manure,[9–11] cattle manure,[10,12–14] cattle manure compost,[15,16] livestock manure compost[17] poultry manure[18] and unspecified compost.[19] Almost all studies reported that NIR predicted moisture highly successfully and organic matter or solids, ammonium-N, organic N and total N well or reasonably well. Phosphorus was generally predicted poorly, except in several of the hog manure studies[9,11] and one compost study,[19] where it was predicted very successfully. The prediction of K was inconsistent. The field-portable instruments, ProSpectra and Corona, show promise for the in-stream, real-time analysis of hog manure during land application.[6] As part of a manure delivery system with GPS and GIS, NIR could accurately monitor nutrient loading to land, balance N : P by the addition of inorganic nutrients and deliver a custom fertilizer application. Challenges to making this work including sample presentation and development of calibrations.

Biosolids

As for manure and compost, applying anaerobically-digested, dewatered municipal sewage sludge, termed biosolids, to agricultural land is a widely-used, cost-effective way of disposing of this waste and making use of the organic matter, N and P that it contains. The UK and the US produce one million and seven million tonnes of biosolids, respectively, every year and apply about half of it to agricultural land.

When biosolids are applied to land, the most important constituents are the macro-nutrients, N, P and K and the heavy metals Cd, Cr, Cu Pb, Ni and Zn that may, potentially, accumulate to toxic levels from repeated application. On the other hand, when biosolids are land-filled, the organic matter content is of concern since C may be biodegraded in the landfill to produce dangerous methane. In a study involving biosolids from the City of Winnipeg, MB, Canada, moisture and organic matter content were predicted successfully, NH_4–N and suspended N were predicted marginally well, but P and K were not predictable.[20] Cadmium, Cu, Pb and Zn were predicted moderately well, likely as a result of binding to the organic matter or clay present in the biosolids.[20]

Aquatic ecosystem monitoring

Water quality

Water quality in lakes can be impacted by the release of wastes by human activities into air, land and water. NIR was found to be highly successful in measuring the concentrations of C, N and P in the suspended particulates (seston) in lakes.[21,22] This technique can monitor changes in the mass of N and P with changes in nutrient loading. NIR analysis of seston has been found to be related to total organic C, total P and pH of the lake water sampled at the same time.[23] NIR spectra from surface (0–1 cm depth) sediments in 58 Swedish lakes reflected water quality over the previous four years for total P, pH and total organic C.[24] This was repeated on Canadian lakes where suspended C and suspended N in lake water in the prior two years were predicted based on NIR analysis of the surface (0–1 cm) sediment.[25] In a study of sediment cores from Swedish lakes, the NIR spectra were used to predict pH in the lake water prior to acidification.[26]

Sediment quality

NIR spectroscopy has been highly successfully used for the prediction of total C, carbonate, organic C and N in lake sediments from four Canadian lakes ranging over more than 18° latitude[27] but P was predicted only marginally well. Phosphorus was predicted very successfully in sediments from eutrophic German Lake Arendsee.[28] Background concentrations of heavy metals, Fe, Mn, Zn, Cu, Pb and Ni in sediments from a pristine Canadian lake were predictable, apparently because of their association with organic matter.[28] Cadmium, experimentally added to the same lake was not successfully predicted in the sediment, either because its concentration was too low or because its association with the organic matter was much shorter than for the naturally-occurring metals.[29]

An elegant study of a small Swedish lake (0.5 km^2) of known bathymetry, involved only intensive sampling of surface sediments, drying and scanning of the samples, ashing of the samples and principal component analysis of the spectral data. Variance due to water depth and organic matter content was removed from the variance in the first principal component. The remaining variance in PC1 reflected organic matter quality. When the PC1 scores were mapped on the lake, the influence of natural inlets and outlets, of the run-off from human habitation and of clear-cutting on the shore on organic matter quality were clearly seen.[29]

Climate change

The most serious environmental problem facing mankind today is global climate change associated with increases in emissions from human activity of greenhouse gases, CO_2, CH_3, N_2O and chlorofluorocarbons (CFCs) into the atmosphere well above natural levels. The Kyoto Protocol, negotiated in Japan in 1997, commits industrialised nations to decrease greenhouse gas emissions between 2008 and 2012. As an alternative to some of the reduction in greenhouse gas emissions, some nations are proposing to take "credit" for large stores of C such as forests and soils that they have within national boundaries.[30] As well, they propose, through resource management, to use forests and soils as sinks to remove or sequester C from the atmosphere.[31]

Carbon quantity

Practical, cost-effective methods operating on a landscape scale are needed to determine the inventories of stored C and changes in C under management strategies. Carbon in soil is highly variable on a small-scale with both horizontal distance and topography and with depth. For example, soil from a 5.6 ha field in the Black Soil Zone of Manitoba, Canada varied in organic C from 0.38 to 4.05% d.w. in all A horizons combined and from 0.92 to 4.05% d.w. in the Ah horizon alone.[32] NIR is probably the only technique that is sufficiently cost-effective to allow the large numbers of samples to be measured that would be required to determine the size of and dynamic changes in C inventories.

Precision of NIR for measuring organic matter or C is comparable to the conventional Walkley Black or Dumas combustion methods. Using NIR, a calibration model developed for the above Manitoba soil samples between the chemically-determined organic C and the NIR spectral data predicted this constituent with an error of ± 0.33% organic C (range 0.38–4.05 %).[33] In a study on US soils, organic C (range 0.6 to 2.8%) was predicted by NIR with an error of ± 0.1% organic C.[34]

Although most analysis for organic matter or organic C with NIR has been on dried samples, the greatest benefit from NIR would be achieved if samples could be analysed on-site without costly sample transport, documentation, storage and analysis in the laboratory. Carbon could be analysed at the time of field sampling and results mapped as they were obtained. Field moist soil has been analysed by NIR. Percentage organic matter (range 0.1 to 40.6%) in field moist clay and peaty soils was predicted with an error of 3.32% organic matter (D.F. Malley and R.S. Currie, unpublished data). Soil organic C (range 0.3 to 2.5%) was measured in a group of US field moist soils with an error of 0.25% organic matter.[1]

Carbon quality

Determining the permanence of the C sequestered or stored in agricultural and forest ecosystems is its chemical nature or quality. Conventional methods for the determination of C quality are even more impractical to apply on a landscape scale than are laboratory methods for organic matter or C content in soil and litter. With NIR, carbon quality can be determined simultaneously with organic matter or C quantity. Total N, lignin and cellulose contents of forest foliage of 13 different species have been successfully determined with NIR.[35] Carbon, N, ash content, acid-detergent fibre and acid–detergent lignin were successfully determined in the litter of eight species of evergreen and deciduous broad-leafed trees, conifers and shrubs using NIR.[36] Measures of decomposition in leaf litter such as C/N ratios, lignin/N ratios have been predicted with NIR.[37] The effects of clear cutting forests with and without prescribed burning on humus quality, total microbial C and substrate-induced respiration were studied with NIR.[38] The usefulness and limitations of NIR to determining various C fractions in litter and soil have been reviewed.[39]

Petroleum hydrocarbons

Petroleum hydrocarbons (PHCs) represent the most wide-spread soil contaminants in many countries. They create a fire or explosion hazard, threaten human health, soil quality, drinking water quality, aquatic ecosystems and can be toxic to plants and animals. NIR is used quite widely for process control and product quality monitoring in oil refineries.[40] If successful for the determination of PHCs in soil, NIR would be very useful for the delineation and assessment of contaminated sites, particularly in remote areas where the need for conventional chemical analyses in laboratories results in costly delays. It may also by useful for monitoring natural recovery or remediation of contaminated sites.

Prediction of petroleum hydrocarbon contamination in soil and sand has been explored using soil and sand spiked with gasoline, diesel, motor oil and synthetic hydrocarbon mixtures. The determination of the hydrocarbons appeared to be more feasible in sand than in soil.[41] Diesel fuel contamination of soil has been predicted by NIR marginally well, using solid, phase micro-extraction–gas chromatography as the reference method for the determination of the hydrocarbons.[42]

Conclusion

A number of features combine to make NIR spectroscopy the most cost-effective method of analysis to emerge during the last three decades. Although NIR is presently poorly known among environmental scientists, consultants, technicians and regulators, the body of literature and experience with the technique is growing quickly. The cost of purchasing instruments, in addition to maintaining conventional chemical analytical capability, is a potential barrier when budgets for environmental monitoring and research are tight. Environmental scientists operating in a multi-disciplinary, cooperative environment with establishments such as agricultural industries or research stations may have an opportunity to try the technology before committing to a major purchase.[43] Nevertheless, the capital cost of an instrument may be roubped in a relatively short period of time, in cases where NIR significantly reduces labour or commercial laboratory costs. As NIR comes into wider use, such as for the analysis of C quantity and quality in relation to C inventories and sequestration, it is expected to lead to improved understanding of the composition, dynamics and variability of NOM. As it is in other fields, NIR in environmental analysis will be a tool to conduct routine analyses more efficiently and a stimulus to the development of new analytical capabilities not presently possible.

References

1. K.A. Sudduth and J.W. Hummel, *Transactions of the ASAE* **36,** 1571 (1993).
2. P.W.R. Doorenweerd and G.L.J. Haaij, International Conference on Contaminated Sediments: Restoration and Management. Rotterdam, The Netherlands, 7-11 September. Preprints, Volume II, p. 1045 (1997).
3. Anonymous, *Case IH Canadian Farming*, Fall, p. 4 (1999).
4. P. Williams, *NIR News* **11(3),** 3 (2000).
5. M. Rode, Abstract #O7-2 in Abstracts and Program of NIR-2001: *Changing the World with NIR*, 10th International Conference on near infrared Spectroscopy, Kyongju, Korea. 10-15 June (2001).
6. P.D. Martin and D.F. Malley, in Proceeding sof the Livestock Options for the Future Conference, Winnipeg, MB, Canada, 25–27 June (2001).

7. L.R. Brown, C. Flavin, H. French and others, *State of the World 2000*. The Worldwatch Institute. W.W. Norton & Co., New York, USA (2000).
8. D.F. Malley and P.D. Martin, in *Proceedings of the Livestock Options for the Future Conference*, Winnipeg MB, Canada 25–27 June (2001).
9. D.F. Malley, R.G. Eilers, E. Rempel, M. Van Walleghem and P.D. Martin, in *Manure Management'99: Proceedings of a Tri-provincial Conference on Manure Management*. Saskatoon, SK, 22–24 June, p. 633 (1999).
10. A. Millmier, J. Lorimor, C. Hurburgh Jr, C. Fulhage, J. Hattey and H. Zhang, *Transactions of the ASAE* **43,** 903 (2000).
11. D.F. Malley, L. Yesmin and R.G. Eilers, *Soil Sci. Soc. Amer. J.* **66,** (in press).
12. T. Asai, S. Shimizu, T. Koga and M. Sato, *Nippon Dojo Hiryogaku Zasshi* **64,** 669 (1993).
13. J.B. Reeves, III and J.S. Van Kessel, *J. Near Infrared Spectrosc.* **8,** 151 (2000).
14. J.B. Reeves, III and J.S. Van Kessel, *J. Dairy Sci.* **83,** 1829 (2000).
15 M. Nakatani and Y. Harada, *Japanese J. Soil Science & Plant Nutrition* **66,** 422 (1995).
16. M. Nakatani, Y. Harada, K. Kaga and T. Osada, *J. Japanese Soil & Fertil. Soc.* **66,** 159 (1996).
17. T. Kinoshita, A. Ichikawa and T. Kotah, *Aichi-ken Nogyo Sogo Shikenjo Kenkyu Hokoku* **29,** 305 (1997).
18. J.B. Reeves, III, *J. Ag. Fd. Chem.* **49,** 2193 (2001).
19. J.-J. Nam, K.-Y. Jung and S.-H. Lee, *Near Infrared Spectroscopy: Proceedings of the 9th International Conference*, Ed by A.M.C. Davies and R. Giangiacomo. NIR Publications, Chichester, UK, p. 613 (2000).
20. D.F. Malley, B. Trybula, R.D. Ross and G. Gay, *Evaluating the use of near infrared spectroscopy for the analysis of biosolids constituents.* Project 99-PUM-6-ET, Water Environment Research Foundation, Alexandria, VA. D00306WW (2000).
21. D.F. Malley, P.C. Williams, M.P. Stainton and B.W. Hauser, *Can. J. Fish. Aquat. Sci.* **50,** 1779 (1993).
22. D.F. Malley, P.C. Williams and M.P. Stainton, *Water Res.* **30,** 1325 (1996).
23. E. Dåbakk, M. Nilsson, P. Geladi, S. Wold and I. Renberg, *Wat. Res.* **34,** 1666 (2000).
24. M. Nilsson, E. Dåbakk, T. Korsman and I. Renberg, *Environ. Sci. Technol.* **30,** 2586 (1996).
25. P.D. Badiou, D.F. Malley, M.J. Paterson and M.P. Stainton, *Proceedings for the 43nd Annual Meeting of the Manitoba Soil Science Society*, January 25–26, p. 54 (2000).
26. T. Korsman, M. Nilsson, J. Ohman and I. Renberg, *Environ. Sci. Technol.* **26,** 2122 (1992).
27. D.F. Malley, L. Lockhart, P. Wilkinson and B. Hauser, *J. Paleolimnol.* **24,** 415 (2000).
28. D.F. Malley, H. Rönicke, D.L. Findlay and B. Zippel, *J. Paleolimnol.* **21,** 295 (1999).
29. D.F. Malley and P.C. Williams, *Environ. Sci. Technol.* **31,** 3461 (1997).
30. T. Korsman, M.B. Nilsson, K. Langren and I. Renberg, *J. Paleolimnol.* **21,** 61 (1999).
31. D. Anderson, R. Grant and C. Rolfe, *Taking Credit: Canada and the Role of Sinks in International Climate Negotiations.* David Suzuki Foundation and West Coast Environmental Law, 86 pp. (2001).
32. G.R. Manning, Master of Science Thesis, Department of Soil Science, University of Manitoba, Winnipeg MB, Canada (1999).
33. P.D. Martin, D.F. Malley, G. Manning and L. Fuller, *Can. J. Soil. Sci.* (accepted for publication).
34. G.W. McCarty and J.B. Reeves, III, in *Assessment Methods for Soil Carbon*, Ed by R. Lal, J.M. Kimble, R.F. Follett and B.A. Stewart, *Adv. Soil Sci.* Lewis Publishers, Boca Raton, USA, p. 371 (2001).
35. T.M. McLellan, J.D. Aber, M.E. Martin, J.M. Melillo and K.J. Nadelhoffer, *Can. J. For. Res.* **21,** 1684 (1991).
36. R. Joffre, D. Gillon, P. Dardenne, R. Agneessens and R. Biston, *Ann. Sci. For.* **49,** 481 (1992).

37 D. Gillon, R. Joffre and P. Dardenne, *Can. J. For. Res.* **23,** 2552 (1993).
38. J. Pietikäinen and H. Fritze, *Soil Biol. Biochem.* **27,** 101 (1995).
39. B. Ludwig and P.K. Khanna, in *Assessment Methods for Soil Carbon*, Ed by R. Lal, J.M. Kimble, R.F. Follett and B.A. Stewart, *Adv. Soil Sci.* Lewis Publishers, Boca Raton, USA, p. 361 (2001).
40. J. Workman, Jr, *J. Near Infrared Spectrosc.* **4,** 69 (1996).
41. H.W. Zwanziger and H. Förster, *J. Near Infrared Spectrosc.* **6,** 189 (1998).
42. D.F. Malley, K.N. Hunter and G.R.B. Webster, *J. Soil Contamin.* **8,** 481 (1999).
43. W.J. Foley, A. McIlwee, I. Lawler, L. Aragones, A.P. Woolnough and N. Berding, *Oecologia* **116,** 293 (1998).

Evaluation of rapid determination of phosphorous in soils by near infrared spectroscopy

Kwan Shig Ryu,[a] Jin Sook Park[a] and Bok Jin Kim[b]

[a]*Department of Agricultural Chemistry, Taegu University, Korea*

[b]*Department of Agronomy, Yeungnam University, Korea*

Introduction

Phosphorous in soil is one of the most difficult elements to assess for plant needs. Phosphorous compounds commonly found in soils are mostly unavailable for plant uptake because they are highly insoluble. When soluble sources of phosphorous, such as those in fertilizer and manure, are applied to soil, they are easily fixed to less available, sparingly soluble forms. Reactions that fix phosphorous in relatively unavailable forms differ from soil to soil and are closely related to soil pH. In acid soils the reactions predominantly involve Al, Fe and Ca either as dissolved ion, as oxides, or as hydrous oxides. Many Korean soils contain such hydrous oxides as coatings on soil particles and as interlayer precipitate in silicate clays. Some of the added phosphorous may also be changed to organic forms and become temporarily unavailable.[1]

Determination of total phosphorous of the soil commonly uses an oxidation and dissolution method with suitable acids to oxidise organic forms and dissolve phosphorous imbedded in the soil minerals. The basic function of a soil phosphorous test depends on both test methods and soil characteristics. The commonly known phosphorous availability indices are dissolution methods with water or unbuffered salt solution, diluted weak acid, diluted strong acid and buffered alkaline. Interpretation of the data from those methods vary with regional preference, consideration of soil characteristics and efficiency of operation.[2]

Davies mentioned the potential of NIR technique and this has been shown a useful technique in determining soil moisture,[3] organic matter and total nitrogen in the soil.[4,5] Krischenko showed a possible near infrared (NIR) technique for soil phosphorous analysis.[6] This study investigated the capability to estimate total and available phosphorous extracted by various extractants from Korean soils.

Materials and methods

Soils

A total of 148 soil samples with a wide range of soil characteristics were collected from paddy, upland, orchard and others over the Kyong book province in Korea. Soil sample were air-dried and prepared to pass 2.0 mm sieve openings.

Determination of soil phosphorous

Total phosphorous in the soil was determined by ICP after four hours digestion with 60% $HClO_4$ – conc HNO_3. Available phosphorous was also determined by ICP after extraction with Bray 1 extractant for one minute and Olsen and Truog extractants for 30 minutes, respectively. All results were expressed on the basis of dry weight.

Measurement of NIR spectra

The NIR reflectance spectra of soils were measured using a scanning NIR spectrometer (Foss, NIR5000) with the wavelength ranging from 1100 to 2500 nm. Functional groups of Ca, Fe and Al bounded phosphates on the NIR spectrum were compared after adding 3% of phosphorous solution with Ca, Fe and Al phosphate to a soil. The difference in the functional groups of the remaining phosphate in the soil was also compared after being extracted by the Bray 1, Olsen and Truog extractants. 99 soil samples were used for the calibration equation and 49 samples were used for prediction. The multiple linear regression was adopted for making the calibration.

Results and discussion

The data in Table 1 shows the various types of phosphorous in the soils used in this study. The average total phosphorus by $HClO_4$ digestion used for calibration was 1384.7 mg kg^{-1} ranged from 227.2 to 3675.9 mg kg^{-1} of the soil and that used for prediction was 1376.0 mg kg^{-1} ranged from 326.1 and 511.3 mg kg^{-1} of the soil. Total phosphorous in the soil was mainly determined by the Na_2CO_3 fusion, H_2SO_4 digestion, $HClO_4$ digestion and NaOBr oxidation methods. Kuo mentioned that the latter two methods were known not to readily dissolve phosphorous in silicate minerals and, therefore, underestimate the total phosphorous in the soils.[2]

Available phosphorous data of the soils used for calibration and prediction by Bray 1, Olsen and Truog extractant showed big differences (Table 1). This seemed mainly due to the different values of the soil pH, time of extraction and different extractants which extracted different types of phosphate. Shin *et al.* showed the composition rate of Ca–P, Al–P and Fe–P in Korean soil.[1] Ca–P was far below average when soil pH was less than 5.5. It was far higher than average when soil pH was more than 5.5. Al–P was increased slightly with the decrease of soil pH and Fe-P had the smallest value when soil pH was below 4.5 and above 6.5. The difference in Al–P and Fe–P to soil pH was smaller compared with Ca-P. After a soil was mixed with Ca, Fe and Al phosphate solution (3% as phosphorous), change in NIR reflectance spectra to the functional group for Ca, Fe and Al–P was not clear as shown in Figure 1. The minor change on raw and 2nd derivatives of the spectrum to these groups was barely identifiable. Therefore, it is difficult to interpret exactly what the main composition for the complicated bands was. After the soil was extracted by Olsen, Bray 1 and Truog extractant, minor changes in NIR reflectance spectra for these functional groups, shown in Figure 2, were of the same order as the functional groups of Al, Ca and Fe bounded phosphates but the magnitude was different near the 1930 nm range. Diluted

Table 1. Total and available phosphorus (mg/kg of soil) by various extractants used for calibration and prediction in the soil.

	Calibration (99)		Prediction (49)	
	Mean (mg kg^{-1})	Range (mg kg^{-1})	Mean (mg kg^{-1})	Range (mg kg^{-1})
Total P	1384.7	227.2–3675.9	1376.0	326.1–3511.3
Bray 1	213.7	16.9–911.6	205.5	8.0–888.3
Olsen	102.0	4.2–317.4	103.7	6.4–292.7
Truog	101.9	0.0–717.6	111.7	0.0–564.1

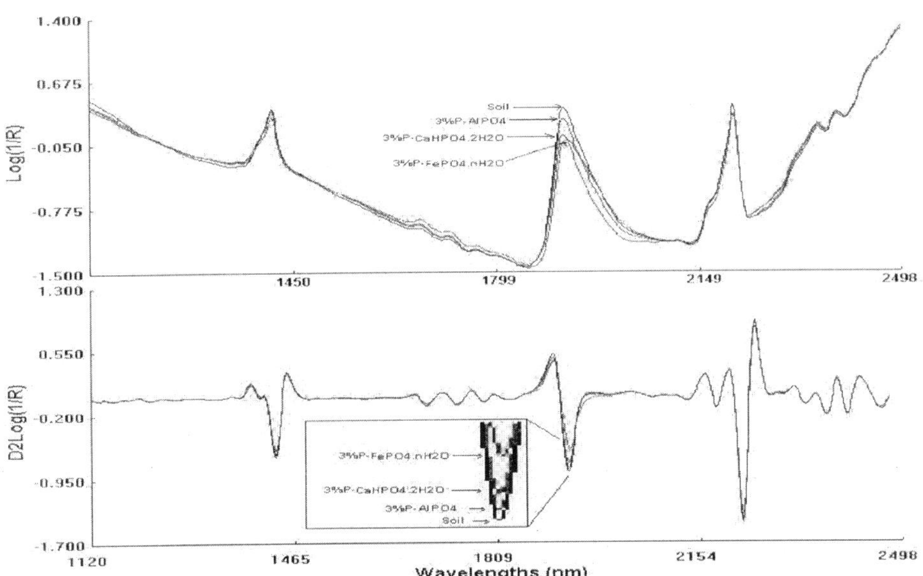

Figure 1. Raw and corresponding 2nd derivative spectra after adding various phosphate (as 3% of phosphorous) compound solutions to a soil and mixed through.

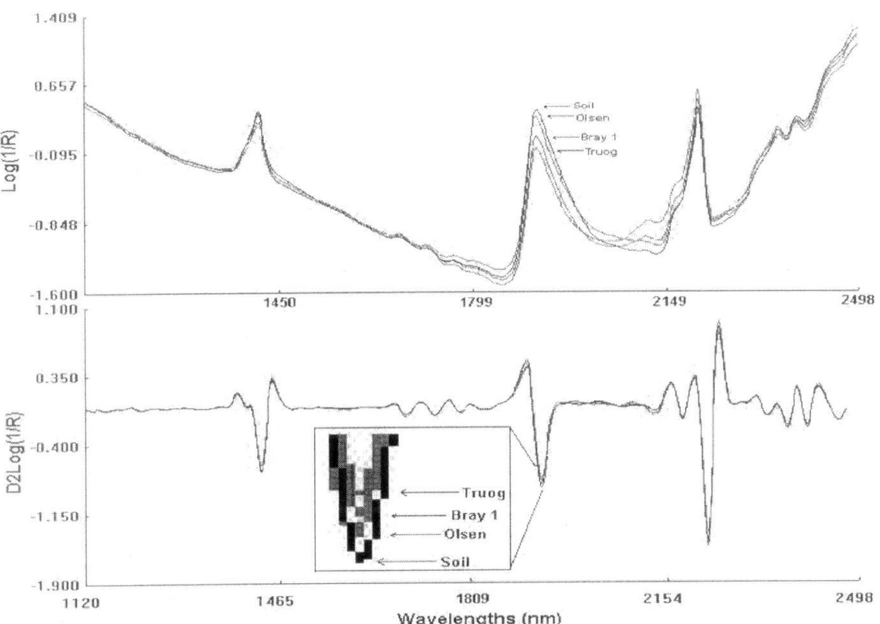

Figure 2. Raw and corresponding 2nd derivative spectra of a soil after extraction by different extractants and washed with distilled water.

strong acid solution such as Bray 1extractant solubilised Ca, Al–P and, to a lesser extent, Fe–P Al–P were the primary soil phosphorous fractions by the Bray 1 test. The Olsen test extracted less phosphorous than the Bray 1 test and correlated well with Al–P P.[2] Mean available phosphate was in the order of Bray 1, Olsen or Truog extractant phosphates but the range of these phosphates in the soil used for calibration and prediction was in the order of Bray 1, Truog and Olsen extractant phosphates. Figure 3 shows the actual measured phosphorous used for this study by means of wet chemistry and predicted value by NIR on a 1 : 1 line.

Scattering of each measurement and prediction on a 1 : 1 line seemed to be a function of a complicated effect of phosphorous compounds on the various extractants, time of extraction and various soil characteristics. As seen, the actually measured values showed relatively good agreement with the predicted value in spite of the complicated soil phosphorous analyses but the Truog P test showed very poor agreement when the P value was less than 200 mg k^{-1} of soil. The calibration developed for phosphorous was the best for total phosphorous with $R = 0.91$, then Bray 1 P with $R = 0.82$ and Olsen P with $R = 0.80$. The poorest one was the Truog P with $R = 0.76$.

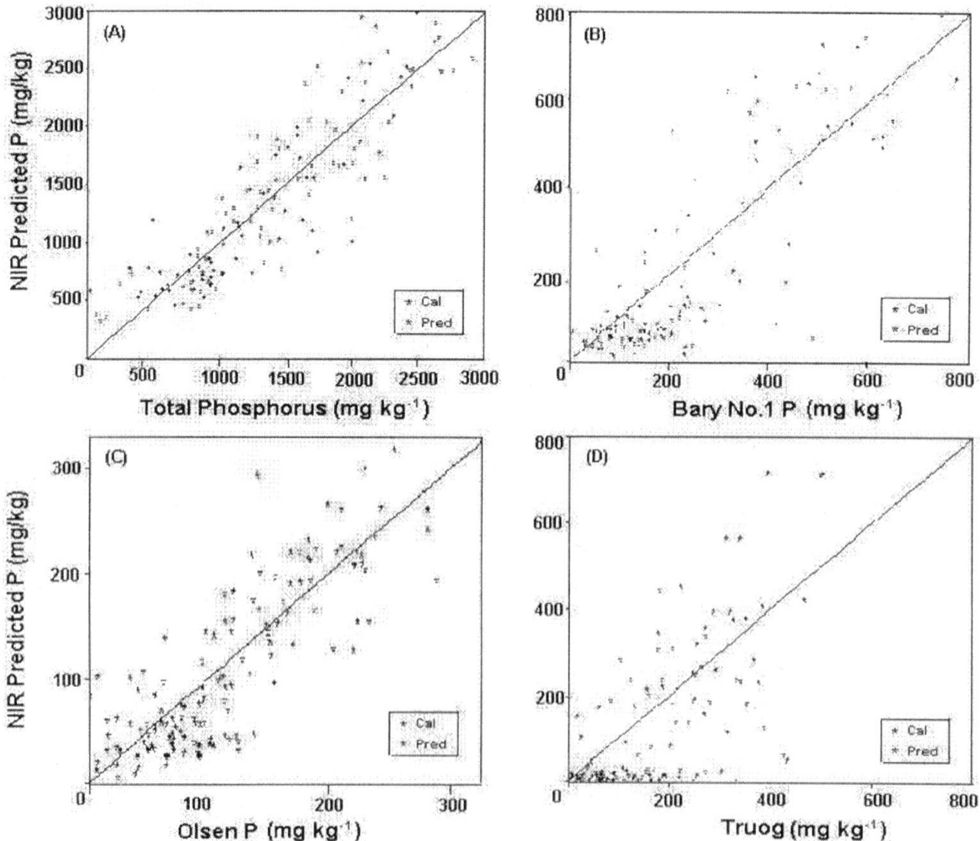

Figure 3. Relationship between measured phosphorous by $HClO_4$ (a) digested total phosphorousand (b) Bray 1, (c) Olsen and (d) Truog available phosphorous and predicted value by NIR in soils.

Table 2. Calibration and prediction statistics of mutiple linear regression for total and available phosphorus(mg kg^{-1} of soil) by the various extractants.

Extaractant	Spectrum	Terms	*R*	*SEE*	*SEP*	Bands used for calibration
Total P	Raw	9	0.91	329.0	429.7	1100,1920,1978,2160,2172,2284,2300,2308,2464
Bray 1	2D	9	0.82	146.4	162.3	1242,1252,1330,1742,1962,1990,2140,2252,2344
Olsen	2D	9	0.80	54.9	54.1	1188,1202,1216,1624,1678,1890,1982,2142,2228
Truog	1D	9	0.76	162.7	152.3	1464.1896,1922,1942,2026,2194,2212,2294,2304

Terms: number of wavelengths used for calibration, *R*: Multiple correlation coefficient
Raw: raw spectrum, 1D: 1st derivative spectrum, 2D: 2nd derivative spectrum
SEE: standard error of estimation, *SEP*: standard error of prediction

Conclusion

NIR spectroscopy is increasingly used as a means of rapid and accurate determination of the properties and qualities of food and agricultural products. However, the NIR technique may not exceed the referenced laboratory analytical method in accuracy due to the calibration equation being based on the referenced analytical data. Phosphorous in the soil is one of the most difficult elements to assess for plant needs mainly due to the various soil characteristics. However, the NIR method could also be used as a routine method to estimate or screen total phosphorous, and Bray 1 and Olsen type of available phosphorous because of ease of operation. Often research requires more accurate soil data for fertilizer recommendations than is achieved by present NIR calibrations. With improvments in accuracy, NIR spectra may become a suitable technique for reliable soil testing.

Acknowledgement

This research was partly funded by the Research Centre of Agricultural R & E Promotion Centre in Korea.

References

1. C.W. Shin, J.J. Kim and I.S. Ryu, *Korean Society of Soil Science and Fertilizer* **23(1),** 15 (1990).
2. S. Kuo, in *Methods of soil analysis*, Part 3. SSSA, p. 869 (1996).
3. T. Davis, "Overview of near infrared spectroscopy", *in The symposium for "Strategies for improvement of agricultural environment and creation of the venture) business from agricultural chemistry.* KSACB and KSEC, pp. 203–214, (1998).
4. N. Inoue, T. Asai, K. Yamada and M. Higuchi. *Hokuriku Gakkaiho* **26,** 104 (1991).
5. K.S. Ryu, B.S. Kim, W.C. Park and R.K. Cho, in *Near Infrared Spectroscopy: Proceedings of 9th International Conference,* Ed by A.M.C. Davies and R. Giangiacomo. NIR Publications, Chichester, UK, pp. 593–597 (1999).
6. V.P. Krischenko, S.G. Samokhvalov, L.G. Fomina and G.A. Novikova, *Making Light Work: Advances in Near Infrared Spectroscopy,* Ed by I. Murray and I.A. Cowe, VCH, Weinheim, Germany, pp. 239–249 (1992).

Evaluation of drainage by near infrared spectroscopy

Hitoshi Takamura,[a,b] Hiroko Miyamoto,[c] Yoshikuni Mori[d] and Teruyoshi Matoba[a,c*]

[a]*Department of Food Science and Nutrition, Nara Women's University, Nara 630-8506, Japan*

[b]*KYOUSEI Science Center for Life and Nature, Nara Women's University, Nara 630-8506, Japan*

[c]*Graduate School of Human Culture, Nara Women's University, Nara 630-8506, Japan*

[d]*Nara Purification Center, Yamato-Koriyama 639-1035, Japan*

Introduction

Near infrared (NIR) absorption is derived from overtones and combinations of the fundamental vibrations of molecules found in the mid-IR region. Of major importance are the absorptions of the C–H, O–H and N–H groups. Therefore, the NIR technique is suitable for the determination of organic compounds. In addition, the determination in aqueous systems is possible since NIR absorption due to water is much lower than that in the mid-IR region. In our laboratory, spectral analysis for protein determination,[1–5] lipid oxidation[6] and moisture content[7] by NIR has been studied for years. However, only a few studies on the application of NIR techniques for environmental pollutions have been reported.[8–10]

Water pollutants in drainage mainly consist of organic compounds. Hence, total organic carbon (TOC), chemical oxygen demand (COD) and biochemical oxygen demand (BOD) were generally used as the indices of pollution. However, these values are determined with a special analyser (TOC), titration method (COD) or microbe culture (BOD), which are time-consuming methods. Therefore, the development of simple and easy-to-use methods for the determination of water pollution is required. The authors reported the evaluation of water pollution by NIR spectroscopy in a model system with food components.[11] In this study, the relationship between NIR spectra and drainage was investigated in order to develop a method for the evaluation of drainage by NIR.

Materials and methods

Sample drainage and river water

Drainage, partially purified drainage, purified drainage and river water at various pollution levels were obtained at the Nara Purification Centre. Approximately 400 samples were used for calibration and 100 samples were used for prediction. The ranges of TOC, COD and BOD were 0–140, 0–100 and 0–220, respectively.

*Author who is responsible for correspondence (E-Mail: matoba@cc.nara-wu.ac.jp)

Determination of NIR spectra

NIR transmittance spectra (680–1235 and 1100–2500 nm) of the drainage were determined with an NIRSystems (Pacific Science) Model 6250 Research Composition Analyzer at 10–40°C. The 10 mm cuvette cell was used as a sample cell for 680–1235 nm and the 0.5 mm cuvette cell for 1100–2500 nm. Statistical analysis was performed using NSAS Ver. 3.27 (NIRSystems).

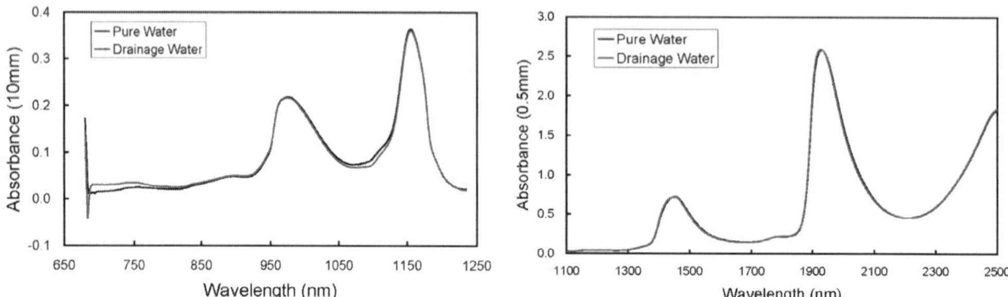

Figure 1. NIR raw spectra of water (TOC = 0) and drainage (TOC = 140) measured at 25°C in the region of (a) 680–1235 nm and (b)1100–2500 nm.

Table 1. Calibration using the short wavelength range of 680–1235 nm for determining TOC/COD/BOD.

Factor	TOC		COD		BOD	
	R	*SEC*	*R*	*SEC*	*R*	*SEC*
1	0.710	24.29	0.775	17.05	0.728	37.33
2	0.837	18.88	0.893	12.17	0.852	28.54
3	0.871	17.01	0.919	10.65	0.883	25.66
4	0.963	9.28	0.954	8.14	0.968	13.73
5	0.968	8.74	0.960	7.63	0.973	12.75
6	0.970	8.43	0.961	7.51	0.976	11.92
7	0.971	8.31	0.963	7.32	0.978	11.56
8	0.972	8.21	0.964	7.26	0.978	11.37
9	0.973	8.10	0.964	7.21	0.979	11.25
10	0.974	7.96	0.967	6.96	0.980	10.91
11	0.975	7.74	0.968	6.83	0.981	10.67
12	0.976	7.59	0.971	6.55	0.982	10.32
13	0.978	7.34	0.972	6.45	0.983	10.20
14	0.979	7.11	0.974	6.24	0.985	9.68
15	0.980	6.90	0.975	6.11	0.985	9.52

Determination of TOC, COD and BOD

TOC was determined with a Shimadzu TOC-5000A TOC analyser. COD was determined by the potassium permanganate titration method. BOD was calculated from the difference in oxygen concentrations of water before and after a 5-day microbe incubation. Oxygen concentration was determined by oxygen electrode.

Regression analysis

A partial least squares (PLS) regression was used for calibration and prediction. All spectra measured at different temperatures were used for the calibration in order to obtain a robust calibration.

Results and Discussion

NIR spectra of water and drainage

Figure 1(a) and (b) show the NIR raw spectra of water (TOC = 0) and drainage (TOC = 140). In the region of 680–1235 nm [Figure 1(a)], the difference between water and drainage could be seen, while there was little difference in the region of 1100–2500 nm [Figure 1(b)].

Calibration for determining TOC/COD/BOD

Tables 1 and 2 show the correlation between the NIR raw spectra and TOC/COD/BOD in the wavelength range of 680–1235 and 1100–2500 nm, respectively. The samples of water and drainage were first determined by NIR spectrometer at 10–40°C and then TOC, COD and BOD were determined.

Table 2. Calibration using long wavelength range of 1100–2500 nm for determining TOC/COD/BOD.

Factor	TOC		COD		BOD	
	R	*SEC*	*R*	*SEC*	*R*	*SEC*
1	0.133	38.61	0.170	27.08	0.121	68.52
2	0.244	37.84	0.308	26.19	227	67.33
3	0.634	30.21	0.686	20.06	0.633	53.62
4	0.757	25.56	0.783	17.18	0.757	45.29
5	0.843	21.10	0.841	14.96	0.848	36.79
6	0.862	19.91	0.863	14.02	0.868	34.54
7	0.913	16.06	0.919	10.95	0.916	28.01
8	0.940	13.40	0.944	9.18	0.942	23.35
9	0.952	12.08	0.955	8.25	0.954	20.88
10	0.959	11.22	0.962	7.63	0.961	19.35
11	0.964	10.55	0.966	7.20	0.966	18.10
12	0.968	9.98	0.970	6.80	0.970	17.13
13	0.972	9.40	0.973	6.44	0.973	16.13
14	0.974	8.97	0.976	6.16	0.976	15.32
15	0.976	8.70	0.977	5.98	0.978	14.74

Table 3. Validation of the calibration equations developed.

Index	680–1235 nm				1100–2500 nm			
	Factor	*R*	*SEP*	Bias	Factor	*R*	*SEP*	Bias
TOC	6	0.960	4.99	3.46	11	0.187	70.0	–309
COD	6	0.892	5.55	8.60	10	0.211	58.8	–309
BOD	7	0.949	9.34	6.54	11	0.253	108.0	–559

Calibration was performed by PLS using 400 spectra measured at different temperatures to get a robust calibration against the change of temperature. As shown in Tables 1 and 2, good correlations between the NIR spectra and TOC/COD/BOD were obtained in both wavelength ranges, which suggests that the NIR raw spectra determined at different temperatures were highly correlated with TOC, COD and BOD.

Validation of the calibration equations developed

Table 3 shows the prediction statistics of NIR raw spectra for TOC, COD and BOD. Though good calibrations were obtained in both wavelength ranges, good predictions were obtained only in the range of 680–1235 nm. This result suggests that NIR spectroscopy can be used to evaluate the quality of drainage water by using the wavelength range of 680–1235 nm.

In this work, we examined the possibility of using NIR spectroscopy to evaluate drainage. NIR spectroscopy has a great advantage for this purpose, since it is simple and easy to use. Although some improvement is necessary for practical use, NIR evaluation of drainage may become useful in the future.

References

1. H. Kamishikiryo, K. Hasegawa and T. Matoba, *J. Jpn Soc. Food Sci. Technol.* **38,** 850 (1991).
2. H. Kamishikiryo, K. Hasegawa, H. Takamura and T. Matoba, *J. Food Sci.* **57**, 1239 (1992).
3. H. Kamishikiryo-Yamashita, M. Tatara, H. Takamura and T. Matoba, *J. Jpn Soc. Food Sci. Technol.* **41,** 65 (1994).
4. H. Kamishikiryo-Yamashita, Y. Oritani, H. Takamura and T. Matoba, *J. Food Sci.* **59,** 313 (1994).
5. H. Yamashita, H. Takamura and T. Matoba, *J. Near Infrared Spectrosc.* **2,** 145 (1995).
6. H. Takamura, N. Hyakumoto, N. Endo, T. Nishiike and T. Matoba, *J. Near Infrared Spectrosc.* **3,** 219 (1997).
7. H. Takamura, N. Endo and T. Matoba, *J. Near Infrared Spectrosc.* **6,** 235 (1998).
8. T. Korsman, M. Nilssonm J. Ohman and I. Renberg, *Environ. Sci. Technol.* **26,** 2122 (1992).
9. M.B. Nilsson, E. Dabakk, T. Korsman and I. Renberg, *Environ. Sci. Technol.* **30,** 2586 (1996).
10. D.F. Malley, P.C. Williams and M.P. Stainton, *Water Res.* **30,** 1325 (1996).
11. H. Takamura, H. Miyamoto, M. Imatani, and T. Matoba, in *Near Infrared Spectroscopy: Proceedings of 9th International Conference*, Ed by A.M.C. Davies and R. Giangiacomo. NIR Publications, Chichester, UK, p. 503 (2000).

Near infrared spectroscopy for measuring purine derivatives in urine and estimation of microbial nitrogen synthesis in the rumen for sheep

Stefka Atanassova,[a] Nana Iancheva[a] and Roumiana Tsenkova[b]

[a]*Thracian University, Agricultural Faculty, 6000 Stara Zagora, Bulgaria*

[b]*Kobe University, Agricultural Faculty, Rokkodai 1-1, Nada, Kobe, Japan*

Introduction

The efficiency of the ruminal fermentation process influences overall efficiency of ruminal production, animal health and reproduction. Ruminants' production systems have a significant impact on the global environment, as well. Animal waste contributes to pollution of the environment as ammonia volatilised to the air and nitrate leached to ground water.

Microbial protein synthesis in the rumen satisfies a large proportion of the protein requirements of animals. Quantifying the microbial synthesis is possible by using markers for rumen bacteria and protozoa such as nucleic acids, purine bases, some specific amino acids, or by isotopic ^{15}N-, ^{32}P- and ^{35}S-labelled feeds.[1–4] All those methods require cannulated animals, they are time-consuming and some methods are very expensive as well. Many attempts have been made to find an alternative method for indirect measurement of microbial synthesis in intact animals. One of them is based on measurement of purine derivatives (allantoin and uric acid in ruminants) excreted in urine and milk. Several authors demonstrated a direct relationship between urinary purine metabolite excretion and microbial protein production in the rumen.[5–7]

Near infrared (NIR) spectroscopy has been demonstrated as a successful toll for compositional analysis of feedstuff and other agricultural products.[8,9] At present there are a number of studies on the use of NIR spectroscopy in animal physiology, for example, brain research, Hb research, bacterial nitrogen in duodenal content[10,11] and in feed residue after ruminal incubation in nylon bags.[12,13]

The present investigation aimed to assess possibilities of usingNIR spectroscopy for the prediction of purine nitrogen excretion and ruminal microbial nitrogen synthesis by NIR spectra of urine.

Materials and methods

Eighty urine samples were collected from 12 growing sheep, six male and six female. The sheep were included in a feeding experiment which consisted of sorghum silage and protein supplements—70 : 30 on a dry matter (DM) basis. The protein supplements were chosen to differ in protein degradability: urea, soybean meal and soybean meal/fish meal (40 : 60 DM basis). The urine samples were collected daily in a vessel containing 60 mL10% sulphuric acid to reduce pH below 3 and diluted with tap water to 4 L. The samples were stored in plastic bottles and frozen at –20°C until chemical and NIR analyses. The samples were analysed for allantoin, uric acid, xantine and hypoxantine content. All methods of analyis were described in detail by Chen and Gomes.[14] Microbial nitrogen synthesis in the rumen was calculated based on the urinary excretion of purine derivatives, according to Chen and Gomes.[14]

NIR transmittance (*T*) spectra of urine samples were obtained by an NIRSystem 6500 spectrophotometer (Foss NIRSystems, Silver Spring, MD, USA). A cell 1 mm thick was used for the NIR measurements. The spectra were taken in the wavelength range from 1100 to 2500 nm at 2 nm intervals and were recorded in the linked computer as absorbance, i.e. log(1/*T*). Prior to spectral analysis each sample was warmed up to 40°C in a water bath.

A commercial software package, ISI NIRS 3 (Infrasoft International, Port Matilda, PA, USA), was used to process the data and to develop models for determination of allantoin, total purine derivatives and purine derivatives nitrogen content in urine and microbial nitrogen synthesis in the rumen.

The calibration was performed using partial least square (PLS) regression. In the development of all calibration models, ten PLS factors were set up as a maximum. The optimum number of PLS factors used in the models was determined by a cross-validation method. In cross-validation, five samples were temporarily removed from the calibration set to be used for validation. With the rest of the samples a PLS model was developed and applied to predict the content of tested parameters of the group of five samples. The results were compared to the respective reference values. This procedure was repeated several times until a prediction for all samples was obtained. Performance statistics were accumulated for each group of removed samples. The validation errors were combined into a standard error of cross-validation (*SECV*). The optimum number of PLS factors in each model was defined to be the one that corresponded to the lowest *SECV*.

Selection of the best calibration equations was made on the basis of the lowest standard error of calibration (*SEC*) and standard error of cross-validation (*SECV*), and the highest calibration coefficient of determination (R^2) and cross-variation coefficient of determination CVr^2, respectively.

Results and discussion

The ranges, mean values and standard deviation of allantoin, total purine derivatives and total purine nitrogen content of urine samples and calculated microbial nitrogen synthesis in the rumen are presented in Table 1. The statistical parameters of calibration and cross-validation procedures for purine derivatives content and microbial protein synthesis determination are shown in Table 2. Figures 1 and 2 graphically illustrate the relationships between determined and NIR predicted values of allantoin content and microbial nitrogen synthesis, respectively.

The results of estimating the purine derivatives, excretion and microbial protein synthesis of urine using NIR spectra showed promising accuracy. The ratio of standard deviation of population and standard error of calibration was 2.96 for allantoin content, 2.65 for total purine derivatives, 2.53 for total purine nitrogen and 2.87 for microbial nitrogen, respectively. These values were considered adequate for fast and noninvasive evaluation of microbial nitrogen synthesis in the rumen.

Table 1. Mean, range and standard deviation (SD) of tested parameters in urine samples.

Parameter	Average	Min	Max	SD
Allantoin, mg L^{-1}	33	849.5	403.68	125.20
Total purine derivatives, mmol d^{-1}	2.36	26.67	13.33	3.75
Purine derivatives nitrogen, mg d^{-1}	165	1867	935.80	259.14
Microbial nitrogen, g d^{-1}	0.77	23.08	11.40	3.36

Table 2. The statistical results of NIR calibration for prediction of purine derivatives and microbial protein synthesis.

Parameter	Math. Transf.	PLS factors	*SEC*	R^2	*SECV*	CVR^2
Allantoin, mg L^{-1}	log(1/*T*)	9	42.25	0.848	52.13	0.770
Total purine derivatives, mmol d^{-1}	first derivative	7	1.42	0.843	1.68	0.776
Purine derivatives nitrogen, mg d^{-1}	log(1/*T*)	8	102.20	0.821	118.63	0.760
Microbial nitrogen, g d^{-1}	first derivative	8	1.17	0.861	1.37	0.810

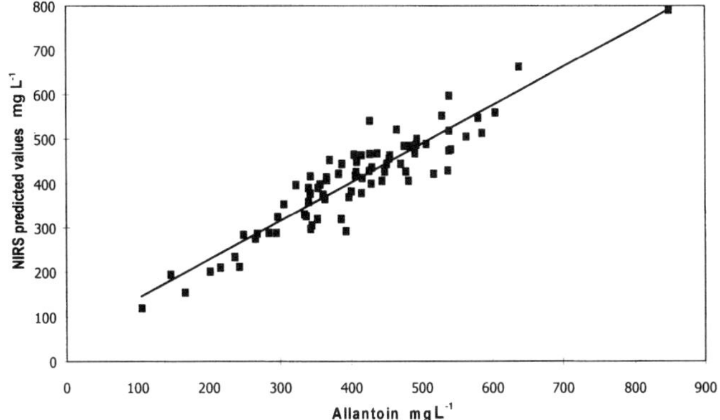

Figure 1. Relationship between actual and NIR predicted values of allantoin content in urine.

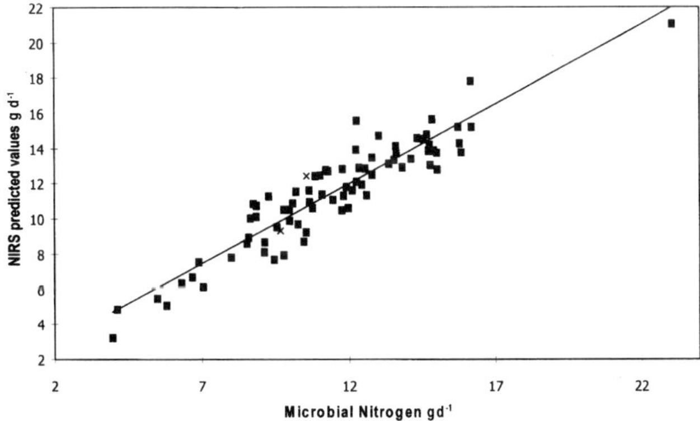

Figure 2. Relationship between actual and NIR predicted values of microbial nitrogen synthesis in the rumen.

The PLS modelling not only aided the development of quantitative models but also was used as a tool for discerning the location of spectral information related to purine derivatives. The regression vector of the calibration equation for allantoin content, based on log(1/*T*) spectral data, was studied. High positive coefficients in the regression vector were found at 1368, 1484, 1838, 1956, 2024, 2190,

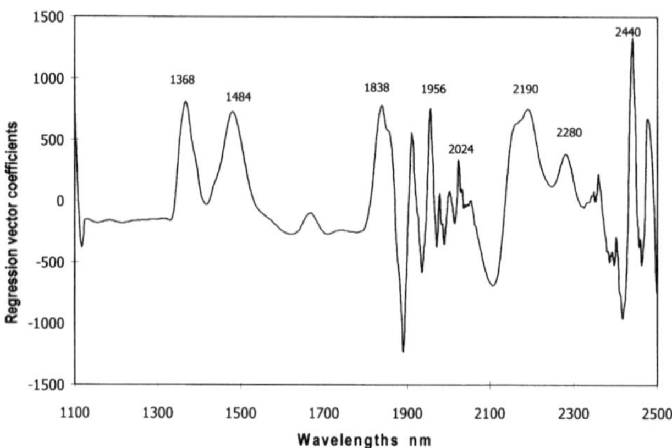

Figure 3. Plot of regression vector coefficients of equation for determination of allantoin content.

2280 and 2440 nm (Figure 3). The NIR absorption around 1484 and 1956 nm may be due to the first overtone in the N–H stretching mode and combination of N–H stretching and amide II vibration, respectively.[8,15] In the regions around 2190, 2280 and 2440 nm dominated absorption of C–H bands and absorption at 2024 nm was connected with C=O absorption.[8,15] It might be concluded that the determination of purine derivatives by NIR spectroscopy was connected with their absorption in the NIR region.

Conclusion

The results indicate that the advantages of the NIR technology can be extended into animal physiological studies. The fast and low cost NIR analyses could be used with no significant loss of accuracy when microbial protein synthesis in the rumen are to be assessed by NIR spectroscopy.

References

1. S.H. Rahnema and B. Theurer, *J. Anim. Sci.* **63,** 603 (1986).
2. G.A. Broderick and N.R. Merchen, *J. Dairy Sci.* **75,** 2618 (1992).
3. D.J. Illg and M.D. Stern, *Anim. Feed Sci. Technol.* **48,** 49 (1994).
4. D. Djouvinov and N.A. Todorov, *Anim. Feed Sci. Technol.* **48,** 289 (1994).
5. J.H. Topps and R.C. Elliot, *Nature.* **205,** 498 (1965).
6. R. Puchala and G.W. Kulasek, *Can. J. Anim. Sci.* **72,** 821 (1992).
7. J.F. Perez, J. Balcells, J.A. Guida and C. Casrtillo, *Animal Science.* **65,** 225 (1997).
8. J.S. Shenk, J.J. Workman and M.O. Westerhaus, in *Handbook of Near-infrared Analysis,* Ed by D.A. Burns and E.W. Ciurczak. Marcel Dekker Inc., New York, USA, p. 383 (1992).
9. I. Murray, in *Sward Measurement Handbook,* Ed by A. Davies, R.D. Baker, S.A. Grant and A.S. Laidlaw. Br. Grassl. Soc., Reading, UK, p. 285 (1993).
10. S. Atanassova, D. Djouvinov, E. Enev and N. Todorov, in *Near Infrared Spectroscopy: The Future Waves,* Ed by A.M.C. Davies and P. Williams. NIR Publications, Charlton, UK, p. 548 (1996).
11. P. Lebzier and Ch. Paul, *Anim. Feed. Sci. Technol.* **68,** 225 (1997).
12. Ph. Lecomte, M. Kamoun, R. Agnessens, J. Beckers, P. Dardenne, A. Thewis and E. Francois, in *Proc. Soc. Nutr. Physiology,* Vol. 3, Ed by D. Giesecke. DLG-Verlag, Frankfurt (Main), Germany, p. 44 (1994).

13. S. Atanassova, N. Todorov, D. Djouvinov, R. Tsenkova and K. Toyoda, *J. Near Infrared Spectrosc.* **6,** 167 (1998).
14. X.B. Chen and M.J. Gomes, *Estimation of Microbial Protein Supply to Sheep and Cattle Based on Urinary Excretion of Purine Derivatives - An Overview of the Technical Details.* Occasional Publication 1992. International Feed Resourses Unit, Rowett Research Institute, Aberdeen, UK (1995).
15. B. Osborne, T. Fearn and P.H. Hindle, *Practical NIR Spectroscopy with Applications in Food and Beverage Analysis.* Longman Scientific and Technical, Harlow, UK (1993).

Non-destructive and fast determination of N, P and K in compound fertiliser by near infrared spectroscopy

Xiangyang Zhou,[a] Ling Liu,[b] Duojia Wang[a] and Tongming Jin[b]

[a]*Shenzhen Agricultural Research Center, Shenzhen, China*

[b]*Beijing Vegetable Research Center, Beijing, China*

Introduction

Fertilisers are an important elemental source for growth and development of crops, whether intended for yield increasing or quality improving, they are indispensable. The amount of compound fertiliser produced in China is more than ten million tons and it is increasing at the rate of 5–7% per year. Chemical analysis methods are widely used for quality testing but these methods are complex and time-consuming. We expect to find an alternative method that is fast, non-destructive and has low running costs.

Near infrared (NIR) spectroscopy has been widely used for fast and non-destructive analysis in agriculture, food, pharmaceuticals, textiles, cosmetic and polymer production.[1–4] Of all the techniques available for fast assessment, NIR is the most suitable method for analysis.

In this study, we explored the feasibility of using NIR spectroscopy for the analysis of N, P and K in compound fertiliser samples.

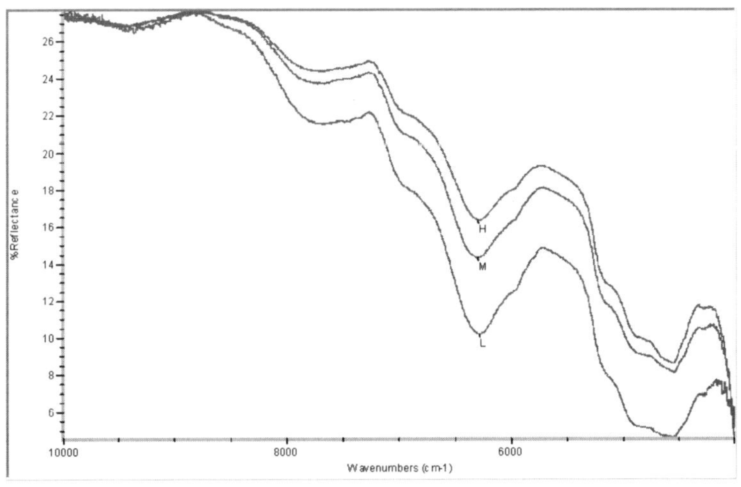

Figure 1. Original spectra of Lanfei (H, M and L).

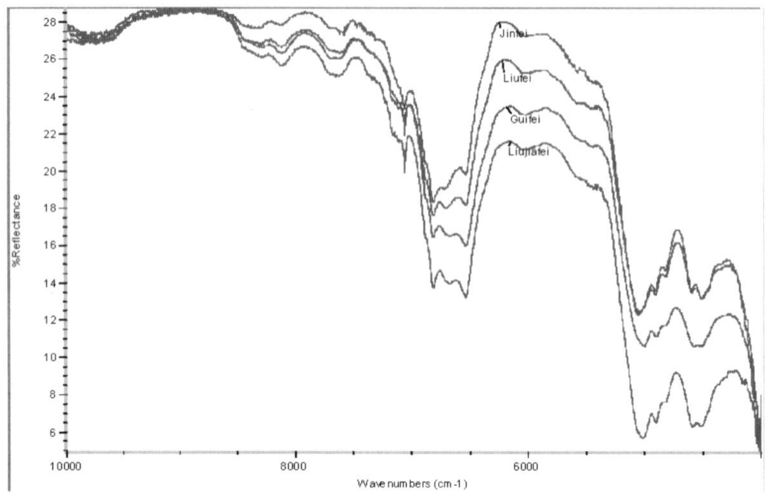

Figure 2. Original spectra of four fertilisers.

Materials and methods

Sample collection and chemical analysis

Seventy compound fertiliser samples were collected in 2000 from Batian Compound Fertiliser Ltd in Shenzhen, China. The five varieties were Lanfei, Jinfei, Liufei, Guifei and Liujiafei, which included 30 from Lanfei and 40 of the other four kinds of fertilisers. The contents of N, P and K were determined chemically and these results were used to develop NIR spectroscopic methods.

NIR spectral acquisition

The NIR spectrometer used in this experiment was a commercially available 360N (Nicolet Co., USA). The spectral data of N, P and K were generated in the wavenumber range from 10000 cm^{-1}–4000 cm^{-1}. The original spectra are shown in Figures 1 and 2.

Table 1. The range of N, P, K contents in compound fertiliser.

	Calibration			Prediction		
	N (%)	P (%)	K (%)	N (%)	P (%)	K (%)
Lanfei range	14.90–16.92	4.43–5.60	8.16–11.63	15.18–16.38	4.55–5.00	8.26–11.36
Average	15.72	4.82	10.08	15.65	—	10.10
Others range	13.27–17.43	—	6.48–10.46	13.82–16.90	—	6.52–10.42
Average	15.29	—	8.30	15.46	—	8.53

Table 2. The main parameters of N, P and K in compound fertiliser.

	Lanfei				others			
	n	Range (cm^{-1})	*R*	*SE*(%)	*n*	Range (cm^{-1})	*R*	*SE* (%)
N	22	5989–5565	0.9925	0.22	3	9750–5250	0.9913	0.50
P	22	10040–5800	0.9731	0.10	—	—	—	—
K	22	8320–7350	0.9973	0.33	3	9950–7780	0.9896	1.20

Data analysis

Partial least square regression (PLSR) analysis was conducted using the Omnic 5.1 data analysis software (Nicolet Co.).

Results and discussion

The analysis characteristics of the samples and original spectra are given in Table 1. The best results were observed using the second derivative and PLSR analysis for N, P and K, which not only had good relationship but also a low standard error of prediction. The main parameters of N, P and K in compound fertiliser using NIR are presented in Table 2.

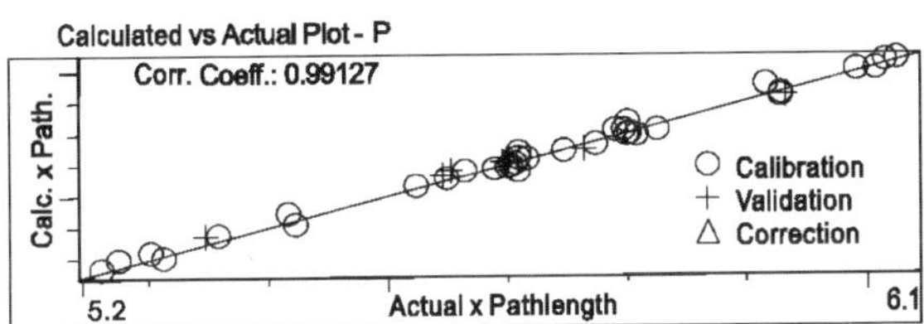

Figure 3. NIR calibration equation for nitrogen in four fertilisers.

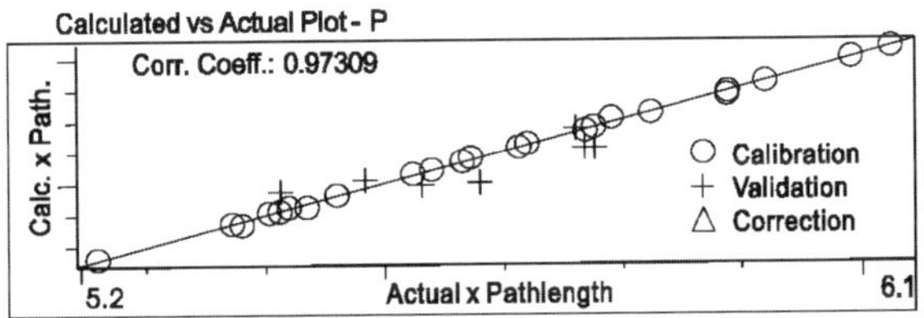

Figure 4. NIR calibration equation for phosphorus in Lanfei.

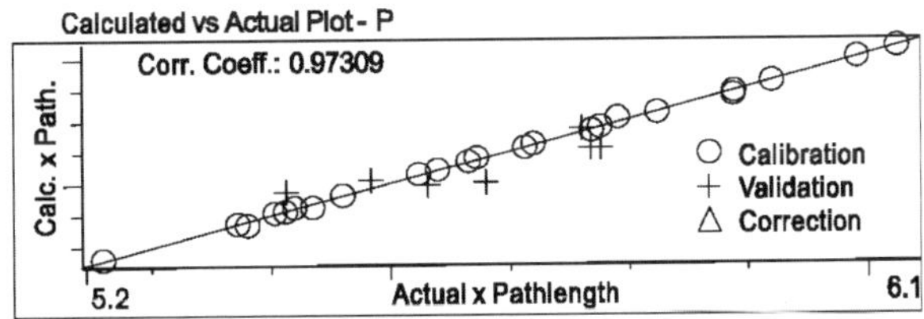

Figure 5. NIR calibration equation for potassium in Lanfei.

Table 3. Prediction results using NIR in Lanfei.

	N (%)			P (%)			K (%)		
No	Lab	NIR	residual	Lab	NIR	residual	Lab	NIR	residual
1	15.68	15.72	0.04	4.73	4.71	0.02	8.26	8.31	0.05
2	15.88	15.84	0.04	4.81	4.79	0.02	8.53	8.56	0.03
3	16.39	16.61	0.22	4.70	4.60	0.10	8.87	8.91	0.04
4	15.18	15.09	0.09	4.82	4.72	0.10	10.76	10.74	0.02
5	15.72	15.69	0.03	4.92	4.90	0.02	11.36	11.31	0.05
6	15.52	15.45	0.07	4.83	4.93	0.10	10.93	10.89	0.04
7	15.37	15.30	0.07	5.00	4.90	0.10	11.13	11.13	0.00
8	15.47	15.46	0.01	4.55	4.68	0.13	11.00	10.97	0.03

Table 4. Prediction results using NIR in the other four kinds fertiliser.

	N (%)			K (%)		
No	Lab	NIR	residual	Lab	NIR	residual
1	16.90	16.62	0.28	8.45	8.47	0.02
2	16.07	16.17	0.10	8.62	8.46	0.16
3	15.40	15.48	0.08	8.71	8.65	0.06
4	15.43	15.44	0.01	6.52	6.67	0.15
5	13.82	13.95	0.13	7.46	7.43	0.03
6	15.10	15.24	0.14	7.85	7.74	0.11
7	15.83	15.61	0.22	8.21	8.21	0.00
8	15.13	15.09	0.03	9.62	9.70	0.08
9	15.08	15.03	0.05	9.48	9.50	0.02
10	15.90	16.14	0.24	10.42	10.38	0.04

Figures 3 to 5 show the NIR regression equations developed for N, P and K in compound fertiliser samples respectively.

Validation statistics from PLSR analysis, comparing the results of chemical analysis with those predicted from NIR analysis, are shown in Tables 3 and 4.

Conclusions

This study has shown that NIR techniques can be used to rapidly and accurately determine the N, P and K nutrient components in compound fertiliser samples. Compared with the national standard methods, the results obtained in these components were multi-correlative coefficients of 0.9913–0.9925; 0.9731; 0.9896–0.9973 correspondingly and the standard errors for the test samples are 0.01–0.28; 0.02–0.13; 0.00–0.16 respectively. The NIR techniques have been successfully used in the Batian fertiliser plant for quality control.

References

1. B.G. Osborne, T. Fearn and P.H. Hindle, *Practical NIR Spectroscopy with Application in Food and Beverage Analysis.* Longman Scientific & Technical, Harlow, Essex, UK (1993).
2. A.M.C. Davies and R. Giangiacomo (Eds), *Near Infrared Spectroscopy: Proceedings of the 9th International Conference.* NIR Publications, Chichester, UK (2000).
3. T. Jin and L. Liu, *J. Near Infrared Spectrosc.* **3,** 89 (1995).
4. T. Jin, L. Liu, X. Zhou and D. Wang, in *Near Infrared Spectroscopy: Proceedings of the 9th International Conference,* Ed by A.M.C. Davies and R. Giangiacomo. NIR Publications, Chichester, UK (2000).

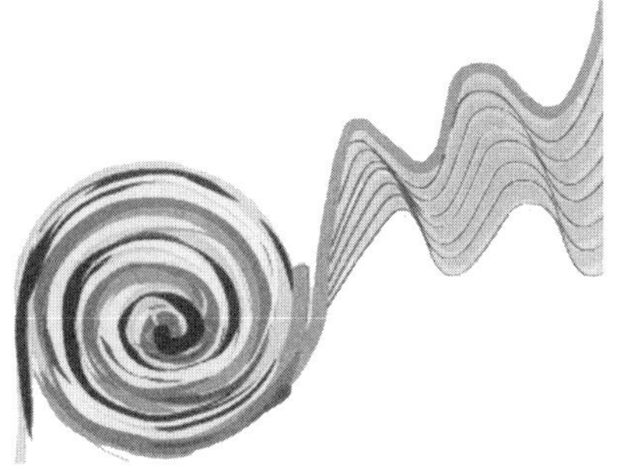

Polymers and Chemicals

Application of time-of-flight near infrared spectroscopy to wood

Satoru Tsuchikawa[a] and Shigeaki Tsutsumi[b]

[a]*Graduate School of Bioagricultural Sciences, Nagoya University, Nagoya 464-8601, Japan*

[b]*Department of Mechanical Engineering, School of Engineering, Fukui University of Technology, Fukui 910-8505, Japan*

Introduction

The behaviour of light propagation in wood is directly affected by the cellular structure. Recently, we constructed the optical characteristic models to provide a background and framework for developing a non-destructive measurement system applicable to such biological materials.[1,2] Furthermore, we examined the non-destructive quantitative assessment of moisture and physical conditions in bulky wood by using a near infrared (NIR) spectrophotometer with low output power. The calibration equations for moisture content, orientation of fibre, etc. with sufficient precision were found from a multi-term linear regression analysis.[3,4] However, the information from such an optical method measuring the diffuse reflectance spectra was confined to part of the subsurface layer of a sample.

To find the wide application of NIR to a wooden article (i.e. timber, lumber or furniture), it is important to establish a high power measurement system in which the measurable sample thickness is larger than that in the traditional apparatus. In this study, an advanced step forward in this investigation, an optical measurement system which was mainly composed of a parametric tuneable laser and a near infrared photoelectric multiplier, was introduced at the viewpoint of the time-of-flight near infrared (ToF-NIR) instrument.[5,6] This system combines the best features of the spectrophotometer and the laser beam, and more advantageously, the time-resolved profile of transmitted output power could be measured sensitively in nanoseconds. The combined effects of the cellular structure of wood samples, the wavelength of the laser beam and the detection position of transmitted light on the time-resolved profiles were investigated in detail.

Materials and method

The wood samples used were Sitka spruce (soft wood) that had an oven-dried density of 0.47 g cm^{-3} and Beech (hard wood) that had an oven-dried density of 0.68 g cm^{-3}.

The wavelength of the pulsed laser λ was tuned from 700 nm to 1300 nm by the optical parametric oscillation of a BBO (β-BaB_2O_4) crystal.[7,8] The transmitted output power was measured by an NIR photoelectric multiplier having a spectral response ranging between 300 nm and 1700 nm through an optical fibre cable having a diameter of 7 mm. The sampling time and the average number of the transmitted output power were 100 ns and 300 times, respectively.

The normalised time-resolved profiles of Sitka spruce (90 mm in width, 34 mm in length) are shown in Figure 1. A time-resolved profile as the reference was selected by complying with the measuring conditions. In this figure, the time-resolved profile at $d = 1$ mm is taken as a reference. As shown Figure 1, the variations of the peak maxima, the time delay of peak maxima and the variation of

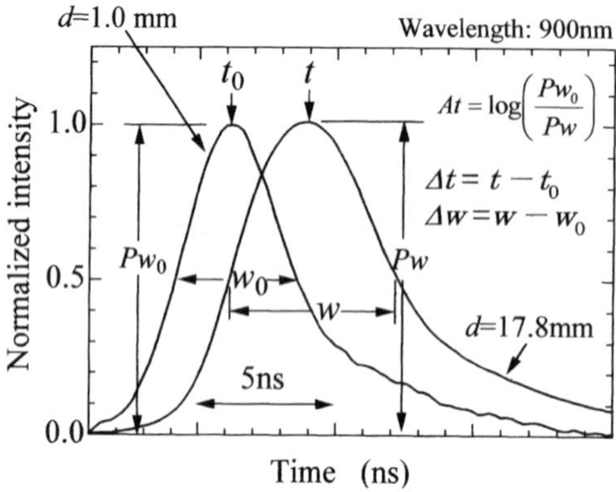

Figure 1. Normalised time-resolved profiles.

full width at half maximum of the profile were examined at the required measuring conditions, respectively.

Results

First, we examined the variation of time-resolved profile with sample thickness d. Figure 2 indicates the spectral variation of each optical parameter at the opposite face. It is commonly known that At increases monotonically as d increases. On the other hand, Δt also increases gradually as d increases. However, its wavelength dependency is inversely related to At. At the absorption band of water around at 1000 nm

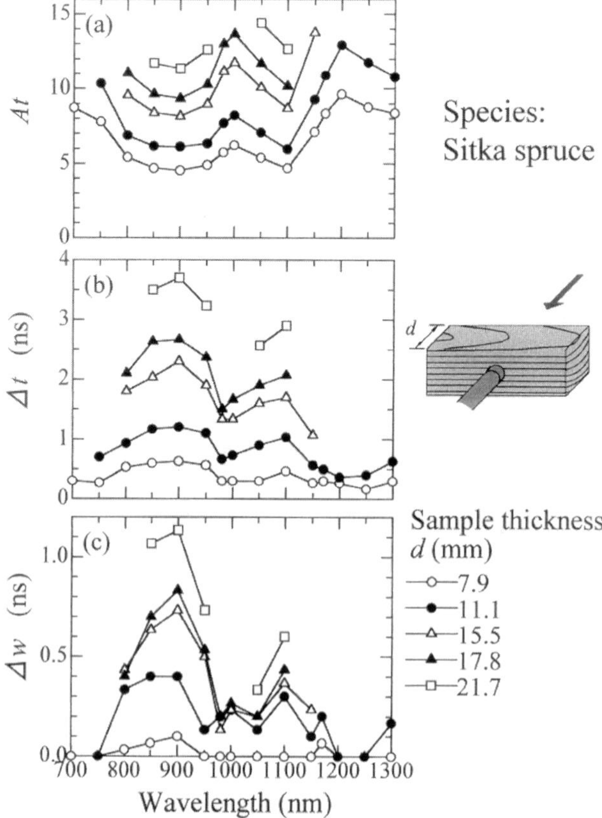

Figure 2. Spectral variation of the attenuance At, the time delay of peak maxima Δt and the variation of full width at half maximum Δw. Wood species: Sitka spruce. Detection position: Centre of opposite face of the sample.

and 1200 nm, the transmitted light is attenuated greatly and the light path time through a sample is becoming faster. As is known from Figure 2(c), Δw also shows the same trend as Δt. These results directly mean that the diffusely-scattered light having relatively longer optical path length is attenuated or absorbed by water and, therefore, the time delay of peak maxima shortens substantially.

The light scattering condition in wood sample was estimated from the product of Δt and Δw. Figure 3 shows the variation of $\Delta t \times \Delta w$ with sample thickness d or irradiation point l at representative wavelengths. In every case, $\Delta t \times \Delta w$ increases exponentially with d or l. However, their absolute value varied greatly with wood species or wavelength. When the transmitted light was detected at the opposite face [see Figure 3(a)], $\Delta t \times \Delta w$ for Beech shows larger value than those for Sitka spruce at every wavelength band. On the other hand, in the case of the measurement at the side face [see Figure 3(b)], $\Delta t \times \Delta w$ for Sitka spruce show larger value. The soft wood (Sitka spruce) is mainly composed of the tracheid with hollow fibre, where the scattered light from wood substance and that from the lumen coexist. In particular, there may be considerable light scattering in the lumen of tracheid, which performs as multiple specular reflection and be easy to propagate along the length of the wood fibre. In the case of Beech with little space between wood fibres, variations in the light propagation along the length of wood fibre may be small, whereas the time delay related to the high density of wood substance is remarkable. When we apply ToF-NIR to the cellular structural materials like wood, it is very important to give attention to the difference in the light scattering within the cell wall and the multiple specular-like reflections between cell walls.

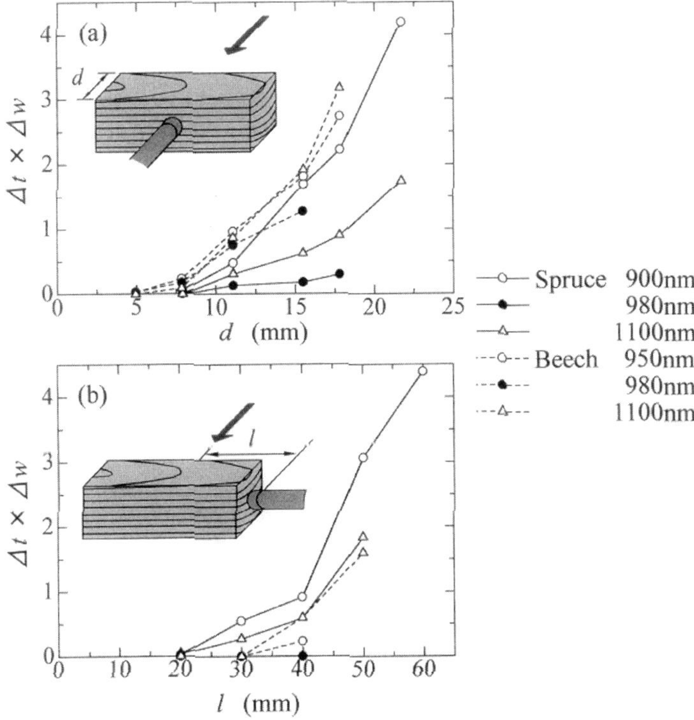

Figure 3. Variation of $\Delta t \times \Delta w$ with sample thickness or detection position.

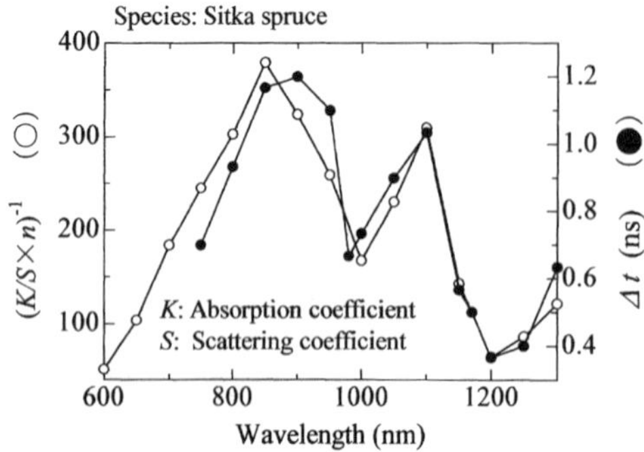

Figure 4. Spectral curves of $(K/S \times n)^{-1}$ and Δt.

We tried to express the characteristics of the time-resolved profile on the basis of the optical parameters for light propagation determined by the previous studies, which were absorption coefficient K and scattering coefficient S from the Kubelka–Munk theory[1] and n from nth power cosine model of radiant intensity.[2] The anisotropy of light scattering in the sample should be expressed by this parameter of n, whereas S and K may be related to the optical characteristics of wood substance, i.e. The cell wall itself. Figure 4 shows the spectral curves of $(K / S \times n)^{-1}$ and Δt. The species used was Sitka spruce. $(K/S \times n)^{-1}$ and Δt showed similar tendency to the variation of wavelength, although the individual variation of K / S or n with the wavelength was not the same as that of Δt. In this way, the time-of-peak maxima reflect directly the interrelationship between the light-scattering and -absorption characteristics of the wood substance and the anisotropic directional characteristics caused by the cellular structure. Such primary analysis suggests that the light propagation, which is governed by the directional characteristics and the absorption/scattering characteristics, can be easily found by the variation of the time-resolved profile synthetically. Thus, our optical characteristics models, which express the light propagation in the cellular structural material, are also supported by ToF-NIR.

Conclusion

The newly constructed optical measurement system, which was mainly composed of a parametric tuneable laser and a near infrared photoelectric multiplier, was introduced to clarify the optical characteristics of wood from the viewpoint of time-of-flight near infrared spectroscopy (ToF-NIR). The combined effects of the cellular structure of wood samples, the wavelength of the laser beam and the detection position of transmitted light on the time-resolved profiles were investigated in detail. The arrangement and orientation of cells were directly related to the time delay of light propagation. Applying ToF-NIR to the cellular structural materials, it is very important to give attention to the difference in the light scattering within the cell wall and that caused by the multiple specular-like reflections between the cell walls. It was also known from the optical parameters for light propagation that the variation of the time-resolved profile was governed by the combination of light-absorbing and -scattering condition and the degree of anisotropy. Thus our optical models were experimentally supported by ToF-NIR. These basic data will be essential for this system to put an in-process measurement system in the wood industry to practical use.

References

1. S. Tsuchikawa, K. Hayashi and S. Tsutsumi, *Appl. Spectrosc.* **50,** 1117 (1996).
2. S. Tsuchikawa and S. Tsutsumi, *Appl. Spectrosc.* **53,** 233 (1999).
3. S. Tsuchikawa, M. Torii and S. Tsutsumi, *Mokuzai Gakkaishi* **42,** 743 (1996).
4. S. Tsuchikawa and S. Tsutsumi, *Mokuzai Gakkaishi* **43,** 149 (1997).
5. L. Leonardi and D.H. Burns, *Appl. Spectrosc.* **53,** 628 (1999).
6. L. Leonardi and D.H. Burns, *Appl. Spectrosc.* **53,** 637 (1999).
7. A. Yariv, *Quantum Electronics,* 2nd Edn. John Wiley & Sons, New York, USA (1975).
8. P.G. Harper and B.S. Wherrett (Eds), *Nonlinear Optics.* Academic Press, New York, USA (1977).

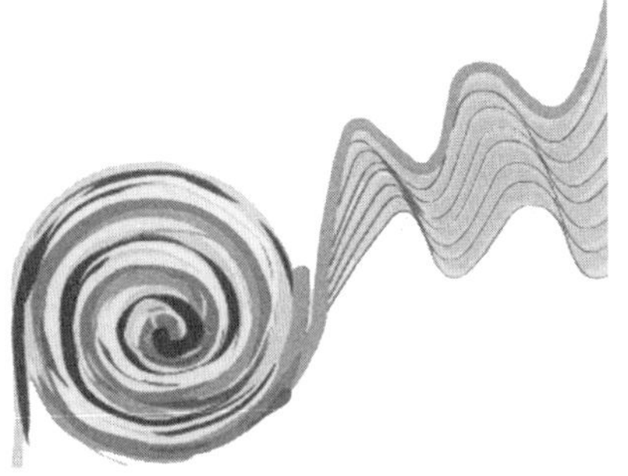

Textiles and Petrochemistry

Evaluation of near infrared spectroscopy for assessing physical and chemical characteristics of linen weft yarn

L. Whiteside, K. Kernaghan and H.S.S. Sharma

The Department of Applied Plant Science, School of Agriculture and Food Science, Queen's University of Belfast and The Department of Agriculture and Rural Development for Northern Ireland, Newforge Lane, Belfast BT9 5PX, UK

Introduction

Conventional chemical and physical methods of assessing yarn quality, although effective, are highly labour intensive. Within the linen industry there is a need for novel, rapid, non-destructive, instrumental means of predicting yarn quality. Current yarn production procedures are based on the blending of highly heterogeneous raw fibre supplies in order to enhance yarn consistency. Changes occurring in response to chemical processing, such as boiling, bleaching and dyeing can be detected and monitored by derivative thermogravimetry (DTG) and near infrared (NIR) spectroscopy. These methods allow sample homogeneity and quality parameters to be monitored during the various stages of processing.[1] Preliminary results suggest that NIR spectroscopy could also be useful during fabric finishing for assessing crease recovery and abrasion resistance performance.[2] The results from conventional physical and chemical analyses have provided supporting evidence for this. NIR spectra can be used to form predictive models allowing rapid assessment of yarn quality for physical and chemical characteristics.

Methodology

The yarn samples supplied for quality analysis were provided by Ferguson's Irish Linen (FIL), Banbridge, N. Ireland. All materials were sub-sampled (50 g) and conditioned for a minimum of 24 h at 20°C and 65% RH before analysis.

Derivative thermogravimetric analysis

Yarn samples were further sub-sampled (10 g) with representative material being cut to less than 1 mm length using a pair of serrated scissors. Thermogravimetric analysis using a robot (TSO 801 RO) controlled TGA/SDTA 851^e (Mettler Toledo) instrument over a temperature range of 35–600°C at a 20°C rise min^{-1} was carried out using 3.0–3.2 mg samples with a minimum of three replicates per treatment. The system was flushed with air at the rate of 20 ml min^{-1} and the results analysed using Star (Mettler Toledo) software to calculate derivative peaks with their weight loss in the primary (220–400°C) and secondary decomposition (400–600°C) bands.

DTG thermograms, when overlaid, revealed that there was a high degree of similarity between individual sub-sample replicates. As DTG analysis can be performed on a variety of materials including raw fibre, roving and yarn, it may therefore be used to predict quality parameters.

Near infrared spectroscopy

Sub-samples of yarn were packed into closed cups comprising a quartz-glass cover and scanned ten times after re-packing with a Visual-NIR spectrophotometer (Foss NIRS 6500) at 2 nm intervals over the range 400–2498 nm producing 1050 data points. A band pass value of 10 nm was used with wavelength accuracy of ± 0.5 nm. Reflectance readings were converted to absorbance (A) values using the equation: $A = \log(1/R)$. Infrasoft International software was used to collect the data and the raw spectral data exported for chemometric analysis (Unscrambler version 7.0, Camo, Trondheim, Norway).

Results and discussion

Derivative thermogravimetry

The most influential factor on yarn quality is processing sequence used during spinning. DTG analysis was performed on yarn samples supplied by three different spinners, the thermal profiles of which are presented in (Figure 1). Although maximum primary peak decomposition temperatures of the three test materials proved to be similar (360–363°C), primary peak weight loss of sample A was much higher (76.6%) than samples B (69.1%) and C (68.4%). Significant differences in terms of secondary peak weight loss, height and decomposition temperature were also observed. Peak two weight loss of Sample A was lower (16.4%) than B (24.1%) and C (25.73%), however, decomposition temperatures of B and C although similar (472°C), were much lower than sample A (492°C). The magnitude of the primary and secondary peak height and width parameters of yarn A compared to B and C would suggest that a greater quantity of non-cellulosic fractions were removed during the processing of A compared to the other two. This would infer that a more efficient processing treatment had been applied whilst spinning by the suppliers of this yarn. These distinct differences in yarn thermal profiles enable the prediction of valuable information on the process history of unknown yarn samples [3].

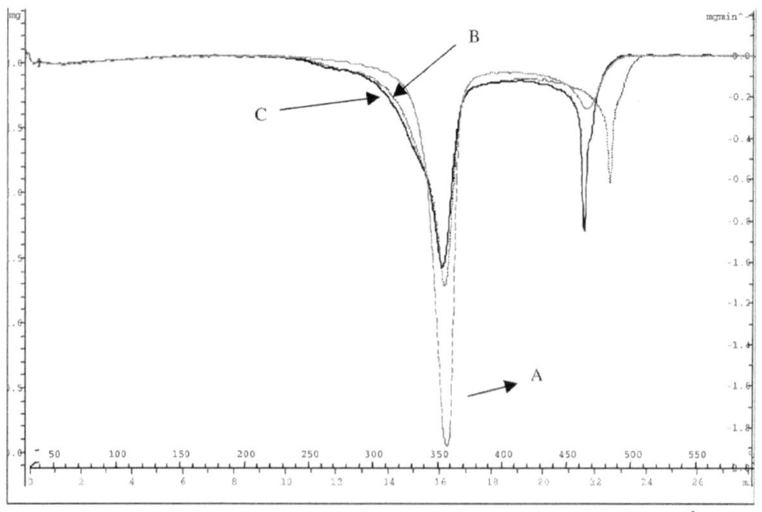

Figure 1. Overlay of 1st derivative thermograms of (a) good, (b) medium and (c) poor weft yarns.

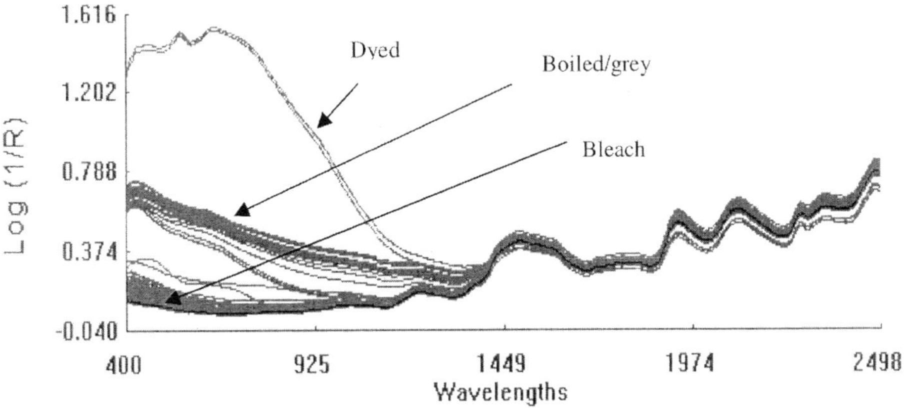

Figure 2. Raw NIR spectra indicating differences between boiled, bleached and dyed yarns.

Near infrared spectroscopy

Differences in yarn treatment and finishing processes have a direct effect on yarn performance characteristics such as tenacity and breaks during weaving. A variety of yarns that had been previously treated by alkali boiling, bleaching and dyeing were assessed using NIR. As shown in Figure 2, there were significant spectral differences between the three treated yarn types—both in the visible and near infrared wavelength bands. Bleached yarns have significantly lower reflectance values compared to dyed yarns in the visible spectral range. Within the near infrared regions the differences are much less clear and further manipulation of the raw spectra, for example by transformation, are necessary in order to highlight the variations.

In order to show how differences in key quality parameters are reflected by the spectral data, an extensive analysis of weft yarns using NIR spectroscopy was performed and a database of yarns with different known quality characteristics developed. Three grades of weft yarn were assessed using NIR

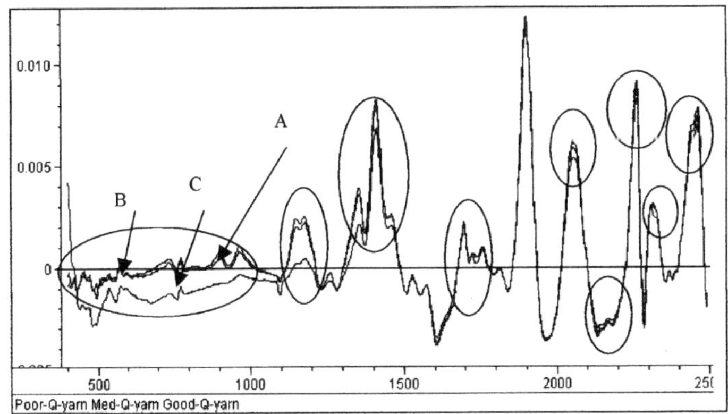

Table 1. Preliminary NIR calibrations for the assessment of weft yarn quality.

Parameters	n	SEC	R^2	SECV	R^2
Tenacity cN/Tex	46	0.01	0.98	1.21	0.75
CV-Mass (%)	48	1.11	0.78	1.41	0.62
Neps 200/km	45	293.60	0.94	505.50	0.81
Elongation (%)	46	0.09	0.89	0.01	0.71
Degree of polymerisation	47	348.30	0.84	471.24	0.70
Cautic weight loss (CWL) (%)	50	2.19	0.89	2.99	0.78
Calcium (ppm)	40	545.00	0.86	678.60	0.78
Nitrogen (%)	32	0.01	0.98	0.001	0.96
Carbon (%)	48	0.36	0.90	0.464	0.83
Ash (%)	36	0.13	0.85	0.72	0.17

SEC: standard error of calibration
SECV: standard error of cross-validation

spectroscopy. The raw spectral data were transformed and the spectra presented in Figure 3. The greatest variation in yarn type was observed in the visible spectral region due to different bleaching treatments used, however additional differences were also observed in the 1100–1200, 1300–1400, 2060, 2146–220, 2312 and 2450 nm wavelengths. Partial least squares calibrations for key physical properties, such as tenacity, CV-mass, neps and elongation, and chemical characteristics, such as degree of polymerisation, caustic weight loss, calcium, nitrogen, carbon and ash, of weft yarn have been developed using the visible and NIR spectra (Table 1). However, r^2 values for both calibration and cross-validation of the parameters need to be improved with additional sample sets. The results have shown that spectroscopy has the potential for assessing yarn quality rapidly.

Conclusion

Quality assessment system using spectroscopy to monitor key yarn parameters is feasible and hence enhancing yarn and fabric quality optimisation during production is possible, provided robust predictive models can be developed. Such a system, supported by additional DTG analysis, would enable rapid, accurate prediction of yarn performance during all stages of processing and thus reducing problems arising from variation in raw material quality.

Acknowledgements

This work is funded by the EU BRITE-EURAM project (BE97-4744). We wish to thank the European Union for supporting the research programme and also CTC (Italy), TDL (France) LCN (Italy) and FIL (UK) for providing samples and some of the test results.

References

1. G. Faughey and H.S.S. Sharma, *J. Near Infrared Spectrosc.* **8,** 61 (2000).
2. D. McCall, K. Kernaghan and H.S.S Sharma, *J. of Appl. Pol. Sci.* **82,** 1886 (2002).
3. H.S.S Sharma and K Kernaghan, *Thermochimica Acta* **132,** 101 (1988).

Quantitative in-line measurements on papers

Angela Schmidt and Klaus Schorb
Bruker Optik GmbH, Rudolf-Plank-Str. 23, D-76275 Ettlingen, Germany

Renate Motsch and Ulli Nägele
Herma GmbH, Fabrikstr. 16, D-70794 Filderstadt, Germany

Introduction

In the paper industry, as well as in the analysis of paper related goods, near infrared (NIR) methods have been described for the control of incoming materials.[1] Quantitative models for the components in papers such as mechanical pulp (TMP, CTMP = chemically/thermally mechanical pulp, containing lignin) or chemical pulp (= classical types of cellulose) and the various detectable filler materials (clay, talc, chalk) or coatings thereof have been shown.[1,2] Much more sophisticated applications involve measurement of additives in papers, for example, sizing agents (resin size, AKD[3]) and wet or dry strength materials.[3] Moreover, the characterisation of paper by their physical properties, including absolute moisture,[4] ash content,[4] basic weight[4] and thickness,[4] as well as tear index,[5,6] tearing strength,[5] bursting strength,[6] tensile stretch,[5] tensile strength,[6] fibre length,[5] debonding energy,[6] hydrophobicity (Cobb by Ultrasound),[6] wettability[6] and printability[6] is possible. Apart from these applications, NIR spectroscopy could be a new tool to classify waste paper during the screening process. In this study, the feasibility to set up in-line NIR methods for products converted from papers is presented. Silicone coated papers are widely used as release liners, with the major application being stick-on labels of all types. Besides the quantification of coat weights from adhesives, it is especially important to control closely a fixed amount of around 1 g m^{-2} silicone applied, because too little will be ineffective and too much is a waste of expensive material.[7,8] To make direct implementations for product consistency and efficiency of the process, in-line monitoring of coat weights by realistic calibration models is required. All the variations from paper, due to the substrate, colour or opacity and basic weight, have to be considered in order for the measurement procedure to be robust enough for paper change at the web. The creation of a comprehensive, but selective, calibration model for silicone (Figure 1) has already been successfully made with an at-line FT-NIR spectrometer based on the VECTOR 22/N, Bruker Optik GmbH (data set C, Table 1). To meet the requirements in-line during label stock production in

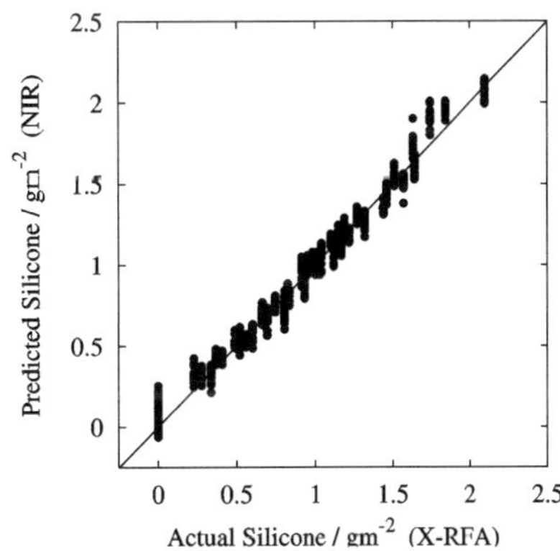

Figure 1. Comprehensive calibration model for silicone (set C).

Table 1. Validation results of quantitative models for the determination of silicone coat weights on glassine papers.

Quantitative Model	Kind of Validation	Silicone Coat weight range gm^{-2}	Number of spectra	Data pre-processing	Frequency cm^{-1}	Number of PLS terms	Coefficient of Determination R^2	*RMS* Error Validation g m^{-2}
At-line Set A Figure 1	Test-Set	0–2	1,320 individual	1st derivative 5 point smoothing	7501–4015	7	97.9 %	0.07
In-line Set B Figure 5(a)	Cross	0.75–1.43	12 average	Vector normalisation + 1st derivative (17 point smoothing)	6101–5446	6	99.8 %	0.01
In-line Set C Figure 5(b)		0.77–1.45	13 average			2	97.2 %	0.03
In-line Set C Figure 6(a)			326 individual			4	96.9 %	0.04
In-line Set C Figure 6(b)			350 individual			9	95.5 %	0.06

the coating machine, a new concept for an FT-NIR spectrometer was designed to perform the measurements of moving paper in a non-contact mode.

Materials and methods

Instrument

The new FT-NIR spectrometer, MATRIX-E, was developed by Bruker Optik GmbH, Ettlingen, Germany. The MATRIX-E is equipped with four lamps to illuminate the moving paper web in the machine. The angle of incidence was chosen so that the contribution due to specular reflection is minimized. The extended illuminated area of 25 diameter (Figure 2) is suitable, in general, to scan large and inhomogeneous samples during movement. Due to the small thickness of the backing papers (= glassine papers) for labels, the light penetrates the sample twice as it is reflected by the roll made from metals such as aluminum or chrome. This reflection introduces a matrix-effect in the NIR spectra and it is important that the kind of metal used is specified. After interac-

Figure 2. FT-NIR spectrometer MATRIX-E illuminating the sampling area.

tion with the sample, the diffusely-scattered light returns to the instrument and is directed to the detector (InGaAs Peltier cooled, operating range 12,800 to 4,000 cm^{-1} or 780 to 2,500 nm). The optimum spatial distance from the measurement window to the sample is 170 mm but slight differences are not relevant to acquire reproducible spectra. Spectralon, built inside the instrument, is used as an internal reference material. The internal background acquisition is performed by closing the instrument with a shutter automatically and making installations in the production environment more stable.

NIR Experiment

The scan time for in-line measurements of the MATRIX-E was set to five seconds per NIR spectrum (32 scans, spectral resolution 16 cm^{-1}, scanner velocity 20 kHz). For internal reference acquisition, the shutter closes at variable time intervals (for example, every 30 minutes) and records a background spectrum. The velocity of the paper machine was 400 m min^{-1}, which corresponds to a length of 33.3 m paper per spectrum. If the machine velocity is different, or becomes faster in the future, the paper will run more quickly through the dryers and changes will result in its temperature or moisture content introducing additional matrix-effects in the NIR spectra. However to develop robust models, "artificial spectral variations" can be created by changing the moisture content in a paper sample but using the same reference value for silicone coat weight (as reported in a previous NIR study albeit from the agricultural sector).[9]

For this in-line experiment, 13 coat weights from silicone ranging between 0.73 to 1.45 g m^{-2} were set at the coating machine supplying one type of glassine paper. The glassine paper was measured before the initial coating with silicone in the machine as well. Influences from vibrations caused by the movement of the paper are kept to a minimum as the scanning is performed on a roll. Moreover, measurements recorded with the MATRIX-E are insensitive to vibrations by the machinery itself. The same observation was made with interference from high voltage equipment. A movement of the MATRIX-E across the web is considered in the future to ensure a flat, uniform coating.

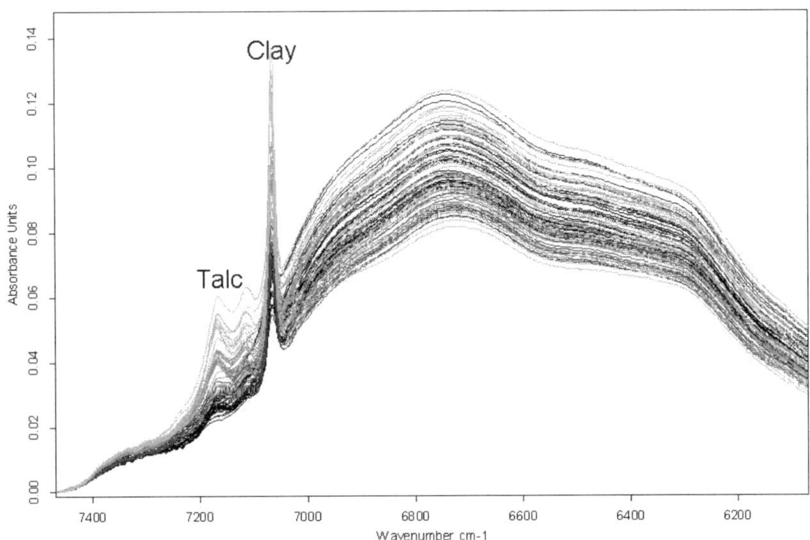

Figure 3. NIR spectra from silicate coated papers showing significant OH absorption bands.

Reference analysis

Sampling methods for the element silicon (*Si*) by X-ray technology are available and require calibrations as well. This XRF analysis is rather time-consuming (two minutes in total) and depends on the surrounding temperature. Moreover, it cannot distinguish between silicate present in clay- or talc- coated papers (Figure 3) and organic silicone compounds provided for the coating. Clay-coated papers are often used as a backing paper for labels and their silicone coat weight is either not examined at all or could be calculated indirectly from the central element {aluminum in clay $Al_2[Si_2O_5](OH)_4$}. In addition, the delay between actual production and off-line XRF-analysis prevents immediate actions onto the process.

Figure 4. MATRIX-E installed in label stock conversion facility at the paper roll.

After in-line measurements with the MATRIX-E, lab sheets from the end product of self-adhesive laminate were collected from each reel produced with a certain setting of silicone. These sheets were taken according to the MATRIX-E measurement position along the web (Figure 4). On the same day, a benchtop XRF analyser (Lab X 3500, Oxford Instruments Analytical, High Wycombe, UK) was used for a single analysis of the silicone value (data set A, Table 2). The same lab sheets were examined again carefully by the same Lab X 3500 a few days later (data set B, Table 2). From each sheet, six replicates alternating with an appropriate standard sample were measured. These six replicates were offset-corrected according to the standard sample and then the averages were calculated. In Table 2 the reference values for the same samples from both XRF analyses are given. Obviously the reference values depend on the time of the XRF measurements as their differences are larger than the absolute deviation between the replicates.

Table 2. Reference values of silicone from two different XRF analyses: 1st set A (single analysis, same day), 2nd set B (averages from six replicates, a few days later).

Machine Setting	1st XRF analysis	2nd XRF analysis
Silicone Coat weight range gm^{-2}	0.75 –1.43	0.77 – 1.45
0	0	0
1	0.75	0.77 ± 0.03
2	0.76	0.78 ± 0.01
3	0.79	0.80 ± 0.01
4	0.84	0.88 ± 0.02
5	0.88	0.92 ± 0.01
6	0.89	0.94 ± 0.02
7	0.97	1.02 ± 0.01
8	0.98	1.02 ± 0.02
9	1.00	1.06 ± 0.01
10	1.07	1.12 ± 0.00
11	1.16	1.18 ± 0.01
12	1.27	1.30 ± 0.01
13	1.43	1.45 ± 0.02

Results and discussion

We compared the cross-validation results from two PLS-1-models (optimised with OPUS NT/QUANT Software by Bruker Optik GmbH) using the same average spectra for 13 settings but the two available reference values (Table 2) either from the first XRF-analysis on the same day and from the second XRF analysis a few

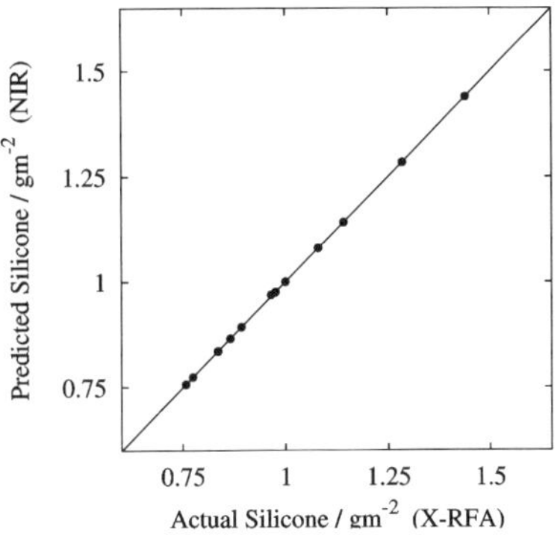

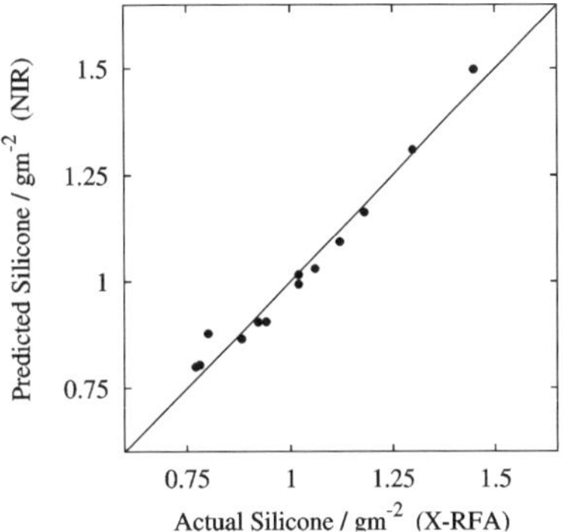

Figure 5. Comparison of PLS models from average spectra using different reference values.

days later (Figure 5). From the first model based on the single reference values, one outlier (Nr.3) was excluded. Nevertheless, this first model proves the feasibility and shows a better correlation R^2 and predictive error *RMSECV* (Table 1) from cross-validation. Because the first XRF measurements were performed for each sample in a quick succession, they are not significantly affected by temperature changes. The second XRF analysis later is more susceptible to the temperature effect and leads to reduced accuracy in the PLS model.

In another comparison, two PLS-1 models from the individual spectra from in-line measurements were built, either with or without consideration of a zero point, from non- coated glassine paper. If the zero point is excluded, the quality of the model will improve, especially in terms of PLS-Vectors used (Figure 6, Table 1). The indication of the non-coated glassine paper spectra as outliers by the software can be explained by two reasons: firstly glassine paper has different scattering properties and no reflective silicone coating, which is even enhancing the signal (Figure 7) and, secondly, the zero point is too remote from the first coat weight setting of 0.75 g m^{-2} to be predicted reliably by the remaining set. As a consequence, the non-coated sample cannot be used for offset adjustments of a paper-type model after reel-to-reel-change.[7,8] However the paper type in one silicone coating machine will be changed up to 20 times per day and is currently supplied manually in a mode called "flying-web-change". This

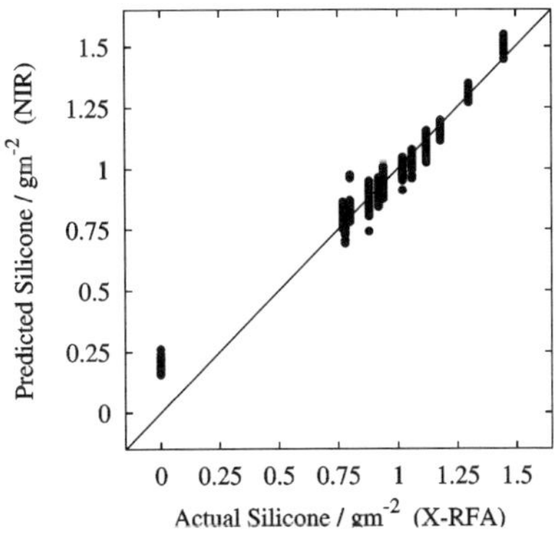

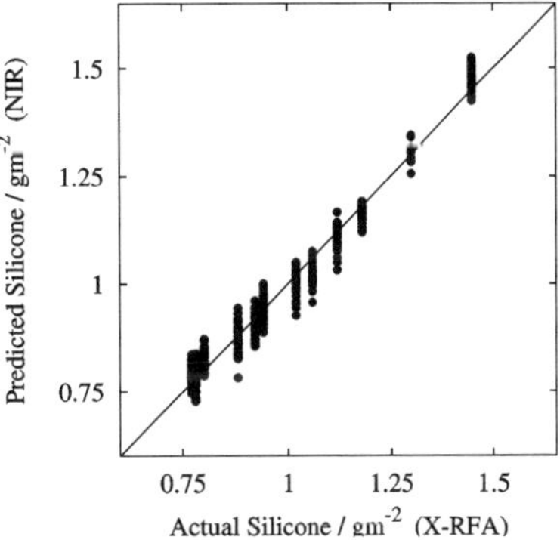

Figure 6. Comparison of PLS-models from individual spectra (a) with and (b) without zero.

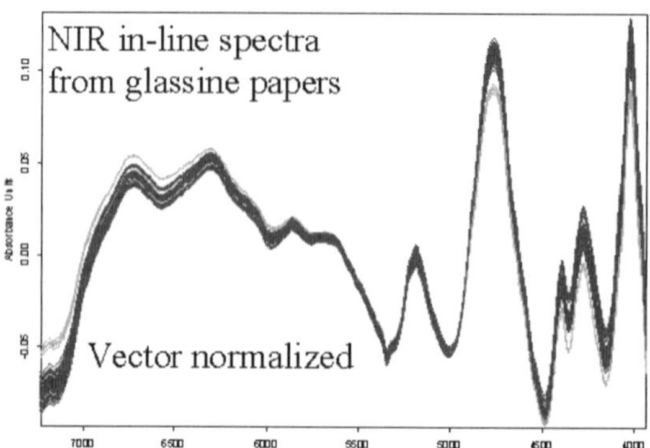

Figure 7. Differences in NIR-spectra from papers with and without reflecting silicone (light coloured spectra).

explains, once again, the demand for comprehensive calibration models without the need for offset adjustments.

Conclusions

The present study shows the possibility of in-line calibrations for one paper type from low coat weights of silicone around 1 g m^{-2} during label stock conversion with a new FT-NIR spectrometer, MATRIX-E, in a non-contact mode. After an immediate reference analysis of lab sheets by X-ray technology required for PLS-evaluation, the statistical performances of the predictive model from averaged in-line spectra are $R^2 = 99.8\%$ and $RMSECV = 0.01$ g m^{-2} for silicone. The inclusion of a zero point from the non-coated glassine paper is not recommended. NIR spectra from other glassine paper types, including all the variations described above, are acquired in-line. The data are used to build one comprehensive in-line calibration model to meet the requirements for frequent paper change at the web, allowing continuous monitoring of an absolute value for coat weights from silicone. A realistic calibration model can be transported[10] to other MATRIX-E instruments to be installed in similar label stock production facilities without the need for instrument standardisation between the FT-NIR spectrometers.

References

1. Chr. Reiter, B. Reinhardt, B. Borchers, P. Plew, in *Das Papier* **10(A),** (1999).
2. H. Furumoto, U. Lampe, H. Meixner, Chr. Roth, in *Das Papier* **4,** (1999).
3. L. Renberg, in European Patent EP 0 760 094 B1, (1995).
4. I. Betz, in *International Munich Paper Symposium*, (2001) Lecture Material available from www.ivp.org/symposium.
5. B. Borchers, P.Plew, in *PTS – Research Report*, PTS-FB 14/2000 (2000).
6. L. Renberg, R. Olsson, in *European Patent EP 0 759 160 B1*, (1997).
7. J. Millard, in *Converter* **33(5),** (1996).
8. I. Benson, in *Converter* **33(8),** (1996).
9. P. Dardenne, G.Sinnaeve, L.Bollen, R. Biston, in *Leaping ahead with NIRS*, Ed by G.D. Batten, P.C. Flinn, L.A. Welsh and A.B. Blakeney, p.154 (1995).
10. J.-P. Conzen, T. Stadelmann, A. Schmidt, H. Weiler, T. Droz-Georget, A. Zilian, in *Near Infrared Spectroscopy,* Ed by A.M.C. Davies and R. Giangiacomo, NIR Publications, p. 221 (2000).

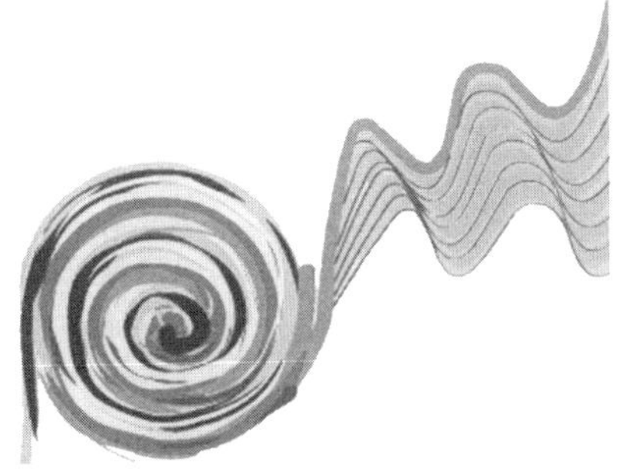

Biomedical

Antioxidative activities of some dietary fibres determined by near infrared emission spectroscopy

Nobutaka Suzuki,[*a] Takeshi Nagai,[b] Kazunari Tokunou,[b] Hiroshi Kusanagi,[b] Iwao Mizumoto,[c] Hiroko Matsuya,[d] Binkoh Yoda,[d] Toshiaki Itami,[e] Yukinori Takahashi,[e] Tateo Nomoto[f] and Akiya Kozawa[g]

[A]*GraduateSchool of Biosphere Sciences, Hiroshima University, Higashi-Hiroshima 739-8528, Japan*

[b]*Deptartment of Food Science, Shimonoseki University of Fisheries, Shimonoseki 759-6595, Japan*

[c]*Toyama National Institute of Maritime Technology, Shin-minato 933-0239, Japan*

[d]*Koriyama Women's University, Koriyama 963-85038, Japan*

[e]*Deptartment of Aquaculture, Shimonoseki University of Fisheries, Shimonoseki 759-6595, Japan*

[f]*Deptartment of Chemistry, Mie University, Tsu 514-8507, Japan*

[g]*ITE/IBA Research Institute, Ichinomiya 491-0806, Japan*

Introduction

We have been working on singlet oxygen, one of the active oxygen species, since 1988.[1] In this paper, we would like to describe the antioxidative activities of some dietary fibres determined by near infrared (NIR) emission spectroscopy as an application of singlet oxygen chemistry.

Singlet oxygen, superoxide, hydroxy radical and hydrogen peroxide are active oxygen species; in the medical field, ROOH, ROO, RO and HOX (X = Cl, Bl, or I) are also included in the term "active oxygen".

Active oxygen species are always generating in living bodies, either during the course of oxidative degradation of nutrients or of photosynthesis. Every living thing protects itself from the harmful effects of the active oxygen species using antioxidants, such as superoxide dismutase (SOD), β-carotene, catalase and vitamins C and E. Otherwise, they suffer cancer, inflammation or mutation and so on. As antioxidants cannot guard bodies perfectly, the result is ageing.
Antioxidants (AH_2) are thought to be radical scavengers or reducing agents.[1]

$$ROO^{\bullet} + AH_2 = ROOH + AH^{\bullet}$$

Dietary fibres are known to prevent cancer formation in the large intestine by their physical effects: (1) absorbing and holding water (faecal bulk increasing; promoting digestion and regulating stools);

(2) absorbing toxic organic compounds (chromatographic action); (3) absorbing metal ions (positive ion-exchanging capability) and (4) gelling capability to result in (a) activation of digestive tracts; (b) increasing bulk of faeces; (c) accelerating faecal passage through digestive tracts; (d) decreasing internal pressure in the digestive tracts; (e) controlling digestion and/or absorption of diet constitution and (f) affecting intestinal bacteria.

Fucoidan, a marine dietary fibre, was also shown by us to prevent infection of some pathogenic virus by its immunological-like activity.[2]

Some dietary fibres have been known to be decomposed by active oxygen species:[3,4] They might have antioxidative properties. Do dietary fibres act as antioxidants? We intend to clarify that some dietary fibres have antioxidative activity, which may prevent cancer. We applied NIR emission spectroscopy to measure the antioxidative activity of some dietary fibres against singlet oxygen.

Materials and methods

Samples

Fucoidan (Sigma, St. Louis, USA), pectin (apple; Sigma), dextran sulfates Na 5000 and 500000 (Wako, Osaka, Japan), ι- and κ-carrageenans (Aldrich, Milwaukee, USA), alginic acid Na (Grade NB-S; Kimitsu Chem. Ind., Tokyo, Japan), NaN_3 (Wako) and vitamin C (Wako) were used as purchased. Fucoidan extracted from Okinawan mozuku *Cladosiphon okamuranus* was generously supplied by Miyako Kagaku Co. Ltd., Tokyo, Japan.

NIR emission spectroscopy

Singlet oxygen is an excited state and gives a dual emission, (1) red light from a dimer and (2) NIR light from a monomer, when deactivated. The red emission is, however, often obscured by emission from the other CL species. Therefore, it cannot be relied on for evidence of singlet oxygen. On the other hand, the NIR emission has no hindering emissions around them. Therefore, when 1270 nm emission is found, you can say safely, singlet oxygen must be present. We have constructed an NIR emission spectrophotometer (Figure 1).[5]

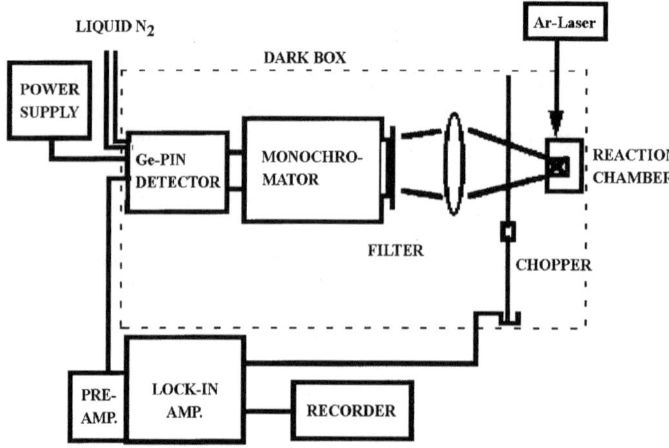

Figure 1. Diagram of the NIR emission spectrophotometer.

$$\text{Dye} + h\nu \rightarrow {}^{1}\text{Dye}^{*} \rightarrow \text{Dye}^{*}$$

$$\underset{\text{mol. oxygen}}{{}^{3}\text{Dye}^{*} + {}^{3}O_2} \rightarrow \text{Dye} + \underset{\text{singlet oxygen}}{{}^{1}O_2}$$

Scheme 1.

There are many ways to make singlet oxygen: the photosensitised oxygenation method is the most convenient way among them. When a dye absorbs light, it is excited to give a singlet excited state, which transforms, sometimes, to give a triplet excited state. The triplet exited state transfers its excited energy to molecular oxygen (triplet ground state) to give excited singlet oxygen (Scheme 1).

Some diseases are known to result from singlet oxygen generation in a living thing as shown below:
(1) a cat eating abalones loses his ears;
(2) some inherited light-hypersensitivity;
(3) as well as a medical treatment, called photodynamic cancer therapy.

Emission spectra of singlet oxygen generated from an aqueous solution of Rose of Bengal under irradiation with a green laser (532 nm) were measured by the NIR emission spectrometer.

The quenching experiments were as follows: intensities of emission spectra were measured in the absence (I_0) and in the presence of the seaweed constituents (I) (Figure 2); ratios of I_0 / I were plotted against every concentration of the quenchers (Stern–Volmer plots) which gives a straight line (Figure 3). The slope of each line gives a $k_q\tau$ value, which gives a quenching constant k_q value (an antioxidative constant against singlet oxygen) when the τ value (half-life time of singlet oxygen in the solvent used) was given.[5]

A solution of a dye [Rose of Bengal (Aldrich); $5\text{–}10^{-4}$ mmol L^{-1}] and a quencher, antioxidant, was introduced into a flow-cell (1.2 mL min^{-1}) and continuously irradiated by a green laser (532 nm; CRGL-1100 (110 mW); CrownEO Co., Ltd.). The generated singlet oxygen was monitored by the

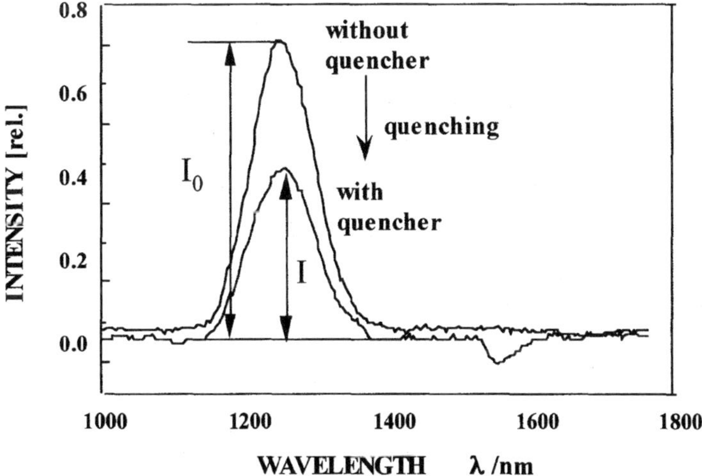

Figure 2. Quenching experiments of singlet oxygen using the NIR emission spectrometer.

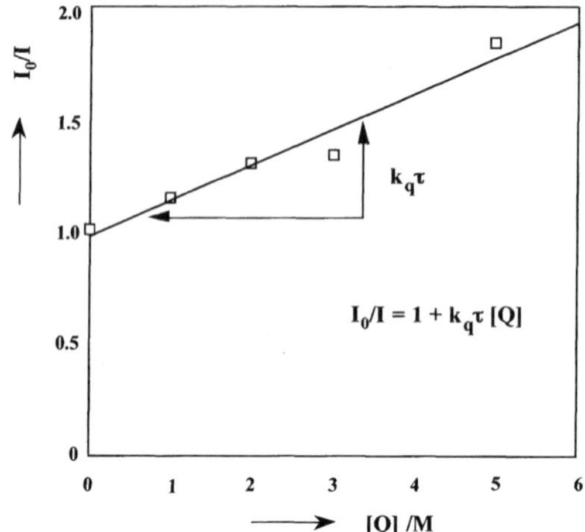

Figure 3. A Stern–Volmer plot for singlet oxygen quenching.

emission spectrophotometer and the spectra were observed. In the Batch type cell, dye could be bleached out and the quencher destroyed quickly.

Empirical calibrations

When measuring the emission intensities for dietary fibres, special consideration is needed to eliminate absorption effects: compensation was made at both 514 and 1100 nm.

Results and discussion

Table 1 shows antioxidative constants (quenching constants) of the dietary fibres. Quenching constants against superoxide, k_3, were also obtained using a method similar to that for singlet oxygen using chemiluminescence of a *Cypridina* luciferin analogue (CLA).[6]

The results obtained for dietary fibres show that fairly large k_q values were obtained from fucoidan, pectin, and Kombu extracts in g L^{-1}. They have also large k_3 values against superoxide oxidation as shown in Table 1.

Some dietary fibres, such as fucoidan (Figure 4), pectin (apple), and Kombu extracts showed antioxidative activity also for auto-oxidation (data not shown). They suppressed peroxidation of linoleic acid.

Several papers are found in the literature dedicated to oxidation of sugars by active oxygen species.[3,4] However, we believe, there are no papers on their quantitative antioxidative activity against active oxygen species.

Hyaluronic acid in connective tissues or synovia, a mucopolysaccharide, was reported to be depolymerised by active oxygen species to give inflammation of a joint or muscle. This reaction is suppressed by either catalase or SOD.[3]

Pectin and dextran are known to be depolymerised by ascorbic acid and Cu ion. Either an OH radical scavenger or quencher of singlet oxygen (DABCO) suppresses the reaction. Therefore, singlet oxygen can be concerned in the reaction as well as the OH radical and superoxide.[4]

Therefore, we examined the antioxidative activities of several representative "dietary fibres," such as fucoidan, pectin, carrageenans, alginic acid, dextran sulfate Na and Kombu extracts using the NIR emission spectrophotometer constructed in our laboratory and found them to quench singlet oxygen

Table 1. Antioxidative constants of dietary fibres.

Sample	MW	for $^1O_2$ $k_q/10^4$ (g L)$^{-1}$ s^{-1} ($k_q/10^{8\,M-1}$ s^{-1})		for O_2^- $k_q/10^4$ (g L)$^{-1}$ s^{-1} ($k_q/10^8$ M^{-1} s^{-1})	
		in EtOH-water (3 : 1) τ = 26.37μs		in water	
Vitamin C	176.13	1900[b] 8.94	34[b] (0.15)	400	(7.1)
NaN_3	65.01	401	(2.61)	—	
Kombu *Laminaria japonica*	—	10.34		4.4	
Fucoidan (from YT)	—	1.74		0.23	
Fucoidan (Sigma)	—	3.96 8.75[b]		[a]	
Alginic Acid Na	—	[a]		0	
ι-Carageenan	—	[a]		0.018	
κ-Carageenan	—	[a]		0.0074	
Pectin (apple)	20000–400000	1.16	(2.32–46.4)	0.32	(0.85–13)
Dextran Sulfate Na	5000	0.39	(0.19)	0.0041	(0.0041)
Dextran Sulfate Na	500000	0.07	(3.27)	0.056	(0.013)

[a]No data
[b]in water; $\tau = 2 \times 10^{-6}$ s

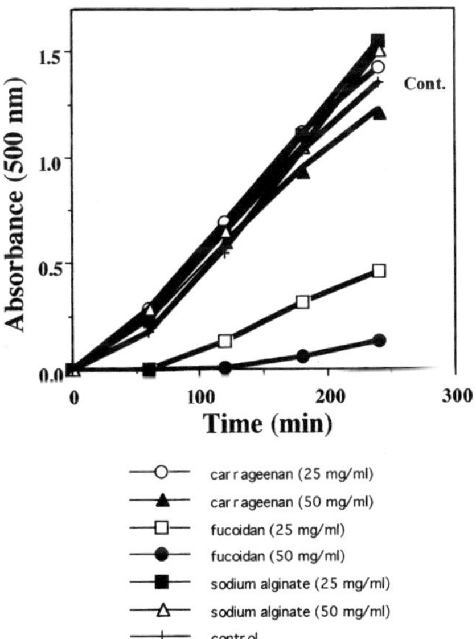

Figure 4. Antioxidative activity of some dietary fibres against autoxidation (POV).

($^1O_2$) as well as superoxide. Constituents of several representative seaweeds, such as wakame *Undaria pinnatifida*; hijikia *Hizikia fusifome* and kombu *Laminaria japonica*, were found to have fairly large reaction rates determined by quenching experiments of emission spectra in the NIR region (λ_{max}: 1270 nm) from $^1O_2$. The determined reaction rates are between 10^3–10^5 $(g\ L^{-1})^{-1}s^{-1}$; the larger ones are as large as that of ascorbic acid, 8.4×10^4 $(g\ L^{-1})^{-1}s^{-1}$. Some of these seaweed constituents also showed antioxidative activity against auto-oxidation and superoxide as well as their immunological enhancing activity.

These results suggest that dietary fibres, which are indigestible in the human body and can reach the large intestine without being hydrolysed, could scavenge either active oxygen species or toxic radical species to prevent carcinogen in the large intestine by their chemical, antioxidative activity.

Conclusions

1. Some "dietary fibres" were found to quench $^1O_2$ as well as O_2^- and to prevent autoxidation effectively. This suggests that they could work as antioxidative compounds in the alimentary canal (digestive organ).
2. Dietary fibres could prevent cancers of the colon by their chemical elimination of carcinogenic compounds as well as by known physical effects.

Acknowledgement

Part of this work was supported by grants from Nakatani Electronic Measuring Technology Association of Japan. The authors are sincerely grateful to the donors. We also acknowledge Optec Ltd., Tokyo, Japan for supplying their facilities and Miyako Kagaku Co. Ltd., Tokyo, Japan and Honpo Co. Ltd., Shimonoseki, Japan for their kind supply of fucoidan and kombu.

References

1. N. Suzuki, R. Tanaka, H. Hatate, T. Itami, Y. Takahashi, I. Mizumoto, T. Nomoto and B. Yoda, *in Recent Research Developments in Agric. & Biological Chemistry,* Vol. 3, Ed by P. Cremonesi, H. Yoshida, K. Mori and H. Tsuge. Research Signpost, Trivandrum, India, pp. 71–83 (1999); N. Suzuki, I. Mizumoto, T. Nagai, T. Nomoto, H. Matsuya, B. Yoda, K. Kozawa and A. Kozawa, in *Bioluminescence and Chemiluminescence 2000*, Ed by J.F. Case, P.J. Herring, B.H. Robinson, S.H.D. Haddock, L.J. Kricka and P.E. Stanely. World Scientific Publishing Co., Singapore, pp. 239–242 (2001).
2. Y. Takahashi, K. Uehara, R. Watanabe, T. Okumura, T. Yamashita, H. Omura, Y. Yomo, T. Kawano, A. Kanemitsu, H. Narasaka, N. Suzuki and T. Itami, in *Advances in Shrimp Biotechnology,* Ed by T.W. Flegel. National Center for Genetic Engineering & Biotechnology, Bangkok, Thailand, pp. 171–173 (1998).
3. J.M. McCord, *Science* **185,** 529 (1974).
4. K. Uchida and S. Kawakishi, *Agric. Biol. Chem.* **50,** 2579 (1986).
5. N. Suzuki, I. Mizumoto, Y. Toya, T. Nomoto, S. Mashiko and H. Inaba, *Agric. Biol. Chem.* **54,** 2783 (1990).
6. N. Suzuki, S. Mashiko, T. Nomoto, Y. Toya, B. Yoda and H. Inaba, *Chem. Express* **5,** 735 (1990).

Application of *in-line* near infrared spectroscopy to the control of industrial fermentation processes

E. Tamburini, G. Vaccari, S. Tosi and A. Trilli
Department of Chemistry, University of Ferrara, Ferrara, Italy

Introduction

Process quality control has become essential for both the agro-food and other industrial sectors. From an analytical perspective, this translates into a requirement for dependable, selective and sensitive analytical techniques. Such a requirement is particularly felt in the biotechnological sector where, in addition to some easily measurable physical and chemical parameters, several other crucial process variables such as biomass, substrates and metabolic products, ideally, should be monitored in real-time in order to maximise process yields and product quality. Modern fermentation technology increasingly relies on interactive process control which can only be optimally applied if a real-time, complete analytical characterisation of the process is available. A key requirement of any process monitoring technique applied to biotechnological processes is that it should not endanger process sterility: in this respect non-invasive analytical techniques which do not need to break the sterility barrier to obtain samples for off-line analysis have a tremendous advantage.

However, the current lack of suitable sensors means that the control of fermentation processes relies largely on the on-line measurement of physical parameters which relate only indirectly to the important biological and biochemical process variables. Real-time monitoring of such parameters is crucial to the optimisation of industrial fermentation processes because conventional off-line monitoring cannot adequately represent a dynamically-changing system. It is clear that any improvement in the response time of process monitoring is bound to translate into process improvements.

A spectroscopy-based approach is possible, in principle, given that the majority of the compounds of interest possess sufficiently differentiated spectroscopic features, but the inherent turbidity of fermentation broths prevents a direct *in-situ* determination by means of conventional UV/vis, or infrared (IR) techniques. As a result, the important biological and biochemical process variables are usually monitored by means of off-line techniques such as chromatography (HPLC, GC, etc.). These are time-consuming, require that samples be withdrawn and, generally, yield their results after a time-lag.

Against this background, near infrared (NIR) spectroscopy, coupled to suitable chemometric techniques, has considerable promise because it can combine real-time determination with non-invasive operation. An additional bonus is the elimination of the potentially toxic and/or environmentally- unfriendly reagents typically required by conventional analytical methods.

In order to demonstrate the applicability of NIR spectroscopy to the control of an industrial bioprocess, we have developed an NIR-based fermentation monitoring system capable of monitoring the concentration of biomass, growth substrates and metabolic products in two fermentations used to produce starter cultures for the industrial production of salami.

Objectives

Our group has utilised NIR spectroscopy in the biotechnological field for several years[1–5] working, in particular, with anaerobic cultures of a homofermentative *Lactobacillus* strain. The work reported here is the ideal continuation of previous research in that it aims at evaluating NIR spectroscopy as a tool for the real-time monitoring of all relevant parameters of two aerobic fermentations based on agitated and aerated cultures of different microorganisms grown in complex cultivation media. Non-invasive monitoring was achieved by means of a sterilisable probe inserted into the fermenter.

Experimental

Acquisition of NIR data

The probe, fitted to the fermenter head-plate, acquired spectroscopic data in the *interactance* mode and was connected to the NIR instrument by means of an optical fibre bundle (Foss NIRSystems 6500). The probe slit was set to 1 mm, thus providing an optical path of 2 mm. The culture broth flowed across the optical path of the probe as a result of mechanical stirring of the culture and was completely exposed to the prevailing hydrodynamic conditions of the fermenter.

Fermentations

The fermenter utilised (BM 3000 CLP-Bioindustrie Mantovane) had a geometrical volume of two litres and was completely computer-controlled. Two microorgansims were utilised, one belonging to the genus *Staphylococcus*, and the other to the genus *Lactobacillus*. The same complex, glucose-based cultivation medium was used with both organisms. The following process variables were monitored: glucose (main C-source), lactic acid and acetic acid (main metabolites) and biomass.

Off-line analysis

The biomass concentration was determined as the dry weight of the culture, measured by the membrane filtration method (cellulose acetate filters, pore size 0.45 µm). The remaining components were determined by direct HPLC analysis of fermentation samples (column: Aminex HPX-87H).

Calibration and validation

NIR spectra were recorded in the 700–1800 nm region and were then transformed into second derivative spectra to reduce baseline offsets. About 170 samples were thus analysed for each culture.

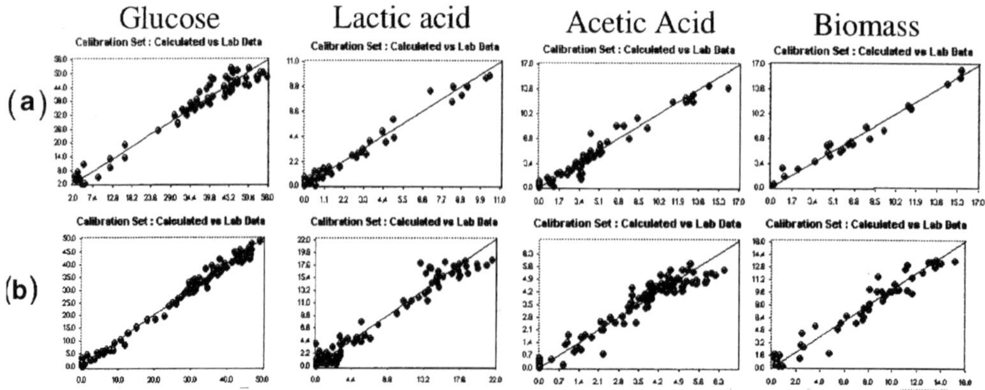

Figure 1. Calibration curves for (a) *Staphylococcus* and (b) *Lactobacillus*.

Table 1. Calibration and external validation parameters.

	Staphylococcus				*Lactobacillus*			
	Calibration		Ext. Validation		Calibration		Ext. Validation	
Glucose	Range R^2	0–58 gL^{-1} 0.9708	R^2 *SEP*	0.9679 2.9123	Range R^2	0–50 gL^{-1} 0.9883	R^2 *SEP*	0.9635 2.2730
Lactic acid	Range R^2	0–23 gL^{-1} 0.9381	R^2 *SEP*	0.9393 1.3669	Range R^2	0–22 gL^{-1} 0.9595	R^2 *SEP*	0.8926 1.8626
Acetic acid	Range R^2	0–19 gL^{-1} 0.9573	R^2 *SEP*	0.9010 0.6280	Range R^2	0–7 gL^{-1} 0.9488	R^2 *SEP*	0.8675 0.4804
Biomass	Range R^2	0–16 gL^{-1} 0.9506	R^2 *SEP*	0.9535 0.8532	Range R^2	0–16 gL^{-1} 0.9514	R^2 *SEP*	0.9578 0.5481

Calibration equations for the four components of interest were obtained by means of PLS regression using an iterative procedure that added the samples used for the *external validation* to the calibration database. The procedure was repeated until the required level of predictive capacity was achieved. The main statistical parameters of the calibration curves and the final *external validations* of all process parameters for both microorganisms are summarised in Figure 1 and Table 1.

Results and discussion

Effects of cultivation enviroment, agitation rate and airflow rate

A preliminary set of experiments was carried out to evaluate the influence of process conditions such as steam sterilisation, fouling by culture components and culture hydrodynamics on data acquisition by means of the immersion probe. Neither *in-situ* steam sterilisation nor fouling appeared to negatively influence probe performance. However, the system was negatively influenced by the air bubbles generated by air sparging and stirring of the culture, required by the aerobic nature of the fermentations being carried out. Stirring speed variations, in particular, were shown to determine major baseline shifts, irrespective of biomass concentration. In fact, it produced a marked signal change, presumably due to its effect on the population of air bubbles suspended in the culture fluid and had to be kept constant throughout the fermentation runs (Figure 2).

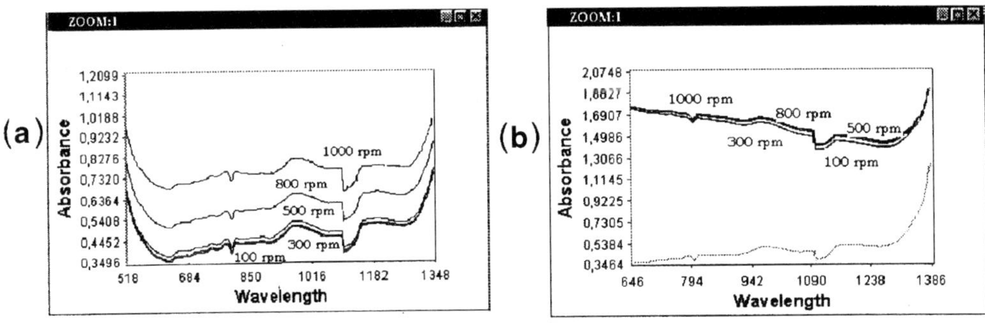

Figure 2. Stirring rate effects on spectrum baseline at (a) low and (b) high biomass concentration.

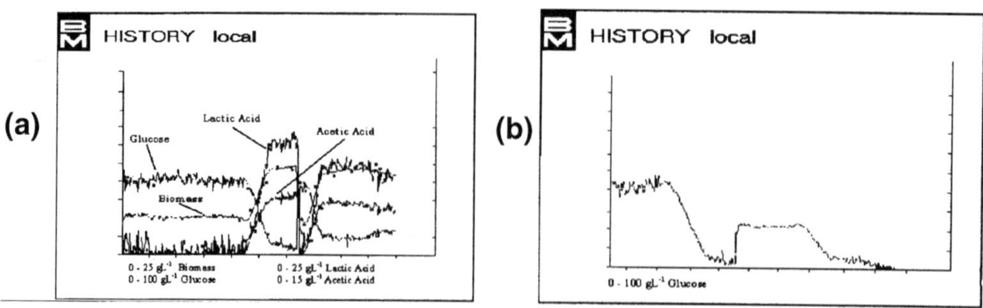

Figure 3. Automatic control of (a) repeated batch and (b) continuous fermentation processes.

Real-time, quantitative monitoring of fermentation parameters

Based on the encouraging results of the external validations for the biochemical parameters of interest, we set out to evaluate the applicability of *in-line* NIR spectroscopy to real-life fermentation monitoring and control. This was achieved thanks to the direct interfacing of the NIR instrumentation with the PC-based control system of the fermenter. The values of the relevant process parameters obtained by means of the NIR technique were thus transferred in real time to the fermenter control system where they were used for immediate numerical and graphical display and for the implementation of various cultivation strategies. As a result, both fed-batch and continuous fermentations employing the two microorganisms were carried out (Figure 3). Simultaneous off-line analysis confirmed the validity of the in-line NIR measurements, demonstrating the ability of this technique to monitor satisfactorily an entire process, while minimising operator interventions.

Conclusions

The NIR-based data acquisition system utilising a steam-sterilisable probe, immersed directly into the culture fluid, has been shown to be a viable alternative to traditional monitoring methods based on sample withdrawal and off-line analysis. The validity of the approach was confirmed even when applied to multicomponent, time-variable matrices such as fermentation broths with varying levels of turbidity. The elimination of sterility hazards connected to sampling is particularly relevant to biotechnological processes. Once the main task of setting up and validating the required calibration curves is completed, a one-probe non-invasive monitoring system is in place and only requires a suitable interfacing with process computers to allow real-time monitoring and control of fermentations. A tool is, therefore, available for the automatic control of fermentation processes on the basis of suitable process models and control strategies. Finally, it should be pointed out that the results reported here are relevant to the monitoring and control of several submerged processes, both inside and outside the scope of biotechnology.

References

1. G. Vaccari, E. Dosi, A. Campi and G. Mantovani, *Zuckerind* **118,** 266 (1993).
2. G. Vaccari, R.A. y Vara-Gonzalez, A. Campi, E, Dosi, P. Brigidi and D. Matteuzzi, *Appl. Microbiol. Technol.* **40,** 23 (1993).
3. G. Vaccari, E. Dosi, A. Campi, R.A. y Gonzalez-Vara and G. Mantovani, *Biotechnol. Bioeng.* **43,** 913 (1994).

4. E. Dosi, G. Vaccari, A. Campi, G. Mantovani, R.A. y Gonzalez-Vara and A. Trilli A, in *Near Infrared Spectroscopy: The Future Waves*, Ed by A.M.C. Davies and P. Williams. NIR Publications, Chichester, UK, p. 249 (1996).
5. R.A. y Gonzales-Vara, G. Vaccari, E. Dosi, A. Trilli, M. Rossi and D. Matteuzzi, *Biotechnol. Bioeng.* **67,** 147 (2000).

Selection of visible/near infrared wavelengths for characterising fecal and ingesta contamination of poultry carcasses

W.R. Windham, B. Park, K.C. Lawrence, R.J. Buhr and D.P. Smith
USDA, ARS, Richard. B. Russell Research Center, PO Box 5677, Athens, GA 30604-5677, USA.
E-mail: rwindham@saa.ars.usda.gov

Introduction

Zero tolerance of feces on the surfaces of animal carcasses during slaughter was established as a standard by the Food Safety Inspection Service (FSIS) to minimise the likelihood of contamination of meat and poultry with microbial pathogens.[1] Compliance with zero tolerance in meat processing establishments is currently verified by visual observation. Three criteria are used for identifying fecal contamination. These are colour, consistency and composition. Inspectors use these guidelines to verify that establishments prevent carcasses with visible fecal contamination from entering the immersion ice water bath (chiller). Real-time visual inspection is both labour-intensive and prone to both human error and variability.

Efforts have been made to develop automated or semi-automated visual inspection systems for detecting the presence of contaminants on food products during processing. These systems utilise a technique in which the food item is illuminated with UV or visible light and emissions of fluorescent light is measured between 660 to 680 nm as an indication of the presence of fecal material.[2,3] Visible and near infrared (vis/NIR) reflectance spectroscopy is a technique that can be used to detect contamination on foodstuffs. Vis/NIR spectroscopic techniques have been used to classify wholesome, speticemic and cadaver carcasses.[4,5] The vis/NIR method showed promise for separation of wholesome and unwholesome carcasses in a partially automated system.

Multispectral and hyperspectral imaging systems also have the potential for inspection of meat products during processing. Hyperspectral imaging has been used for the identification of surface contaminates on poultry carcasses.[6,7] Hyperspectal and multispectral imaging techniques were used to detect chicken skin tumors.[8] Park *et al* reported a hyperpspectral imaging system for detecting feces and ingesta on the surfaces of poultry carcasses.[9] The objectives of this research are to investigate the use of visible/ NIR reflectance spectroscopy to discriminate between uncontaminated poultry breast skin and pure feces and to select optimum or key wavelengths.

Materials and methods

Broilers and processing procedures

Male broilers were obtained from a local commercial farm at 38 days of age and transported to grow-out facilities. Broilers were provided a non-medicated, corn-soybean meal pelleted growers diet (3,200 kcal ME kg^{-1}, 19% crude protein) *ad libitum*. Feed and water were withdrawn eight hours and four hours prior to processing the birds, respectively. In Experiment 1, two replicates with 20 broilers were stunned, bled in shackles for a total of 120 sec, scalded at 57°C (hard scald) for 2 min and defeathered in the Russell Research Center pilot scale processing facility. In Experiment 2, two replicates of 16 broilers were scalded at 57°C and 52°C (soft scald). To collect intestinal content, broilers were eviscerated to obtain fecal material from the duodenum, ceca and colon portion of the viscera. Ingesta was collected from the proventriculus and/or gizzard. Fecal and ingesta contents were pressed from the dissected viscera into sample vials for analysis by visible/NIR spectroscopy. Circular samples of skin (38 mm diameter) were taken from the breast area of the carcass and placed in plastic bags.

Spectroscopic and multivariate analysis

Fecal, ingesta and breast skin samples were scanned with an NIRSystems 6500 monochromator (NIRSystems, Silver Spring, MD, USA). Spectra were recorded from 400 nm to 2500 nm at 2 nm intervals and analysed from 400 nm to 950 nm to correspond with the wavelength range of the hyperspectral imaging system described by Park *et al.*[9] Three replicates of uncontaminated breast skin from each carcass were presented in cylindrical sample cells. Samples of pure feces were presented in cylindrical sample cells (internal diameter: 38 mm; depth: 0.1, 0.2, or 0.3 mm) with an optical quartz surface and a locking back.

A commercial spectral analysis program (NIRS3, Infrasoft International, Inc., Port Matilda, PA, USA) was used to analyse the spectra of pure feces, ingesta and uncontaminated carcass skin and for multivariate analysis. The spectral data set (N = 140 and 48 uncontaminated hard and soft scald, respectively, breast skin; N = 42 duodenum; N = 37 ceca; N = 25 colon and N = 21 ingesta) was transformed with multiplicative scatter correction (MSC).[10] Spectra were mean centered and reduced by principal component analysis (PCA).[11]

Results and discussion

Immersion of poultry carcasses in hot water (scalding) aids in the removal of feathers. Scalding temperature is one of the factors that affect the appearance of carcass skin. Hard scald (57°C) removes the cuticle resulting in a whiter carcass, whereas soft scald (52°C) keeps the cuticle intact resulting in a yellow carcass. Figure 1 shows the average log (1/*R*) spectra of uncontaminated hard and soft breast skin. Skin from soft scalding had higher absorbance than hard scald immersion due to the yellow pigmentation of the cuticle. The visible region (400–700 nm) represents the colours of the skin as well as the muscle pigments of myoglobin. The colour of meat is largely determined by the relative amount of three forms of myoglobin, i.e. deoxymyoglobin, oxymyoglobin, and metmyoglobin[12] at the meat surface. Liu and Chen[13] reported at least seven absorption bands at 430, 440, 455, 545, 560, 575 and 585 nm associated with the changes in the oxidation and denaturation of myoglobin due to cooking and cold storage. Soft scald breast skin had absorption bands at 432 and 556 nm attributed to oxymyoglobin and the shoulder at 485 nm to metmyoglobin (Figure 1). Hard scalded skin also had bands attributed to oxymyoglobin at 544 and 570 nm.

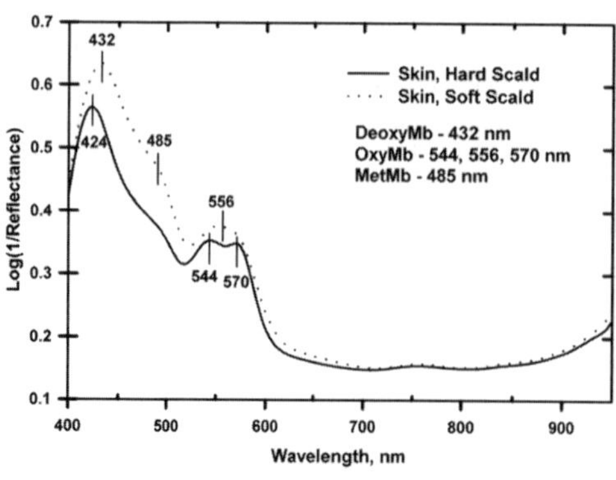

Figure 1. Average absorbance (log 1/*R*) spectra of hard and soft scald uncontaminated breast skin.

The appearance of two bands compared with one oxymyoglobin band in soft scalded skin was possibly due to a change in the secondary structure of the heme groups in myoglobin due to higher immersion temperature. The band at 424 nm can be assigned to the Soret absorbance band for oxymyoglobin.[14]

The feces spectra represent a mixture of plant pigments with broad absorbance, which increased with decreasing wavelength (Figure 2). Fecal and ingesta obtained from the four sampling sites varied greatly in colour. In general, cecal and colon feces were brown to dark brown in colour compared to duodenum feces and ingesta that was yellow–orange to light brown. Feces and ingesta spectra decreased in absorbance at 458 nm as fecal material became browner. In contrast, feces and ingesta had a broad absorbance characteristic (525–700 nm), which increased as the sample became darker (yellow–orange to dark brown).

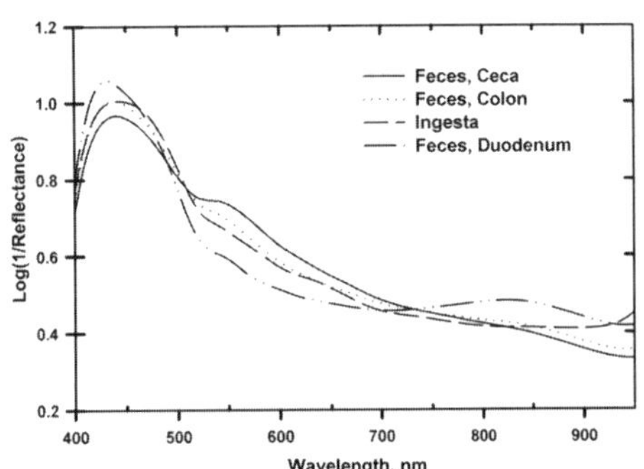

Figure 2. Average absorbance (log 1/*R*) spectra of pure feces and ingesta.

The ceca samples were the brownest of the feces and had the greatest absorbance (525–700 nm), whereas the duodenum samples were the lightest in colour of the feces and had the lowest absorbance. In general, the feces and ingesta spectra had greater absorbance than uncontaminated breast skin samples in this region (Figure 1).

Prior to PCA, the spectral dataset was transformed with MSC to remove interferences of light scatter from the skin and differences in pathlength due to pure feces and ingesta sample thickness variations. The first four components accounted for 98% of the total spectral variation, with components 1, 2 and 4 expressing 92% of the variation. Scores from the first PC separated the uncontaminated breast skin from contaminates (Figure 3). Uncontaminated breast skin had negative scores, whereas feces and ingesta had positive scores. Scores from PC2 and 4 partially separated fecal type with some overlap, whereas ingesta was clearly separated from feces. Fecal type was distributed along the axes of the second PC, indicating its wide variation. In addition, there was less spectral variation in feces from the duodenum compared to ceca and colon feces. In an overall evaluation, the first PC is clearly related to the uncontaminated skin, while the second and fourth PCs describe the variations of fecal type and ingesta.

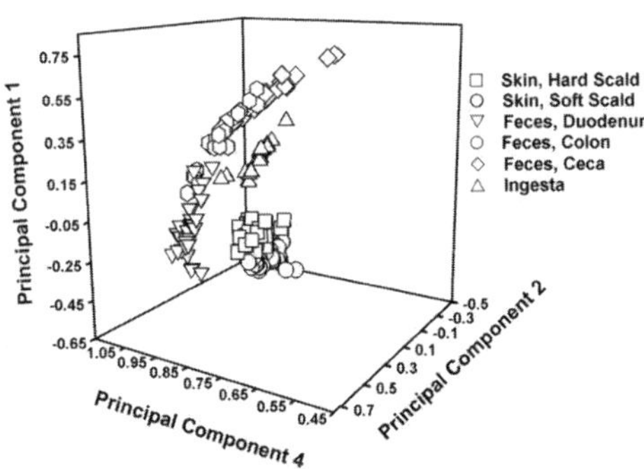

Figure 3. PCA score plot of components 1, 2 and 4 for uncontaminated hard and soft scald breast skin and duodenum, ceca, colon feces and ingesta.

These results demonstrate the ability to discriminate between uncontaminated skin and feces due to the variations in the visible/NIR spectra, which arise from the intrinsic differences in colour and chemical constituents of the samples.

Loadings are the regression coefficients of each variable (wavelength) for each PC. Loadings often resemble the spectra of samples and the spectra of constituents and thus offer scope for interpretation of maximum weighting matching known absorbance bands. The loadings of the

PCs, used to discriminate between uncontaminated breast skin and contaminates are shown in Figure 4. The loading plot indicates how the variance is accounted for in a PC across the wavelength scale. Numerically higher (+/-) weights indicate a relative high contribution of the wavelength area to that PC.

The shape of the plot for the first PC showed the broad absorbance (628 nm) characteristic and the inverse relationship of feces and ingesta colour compared to the colour of uncontaminated breast skin. Uncontaminated breast skin had lower absorbance (Figure 1) than contaminates (Figure 2) in this wavelength region. Weights 2 and 4 had large intensities at 565 and 434 nm, respectively, related to the myoglobin and/or hemoglobin of the breast skin. Weight 4 also had significant intensity at 517 nm possibly related to colour differences.

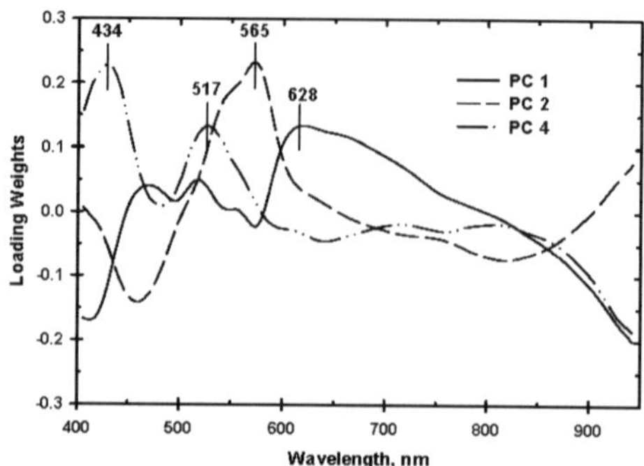

Figure 4. Loading weights of principal components 1, 2 and 4, for uncontaminated poultry breast skin and contaminates.

Conclusions

Visible/NIR spectroscopy can be used to discriminate between uncontaminated poultry breast skin and feces and ingesta from different portions of the digestive tract. Visual assessment of the loadings suggests that discrimination was dependent on the spectral variation related to contaminate colour and myoglobin and/or hemoglobin content of the uncontaminated breast skin. Key wavelengths were identified by intensity of loadings at 628 nm for PC1, 565 nm for PC2 and 434 and 517 nm for PC4. These key wavelengths were selected and applied to hyperspectral images of uncontaminated and fecal contaminated poultry carcasses as described by Park *et al.*[9]

References

1. USDA, *Enhanced Poultry Inspection. Proposed Rule. Fed. Reg.* **59,** 35659 (1994).
2. A. Waldroup and J. Kirby, US Patent No. 5, 621, 215 (1997).
3. T.A. Casey, M.A. Rasmussen and J.W. Petrich. US Patent No. 5, 914, 247 (1999).
4. Y.R. Chen, R.W. Huffman, B. Park and M. Nguyen, *Appl. Spectrosc.* **50,** 910 (1996).
5. Y.R. Chen and W.R. Hruschka, ASAE Paper No. 983047. American Society of Agricultural Engineers, Orlando, FL, USA (1998).
6. R. Lu and Y.R. Chen, *Proc. SPIE* **3544,** 121 (1998).
7. J. Heitschmindt, M. Lanoue, C. Mao and G. May, *Proc. SPIE* **3544,** 134 (1998).
8. K. Chao, P.M. Mehl and Y.R Chen, ASAE Paper No. 003084. ASAE, St Joseph, MI, USA (2000).
9. B. Park, K.C. Lawrence, W.R. Windham and R.J. Buhr, ASAE Paper No. 01-3130. ASAE St. Joseph, MI, USA (2001).
10. T. Isaksson and T. Naes, *Appl. Spectrosc.* **42,** 1273 (1988).
11. I.A. Cowe and J.W. McNicol, *Appl. Spectrosc.* **39,** 257 (1985).
12. J. Price and B. Schweigert, in *The Science of Meat and Meat Products.* Food and Nutrition Press, Westport, CT, USA, pp. 155–343 (1987).
13. Y. Lu and Y.R. Chen, *Appl Spectrosc.* **54,** 1458 (2000).
14. H.J. Swatland, *Can. Inst. Food Sci. Technol. J.* **22,** 390 (1989).

Imaging spectrometry for detecting feces and ingesta contamination on poultry carcasses

B. Park, W.R. Windham, K.C. Lawrence, R.J. Buhr and D.P. Smith
USDA, ARS, Richard B. Russell Research Center, Athens, GA 30604-5677, USA.
E-mail: bpark@saa.ars.usda.gov

Introduction

With the implementation of the Hazard Analysis Critical Control Point (HACCP) System,[1] industry was mandated to establish science-based process controls. The detection of fecal and ingesta contamination by visual observation is far from a science-based approach to process control. In addition, there has been a dramatic increase in water usage in most plants as a result of the zero tolerance standard. Plants have nearly doubled their previous water usage and, nationwide, the usage has increased by an estimated two billion gallons.[2] Development of image sensor technology that retains individual carcass inspection for on-line detection of fecal and ingesta contamination, would provide a science-based process control and decrease water usage.

Recently, hyperspectral imaging (or imaging spectrometry) has emerged as a powerful technique in earth remote sensing, but is also being utilised in medical,[3–5] biological,[6–8] agricultural[9–14] and industrial[15–18] areas as well. Hyperspectral imaging is an imaging technique that combines aspects of conventional imaging with spectrophotometry and radiometry. This technique is capable of providing an absolute radiometric measurement over a contiguous spectral range for each and every pixel of an image. Thus, data from a hyperspectral image contains two-dimensional spatial information as well as spectral information. These data can be considered as a three-dimensional "hypercube" which can provide physical and/or chemical information of a material under test. This information can include physical and geometric observations of size, orientation, shape, colour and texture, as well as chemical/molecular information such as water, fat, proteins and other hydrogen-bonded constituents. Hyperspectral imaging is an extremely useful tool to thoroughly analyse the spectra of inhomogeneous materials that contain a wide range of spectral and spatial information. Therefore, it can be an effective technique for identifying surface contaminates on poultry carcasses. The objective of this research is to develop a hyperspectral imaging technique for the identification of fecal and ingesta contamination on the surface of poultry carcasses.

Materials and methods

Materials

A total of 80 fecal and ingesta samples from 20 poultry carcasses were obtained from a local broiler house, transported to the grow-out facilities located at Athens, Georgia and held for four days. Many random variables affect the fecal and ingesta content of the digestive tract. However, the variables were fixed as follows. The samples used were six-week old male birds. Corn with soybean meal were fed for diet. Food and water were withdrawn eight hours and four hours prior to processing the birds, respectively.

The feeding regime was scheduled for meal feeding to provide a consistent amount of fecal material in the digestive tract among the birds. The birds were stunned (12 VAC), bled for 90 s, scalded at 57.5°C (hard scald) for 2 min and picked in the Russell Research Center pilot scale processing facility. In order to collect feces and ingesta, four replicates of 20 birds were processed and eviscerated to obtain fecal materials from the duodenum, ceca, colon and ingesta from the proventriculus and gizzard.

Methods

A transportable hyperspectral imaging system was designed to provide both portability and flexibility in positioning both the lights and the camera system. The imaging system consists of an imaging spectrograph, a high resolution CCD camera with a 1280 × 1024 pixel resolution, compact C-mount lens, motor for lens motion control, frame-grabber and computer. The prism–grating–prism spectrograph has a nominal spectral range of 400 nm to 900 nm with 6.6 mm axis and attaches to the camera for generating line-scan images. The lighting system consists of a 150-watt quartz halogen DC stabilised fibre-optic illuminator, lamp assembly, fibre-optic cables and 10-inch illuminating size of quartz halogen line lights.

An experiment was conducted to collect hyperspectral images from hard-scald poultry carcasses. Hard scalding removes the skin cuticle, resulting in a white carcass, whereas a soft scald keeps the cuticle intact, resulting in a yellow carcass. Then, fecal and ingesta detection algorithms were developed from uncontaminated and contaminated hard scald carcasses.

To fully characterise the spectral and spatial nature of the hyperspectral imaging system, a full calibration, considering both spectral and spatial information, is required. This allows comparison of images collected from the hyperspectral imaging system with the key wavelengths selected by the visible/NIR reflectance spectrometer. Wavelength calibration was conducted with several laser lights to determine the wavelength value of each vertical pixel in the line-scan camera system. Also, two pencil lights, Mercury–Argon (HgAr) and Krypton (Kr); lasers (543.5, 594, 612, and 632.8 nm) and a spectral lamp power supply were used to identify peaks from known wavelengths.

For the hyperspectral image acquisition, line-scan images were collected using SensiCam software based on the following control settings. Image pixels of high resolution CCD detector were 1280 by 1024; however, actual image size and number of wavelength bands are defined by binning size. For our experiment, four by two binning of the line-scan image created an actual image size of 320 (horizontal) × 340 (vertical) spatial resolution with 512 wavelengths spectral images. In this case, the number of line scans, which depend upon the size of carcass and speed of motor to move lens, was 340. The exposure time and time delay of camera control during image acquisition were 50 msec and zero, respectively. The spectral resolution of hyperspectral images was approximately 0.9 nm and a total file size of each image was 106 Mbytes. Even though scanning time depends on the size of a carcass, the average time to scan the whole carcass was about 34 s.

Hypercube image files were created from line-scan image data using HyperVisual software (ProVision Technologies, Stennis Space Center, MS, USA), which can convert 16 bits of binary data into binary sequence mode data for hyperspectral image processing.

After hyperspectral image files were created, the data size was reduced and the features in the image were enhanced. Dimensionality reduction is an important step in hyperspectral image processing because hyperspectral imaging creates a large amount of data containing enormous spectral and spatial information. Specific wavelength bands of hyperspectral images selected by the results of spectrophotometry were processed and analysed for finding algorithms to detect fecal and ingesta contamination on the surface of carcasses.

The following four steps of image processing algorithms were executed for the identification of fecal and ingesta contaminated spots on the poultry carcasses. Based on the finding of dominant wavelengths from spectroscopy and band selection from the wavelength calibration, four different wavelength spectral images were selected from hypercube image data. The band selection obtained from wavelength calibration was followed by the calculation of band ratios among the four selected spectral images. After the algorithm to calculate the band ratio of individual spectral image has been conducted, a masking template, which is created by the spectral image from one of the 512 spectral image data, masked the images to eliminate background noise from the carcasses. Finally, the algorithm of histogram stretching was applied to all masked images to visually segregate individual fecal and ingesta contaminants.

Results and discussion

Spectral calibration was conducted to correlate absolute wavelength data from known spectral light sources (HgAr, Kr and Green, Red Laser) to the 512 hyperspectral image bands (1024 pixels with a binning of two) obtained from the CCD sensor. The calibration equation for this hyperspectral imaging system is as follows, where X is the band number ranging from 0 to 511, which can be obtained by hypercube image data converted from line scan image data:

$$\text{Wavelength (nm)} = 380.277 + 0.905\,X + (4.369 \times 10^{-4})\,X^2 - (4.356 \times 10^{-7})\,X^3 (r^2 = 0.9999)$$

Based on the calibration equation, we selected four bands: 58, 143, 190, and 251. Each band corresponded with the wavelength of 434 nm, 517 nm, 565 nm and 628 nm, respectively.

Band-ratio images were calculated from the selected four wavelengths spectral images. Six band-ratio images were obtained from the combination of different wavelength selections. As shown in Figure 1, among band ratio images, the band ratio of 565/517 could identify all different types of feces (duodenum, ceca, colon) and ingesta contaminants including colon feces located at the vent area of

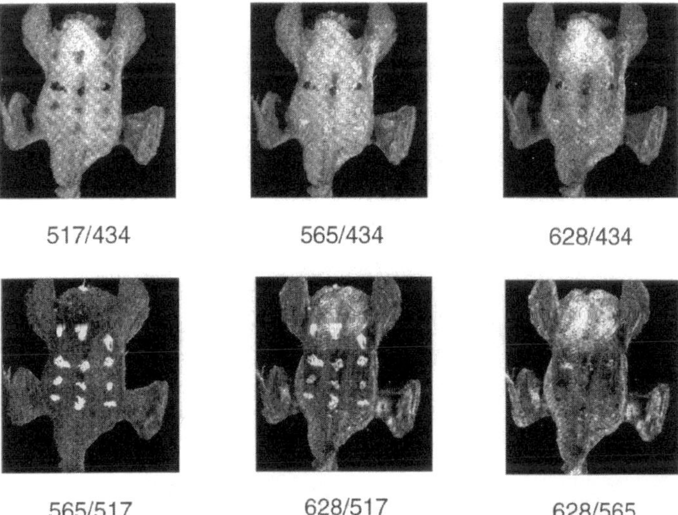

Figure 1. Band ratio images calculated by the combination of two selected spectral images for the identification of feces and ingesta on the poultry carcasses.

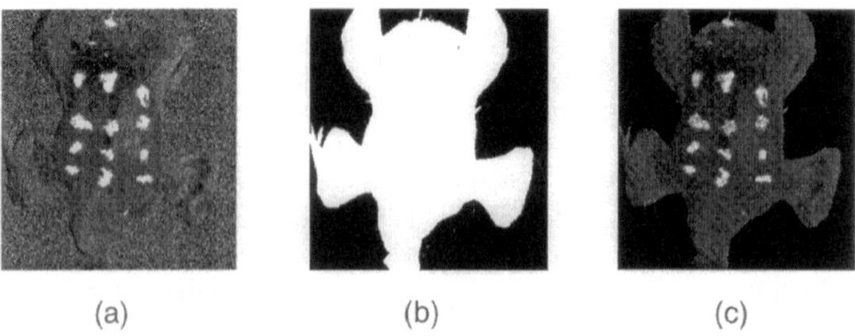

Figure 2. Masking process to eliminate background noise or band ratio processed spectral images. (a) band ratio image (565/517); (b) masking template; (c) band ratio image after masking process.

the tail, which was naturally contaminated during the slaughter of the chicken. The image ratios of 517/434, 565/434 and 628/434 show distinctive ceca (dark spots on the body) contamination.

However, other contaminated spots of duodenum, colon and ingesta were not readily apparent. Even though the ratio image of 628/517 shows contaminated spots on the body, other white spots under the wings and the area between the legs caused false positive errors for feces from the duodenum. Similarly, as seen on the ratio image of 628/565, false positive spots between the legs on the image were actually caused by cuticles or blood hemorrhages on the skin of a carcass.

The background of the original band ratio image was noisier than the chicken body. To eliminate the noisy background, the masking process was implemented for further image processing to segregate the ratio image of a carcass from the background. Figure 2 shows the band ratio (565/517) image before and after the masking process. It was obvious that the masking process made the spots of contamination on the carcass more visually distinctive. To build a masking template, the single spectral image [Band 257 (634 nm)] was selected from 512 hyperspectral images. The template [Figure 2(b)] was created by thresholding the image by choosing minimum thresholding value as zero and maximum value as120, respectively.

After the masking process was applied to ratio images to eliminate background noises, the histogram stretching algorithm was executed to separate feces and ingesta contaminants from a poultry carcass. Both linear and nonlinear histogram stretching algorithms were tested. As shown in Figure 3, the top center white portion of vent area indicates natural contamination of colon feces the second row represents duodenum feces, the third row repersents cecal feces, the fourth row represents colon feces and

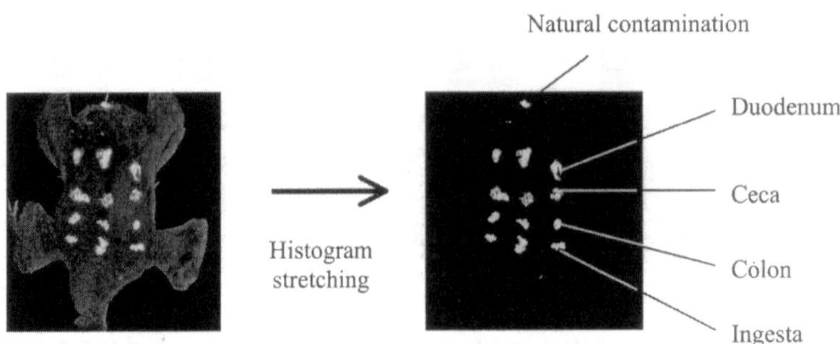

Figure 3. Histogram stretching to separate feces and ingesta from the carcass.

the fifth layer (bottom) represents ingesta contaminants, respectively. The parameter values of histogram stretching algorithm for the sample in Figure 3 were determined as follows: minimun input 1.28; maximum input 1.60; minimum output 0.75; maximum output 2.38.

Conclusions

A hyperspectral imaging system, designed and constructed for this study, provided high resolution multiple spectral images. In conjunction with image processing algorithms (band ratio, histogram stretching, thresholding), the hyperspectral imaging system is an effective technique for the identification of fecal and ingesta contaminants on poultry carcasses. Spectral image band ratio of 565 to 517 performed very well for identification of both feces (duodenum, ceca, colon) and ingesta. Further research into the development of classification algorithms of individual feces and ingesta will enable us to implement a hyperspectral imaging system for HACCP application. Also, the algorithms developed in this study can be effective for the selective multispectral imaging system, in conjunction with batch image-processing algorithms for the real-time, on-line poultry processing applications. By implementing hyperspectral/multispectral imaging techniques, all fecal and ingesta contaminated carcasses can be removed before the chiller tank to prevent cross-contamination at the chiller.

However, even though the hyperspectral imaging technique performed very well, the samples used in this study were limited and different sample conditions need to be considered for the robustness of the system. Therefore, different diet feeding studies should be conducted to investigate the effect of feed ingredients used in broiler finishing diets on the vis/NIR fecal/ingesta spectra and hyperspectral images of contaminated and uncontaminated carcasses, because this is an important factor to select the optimum wavelength to differentiate the location of each feces and ingesta in the digestive tracts of poultry birds.

References

1. USDA, *Fed. Reg.* **61,** 28805 (1996).
2. F.T. Jones, *FSIS Poultry* **6,** 38 (1999).
3. R.M. Levenson, E.S. Wachman, W. Niu and D.L. Farkas, *Proc. SPIE.* **3438,** 300 (1998).
4. M.A. Afromowitz, J.B. Callis, D.B. Heimbach, L.A. DeSoto and M.K. Norton, *IEEE Trans. on Biomedical Engineering* **35,** 842 (1988).
5. C.C. Hoyt, R.R. Richards-Kortum, B. Costello, B.A. Sacks, C. Kittrell, N.B. Ratliff, J.R. Kramer and M.S. Field, *Lasers in Surgery and Medicine* **8N,** 9 Jan (1988).
6. L. Chaerle and D.V.D. Straeten, *Biochemica et. Biophysica Acta* **1519,** 153 (2001).
7. E.B. Chase, C.E. Cooper, D.T. Delpy and E.O.R. Reynolds, *Phil. Trans. R. Soc. Lond.* **B352,** 649 (1997).
8. P. Robert, D. Bertrand, M.F. Devaux and A. Sire, *Anal. Chem.* **64,** 664 (1992).
9. J. Heitschmidt, M. Lanoue, C. Mao and G. May, *Proc. SPIE.* **3544,** 134 (1998).
10. R. Lu, Y.R. Chen, B. Park, and K.H. Choi, *ASAE Technical Paper No.* **993120,** (1999).
11. R. Lu and Y.R. Chen, *Proc. SPIE.* **3544,** 121 (1998).
12. M.S. Borhan, S. Panigrahi, J. Lorenzen and H. Gu, *ASAE Technical Paper No.* **011153,** (2001).
13. A.Y. Muir, I.D.G. Shirlaw and D.C. McRae, *Agricultural Engineer* **Autumn,** 79 (1989).
14. M. Tsuta, J. Sugiyama and Y. Sagara, *J. Agric. Food Chem* **1,** (2001).
15. C.T. Willoughby, M.A. Folkman and M.A. Figueroa, *Proc. SPIE.* **2599,** 264 (1996)
16. W.F. McClure. *NIR news* **2(2),** 8 (1991).
17. Y. Sakamoto, K. Tajiri, T. Sawai and Y. Aoki, *IEEE Trans. on Geoscience and Remote Sensing* **26(4),** 430 (1988).
18. D. Wienke, W. v. d. Broek, W. Melssen, L. Buydens, R. Feldhoff, T. Huth-Fehre, T. Kantimm, F. Winter and K. Cammann, *J. Anal. Chem.* **354,** 823 (1996).

Preliminary study on the use of near infrared spectroscopy for determination of plasma deuterium oxide in dairy cows

Agung Purnomoadi,[a] Itoko Nonaka, Koji Higuchi, Osamu Enishi, Masahiro Amari, Fuminori Terada

National Institute of Livestock and Grassland Science, Tsukuba Norindanchi, PO Box 5, Tsukuba 305-0901, Japan

Introduction

Information on body composition (fat and protein) in live animals is important in the determination of the nutrient requirement of the animal. The change of body composition due to mobilisation and reserves of body tissues in dairy cows occurs during the whole lactation period.[1] Significant changes in body composition especially occurs during early lactation because of the insufficiency of nutrient intake to meet the requirement for lactation, and it is reported[1–4] that empty body fat was reduced by 42.4 kg for early lactation cows compared with that of prepartum cows.

The conventional method for measuring body composition is direct slaughter, a method that is expensive, laborious and unrepeatable. Non-destructive methods such as various body water dilution procedures, have been proposed to estimate body composition and one of them is the deuterium oxide (D_2O) dilution technique.[5–7] The D_2O dilution technique has also been validated as a direct method, and shown its ability to detect changes in body fat and protein across physiological stages in dairy cows[6] and in fat-tailed Barbary ewes.[8] The usefulness of this technique, however, is limited by the time consuming use of special equipment for liophilisation to extract water from plasma prior to the D_2O concentration measurement. This study was carried out to determine if the rapid analytical method of near infrared spectroscopy had the potential to solve this problem.

Materials and methods

Four dairy cows (Cows #474, 478, 550, 942; mean body weight 575 kg) in early lactation were used. They were fed total mixed rations consisting of corn silage, timothy hay and concentrates to make 17.0% CP and 14.0 MJDE kg DM^{-1}. At weeks 1, 3 and 5 after parturition, D_2O with 0.9% NaCl was infused into the jugular vein at a dosage rate of 250 mg kg BW^{-1}. Blood samples were collected at 0, 5, 10, 15, 20, 25, 30, 40, 50, 65, 80, 100, 120, 150 min, 3, 4, 6, 8, 10, 12, 24, 36, 48 and 72 h after infusion. The samples were then centrifuged at 3,000 rpm for 10 minutes to obtain plasma that was then stored in a freezer at –35°C for chemical and spectral analysis.

---

[a]Home address: Faculty of Animal Science, Diponegoro University, Semarang 50275 Indonesia

D_2O analysis, chemically

Plasma samples were thawed at room temperature (23°C) prior to the determination of D_2O. The D_2O concentration was analysed by gas chromatography (deuterium oxide analysis system, HK102, Shokotsusyou) after extraction from plasma by liophilisation.

D_2O analysis with near infrared spectroscopy

The rest of the blood plasma sample remaining after chemical analysis was used for near infrared analysis. The NIR spectra of plasma samples were recorded by a Pacific Scientific (Neotec) model 6500 (Perstorp Analytical, Silver Spring, MD, USA) using transmittance cell samples (1 mm thickness). Spectra were read at 2 nm interval over the wavelength range from 1100 to 2500 nm, which were then converted to second derivative of log A^{-1}; where A is the absorbance, using ISI software (InfraSoft international, Port Matilda, PA, USA). A calibration equation was developed by multiple linear regression using a combination of four wavelengths.

The calibration equation and wavelength selections were developed using samples from one animal (calibration set; cow #550; *n*: 74), while the rest of the samples from three animals (cows #474, $n = 44$; #478, $n = 62$; #942, $n = 68$) were used for validating the developed calibration equation (validation set). This methology was used to make it possible to judge whether the calibration equation (including the selected wavelengths) was valid for D_2O determination while allowing for variation between individual animals. Judgement of accuracy was done on averaged values from three weeks collection for each cow for replication of NIR measurements, ignoring the effect of the sample collection period.

Results and discussion

The range and average of D_2O concentration in blood samples collected in this study is presented in Table 1. The lowest values in each animal is considered as the base level of D_2O in blood at zero time collection, while the highest concentration of D_2O in blood occurred in five or ten minutes after infusion. Development of the calibration equation using four wavelengths in the 1100–2500 nm range wavelengths from the spectra of blood samples from cow #550 showed a high correlation ($R = 0.93$) with a standard error 48.1 ppm. The four wavelengths used were 2128, 1636, 1190 and 2210 nm. These wavelengths were correlated with bonds of N–H (amide) and C–H.[9] This calibration result is presented in Figure 1.

Table 1. The deuterium oxide (D_2O) concentration in blood samples from four cows used in this study and the statistical summary of calibration and validation set on average value of NIR predicted value from three weeks.

Cow no	*n*	D_2O (ppm)	Average (ppm)	*R*	*SEP*	RPD
Calibration set sample s						
#550	74	135 – 925	535	0.93	48.1	—
Validation set samples						
#474	44	303 – 840	530	0.94	53.1	2.34
#478	62	303 – 984	544	0.98	23.5	6.64
#942	68	283 – 919	525	0.95	37.2	3.36

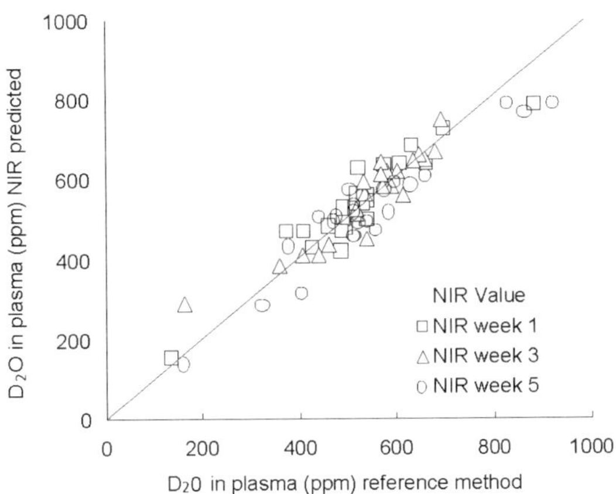

Figure 1. Correlation between deuterium oxide (D_2O) in blood plasma determined by conventional method and the NIR predicted in the calibration set samples using cow #550 at three weeks collections; NIR predicted at week 1, 3 and 5.

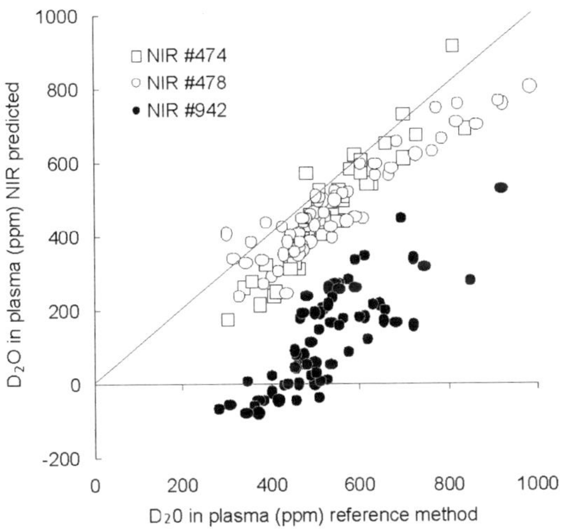

Figure 2. Correlation of NIR predicted value and Lab value of three cows used for validating the calibration equation.

Validation of the calibration equation for three individual cows was done on the average of NIR predicted value of D_2O at each collection time from the three weeks injection. The results showed a high correlation as presented in Table 1 and Figure 2. The *r* and *SEP* for plasma from cows in #474 were 0.94 and 51.8 ppm; cow #478 were 0.98 and 23.5 ppm; cow #942 were 0.95 and 37.2 ppm, respectively. Judgement of the accuracy based on the ratio of standard deviation and standard error in validation set samples (RPD) for cows #474, #478 and #942 were 2.3, 6.6 and 3.4, respectively. Based on RPD values, the results were higher than 2.5, the limit value available for screening[10] while cow # 474 was just below the limit. From this figure, the most interesting is cow #942. The predicted value

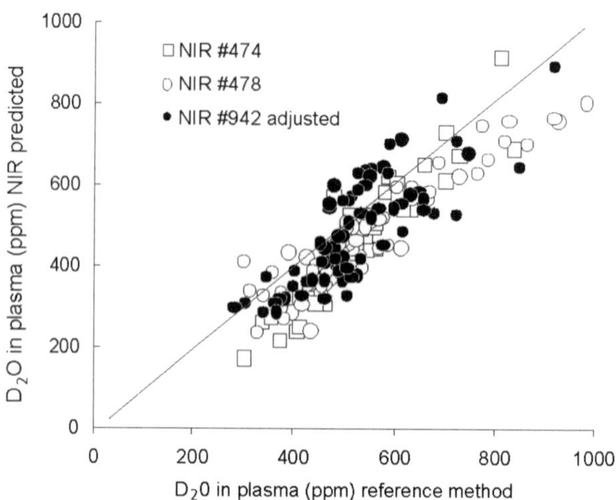

Figure 3. Correlation of NIR predicted value and Lab value of three cows used for validating the calibration equation. The value from cow #942 was adjusted using the difference value on deuterium oxide (D_2O) concentration at the 72 h collection period.

was much lower that the Lab values that resulted in a bias in prediction because the *r* and *SEP* showed high accuracy. This was considered as an effect of the lower level in NIR absorbance compared to those of the other three cows. By converting the lowest value (concentration at 72h) of cow #942 to the point of Lab determined value, the predicted values were lying in the area of NIR predicted values, as shown in Figure 3. These facts showed that the existence of variation between animals in their blood composition possibly biases the prediction values. However, this problem can be overcome by providing standard samples at known concentrations to allow calculation of a correction factor.

Conclusion

The high accuracy in prediction found in this preliminary study on the use of NIR spectroscopy for determining deuterium in plasma suggests that this is a very promising application. The bias measurement may come from the variation between individuals, but this can be overcome with provision of a correction. Further studies on various physiological stages of animals should be done to measure any other factors that biases the measurement.

Acknowledgements

The authors gratefully acknowledge the staff of cattle management, National Institute of Livestock and Grassland Science, Japan for taking care of the animals during the experiment. This work was supported by the Programme for Promotion of Basic Research Activities for Innovative Biosciences (PROBRAIN), Japan.

References

1. S.M. Andrew, D.R. Waldo and R.A. Erdman, *J. Dairy Sci.* **77,** 3022 (1994).
2. W.P. Flatt, P.W. Moe, A.W. Munson and T. Cooper, in *Energy Metabolism of Farm Animals, Proc. 4 th Symp. Energy Metab. Publ,* No. 12. European Association of Animal Production. Warsaw, Poland. p. 235 (1969).

3. B.W. Butler-Hogg, J.D. Wood and J.A. Bines, *J. Agric. Sci. (Camb.)* **104,** 519 (1985).
4. R.L. Belyea, G.R. Frost, F.A. Martz, J.L. Clark and L.G. Forkner, *J. Dairy Sci.* **61,** 206 (1978).
5. S.S. Waltner, J.P. McNamara, J.K. Hillers and D.L. Brown, *J. Dairy Sci.* **77,** 2570 (1994).
6. S.M. Andrew, R.A. Erdman and D.R. Waldo, *J. Dairy Sci.* **78,** 1083 (1995).
7. M.V. Komaragiri and R.A. Erdman, *J. Dairy Sci.* **80,** 929 (1997).
8. N. Atti, F. Bocquier, M. Theriez, G. Khaldi and C. Kayouli, *Livest. Prod. Sci.* **65,** 39 (2000).
9. B.G. Osborne and T. Fearn, *Near infrared spectroscopy in food analysis*. Longman Scientific and Technical Group, Harlow, Essex, UK, pp. 20–42 (1986).
10. P.C. Williams, *Proc. Symp. the 12th non-destructive measurements*, Tsukuba, Japan. p. 1 (1996).

Discrimination analysis of gallstones by near infrared spectrometry using a soft independent modelling of class analogy

Sang Hak Lee,[a] Bum Mok Son,[a] Ju Eun Park,[a] Sang Seob Choi[b] and Jae Jak Nam[c]

[a]*Department of Chemistry, Kyungpook National University, Taegu 702-701,Korea*

[b]*Department of Environmental Management, Andong Science College, Andong 760-820, Korea*

[c]*National Institute of Agricultural Science and Technology, Suwon 441-707, Korea*

Introduction

Classification of gallstones has been based mostly on morphology by visual inspection.[1] Classification by chemical analysis has improved the results but it has been limited by the dissolution of the sample and the insolubility of many of the components. Discrimination of gallstones presented a problem primarily because of the presence of the bile pigment and inorganic compounds. Various instrumental methods have been proposed but are not complete in themselves alone. However, the application of solid state analytical methods such as infrared spectroscopy and X-ray diffraction has permitted the identification of the type of gallstones and to determine the contents of insoluble components.[2,3]

Recently, the application of near infrared (NIR) spectroscopy to quantitative and qualitative analysis of organic compounds is finding increasing use since the NIR technique allows rapid analysis of powdered samples with little sample preparation.[4] The analysis and interpretation of NIR spectra require a variety of chemometric tools. The identification of the NIR spectra needs a suitable chemometric classification method to correctly identify unknown samples. Several methods for this purpose have been reported.[5–8] A soft independent modelling of class analogy (SIMCA) is a classification technique which gives a distinct confidence region around each class after applying principal components analysis (PCA).[9–12] New measurements are projected into each principal component's (PCs) space that describes a certain class to evaluate whether they belong to it or not.

In the present work a method to discriminate human gallstones by NIR spectrometry using SIMCA has been studied. The NIR spectra of 150 gallstones in the wave number range from 4500 to 10,000 cm^{-1} were measured. The 150 gallstone samples were classified to three classes (cholesterol stone, calcium bilirubinate stone and calcium carbonate stone) according to the contents of major components in each gallstone. The training set which contains objects of the different known classes was constructed using 120 NIR spectra and the test set was made with 30 different gallstone spectra. The number of important PCs to describe each class was determined by cross-validation in order to improve the decision criterion of the SIMCA for the training set. For each class the score plot of the objects in the

Table 1. The construction of class data set.

Class	Type of gallstone	Number of samples (Training set)	Number of samples (Test set)
A	calcium bilirubinate stone	40	10
B	calcium carbonate stone	40	10
C	cholesterol stone	40	10

training set belonging to the other classes was inspected. The critical distance for each class was computed using both Euclidean distance and Mahalanobis distance at an appropriate level of significance (α). Two methods were compared with respect to classification and their robustness towards the number of PCs selected to describe different classes.

Experimental

Sample and instrumentation

The gallstone samples studied in this work were kindly provided by the Kaemyung University Hospital (Taegu, Korea). The NIR spectra were collected in the reflectance mode with an InfraProverII FTIR spectrometer (Bran+Luebbe). Before the data acquisition, a successful system suitability test (wavenumber scale, absorbance scale and noise) was performed. Each spectrum used for SIMCA is the average spectrum of 32 scans. The spectral range used for the data analysis is from 4500 to 10,000 cm^{-1}. The samples were classified into three classes; cholesterol stone, calcium bilirubinate stone and calcium carbonate stone. The composition of experimental design employed in this study is listed in Table 1.

Procedure of SIMCA

A training matrix contains objects of different known classes and the submatrix contains *n* training objects belonging to a class that were measured at *p* variable. Each class is modelled separately, based on the similarity of objects within the class. The singular value decomposition (SVD) can be used to

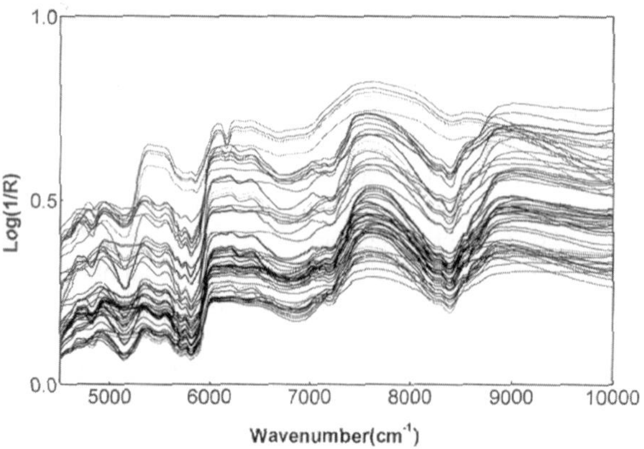

Figure 1. NIR spectra of gallstones taken from Korean patients.

perform PCA after column centering. In this work the number of PCs was selected according to the percentile of the total variance that is expressed by each PC. The PCs containing more than 1% of total variance were arbitrarily chosen for modelling.[12] The class boundaries or confidence limits are then constructed around the PC model. They are based on the distribution of distances (Euclidean distance, ED or Mahalanobis distance, MD) between the objects and the origin in the space of the residual PCs. With the help of an F-Test the critical distance can further be computed at a certain level of significance (a). To predict whether a certain object belongs to a certain class, it is projected on the space defined by the selected PCs of the training set of that class. After the model has been developed on the training set, new objects can be classified. For the identification of such a new object it is projected into the PC space defined by the PCA model and its distance from the class model is compared to the critical distance. If the distance is large than the critical distance, the object is considered a part of the class for which the model was established.

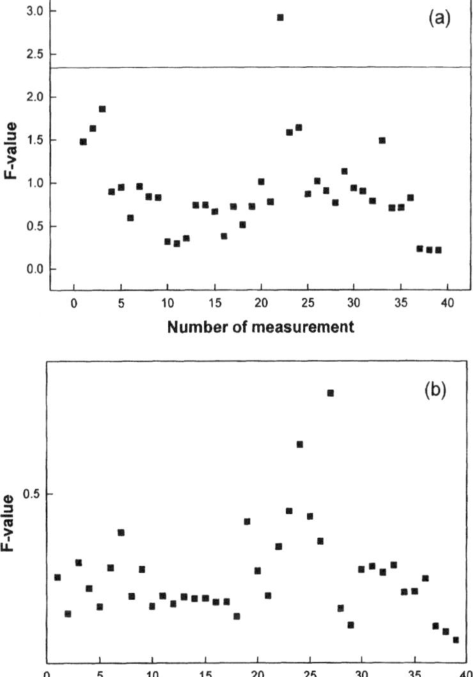

Figure 2. Calculated F-values based on Euclidean distance (a) and Mahalanobis distance (b) for the class A. Solid line indicates the critical F-value.

Result and discussion

Figure 1 shows the mean spectra of all classes. To determine the number of significant PCs for the training set, the SCREE and LEV plot method and the Malinowski IND function were used. The SCREE and LEV plot indicated eight significant PCs. Malinowski's IND function selected 11 PCs. The selection of all PCs that explain more than 1% of total variance within the data gave five significant PCs. The loadings on the early number PCs attribute weight to the different ranges of spectra. This indicates that the first eight PCs contain information that needs to be included in the model. From PC Number 9, the loadings are all

Table 2. Classification results of SIMCA based on Euclidean distance (ED) and Mahalanobis distance (MD).

Class	ED				MD			
	Training set		Test set		Training set		Test set	
	Sample	Outlier	Sample	Outlier	Sample	Outlier	Sample	Outlier
A	40	1	10	0	40	0	10	1
B	40	2	10	1	40	1	10	2
C	40	0	10	1	40	1	10	2

near to zero. From the result of PC selection methods and the investigation of the loadings, the selection of five to eight significant PCs seems to be reasonable. In this work we have decided to retain the PCs containing more than 1% of the total variance for our modelling. The remaining, residual PCs were used to build the confidence interval around the model.

Figure 2 shows the SIMCA results of class A for F-values obtained by using ED(a) and MD(b) for the training set at a significance level of 0.05. The SIMCA using MD seems to lead to fewer outliers. Table 2 shows the results of discrimination analysis of the test set. The SIMCA results using ED seem to have fewer outliers. The results of this work indicate that human gallstones can be successfully classified by NIR spectrometry using SIMCA.

Acknowledgement

This research was supported by Kyungpook National University Research Fund, 2000.

References

1. J.Y. Chu, I.H. Kim, T.J. Lim, H.J. Ryu, S.B. Kim and S.H. Lee, *J. of Korean Surgical Society* **51,** 88 (1996).
2. P.F. Malet, M.A. Dabezies, G. Huang, W.B. Long, T.R. Gadacz and R.D. Soloway, *Gastroenterology* **94,** 1217 (1988).
3. R.D. Soloway, B.W. Trotman and J.D. Ostrow, *Gastroenterology* **72,** 167 (1977).
4. B.M. Smith and P.J. Gemperline, *Anal. Chim. Acta* **423,** 167 (2000).
5. D. González-Arjona and A.G. González, *Anal. Chim. Acta* **363,** 89 (1998).
6. D. Jouan-Rimbaud, B. Walczak, D.L. Massart, I.R. Last and K.A. Prebble, *Anal. Chim. Acta* **304,** 285 (1995).
7. W.J. Welsh, W. Lin, S.H. Tersigni, E. Collantes, R. Duta, M.S. Carey, W.L. Zielinsk, J. Brower, J.A. Spencer and T.P. Layloff, *Anal. Chem.* **68,** 3473 (1996).
8. R. De Maesschalck, A. Candolfi, D.L. Massart and S. Heuerding, *Chemom. Intell. Lab. Syst.* **47,** 65 (1997).
9. A. Candolfi, R. De Maesschalck, D.L. Massart, P.A. Hailey and A.C.E. Harrington, *Chemom. Intell. Lab. Syst.* **19,** 923 (1999).
10. S. Wold and M. Sjöström, in *Chemometrics: Theory and Application*, Ed by American Chemical Society, Washington, DC, USA, p. 243 (1977).
11. M.P. Derde and D.L. Massart, *Chemom. Intell. Lab. Syst.* **4,** 65 (1988).
12. B. Mertens, M. Thompson and T. Fearn, *Analyst* **119,** 65 (1988).

Near infrared analysis of Iberian pig fat: repeatability file effect

M.D. Pérez-Marín, E. De Pedro, J. García-Olmo and A. Garrido-Varo

Escuela Técnica Superior de Ingenieros Agrónomos y Montes, University of Córdoba, Avda Menéndez Pidal s/n, E-14080 Córdoba, Spain

Introduction

Previous studies have demonstrated the viability of near infrared (NIR) technology for predicting the fatty acid content of Iberian pig fat.[1,2] However, despite the high degree of precision afforded by the equations obtained, considerable fluctuations were detected when predicting new samples; these fluctuations became more marked as the time interval between obtaining the equation and testing the product increased, thus hindering routine laboratory use of the equation. Moreover, in our hands, the problem was encountered only in products such as fat, which display an absorption pattern with well-defined peaks at given wavelengths; this prompts acute sensitivity to the slight instrumental changes which escape detection during routine instrument monitoring using the standard check cell. To correct this drawback, WinISI 1.04 software offers a mathematical algorithm consisting, briefly, in a "Repeatability File" which, when used for calibration purposes, enables minimisation of the sources of variation that might affect NIR predictions.[3]

The purpose of the present study was to assess the effect of using a repeatability file in the quantitative NIR analysis of subcutaneous Iberian pig fat.

Material and methods

Experimental material

A total of 188 samples of subcutaneous Iberian pig fat were taken from the carcass rump of batches of pigs reared on different feeding systems. Samples were obtained at COVAP, a Spanish pork- producing cooperative taking part in R+D Project IFD-0990, over two production seasons: 1997–98 ($N = 97$) and 1999–2000 ($N = 91$).

After removal of remnant skin, lean tissue and the surface fat layer, fat samples were melted in a microwave oven following De Pedro *et al.*[4]

NIR spectrum collection and reference data

Melted subcutaneous fat samples were analysed in a Foss NIRystems 6500 SY-I scanning monochromator equipped with a spinning cup, working in reflectance mode in the spectral range 400–2500 nm.

Measurements were made using folded-transmission gold reflector cups with a pathlength of 0.1 mm. Two spectra were measured per sample, the mean spectrum being used for subsequent analysis.

The fatty acid composition of each sample was determined by gas chromatography (GC). The methyl esters of fatty acids were extracted with hexane and were determined using a Perkin Elmer Sigma 3 D chromatograph with FID detector.

Chemometric treatment of data

Spectroscopic and chemical data were subjected to chemometric treatment using WinISI ver. 1.04 software.[5] Initially, the "Center" algorithm was applied to the sample set and six anomalous spectra were detected and eliminated. Calibration equations were then obtained for the 182 remaining samples, using the following: modified partial least squares (MPLS) for regression purposes; wavelength range 1100–2500 nm (at 2 nm intervals); SNV and Detrending treatments to correct for diffuse radiation phenomena. Several mathematical derivation treatments were also tested.

Calibration equations were then obtained using what the WinISI chemometric package termed a "repeatability file", composed of 128 spectra from a single sample, obtained weekly over a nine-month period.

The best equations obtained, with and without the repeatability file, were validated using a validation set composed of 12 fat samples not included in the calibration; sample spectra were collected some time after calibration.

The following statistical parameters were used to evaluate the predictive capacity of the equation: standard error of cross-validation (*SECV*), standard error of prediction corrected for bias [*SEP*(*c*)], coefficient of determination for the cross-validation process (r^2) and for the external validation process (R^2), and bias or mean residual error for the external validation set.

Results and discussion

NIR calibration equations were obtained for predicting the content of six fatty acids in melted Iberian pig fat: myristic acid (C14 : 0), palmitic acid (C16 : 0), palmitoleic acid (C16 : 1), stearic acid (C18 : 0), oleic acid (C18 : 1) and linoleic acid (C18 : 2). Resulting statistics are shown in Table 1.

The calibration statistics obtained (Table 1) suggest that these equations afford a high degree of precision for predicting content of the six fatty acids under study, yielding *SECV* values similar to, and in some cases lower than, those reported by García-Olmo *et al.*[2], using folded transmission aluminium reflector cups.

Table 1. Calibration and validation statistics obtained for predicting fatty acid content in subcutaneous Iberian pig fat with and without repeatability file.

			C14 : 0	C16 : 0	C16 : 1	C18 : 0	C18 : 1	C18 : 2
Without repeatability file	Calibration	*N*	172	176	174	175	166	175
		r^2	0.76	0.97	0.94	0.97	0.99	0.98
		SECV	0.07	0.28	0.08	0.27	0.20	0.16
	Validation	*N*	12	12	12	12	12	12
		R^2	0.52	0.92	0.74	0.94	0.98	0.99
		SEP(*c*)	0.09	0.43	0.13	0.27	0.47	0.13
		Bias	–0.05	-0.42	–0.03	0.47	0.14	0.25
With repeatability file	Calibration	*N*	179	168	175	180	171	163
		r^2	0.65	0.98	0.92	0.96	0.99	0.98
		SECV	0.08	0.24	0.09	0.29	0.20	0.16
	Validation	*N*	12	12	12	12	12	12
		R^2	0.90	0.97	0.79	0.99	0.99	0.94
		SEP(*c*)	0.06	0.26	0.10	0.17	0.32	0.26
		Bias	–0.04	–0.11	0.03	0.28	–0.04	–0.20

The validation statistics obtained suggest that the repeatability file is highly efficient for minimising sources of variation which may undermine the precision of predictions, particularly in products such as fat, which display well-defined peaks in the near infrared range. As Table 1 shows, differences with reference to calibration are negligible and, indeed, better results were sometimes obtained without applying the repeatability file. The positive effect of using the repeatability file is chiefly appreciable when predicting new samples not involved in the calibration process, whose spectra were collected some time after calibration; bias values were much lower than those recorded when not using the repeatability file.

Conclusions

The results obtained here support the use of a repeatability file when obtaining NIR calibration equations; this file ensures the required precision in routine prediction of fatty acid content in new subcutaneous Iberian pig fat samples.

Acknowledgements

This study formed part of Project CICYT-Feder IFD1997-0990 and was performed using equipment and infrastructure belonging to SCAI (Unidad NIR/MIR), University of Córdoba (Spain). GC data were obtained at the Laboratorio Agrario de Córdoba (Junta de Andalucía). The authors thank Ms Francisca Baena, Mr Alberto Sánchez de Puerta, Mr Antonio López and Ms Isabel Leiva of the Animal Production Department (ETSIAM-UCO) for technical assistance.

References

1. E. De Pedro, A. Garrido, I. Bares, M. Casillas and I. Murray, in *Near Infrared Spectroscopy: Bridging the Gap between Data Analysis and NIR Applications*. Ed by K.I. Hildrum, T. Isaksson, T. Næs and A. Tandberg. Ellis Horwood, Chichester, UK, p. 341 (1992).
2. J. García-Olmo, A. Garrido and E. De Pedro, in *Near Infrared Spectroscopy: Proceedings of the 9th International Conference*, Ed by A.M.C. Davies and R. Giangiacomo. NIR Publications, Chichester, UK, p. 253 (2000).
3. ISI, *A collection of new NIR topics*. Foss NIRystems/Tecator, Infrasoft International, LLC, Silver Spring, MD, USA (1999).
4. E. De Pedro, M. Casillas and C.M. Miranda, *Meat Sci.* **45(1),** 45 (1996).
5. ISI. The complete software solution for routine analysis, robust calibrations, and networking manual. Foss NIRystems/Tecator, Infrasoft International, LLC, Silver Spring, MD, USA (1998).

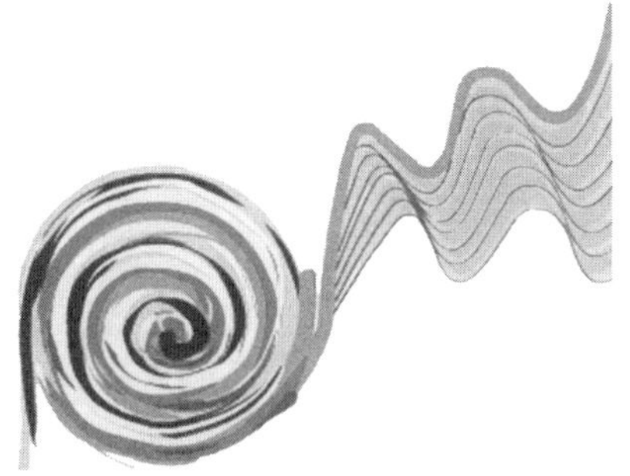

Pharmacy and Cosmetics

Application study of chemometric near infrared spectroscopy in the pharmaceutical industry

Makoto Otsuka*

Department of Pharmaceutical Technology, Kobe Pharmaceutical University, Higashi-Nada, Kobe 658-8558, Japan

Introduction

The polymorphs of pharmaceutical solids exhibit different physicochemical stability, processing characteristic, dissolution rate, etc. Particularly, dissolution rate may be affected which, in return, may significantly affect drug absorption for oral dosage form in the gastro-intestine resulting in a variation of bioavailability for the pharmaceutical compounds.[1–4] Therefore, an accurate assessment of polymorphism and solvate of bulk materials are required for reproducible preparation of pharmaceutical products. There are various analytical methods for polymorph determination, including powder X-ray diffraction,[5] differential scanning calorimetry (DSC),[6] thermal gravimetric analysis (TGA), microcalorimetry,[7] infrared (IR) spectroscopy,[8] Raman spectroscopy[9] and dissolution kinetics.[10] However, these methods are too time-consuming in the preparation of samples and/or their measurements. In contrast, near infrared (NIR) spectroscopy is simple due to its method of non-destructive sample preparation. Consequently, NIR spectroscopy is fast becoming an important technique used for pharmaceutical analysis in the industry.

On the other hand, chemometrics provides an ideal means of extracting quantitative information from UV-vis, IR and NIR spectroscopy, chromatography, mass spectrometry and NMR[11,12] spectra of multi-component samples. A number of chemometric and statistical techniques are employed in NIR quantitative and qualitative analysis because these approaches have been proved successful in extracting the desired information from unprocessed NIR spectra. Calibration methods such as multiple linear regression (MLR), principal component analysis/principal component regression (PCA/PCR) and partial least squares regression (PLS) are in common use.[13–18] Norris *et al.* and the others[19–22] reported that polymorphism of pharmaceuticals were evaluated based on their NIR spectra by MLR, PCR and PLS.

The purpose of this study is to investigate the application of the PCR method on analysing NIR spectroscopy for the quantitative determination of indomethacin (IMC) polymorphism. A direct comparison with the accuracy and experimental advantages with the conventional powder X-ray diffraction method was also explored.

*Corresponding address: Makoto Otsuka, PhD, Department of Pharmaceutical Technology, Kobe Pharmaceutical University, Motoyama-Kitamachi 4-19-1, Higashi-Nada, Kobe 658-8558, Japan. E-mail: m-otsuka@kobepharma-u.ac.jp

Experimental

Materials

The bulk powder of IMC was obtained from the Yashiro Co., Japan. The α form of IMC was prepared by using the following method. Ten g of IMC bulk powder was dissolved in 10 mL of ethanol at 80°C. The undissolved drug was filtered off. Then, 20 mL of distilled water at room temperature was added to the IMC-saturated ethanol solution at 80°C. The precipitated crystals were removed by filtration using a glass funnel and then dried under a vacuum at room temperature. The γ form of IMC was prepared by recrystallisation from ethyl ether at room temperature.[6]

X-ray powder diffraction analysis

X-ray powder diffraction profiles were obtained using X-ray diffractometer (XD-3A, Shimadzu Co., Japan). The measurement conditions include (1) scan mode: step scan, (2) target: Cu, (3) filter: Ni, (4) voltage: 20 kV, (5) current: 20 mA, (6) receiving slit: 0.1 mm, (7) time constant: 1 s, (8) scan width: 0.1 degree/step. The X-ray powder diffraction profiles were measured using the following method. Known quantities of standard mixtures were obtained by physically mixing α and γ forms of IMC powders at various ratios (0, 20, 40, 60, 80 and 100 w/w% γ form content) in a V-type mixer for one hour. About 80 mg of each sample powder were carefully loaded in a glass holder without particle orientation using a spatula and glass plate. After the powder X-ray diffraction profiles of samples had been measured under the above conditions, the intensity values were normalised against the intensity of silicon powder ($2\theta = 28.8°$) which was the external standard. The calibration curves for quantification of crystal content were based upon the total relative intensity of four diffraction peaks, $2\theta = 11.6$, 19.6, 21.8, 26.6°, of the γ form crystal. All data were reported as the average of five runs.

Fourier transform near infrared (FT-NIR) spectroscopy

FT-NIR spectra were taken using an NIR spectrometer (InfraProver, Bran+Luebbe Co., Norderstedt, Germany). Briefly, a fibre-optic probe was inserted into the sample powder (2 g) in a 20 mL glass bottle. Five scans per sample were recorded in the spectral range of 4500 to 10000 cm^{-1}. A ceramic (Coor's Standard) reference scan was taken for each set of samples. FT-NIR spectra of six calibration sample sets were recorded five times with the NIR spectrometer. A total of 30 spectral data were analysed by the various methods and chemometric analysis was performed using the PCR program associated with the SESAMI software (Bran+Luebbe Co.).

Results and discussion

Characterisation of α and γ forms of IMC

The results of powder X-ray diffraction profiles and the DSC profiles of the pure α and γ forms of IMC suggested that the main X-ray diffraction peaks of the α form were at 8.4, 14.4, 18.5, 22.0° (2θ) and those of the γ form were at 11.6, 16.8, 19.6, 21.9 and 26.7° (2θ), as reported prevously.[5] The DSC curves of the α and γ forms showed corresponding endothermic peaks at 155 and 162°C, respectively, which are attributable to sample melting. These results suggested that the α and γ forms of IMC used in the present study were highly purified.

Measurement of the polymorphic content of the γ form of IMC by conventional X-ray powder diffractometry

The calibration curve for measuring the content of the γ form by conventional X-ray diffraction method was based on the total intensity of the four specific diffraction peaks. The X-ray diffraction profiles showed two main causes in fluctuation in the determination of crystal content, one is a inten-

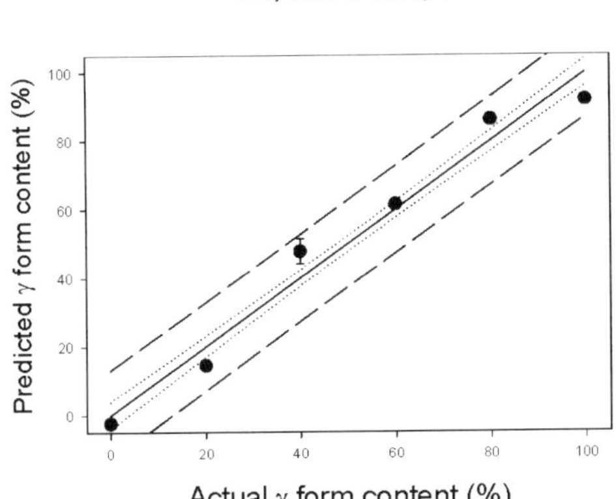

Figure 1. Relation between the actual and predicted content of form γ IMC obtained by conventional X-ray powder diffractometry. Bars present standard deviation. The symbols and error bars present average and standard deviation (n = 5). The solid line, long dash line and dotted line are represented a regression line, 95% predicted interval and 95% confidential interval, respectively.

sity fluctuation of the X-ray direct beam during measurement and the other is crystal orientation when the sample powder was loaded in the sample holder. In order to avoid fluctuation of direct beam intensity, the peak at $2\theta = 28.8°$ of silicon powder was measured as an external standard for correction of the value of crystalline content. The four diffraction peaks with the highest intensity were measured to minimise a systematic error due to crystal orientation.

Figure 1 shows a plot of the relationship between the actual and predicted polymorphic contents of the γ form of IMC measured using the X-ray diffraction method. This plot shows a linear relationship. It has a slope of 0.9983, an intercept of 0.7739×10^{-3} and a correlation coefficient of 0.9699. However, it has slightly higher, 95%, confidence levels for the prediction of individual y-values and 95% confidence intervals of regression, indicating that the X-ray diffraction method has relatively low accuracy in the determination of crystalline content.

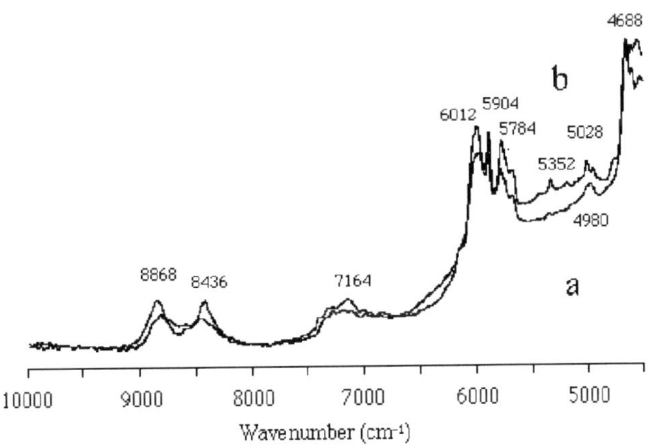

Figure 2. FT-NIR spectra of form α and form γ indomethacin. (a) form α; (b) form γ.

Measurement of content of the γ form by chemometric FT-NIR spectroscopy

Figure 2 shows the FT-NIR spectra of the α and γ forms of IMC. The α and γ forms of IMC showed significant NIR spectral peaks. The NIR absorption peaks of IMC were identified.[23]

In this study, the NIR spectra consist of 459 data points between 4500 to 10000 cm^{-1} at intervals of 12 cm^{-1}. Even batches of standard samples with various content of the γ form of IMC were prepared; four spectra were collected per batches. A total of 24 spectra were selected for the calibration (calibration set) and six spectra were removed and used for prediction of the calibration (prediction set). A pre-treatment was performed on the NIR spectra of the samples to minimise experimental error by using transformations of absorbance, normalised absorbance and second derivative. The best conditions were determined to minimise the root mean squared error of prediction (*RMSEP*, Equation 1).

$$RMSEP = \sqrt{\frac{\sum (y_p - y_r)^2}{n}} \tag{1}$$

Table 1 shows *RMSEP* of the correlation curves were calculated based on the spectral data corrected by normalisation. The *RMSEP* value decreased with an increase in the number of principal component factors, but this was almost constant after three PCs. Table 2 shows *RMSEP* of the correlation curves, which were calculated based on the spectral data corrected by three transformations. As a result, the minimum *RMSEP* value was the normalised NIR spectra based on a three-principal component model, so, the three-principal component model based on normalised NIR spectra was taken for the later analysis.

Figure 3 shows loading vectors corresponding to the principal components (PCs), respectively. The peak at 4560 cm^{-1} was the highest value and the peaks at 6048, 5772, 5352, 8836 and 8486 cm^{-1} were lower on PC1, because there were large spectral intensity differences between α and γ forms at the peaks. The loading vector of PC1 was similar to that of PC2, but not to that of PC3. The result suggested that the loading vectors reflected the spectral difference between α and γ forms.

Figure 4 shows a plot of the calibration data obtained by the NIR method between the actual and predicted contents of the γ form of IMC. The predicted values were reproducible and had a smaller standard deviation. The multiple correlation coefficient, the standard error of estimate (*SEE*) and the *RMSEP* were evaluated to be 0.998 2.559, 3.507, respectively. Since the purpose of this study is to compare the accuracy of the chemometric NIR method with that of conven-

Table 1. *RMSEP* of correlation calculated by PCR based on number of PC.

Transfomation	Number of PC	*RMSEP*
Abs.	2	7.401
Abs. + Nor.	3	3.507
Abs. + 2nd deriv.	2	5.680

Abs: Absorbance
Nor: Normalise
2nd deriv: second derivative.

Table 2. *RMSEP* of correlation calculated by PCR based on various transformations.

Number of factor	*SEP*
0	35.642
1	12.802
2	6.027
3	3.208
4	2.665
5	3.281
6	2.282
7	2.476
8	2.510
9	2.349
10	2.361

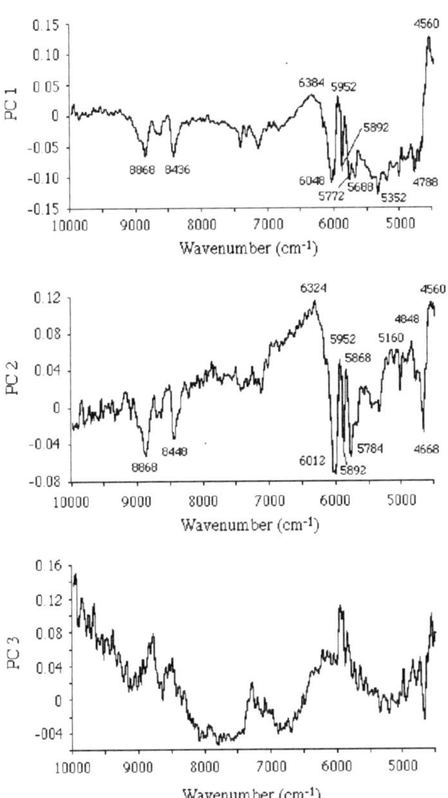

Figure 3. Loading vectors of PCs 1, 2 and 3 based on normalised NIR spectra calculated by PCR.

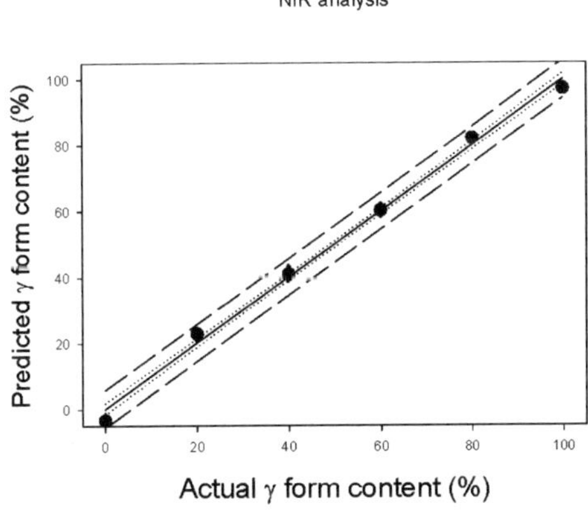

Figure 4. Correlation between actual and predict content of form γ IMC obtained by FT-NIR spectroscopy. The symbols and error bars present average and standard deviation (*n* = 5). The solid line, long dash line and dotted line are represented a regression line, 95% predicted interval and 95% confidential interval, respectively.

tional X-ray powder diffraction, the mean bias and the mean accuracy were determined by Equations 2 and 3, respectively.

$$B_m = \frac{\sum_{i=1}^{n} \frac{(X_c - X_t)}{X_t}}{n} \times 100 \tag{2}$$

$$A_m = \frac{\sum_{i=1}^{n} \frac{|X_c - X_t|}{X_t}}{n} \times 100 \tag{3}$$

B*m* is percentage mean bias, A*m* is percentage mean accuracy, *Xc* is the predicted value of content of the γ form of IMC, *Xt* is actual value of content of the γ form of IMC and *n* is number of experiments.

The mean bias for the NIR and X-ray powder diffraction method were calculated to be 2.95% and –0.94% and the mean accuracy were 4.29 and 10.80%, respectively. The confidence levels for the prediction of individual y-values for the NIR method were much narrower than that for using the conventional X-ray method, but the result was consistant with the X-ray method. These results indicate that the NIR method was more accurate than the X-ray method. Thus, this assay is found to have significant advantages for quantitative analysis of IMC polymorphs.

Reference

1. FDA papers, *Pharm. Tech. Japan* **1,** 835 (1985).
2. J.K. Haleblian, *J. Pharm. Sci.* **64,** 1269 (1975).
3. M. Otsuka and Y. Matsuda, in *Encyclopedia of pharmaceutical technology*, Volume 12, Ed by J. Swarbrick and J.C. Boylan. Marcel Dekker, New York, USA, pp. 305–326 (1995).
4. T.L. Threlfall, *Analyst* **120,** 2435 (1995).
5. H. Yoshino, Y. Hagiwara, S. Kobayashi, M. Samejima, *Chem. Pharm. Bull.* **32,** 1523 (1984).
6. N. Kaneniwa, M. Otsuka and T. Hayashi, *Chem. Pharm. Bull.* **33,** 3447 (1985).
7. H. Ahmed, G. Buckton and D.A. Rawlins, *Int. J. Pharm.* **130,** 195 (1996).
8. D.B. Black and E.G. Lovering, *J. Pharm. Pharmacol.* **29,** 684 (1977).
9. L.S. Taylor and G. Zografi, *Pharm. Res.* **15,** 755 (1998).
10. J.C. Berridge, P. Jones and A.S. Roberts-McIntosh, *J. Pharm. Biomed. Anal.* **9,** 597 (1991).
11. U. Edlund and H. Grahn, *J. Pharm. Biomed. Anal.* **9,** 655 (1991).
12. D. Lincoln, A.F. Fell, N.H. Anderson and D. England, *J. Pharm. Biomed. Anal.* **10,** 837 (1992).
13. M. Otsuka, F. Kato and Y. Matsuda, *Pharmsci.* **2,** (2000).
14. K.M. Morisseau and C.T. Rhodes, *Pharm. Res.* **14,** 108 (1997).
15. J.K. Drennen and R.A. Lodder, *J. Pharm. Sci.***79,** 622 (1990).
16. B.R. Buchanan, M.A. Baxter, T.S. Chen, X.Z. Qin and P.A. Robinson, *Pharm. Res.* **13,** 616 (1996).
17. P. Frake, I. Gill, C.N. Luscombe, D.R. Rudd, J. Waterhouse and U.A. Jayasooriya, *Analyst* **123,** 2043 (1998).
18. H. Martents and T. Næs, *Multivariate Calibration*, John Wiley & Sons, Chichester, UK (1989).
19. T. Norris, P.K. Aldridge and S.S. Sekulic, *Analyst* **122,** 549 (1997).
20. R.W. Saver, P.A. Meulman, D.K. Bowerman and J.L. Havens, *Int. J. Pharm.* **167,** 105 (1998).
21. A.D. Patel, P.E. Luner and M.S. Kemper, *Int. J. Pharm.* **206,** 63 (2000).
22. M. Blanco, J. Coello, H. Iturrriaga, S. Maspoch and C. Perez-Maseda, *Anal. Chim. Acta* **407,** 247 (2000).
23. M. Iwamoto, S. Kawano and J. Uozumi, *Introduction of Near Infrared Spectroscopy.* Sachi Syobou Co., (1994).

Fast quantitative and qualitative analysis of pharmaceutical tablets by near infrared spectroscopy

Line Lundsberg-Nielsen,[a*] Charlotte Kornbo,[a] Mette Bruhn[a] and Marianne Dyrby[b]

[a]*H. Lundbeck A/S, Ottiliavej 9, DK-2500 Valby, Denmark*

[b]*The Royal Veterinary and Agricultural University, Rolighedsvej 30, DK-1958 Frederiksberg C, Denmark*

Introduction

The implementation of near infrared (NIR) spectroscopy and chemometrics in the pharmaceutical industry is still progressing strongly, both regarding qualitative and quantitative applications and beneficial results are seen. Looking at the development so far, NIR will change the pharmaceutical industry even more in the future.

This paper addresses the experiences and progress achieved regarding the application and implementation of quantitative methods for determination of content uniformity of tablets with less than 10% w/w (weight percent) of active substance, using NIR transmittance spectroscopy in combination with chemometric/multivariate data analysis such as partial least squares (PLS) regression. Also, qualitative methods for identification of the same tablets by NIR reflectance spectroscopy will be discussed.

Four commercial tablet strengths are formulated (5, 10, 15 and 20 mg) and produced from two different compositions by direct compression. Three strengths (10, 15 and 20 mg) are dose proportional, i.e. fixed concentration but varying in tablet size. The concentration of active drug substance within the 5 mg tablet is 5.5% w/w, whereas the 10, 15 and 20 mg are 8.0% w/w. The aim was to replace the conventional primary methods for analysing content uniformity and identification by NIR, whereby the lead-time could be reduced by more than a factor of 300, sample preparation was unnecessary, a non destructive method could be used and no chemical reagent consumption was needed.

The task was therefore to:

- develop an NIR calibration for quantitative determination of the active substance within coated as well as uncoated tablets.
- develop an NIR calibration for qualitative identification of the tablets, including the active drug salt as well as coating.

Equipment

All tablet analysis was performed on an FT-NIR instrument (MB160 Pharma, ABB Bomem) equipped with a tablet sampler for transmittance measurements and a powder sampler for reflectance measurements. The reference method for the quantitative measurements was high performance liquid

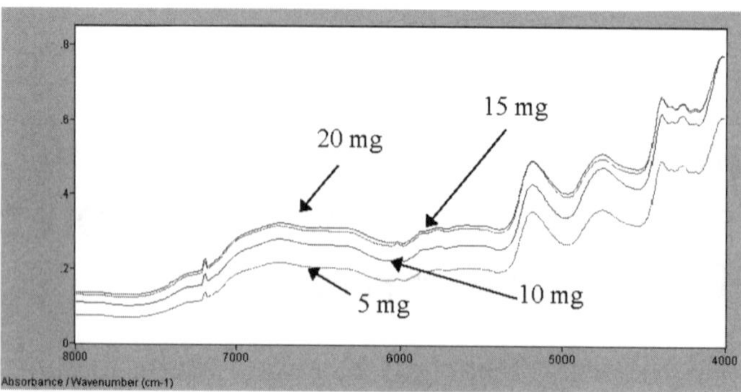

Figure 1. Diffuse reflectance spectrum of 5, 10, 15 and 20 mg tablets.

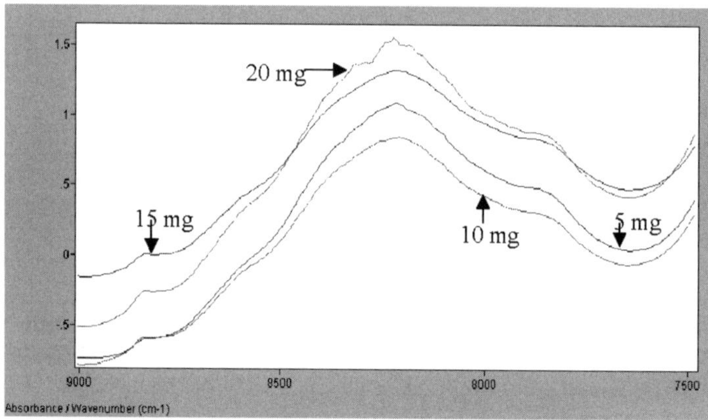

Figure 2: Diffuse transmittance spectrum of 5, 10, 15 and 20 mg tablets.

chromatography (HPLC) (Merck and Waters equipment) and for the qualitative identifications, traditional wet chemical methods were used.

Analysis of active content

Preliminary studies

A preliminary study was carried out to investigate which kind of NIR measurement and which type of chemometric/multivariate calibration was the most suitable for these tablets. The study included:

- reflectance or transmittance spectroscopy
- absolute calibration units (mg tablet^{-1}) or weight concentration units (% w/w).
- a global calibration covering all four table strengths or one calibration per tablet or formulation.

The reflectance spectra and transmittance spectra of the four tablet strengths are shown in Figures 1 and 2, respectively.

Table 1. Result of the preliminary study for determination of drug substance.

Measurement technique	Calibration	Pre-treatment	Unit	Correlation $(R^2$	Standard error of cross-validation (*RMSEC*)	Actual relative prediction error (*RSEP*)
Reflectance	PLS, global	MSc	% w/w	0.91	0.41%	5.117.4%
Transmittance	PLS, global	MSc	% w/w	0.96	0.33%	4.115.9%
Transmittance	PLS, global	1st deriv.	Absolute	0.99	1.46 mg tablet^{-1}	7.3–30%
Transmittance	PLS, local	MSC	% w/w	0.88–0.92	0.22–0.31%	< 4%

Where *RSEP* = *RMSECV* · (nominal concentration)$^{-1}$ · 100%. The study is described in detail by M. Dyrby *et al.*[1]

The study was performed based on tablets from nine laboratory batches, 12 pilot batches and seven full-scale batches. The results are shown in Table 1:

The study showed that transmittance should be used together with local models, developed on weight concentrations and, furthermore, coated as well as uncoated tablets. We decided, therefore, to develop one PLS calibration per formulation, using MSC as data pre-treatment.

Quantitative models

The final calibrations were made based on four production batches and three pilot batches (292 spectra in total) for the 5 mg model and 14 production batches and nine pilot batches (724 spectra in total) for the 10, 15, 20 mg model (Figures 3 and 4).

The specification for content uniformity is that the active drug content within a tablet should be the nominal value ± 15%. In real production, the variation of the content is typically within ± 8%. In order to extend the calibration range to include the whole specification range, pilot batches with increased or

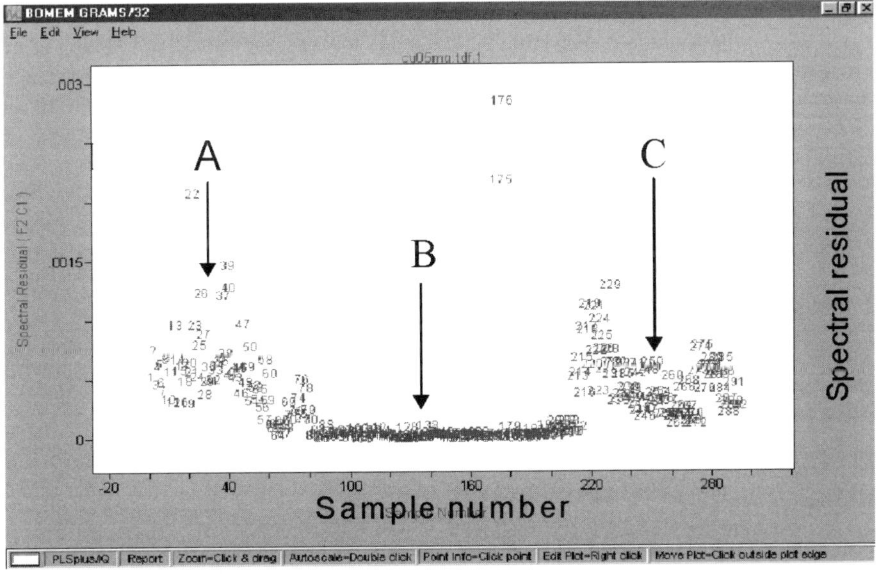

Figure 3. PLS calibration, 5 mg tablets. Weight concentration: 5.5%, as well as residual plot.

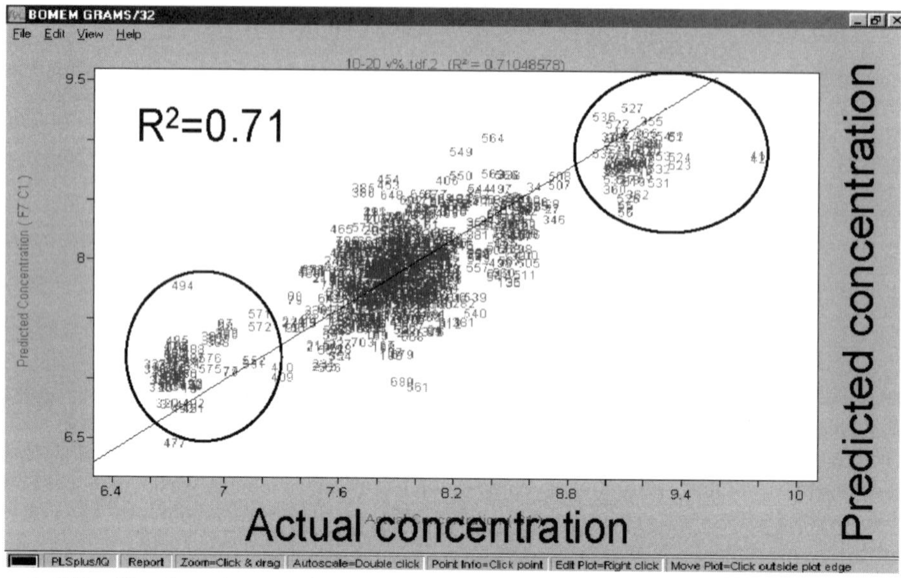

Figure 4. PLS calibration 10, 15 and 20 mg tablets. Weight concentration: 8%, as well as a typical scoreplot.

reduced content of the active substance were produced. In Figure 4, these samples are marked by the two ellipsoids. One major problem of varying the active drug content is that the formulation is changed, since increasing the drug content results in decreasing one of the major excipients. Even though the calibration window was chosen around the 2nd overtone of the C–H stretch at 8830 cm^{-1}, the responses from the other excipients are still contributing to the calibration as well as the physical properties of the tablets, which is seen from the relatively low correlation ($R^2 = 0.71$) in Figure 4. As can be seen from both Figures 3 and 4, these calibrations also reflect real production, i.e. includes variations in coating thickness, tablet sizes, tablet presses, different pounces, as well as the slight variation in the formulations. The model is not able to separate the three different tablet strengths (10, 15, 20 mg) from one another as expected since the concentrations are the same.

Table 2. Result of the preliminary study for identification of tablets.

Measurement technique	Calibration	Pretreatment	Correlation (R^2)	Standard error of cross-validation (*RMSECV*)
Transmittance	PLS-1 discriminatior	2nd deriv.	0.98	7.2
Transmittance	PLS-1 discriminatior	MSC	0.94	17
Reflectance	PLS-1 discriminatior	1st deriv.	0.99	5.7
Reflectance	PLS-1 discriminatior	2nd deriv.	0.99	5.4

PLS-1 discrimination: In the models the true samples have been assigned the value 100 and the false samples 0 for the "dummy" Y-variable. Further details are described by M. Dyrby *et al.*[1]

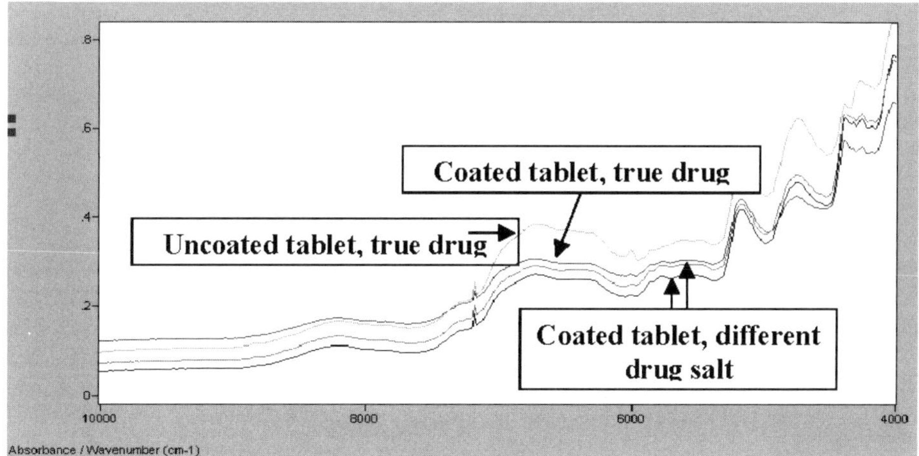

Figure 5. Reflectance spectrum of different 10 mg tablets.

The result of the validation of these calibrations with a new test spectra is that the relative prediction error,[2] is 3.8% for the 5 mg model and 3.9% for the 10, 15, 20 mg model. Both are in the same order-of-magnitude as the error of the HPLC reference measurements (typically 3.5%) and, therefore, suitable for its extended purpose of measuring the relative active drug content.

Identification of the tablets

Preliminary studies

Another preliminary study was carried out to investigate how to identify the tablets. The study considered:

- using reflectance or transmittance spectroscopy
- how to develop the discrimination models for identification of drug, salt and coating

The results of the study is presented in Table 2.

Reflectance spectra of the 10 mg tablets, coated and uncoated, as well as special tablets produced with the different drug salts, are shown in Figure 5.

The study shows that reflectance should be used and that one global calibration, including drug salts, coating and the four tablet sizes, is possible. Due to GMP reasons, we decided to develop two different PLS-1 discrimination qualitative models, one per formulation, using derivatives as data pre-treatment. In the following, only the model for the 10, 15, 20 mg formulation will be presented, as the 5 mg model is very similar.

Qualitative model

Based on 14 production batches and six laboratory batches (556 spectra in total) for the 10, 15, 20 mg model, the final calibrations were made.

In order to be able to distinguish between the different drug salts possible within our production, tablets with different drug salts were produced, coated as well as uncoated. The calibration result showed $SECV = 7.47$. The model (Figure 6) is able to distinguish the three different classes from one another and as the separation in prediction is clear, it can be used for identifying the tablets.

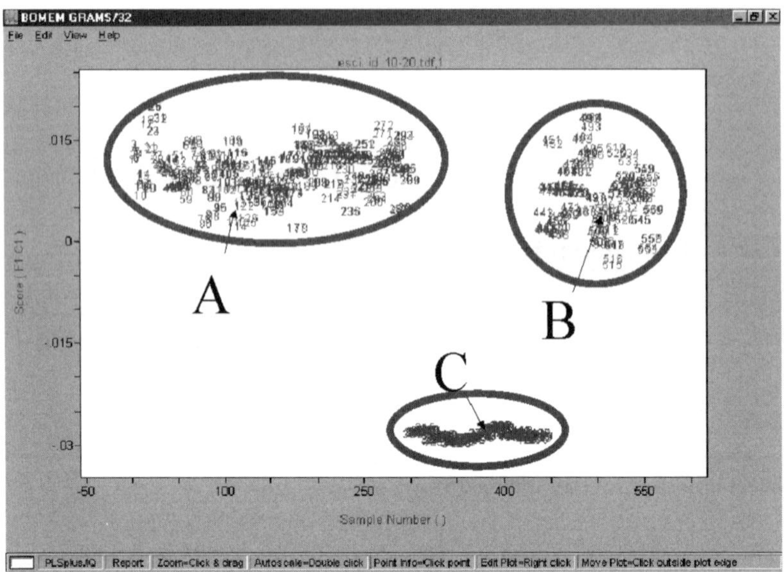

Figure 6: PLS-1 discrimination model for identification of 10, 15 and 20 mg tablets and factor 1 scoreplot, visualising the three different classes: True coated tablets, uncoated tablets with right drug salt, coated as well as uncoated tablets with different drug salt.

Concluding remarks

It has been shown that it is possible to qualify and quantify dose proportional tablets in real production, but one calibration per formulation is needed for the quantitative methods. The impact of physical production parameters can not be eliminated completely in the models by pre-treatment, so effects from different tablet presses, pounces, coating thickness etc. do influence the prediction accuracy of the model. However, the total uncertainty of the model is of the same order-of-magnitude as that of the reference method itself, so the models can be used in the laboratory or directly at-line or on-line in the production chain.

The use of the quantitative chemometric/multivariate models directly in production can be used not only to control the tablet quality in real-time, but also for different production trials, such as investigation of the impact on the active drug content by variation of particle size, tablet presses and new raw material suppliers.

References

1. M. Dyrby, S.B. Engelsen, L. Nørgaard, M. Bruhn and L. Lundsberg-Nielsen, submitted to *Applied Spectroscopy.*
2. A. Eustaquio, M. Blanco, R.D. Jee and A.C. Moffat, *Anal. Chim. Acta* **383,** 283 (1999).

Advances in near infrared spetroscopy in phytochemistry

C.W. Huck

Institute of Analytical Chemistry and Radiochemistry, Leopold-Franzens University Innsbruck, Innrain 52a, 6020-Innsbruck, Austria.

e-mail: christian.w.huck@uibk.ac.at

Introduction

During recent years the pyhtopharmaceutical industry has continued to be an area of growth—between 1994 and 1997 sales of phytopharmacia in the United States increased by 93%, in Europe by15% and in Asia by 15%. Due to this increase in demand, new analytical methods for quality control are required and the analysis of plants, their extracts and natural compounds has become a demanding challenge. Therefore, near-infrared (NIR) spectroscopy, supported by sophisticated statistical software, offers a fast, reliable and non-destructive method for controlling not only chemical but also physical parameters at once. As a continuation of our earlier developed method for the quality control of *Flos Primulae veris* extracts[1,2] and its phenolic ingredients, an NIR method for the quantitative analysis of hypericin in *Hypericum perforatum* L. extracts was established. *St. John´s Wort* extract is used for the treatment of skin injuries, burns, neuralgia, antibacterial activity and as a treatment of mild to moderate depression.[3–9] In recent years, scientific studies to examine ingredients and quality of red wine also have become very important. The reason for this can be found in its health benefits.[10-12] Some papers have already discussed the problem of finding out the geographical origin of a wine without sensorial tests, for example with high-performance liquid chromatography (HPLC),[13] gas chromatography (GC) or pyrolysis mass spectrometry.[14] These classifications are only possible by the use of multivariate analyses. Due to the high demand for a fast analytical tool for the analysis of red wine, an NIR method to distinguish between different varieties and age groups was established. Correlation with results received from HPLC and HPLC coupled to mass spectrometry (HPLC-MS) also enabled the content of antioxidative phenolic ingredients to be predicted. Finally, based on these experiments, a qualitative model for the differentiation between *arabica* and *robusta* beans was established. This method has proved to be an interesting and powerful analytical tool for the coffee producing industry as quality and prices are fixed by the type of beans.

Materials and methods

The near infrared spectra used in this work were collected with a FT-NIR universal spectrometer from Büchi Labortechnik AG (Uzwil, Switzerland), which is equipped with a tungsten–halogen source and a PbS-detector. The absolute wavelength accuracy is $\pm$ 2 cm^{-1} over a wavelength range from 1000 to 2500 nm. The sample measurement was performed using fibre optics with ten scans for one average spectrum to eliminate inhomogenities. Software from Büchi Labortechnik was used: BCAP 5.0 for controlling the spectrometer and NIRCAL 3.0 for the processing of spectral data and chemometrics. All samples were heated to 23°C and scanned without any sample pretreatment. For reference analyses HPLC, HPLC-MS and capillary electrophoresis (CE) were used. All individual parameters can be found in the cited literature.[15–17] The determination of the ethanol/water content in red wine was carried out by gas chromatography with flame ionisation detection (GC-FID) and Karl–Fischer titration. GC-FID: HP1 fused silica (50 × 0.32 mm I.D.); gas, hydrogen 0.7 bar; injector temperature, 30°C; temperature from 50°C (4 min) up to 220°C (8°C min^{-1}) and up to 300°C (20°C min^{-1}); detector temperature, 300°C; split, 35 mL min^{-1} 1 : 120; sample size, 5 µl. Karl-Fischer titration: 20 µl of each extract were titrated on a 684 KF Coulometer (Metrohm, Filterstadt, Germany).

Quantitative analysis of hypericin in *Hypericum perforatum* L. extracts

The otpical pathlength used in the transflectance mode was 3 mm. Eighty three samples were measured over a wavelength range from 4500–9996 cm^{-1}. First derivation was calculated and the 332 spectra were divided into a learning (60%) and a validation set (40%). The smoothed average of the spectra was three points. For partial least square regression (PLSR) 15 factors were selected.

Qualitative wine analysis

The otpical pathlength used in the transflectance mode was 0.5 mm. After performing a second derivation of the recorded 172 spectra they were randomly partitioned into a calibration set (70% of total samples) and a validation set (30% of total samples). The smoothed average of all spectra was three points and six factors were used. With these data sets a principal component analysis (PCA) over a wavelength range from 4000 to10000 cm^{-1} was carried out.

Qualitative coffee analysis

Eighty five samples were measured in the reflectance mode over a wavelength range from 4500 to 9996 cm^{-1}. A first derivation and three points average smoothing were performed. Five factors were used for the calculation. 67% of the recorded 255 spectra were used for the calibration and 33% for the validation. With these data sets a PCA was carried out.

Results and discussion

The strategy scheme depicted in Figure 1 for the quantitative analysis of hypericin in *St. John´s Wort* can also be used as a general scheme. At the beginning, the extraction procedure and reference

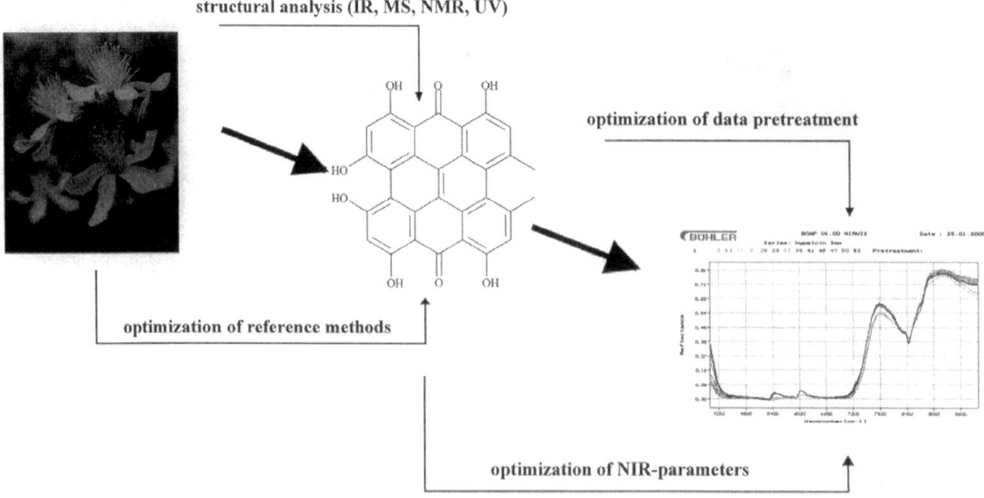

Figure 1. Strategy scheme for the analysis of hypericin in *Hypericum perforatum* L. extracts.

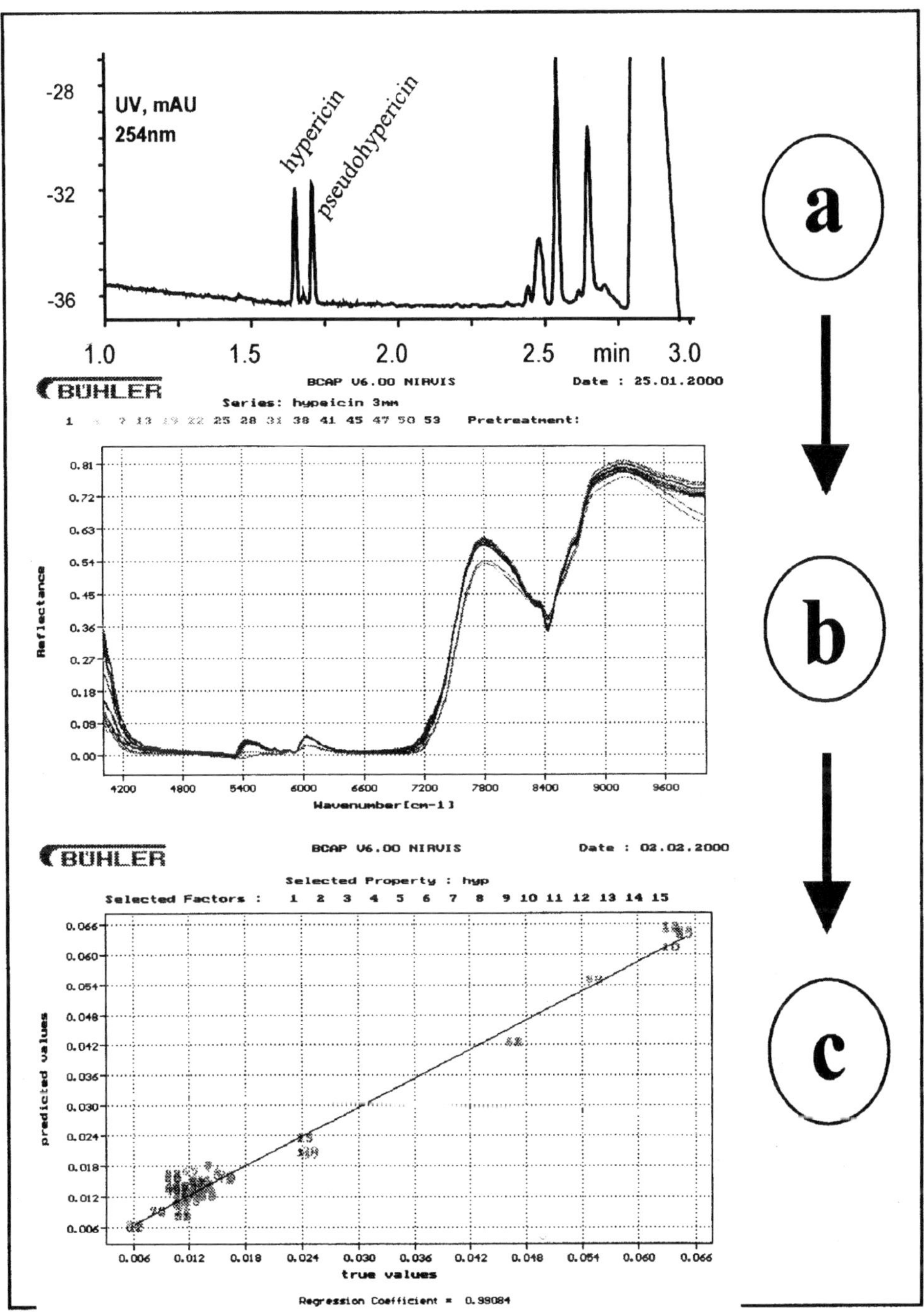

Figure 2. Quantitaive analysis of hypericin in *Hypericum perforatum* L. extracts. (a) Capillary electrophoresis (reference method); (b) NIR-spectra; (c) calibration curve.

method have to be established, optimised, evaluated and validated. After that, NIR parameters such as temperature and optical thin-layer thickness must be optimised. Furthermore, the best data pretreatment must also be found. For the control of the hypericin content in *Hypericum perforatum* L. extracts, CE was used as the reference method. Each extract was measured three-fold and cross-wise by CE [Figure 2(a)]. For the separation, a buffer system consisting of 26 mM phosphate, 0.001% HDB, 4.5% butanol and 20% acetonitrile (pH 2.65) was used. The calibration calculated the content of hypericin. After optimisation of the temperature and the optical thin layer, 332 spectra of 83 extracts were recorded in the transflectance mode with a NIRVIS instrument. Mathematical pretreatment and statistical analysis were carried out by performing PLS. By recording the NIR spectrum [Figure 2(b)] and calculation of its first derivation, characteristic absorption bands were identified. The most intensive band belonged to the vibration of the second overtone of the carbonyl group (5376 cm^{-1}), followed by the C–H stretch and C–H deformation vibration of ethanol (7212 cm^{-1}), the –OH vibration of water and ethanol (4440 cm^{-1}), the $-CH_2$ overtone (5742 cm^{-1}) and the $-CH_2$ / $-CH_3$ overtone (5808 cm^{-1}). All recorded spectra were transformed to their first derivative before being calculated in the linear PLS model. Fifteen principal components were necessary in order to reach the best calibration equation. The multivariate statistical method, PLS, is a full spectrum method. Therefore, the information of the whole recorded spectral range can be used for the calibration. In the course of model optimisation the best statistical results were obtained when the spectral information in the interval between 4500 and 9996 cm^{-1} was used for calculating the PLS. Finally, a correlation coefficient of 0.99084 for the calibration curve of NIR values against CE values helped to assess the linearity of the model [Figure 2(c)]. For a higher robustness of the calibration system, extracts with spiked hypericin were included in the curve. This method was proven to be a highly suitable spectroscopic tool to quantify hypericin in *Hypericum perforatum* L. extracts in the low percentage range.

As the variety of wine has a big influence on the quality of a red wine, a method to distinguish between different bottles made of different grapes was established. Fifty bottles of red wine of three different wine variety samples were inspected: *Cabernet Sauvignon*, *Lagrein* (both pure wines) and *Sangiovese* (Chianti). After recording the spectra, partitioning into a calibration and a validation set, average smoothing of three points and performance of a second derivation, PCA, over a wavelength range from 1010 to 2222 nm, was carried out. Figure 3(a) shows that

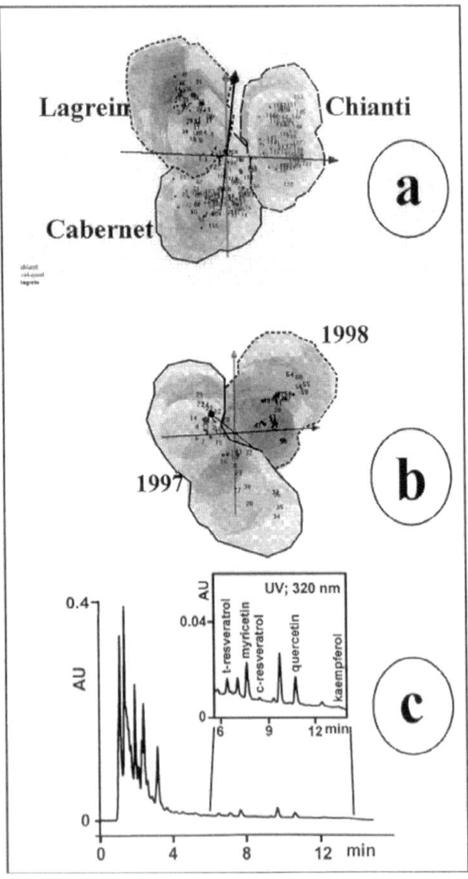

Figure 3. Qualitative analysis of red wine. (a) 3d-scatter plot of 3 different varieties; (b) 3d-scatter plot of two age groups of Cabernet; (c) HPLC-analysis (reference method).

the inspected wines (*Cabernet Sauvignon*, *Lagrein* and *Chianti*) can be classified very clearly by a 3-d scatter-plot. Each property (wine variety) could be assigned to one cluster. Due to the fact that the inspected wines come from very close geographical origins and two of them were produced using the same method in the same wine cellar at the same time, it is possible to estimate that the reason for this separation is the different wine variety. Furthermore, it was possible to distinguish between *Cabernet* 1997 and 1998 [Figure 3(b)]. One reason for this possible classification can be found in the different fingerprint caused by the phenolic ingredients, which was analysed by HPLC [Figure 3(c)]. Correlation of the results received from these HPLC analyses with those from NIR spectroscopy allows us to give a statement about the amount of phenolic ingredients found. If NIR spectroscopy identifies a wine as a *Cabernet Sauvignon* 1997, for example, the content of quercetin is between 18 and 30 µg mL^{-1}.

Based on these results, investigations in the field of coffee analysis were carried out. The development of an NIR spectroscopic method to distinguish between *arabica* and *robusta* green coffee beans is an important quality control tool, as the international coffee trade is conducted almost exclusively with these two varieties. With NIR in reflectance mode it is possible to perform a complete profile of the whole bean. At first, different extracts of roasted *arabica* and *robusta* beans were investigated using HPLC analysis of the three main ingredients which are necessary for the taste of coffee: caffeine, theobromine and theophylline [Figure 4(a)]. By this method, the different patterns of *arabica* and *robusta* could be confirmed. These differences were used in NIR to distinguish between the two varieties. Two hundred and fifty five spectra of 85 samples were recorded in the reflectance mode with a NIRVIS instrument. Mathematical pretreatment and statistical analyses were carried out by performing PCA. Recording the NIR spectrum of a green bean and calculating its first derivative again identified the characteristic absorption bands of xanthines [for structure see, for example, caffeine in Figure 4(a)]. The most intense band belonged to the vibration of the second overtone of the carbonyl group (5268 cm^{-1}), followed by the C–H stretch and C–H deformation vibration at 7116 cm^{-1}. All spectra were transformed to their first derivative before calculating in the PCA. Finally, the *arabica* and *robusta* beans could be classified very clearly by 3-d scatter plot [Figure 4(b)]. Each property (bean variety) could be assigned to one cluster. Due to the fact that *arabica* and *robusta* beans come from different cultivation areas, this method is highly suitable for a fast differentiation of green coffee varieties.

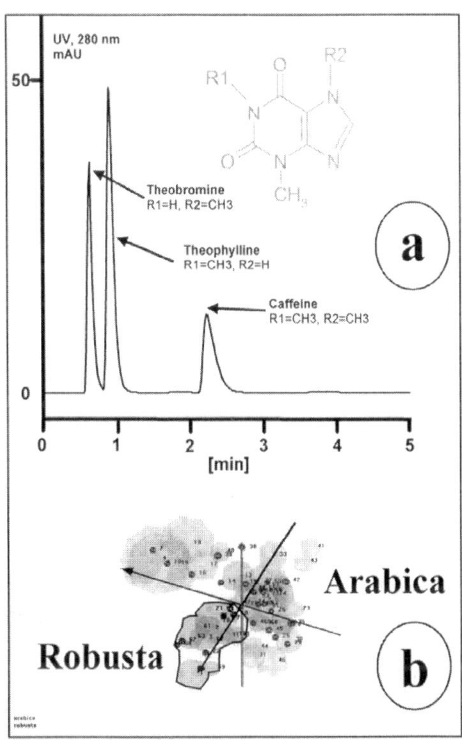

Figure 4. Qualitative differentiation between green coffee varieties. (a) HPLC analysis (reference method); (b) 3d-scatter plot.

References

1. C.W. Huck, R. Maurer, G.K. Bonn, M. Popp and N. Basener, *Near Infrared Spectroscopy: Proceedings of the 9th International Conference*, Ed by A.M.C. Davies and R. Giangiacomo. NIR Publications, Chichester, UK, p. 487 (2000).

2. C.W. Huck, R. Maurer, M. Popp, N. Basener and G.K. Bonn, *Pharm. Pharmacol. Lett.* **9,** 26 (1999).
3. C.M. Hawken, *St. John's Wort.* Woodland Publishing Inc., Pleasant Grove, USA (1997).
4. G. Lavie, Y. Mazur, D. Lavie and D. Meruelo, *Med. Res. Rev.* **15,** 111 (1995).
5. Ch.M. Schempp, K. Pelz, A. Wittmer, E. Schöpf and J.C. Simon, *The Lancet* **353,** 2129 (1999).
6. N. Rosenthal, *St. John's Wort The Herbal Way to Feeling Good.* Harper Collins Publisher Inc., New York, USA (1995).
7. J. Deltito and D. Beyer, *J. Affect. Disord.* **51,** 345 (1998).
8. J. Chang, *Biochem. Pharmacol.* **59,** 211 (2000).
9. E. Celotti, R. Ferrarini, R. Zironi and L.S. Conte, *J. Chromatogr. A* **730,** 47 (1996).
10. F. Fauconneau, P. Waffo-Teguo, F. Huguet, L. Barrier, A. Decendit and J.M. Merillon, *Life Sciences* **61,** 2103 (1997).
11. N. Worm, *Täglich Wein*, 5th Edn. Hallwag AG, Bern, Switzerland (1998).
12. S. Renaud and M. De Lorgeril, *The Lancet* **339,** 1523 (1992).
13. K. Sivertsen, B. Holen, F. Nicolaysen and E. Risvik, *J. Science of Food and Agriculture* **79,** 107 (1999).
14. L. Montanarella and M.R. Bassani, *Rapid Commun. Mass Spectrom.* **9,** 1589 (1995).
15. G. Stecher, C.W. Huck, M. Popp and G.K. Bonn, *Fresenius J. Analyt. Chem.*, in press (2001).
16. N. Schelosky, *Analytische Untersuchungen an Johanniskraut*, Ph.D. thesis, University of Innsbruck (1998).
17. R. Pirker, C.W. Huck and G.K. Bonn, *J. Chromatogr. B*, in press (2001).

Quantification of an active ingredient in tablets using near infrared transmission measurements

Andreas Niemöller, Angela Schmidt, Helmut Weiler and Aaron Weis
Bruker Optik GmbH, Rudolf-Plank-Straße 23, D-76275 Ettlingen, Germany
E-mail: optik@bruker.de

Introduction

In the pharmaceutical industry spectroscopic techniques are suitable for monitoring different steps of the production process (for example, receipt of goods, blending and mixing of components) and the QC of final products.[1] Near infrared (NIR) spectroscopy offers a lot of options to do fast, easy, non- invasive and non-destructive analyses of liquids, powders and solids without time-consuming sample preparation. The capability of NIR for non-destructive analysis makes it highly qualified for quality control of tablets, for example, batch uniformity. Several chemical and some physical parameters can be checked, but the most important factor is the content of the active ingredient which usually has to match a narrow range around the designated content.

In principle NIR analyses of tablets can be performed using diffuse reflectance and transmission measurements.[2–5] Depending on the tablet composition and the property of interest (coating, excipients, active ingredients, hardness etc.) both methods have advantages and disadvantages. In general, reflection is preferred for extrinsic properties like thickness and composition of coatings whereas using transmission permits analysing the tablet core.

In this work tablets for a clinical study (placebo/verum studies) with very low concentrations of the active ingredient were analysed in transmission and in reflectance using different accessories. The dosage range was 0 to 6 mg with a total tablet weight of 105 mg, leading to a highest concentration of the active component of 5.7% by weight. In particular, the spectroscopic distinction between the placebo and the low dosage forms with 0.25 and 0.5 mg active ingredient requires an extraordinary accuracy.

Experimental

The sample set was made for a clinical study and provided by Janssen Research Foundation (Beerse, Belgium). It consists of 45 coated tablets with a thickness of 3 mm and a diameter of 7 mm: three tablets of two batches with 0, 0.25, 0.5, 1, 2, 3 and 4 mg dosage and three tablets of one batch with 6 mg. All experiments were carried out using different Bruker FT-NIR-spectrometers (Bruker Optik GmbH, Ettlingen, Germany). Details are shown in Table 1.

Reflectance experiments

For the UpIR-experiment an upward-looking diffuse reflectance accessory from Pike (Madison, USA) was placed in the sample compartment of an IFS 28/N spectrometer equipped with a wide range InAs-detector. The integration sphere and the fibre probe (random fibre bundle) are standard accesso-

Table 1. Instrument set up and measurement settings.

	Instrument	Detector	Spectral range cm^{-1}	Resolution cm^{-1}	Scan time s
UpIR	IFS 28/N	InAs	11,500 – 3,300	8	30
Integration sphere	VECTOR 22/N-I	Ge	11,500 – 5,300	8	30
Fibre probe	VECTOR 22/N-F	InGaAs	11,500 – 3,800	8	10
Transmission	VECTOR 22/N-T	InGaAs	11,500 – 7,000	8	30

ries of the corresponding instruments. For the UpIR and integration sphere tests, the tablets were centered on the window of the accessory. The fibre probe and the tablets were fixed during the measurements to minimise variations due to manual handling of the probe.

Transmission experiments

A sample holder, furnished with an iris, was used to place the tablets in the focus of the recently developed transmission accessory (Figure 1). Here the NIR radiation coming from the interferometer is guided from the bottom up through the tablet to the detector mounted about 15 mm above the tablet. Because of scattering effects on its way through the tablet the NIR radiation is collected by a lens in front of the detector.

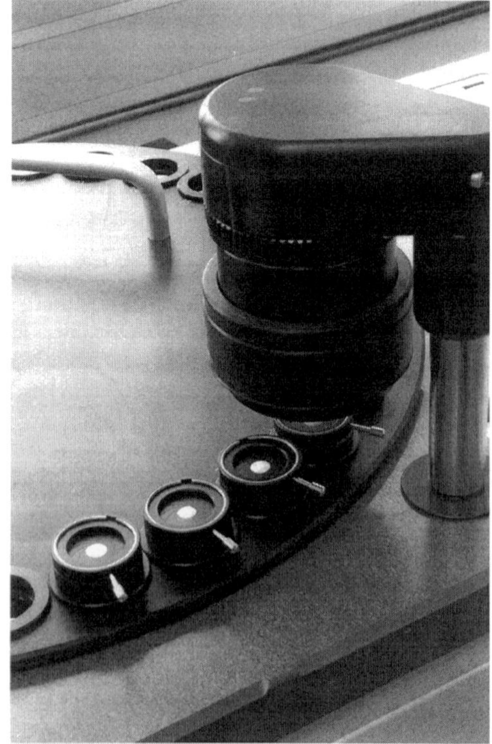

Figure 1. Accessory for transmission measurements using the Bruker VECTOR 22/N-T spectrometer.

Evaluation

The regression models were calculated by PLS using the Bruker OPUS software and the implemented optimisation option. Because of the limited number of samples a full cross validation (leave one individual sample out) was performed leading to the root mean square error of cross-validation values (*RMSECV*) for comparing the results.

Results and discussion

To compare the spectra obtained with the various accessories (Figure 2) the different spectral ranges have to be considered. In particular the useful spectral range for transmission experiments is limited to higher wavenumbers, depending on the thickness and hardness of the tablets. The tablets analysed in this study are rather thin. Tests done with other pressed tablets

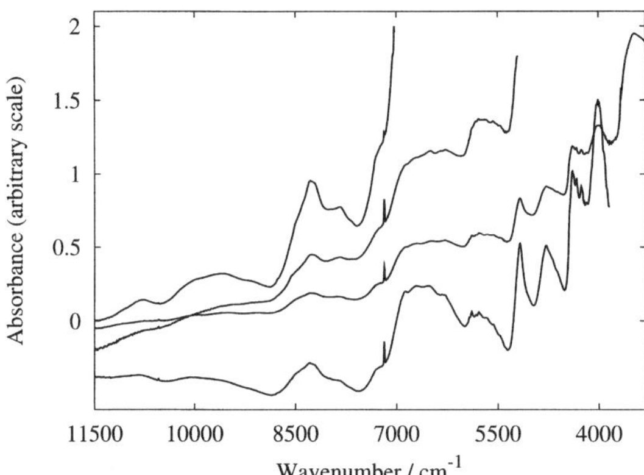

Figure 2. FT-NIR spectra of tablets using different techniques (from bottom to top at 8.500 cm^{-1}: fibre probe, UpIR, integration sphere and transmission).

showed that thicknesses up to 8 mm can be measured.

The basic shape of the spectra is very similar apart from peak intensities and baseline slopes, but there are details where the reflection spectra differ from the transmission spectrum. In all spectra narrow peaks typical of talc—which is part of the coating—can be found at 7185 cm^{-1} and 10534 cm^{-1}, but in the transmission spectrum the intensities are noticeably weaker. That indicates a smaller contribution of the coating proportional to the tablet bulk. In addition, the broad band at 8280 cm^{-1} is most intense in the transmission spectrum compared to the reflection spectra and represents mainly tablet bulk material.

The calibration results are represented in Figure 3 and summarised in Table 2. Obviously the results of the reflection experiments for the UpIR and the integration sphere with *RMSECV* values of 0.41 and 0.35 mg, respectively, are different from those obtained with the fibre probe (*RMSECV* 0.15 mg). For the first two models the predicted vs actual plots show high variations for measurements of tablets containing the same amount of the active ingredient. Especially for the lower dosage forms, these inaccurate results make a distinction between the lowest dosages impossible. Here, the drawback of reflection techniques becomes visible: only the surface can be analysed. In this case, it mainly consists of the coating. In addition, the amount of information about the content of the active ingredient depends on its concentration in the core close to the surface. Basically the homogeneity or distribution of the compound of interest in the tablet is very important for the accuracy of reflection measurements.

But why are the results for the third reflection experiment with the powder probe better than for the two others? To answer this question a closer look at the spectra in Figure 2 is helpful. In the three reflection spectra the talc peaks of the coating at 7185 cm^{-1} have comparable intensities. Apart from this the intensities in most other parts of the spectra differ a lot. Relative to the talc peaks, the intensities of the broad bands around 8280 cm^{-1} and 6600 cm^{-1}—representing mainly the tablet core—are weaker in the UpIR and integration sphere spectra compared to the fibre probe spectrum.

Obviously the better calibration results obtained with the 4 mm diameter fibre bundle are caused by a more effective penetration of the tablet bulk by the NIR radiation. The other two accessories are worse because they were designed for analysing solids and powders in vials or sample cups and not primarily for analysing tablets. The sampling spot of the integration sphere of 20 mm diameter is much bigger than the tablet and so a lot of light is lost for the analysis. In the UpIR accessory a certain amount of the primary light is reflected by specular reflection without any interaction with the tablet core.

Table 2. Calibration parameters and results.

	Data pre-processing	Spectral range cm^{-1}	R^2 %	PLS factors selected	*RMSEE* mg	*RMSECV* mg
UpIR	17pt 1st-derivative	6,100 – 5,968; 5,906 – 5,700	94.46	2	0.38	0.41
Integration sphere	MSC	9,900 – 6,400	95.93	5	0.29	0.35
Fibre probe	17pt 1st-derivative + MSC	9,542 – 7,900; 7,082 – 5,438	99.28	5	0.08	0.15
Transmission	vector normalisation	10,850 – 10,298; 9,200 – 8,096	99.83	5	0.05	0.07

In terms of variance and *RMSECV* of 0.07 mg the transmission experiment leads to the best results, because a larger part of the tablet volume is registered and, therefore, the spectra provide much more information about the tablet core and its composition. The improvement with respect to the powder probe results is noticeable and important for the accurate discrimination between the low dosage forms

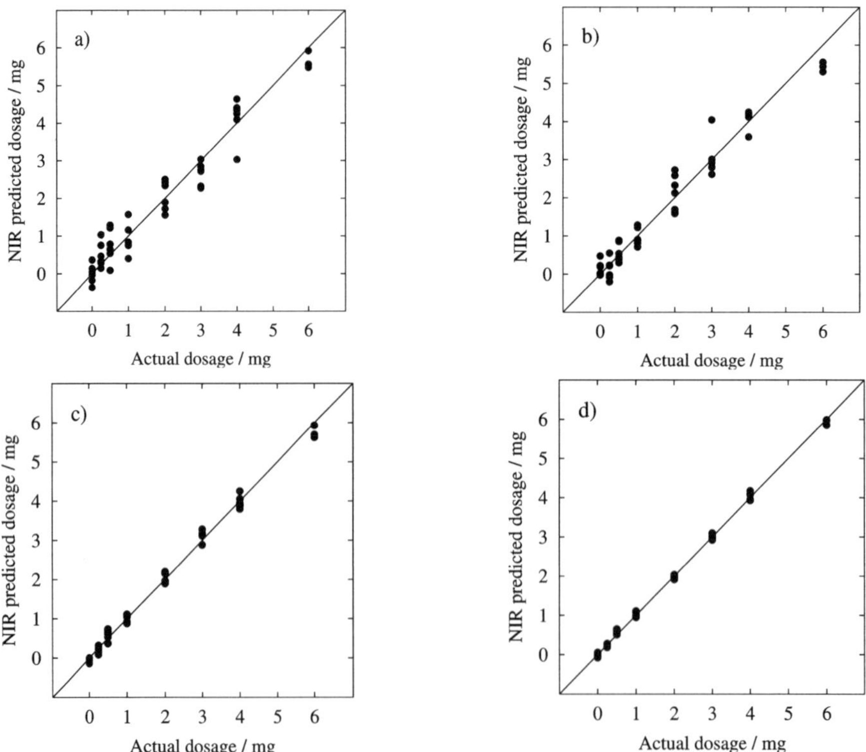

Figure 3. Calibration results for (a) UpIR, (b) integration sphere, (c) fibre probe and (d) transmission experiments.

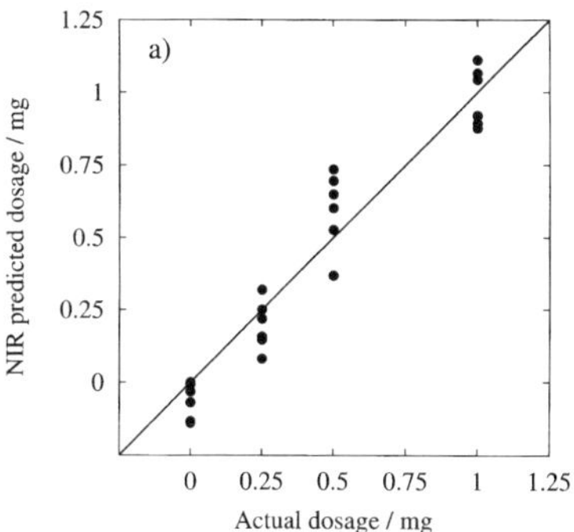

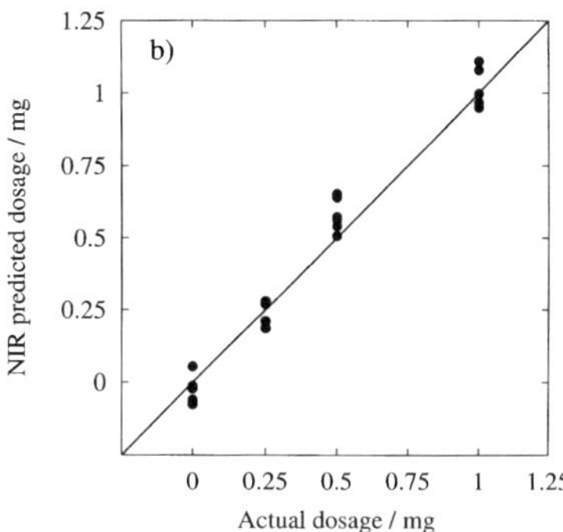

Figure 4. Detailed calibration results for low dosage forms (a) fibre probe and (b) transmission experiments.

with 0 mg (placebo), 0.25 mg and 0.5 mg active ingredient. As shown in Figure 4, only the transmission measurements allow a reliable distinction in that concentration range.

Conclusions

In this study tablets with very low dosages of the active ingredient were analysed by NIR with different accessories for reflection and transmission measurements. Since a property of the tablet core was investigated, only the transmission experiments yield results which allow discrimination of all dosage steps. The practicality of transmission measurements depends on the tablet type and much more on their thickness and hardness, which determine the accessible spectral window. Whether the needed spectral information for the modelling can be found in that window or not is the critical point of such applications and has to be tested in advance. Some types of tablet, for example extruded tablets, have a much higher transparency for NIR radiation than pressed tablets. Here the detector is easily overloaded and a smaller aperture has to be used to reduce the beam intensity.

This study shows just one particular application and does not have the intention to devalue the importance of reflection experiments for tablet analyses. In contrast to the problem discussed here, the diffuse reflection technique is useful for other tasks, for example coating analysis. The optimal solution for characterising tablets by NIR would be a combination of transmission and reflection data.

Acknowledgements

We gratefully acknowledge the kind support of Roy de Maesschalck (Janssen Research Foundation, Beerse, Belgium).

References

1. S.V. Hammond in *Leaping Ahead With Near Infrared Spectroscopy*, Ed by G.D. Batten, P.C. Flinn, L.A. Welsh and A.B. Blakeney. NIR Publications, Chichester, UK, p. 394 (1993).
2. R.A. Lodder and G.M. Hieftje, *Appl. Spectrosc.* **42,** 556 (1988).
3. J. Gottfries, H. Depui, M. Fransson, M. Jongeneelen, M. Josefson, F.W. Langkilde and D.T. Witte, *J. Pharm. Biomed. Anal.* **14,** 1495 (1996).
4. I. Jedvert, M. Josefson and F. Langkilde, *J. Near Infrared Spectrosc.* **6**, 279 (1998).
5. H. Frickel and G. Reich, *Proceed. Intern. Symp. Control. Rel. Bioact. Mater.* **27,** 740 (2000).

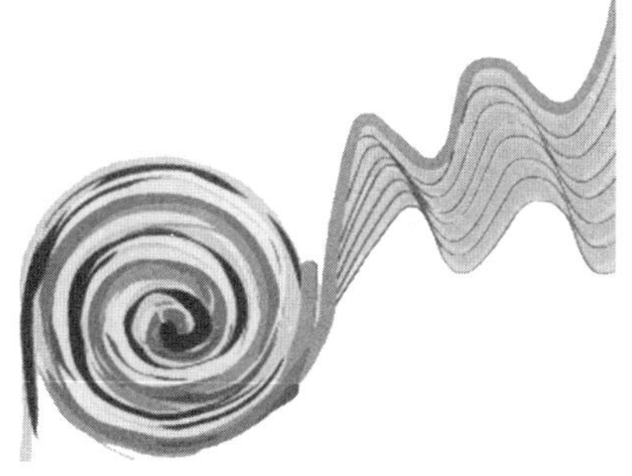

Emerging Aspects

Monitoring the effect of the "ultra-high pressure" preservation technology by near infrared reflectance spectroscopy

Karoly J. Kaffka,[a] József Farkas,[a] Zsolt Seregély[b] and László Mészáros[a]

[a]*Szent István University, Ménesi út 45, H-1118 Budapest, Hungary*

[b]*Metrika R & D Co., Petzvál J. u. 25 H-1119 Budapest, Hungary*

Introduction

Ultra-high pressure (UHP) technology for the preservation of foods is under intense research to evaluate its potential as a complementary or alternative process to traditional methods of food preservation. UHP is emerging as one of the most promising new, non thermal technologies in the food processing industry.[1–4] Traditional processing methods require large amounts of energy which may cause unwanted reactions in the food, leading to lack of flavour and loss of vitamins. The application of UHP requires that the food be subjected to pressures in the range of 50–800 MPa. Such high pressure will usually extend the shelf-life of foods by inactivating vegetative microorganisms, enzymes while promoting the germination of bacterial spores into heat-sensitive cell states, while retaining vitamin content and preserving natural flavours.[5,6] This new technology follows the "minimal processing" concept of minimising the quality degradation utilising less energy. At the same time, UHP technology has the unique ability to create diverse textures and gels. We joined the research team at our university involved in the mentioned technology using an ultra-high pressure equipment, recording the near infrared spectra of meat samples exposed to different pressures. The objective of this investigation was to see whether changes in minced meat samples due to UHP could be followed by near infrared spectroscopy.

Materials and methods

Minced beef and pork were prepared from *longissimus dorsi* muscle. Three experiments were designed to study raw beef samples and two experiments were designed to study raw pork. A total of 102 samples of raw beef and 75 samples of pork were analysed. The raw meat was minced and vacuum packed in PE-PA-PE foil. The samples were UHP treated with pressures ranging from 50 to 800 MPa for 20 minutes. All samples had an initial temperature of 8°C. A Food Lab Model SFL850 (Stansted Fluid Power Ltd, UK) was used in batch mode to induce UHP. The equipment had a chamber size of 40 mm diameter × 240 mm length. UHPs were attained within two minutes. Temperature was maintained by circulating water through the cylinder wall of the pressure vessel. Ethyl alcohol, containing 15% castor oil for lubrication and anticorrosion purposes, was used as the pressure-transmitting medium in the UHP vessel. Since the liquid pressure-transmitting medium changed its volume slightly at compression, the UHP vessel did not present operating hazards.

Near infrared spectra of all samples were recorded immediately after the UHP treatment. A "Spectralyzer 1024" (PMC, LaborChemie, Vienna, Austria). was used to measure and store the NIR spectra (range 1000–2500 nm in 2 nm steps). Quality points were calculated using the polar qualification system (PQS)[7,8] data reduction and qualification software along with the "wavelength range optimisation" program. The optimisation goal was to determine the optimum wavelength range that would give the best separation between UHP treatments. The quality of separation was expressed as sensitivity (S), where sensitivity was defined as the distance between the centres of the quality points of two samples divided by the sum of the standard deviations of the quality points of the two samples exposed

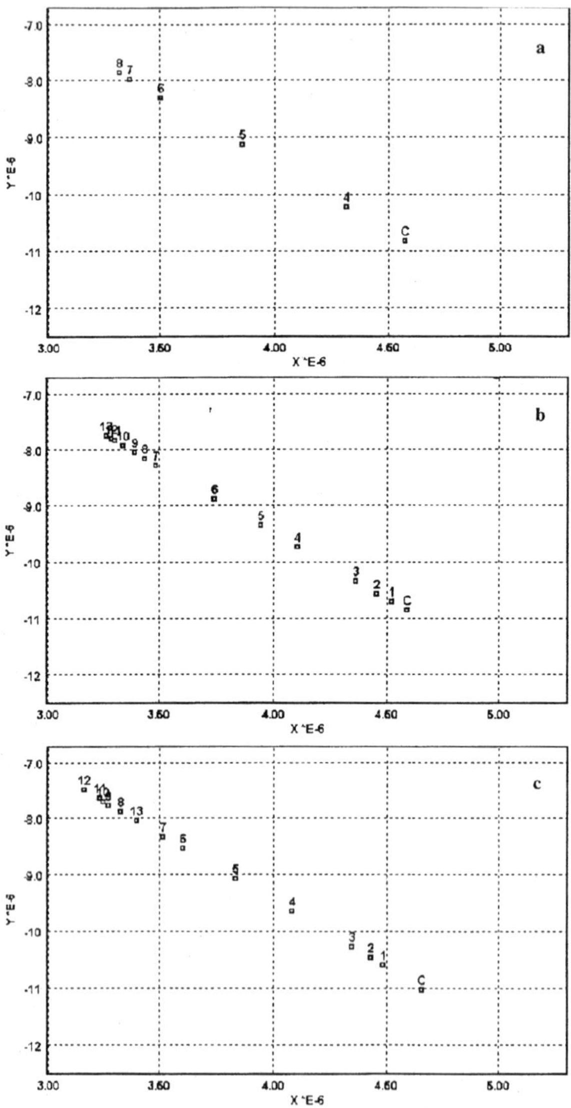

Figure 1. Quality points of the minced beef samples exposed to different high pressures. Quality points were determined from the second derivative spectra in the 1250–1300 nm wavelength range using PQS software [first (a), second (b) and third (c) experiments].

to different pressures. After treatment, the samples were stored at 4 ± 1°C for a week, then they were measured with an "electronic nose" (Daimler–Chrysler Aerospace Model SamDirect, Germany, made chemosensor array in order to see the differences in odour caused by extended shelf-life for the various pressures applied. Evaluation of the electronic nose measurements was performed by using principal component analysis.

Results and discussion

Maximum sensitivity S was achieved using the 1250–1300 nm range. Figure 1 shows the results of the three beef experiments. Figure 2 shows the results of the two pork experiments. Quality points of samples exposed to the same pressure are marked with the same number in the figures. Quality points of the control samples are marked with the letter "c" while quality points of the samples exposed to different pressures are marked with numbers. The numbers and the corresponding pressures were as follows: 1 ÷ 50 MPa, 2 ÷ 100 MPa, 3 ÷ 150 MPa, 4 ÷ 200 MPa, 5 ÷ 250 MPa, 6 ÷ 300 MPa, 7 ÷ 350 MPa, 8 ÷ 400 MPa, 9 ÷ 450 MPa, 10 ÷ 500 MPa, 11 ÷ 600 MPa, 12 ÷ 700 MPa, 13 ÷ 800 Mpa.

Location of the quality points in Figures 1 and 2 clearly show an existing relationship between spectral data and the changes in meat samples caused by the UHP treatment. Shift of the quality points

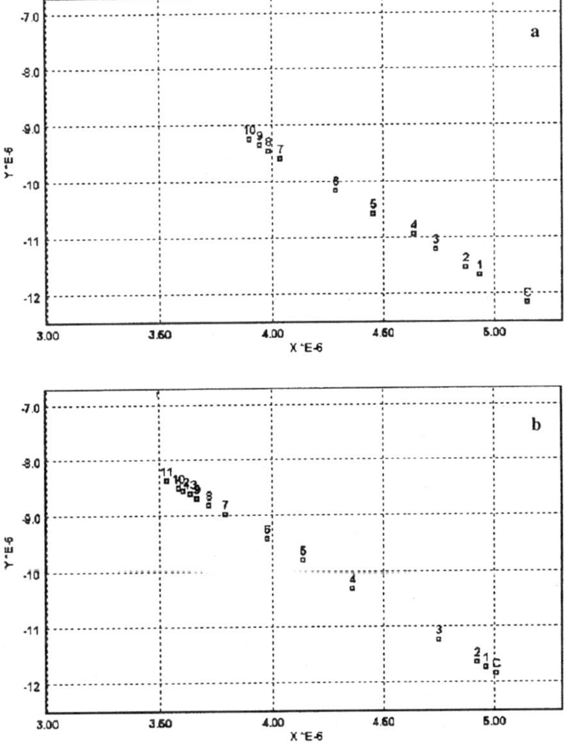

Figure 2. Quality points of the minced pork samples exposed to different high pressures. Quality points were determined from the second derivative spectra in the 1250–1300 nm wavelength range using the PQS software [fourth (a) and fifth (b) experiment].

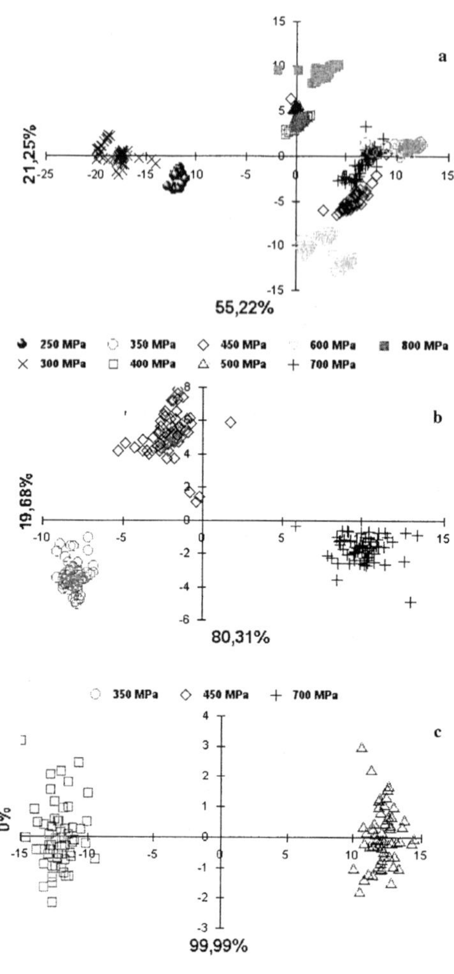

Figure 3. Quality points ("the odour") of nine minced beef samples after a week storage at 4 ± 1°C. The samples were exposed to different high pressures before storage. After storage they were measured repeatedly by an electronic nose. The quality points were determined using principal component analysis, and represented on the projection planes (a, b, c) of the first two principal components of these nine and the a bit overlapping samples.

as a function of the UHP shows that significant changes in meat samples occurred between 200–300 MPa. This change can be assigned to the change in protein structure as an absorption peak of protein can be found around 1276 nm. This wavelength is the middle of the wavelength range found by the wavelength range optimisation program. The sensitivity achieved between the control sample and the sample exposed to 800 MPa was higher than 100. This means that the distance between the two quality points is at least 100 times higher than the sum of the standard deviation of the quality points of these two samples. Repeatability of the measurements was very good, in spite of the fact that it was extremely difficult to produce homogeneous samples.

Immediately after treatment, odour change was not observed among the control sample and the samples exposed to different pressures. Odour measurements, using sensory assessment of the samples after one week storage at 4 ± 1°C, were quite different. Samples exposed to 350 MPa or more tended to maintain their original smell while samples exposed to less than 350 MPa were definitely stinking. Samples exposed to less than 250 MPa were definitely unbearable.

Figure 3(a) shows the quality points of nine beef samples exposed to different high pressures. These samples were stored for one week and measured repeatedly with the electronic nose. The projection plane shown in Figure 3(a) for these nine samples was defined by the first and second principal components for these nine samples. For the three samples exposed to 350, 450 and 700 MPa, a slight overlap exists. However, a new projection plane can be found [see Figure 3(b)] where their separation is perfect. Also note that for the two samples exposed to 400 and 500 MPa—also having overlapping quality points—a new (third) projection plane can be found [see Figure 3(c)] providing excellent separation.

The effect of the pressure on the raw meat is very different depending on the magnitude of UHP. Lower pressures do not inactivate enzymes while the higher pressures definitely inactivate the enzymes. Microorganism behaviour varys depending on the magnitude of UHP. Thus, sample deterioration develops differently during the one week stored at

4 ± 1°C. This seems to explain why the quality points in Figure 3 do not change their position along a strait line or curve as a function of UHP treatment.

Conclusion

Near infrared reflectance spectroscopy is suitable for monitoring changes in meat samples caused by UHP treatments. Both sensory and electronic nose (chemosensor array) evaluations show that the UHP treatment has a significant effect on food quality and on shelf-life of raw meat samples.

Acknowledgment

Supported by grants from the National Scientific Research Fund (OTKA) No: T 023020 and No: T 032814. Thanks are due to G. Bayer and K. Aschenbrenner for their conscientious scientific and technical assistance in performing measurements and data processing.

References

1. B. Mertens and G. Deplace, *Food Technol.* **47(6),** 164 (1993).
2. F. Zimmerman and C. Bergman, *Food Technol.* **47(6),** 162 (1993).
3. D. Knorr, *Food Technol.* **47(6),** 156 (1993).
4. D.G. Hoover, *Food Technol.* **47(6),** 150 (1993).
5. I. Hayakawa, T. Kanno, M. Tomita and Y. Fujio, *J. Food Sci.* **59(1),** 159 (1994).
6. N. Homma, Y. Ikeuchi and A. Suzuki, *Meat Sci.* **38,** 219 (1994).
7. K.J. Kaffka and L.S. Gyarmati, *J. Near Infrared Spectrosc.*, **6(A),** 191 (1998).
8. K.J. Kaffka and Zs. Seregély, in *NIR Spectroscopy: Proceedings of the 9th International Conference*, Ed by A.M.C. Davies and R. Giangiacomo. NIR Publications, Chichester, UK 259 (1999).

Factors concerning development of electronic systems for grading grains and seeds in Canada

Phil Williams

Canadian Grain Commission, Grain Research Laboratory, 1404-303, Main Street, Winnipeg, Manitoba, Canada, R3C 3G8

philwilliams@pdkgrain.com (present e-mail address)

Introduction

This paper will document the concept of grading grains for value determination and marketing by electronic methods, with particular reference to near infrared (NIR) spectroscopy. The factors necessary to the development of an electronic grading system are summarised in Table 1 (not in order of importance). The paper will provide a short commentary on each factor, together with the methods used to date. Most of the work to be described concerns wheat, but extensive research has also involved barley, oats and the oilseed and pulse crops.

Establish reasons for development of electronic grading *vis-à-vis* "Status Quo" systems

The present visual system works, so why change it?

Grain will increasingly be purchased on quality specifications, rather than on grades. Add to this the increasing proclivity of Canadian marketing agencies to seek and service "Niche" markets and the looming threat of segregation of Genetically Modified wheat. The complexities of these changes in marketing strategy, combined with the already progressing changes in the elevator system and grain

Table 1. Factors to consider in the development of electronic systems for grading grains and seeds.

1	Establish reasons for development of electronic grading *vis-à-vis* "Status quo" systems
2	Identify factors that affect end-product quality and the value of commodities
3	Identify factors that should be predictable by NIR spectroscopy
4	Define method of expression of efficiency of NIR estimations and establish limits of acceptable accuracy
5	Identify instruments for application in appropriate settings
6	Evaluate software for calibration development and monitoring
7	Develop calibrations for grading of grains and seeds on the basis of functionality and end-value
8	Evaluate calibrations for electronic grading for application in grain-handling and processing
9	Advertise proposed electronic grading systems among potential user and their clients
10	Implement systems for electronic grading

transportation, will make it extremely difficult for the present system of visual grading to meet the demands of future grain marketing.

Canada's Prairie provinces annually produce about 50 million tonnes of grain, of which about 80% is exported. This volume embraces 17 different crops (crops grown to a significant extent). Each crop is graded visually into at least two grades, so that a country (called Primary) elevator would require nearly 40 bins or cells to enable it to segregate all grades of all grains and seeds. Add to this the need to segregate the top two millings of CWRS wheat into at least two protein levels. In times of a wet harvest period, still further cells are needed to keep high moisture grains separate until the grain could be dried, or efficiently blended with other grain of the same type and grade of moisture content sufficiently low as to render it safe for shipping or further storage.

Binning of the eight classes of wheat is further complicated by the fact that some classes, such as western Canadian Red Winter (WCRW) and white Canada Prairie Spring (CPSW) wheat, are only produced in relatively small volumes. This means that an elevator may be obliged to tie up a 2000 tonne cell with only two or three railcar-loads (about 270 tonnes) of one of these classes until sufficient has been accumulated to market it. Combining the present six grades of western Canadian red and white classes [all classes other than Canada Western Red Spring (CWRS) and Canada Western Amber Durum (CWAD)] into no more than four classes would ease some of these binning stresses.

The need for electronic grading in western Canada (as well as testing for composition) was recognised as long ago as 1994, around the time when the western Canadian Grain-handling industry was exhorted to double grain-handling throughput within the following 10 years or so. At that time it was recognised that such a metamorphosis would not be possible without the application of electronics in grain-handling and the concept of Electronic Grading was spawned at the Canadian Grain Commission (CGC) Grain Research Laboratory (GRL) mainly in recognition of the above mandate.

The portended introduction of new wheat classes, such as a Hard White Spring class, will create even more demands for bin-space on the elevator system. During the past 20 years, the elevator system has been undergoing major changes, epitomised by reduction in the number (from over 5,200 in the mid-1960s to less than 550 at present) and increases in capacity (of individual elevators) and throughput potential. This development has coincided with a gradual emergence of extensive storage capacity on farms, sufficient in most cases to accommodate the full year's harvest.

These items formed the basis for the concept of an electronic system for identifying, at farm level, grains and seeds of the grades and composition required for shipments internationally and domestically. The concept is based on retaining the present visual system for the Hard Red Spring (CWRS) and durum (CWAD) classes and "streamlining" the classes of all other types of wheat by electronic grading. The electronic system would differ fundamentally from the present method, which is based on visual evaluation. It would be based on functionality—how will the grain **perform**, rather than what does it look like. For example, the main "grade" characteristics of wheat would stem from colour, texture, water absorption and gluten strength (protein content would still be used to segregate wheat of different classes). Visually assessed factors, such as sprouting and frost damage, would also be incorporated. Electronic systems for assessing this type of damage could also be used to improve the precision and reduce the time per test of grading the CWRS and CWAD classes.

A second objective of the electronic grading concept has been the development of systems for identification of wheat gluten "strength" for use in wheat-breeding. Prediction of recognised "strength" parameters, such as Farinograph development and stability times, Extensigraph peak height and area and Alveograph "W" value on whole kernels, would enable breeders to screen material in early generations, to avoid the expense of maintaining inferior lines. In later generations that are tested for environment/genotype interaction by growing them on several growing locations, the calibrations would identify lines that were most susceptible to genotype/environment interactions. Lines of this type should be rejected.

Identify factors that affect end-product quality and the value of the commodity

Weather-induced factors such as vitreous kernel %, frost, wet harvest conditions and growing conditions that favour the prolification of fungi and other hazards have been thoroughly researched, for example references 1 and 2. Functionality factors, such as kernel colour and texture and gluten "strength" would be added to these electronic gradings.

Identify factors that should be predictable by NIR spectroscopy

Factors such as sprouting, frost damage and *Fusarium* head blight affect kernel texture (degree of hardness or softness) in different ways. Sprout and *Fusarium* cause slight softening of the kernels, while frost makes the kernels appreciably harder. NIR spectroscopy has been found to be capable of detecting these subtle differences in texture. Significant differences in texture have been demonstrated among some wheat classes, such as Red Canada Prairie Spring (CPSR) and Canada Western Extra Strong (CWES) wheat. Both the texture and gluten characteristics of Canada Western Red Winter (CWRW) and CPSR wheat types are sufficiently close to each other that they could be incorporated into a single class for marketing. The white classes, Canada Prairie Spring White (CPSW), Canada Western Soft White Spring (CWSWS) and Canada Western Hard White Spring classes can be differentiated from the red classes on the basis of colour and from each other on the basis of kernel texture.

Both colour and texture can be measured by modern NIR instruments. Figure 1 illustrates spectra of four wheat classes of different kernel texture. Durum wheat is the hardest and Soft White Spring the softest.

The percentage of hard vitreous kernels (HVK) is an important parameter in establishing grades of durum wheat. The assessment of this by visual methods is subjective. A "pinpoint" of starchiness is sufficient for a grain inspector to regard the kernel as non-vitreous when, in fact, that pinpoint is not likely to have any influence on the functionality of the wheat. The HVK percentage can be predicted with accuracy by NIR as well as by digital imaging. Figure 2 illustrates spectra of six samples (three pairs) of durum wheat, each pair having different HVK % from the other two. The HVK % of each pair was clearly distinguishable.

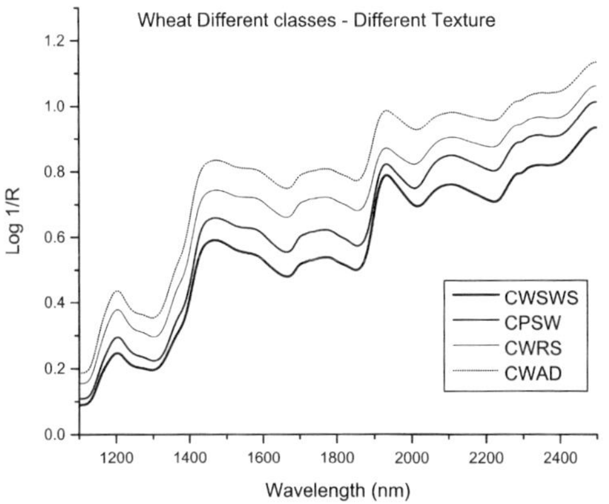

Figure 1. Spectra of four wheat classes of different kernel texture.

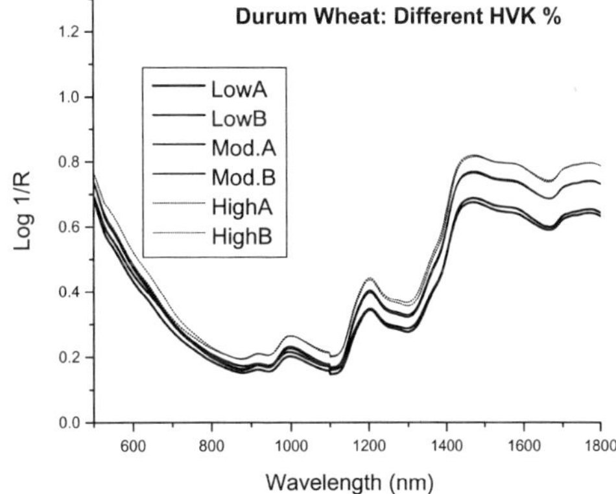

Figure 2. Spectra of six samples (three pairs) of durum wheat with different HVK % values.

Define method of expression of efficiency of NIR estimation and establish limits of acceptable accuracy

The time-honored statistics of coefficient of correlation (*r*), standard error of prediction (*SEP*) and bias have been used in the evaluation of calibrations developed for prediction of electronic grading parameters. Other important statistics include the slope and the relative standard deviations of the reference and NIR-predicted data of the validation sample set, which should be close to identical. The RPD statistic (ratio of *SEP* to standard deviation of reference data of the validation sample set) has also been found useful in "summarising" calibration validation statistics. Several parameters used as reference data (in calibration development) were based on visually-assessed characteristics and such parameters as the Falling Number values. Most of these parameters lack the degree of precision in reference data normally encountered with composition factors, such as protein content. Accordingly, RPD values upwards of 2.5 have been regarded as adequate for the establishment of electronic grading systems.

Identify instruments for application in appropriate settings (including on-farm)

Identification of suitable instruments for application at elevator and farm level have to be evaluated from the aspects of economic as well as technical efficiency. It would be inefficient to assemble and scan several thousand samples to calibrate an instrument that is likely to be superseded by one that is more efficient and/or of lower cost, for the same efficiency. The Foss/NIRSystems Model 6500 visible/NIR scanning spectrophotometer is regarded as the standard by which other instruments can be evaluated. This instrument operates over a wavelength range of 400 – 2500 nm. Several other instruments are under evaluation for possible application in electronic grading. The ideal NIR instrument will combine suitable accuracy and precision with consistent and reliable sample presentation, compact size, simple operation (from the aspect of software and transferability of calibrations and moderate cost). The capability of networking to assist with monitoring would be another asset.

Evaluate software for calibration development, regular analysis and monitoring

Software should be evaluated from aspects of user-friendliness in operation and effectiveness in resolving non-linearity, outliers and other features that detract from accurate and reproducible analysis. Evaluation includes generic software, such as Unscrambler and Grams, as well as dedicated software developed by instrument companies for calibration, operation and instrument diagnostics. Too often, the manufacturing companies' operators' manuals fall short of efficiency. Essential steps in the operation are overlooked by compilers of manuals and it is left to the user to locate such deficiencies and insert the missing steps.

Develop calibrations for electronic grading of grains and seeds on the basis of functionality and end-value

Modern software options, such as artificial neural networks (ANN) and the "Local" option of InfraSoft International's WINISI are most effective with large sample sets. The assembly of sample sets with several thousand samples can best be achieved by arranging for samples of railway carloads or farmers' deliveries to be forwarded to the central laboratory responsible for calibration development. Electronic grading is aimed at all levels of elevator and farm operation. Large terminal elevators may process over 1000 railway carloads in a week, while Primary elevators may process hundreds of farm deliveries in the same time-frame. A very wide range of diversity is represented by both railway and farm deliveries. All of this variance has to be incorporated into reliable calibrations. This is further complicated by the variance in the reference data. The "frowned-upon" practice of splitting large populations into calibration and validation sample sets is acceptable in this type of calibration, because of the wide variance inherent in the very large populations of railway carloads and, on a more regional basis, farmers' deliveries to Primary elevators.

In Canada, because of the handling system, the abrupt changes in grain quality that can occur from season to season usually have less immediate impact than in countries where farmers are able to deliver all of their newly harvested grain immediately. Grain arriving at terminal elevators from Primary elevators has undergone variable degrees of blending and individual carloads may contain grain from more than one season. Similarly, farmers may elect to retain new crop grain when it is of low moisture content and deliver grain from their storage bins from a previous season.

After calibration development, the efficiency of the calibrations has to be tested by setting up the instruments on-location at elevators. Here, the operating conditions may reveal areas where modifications to the operating procedure have to be introduced. Modern elevators can discharge railway hopper-cars in about 2–3 minutes and most elevators have more than one receiving line. At the same time, the elevator may be loading one or more vessels and operating conditions are very intensive. Systems have to be developed for monitoring instrument performance. Above all, the NIR instruments must be capable of efficient throughput with no inconvenience to the elevator staff, while retaining the integrity of the grades.

The chief difficulty in development of NIR calibrations for grade factors lies in the subjectivity of the data used as reference for calibration development. Most of the factors rely on the opinions of individual grain inspectors as to the degree of damage caused by factors such as frost and *Fusarium*. The Falling Number (FN) test is used in place of % sprouted kernels, but even this test has been found to show a standard error of about 30 seconds (including sample preparation). To put this in context, the 95% confidence limits stretch to +/– 60 seconds, so that an apparent result of, for example, 340 seconds for the FN of a sample used in calibration could, in fact, lie between 280 and 400 seconds. Precision of prediction of FN by NIR is only 14.7 seconds, including sample cell reload. The precision of the instrument on re-scanning a sample without re-loading is only six seconds.

Evaluate calibrations for electronic grading for application in grain-handling and processing

The development and evaluation of calibrations for electronic grading has been in process at the Grain Research Laboratory (GRL), Winnipeg, during the past four years. The process has been carried out as a collaborative project between the CGC Industry Services division and the GRL. Methods are as follows:

Sample accumulation

Samples of railway carloads have been forwarded from terminal elevators, together with cargo-loading increments and cargo composite samples. These have been augmented with samples submitted by farmers in response to the annual harvest surveys conducted by the GRL and further augmented by samples of pure varieties and breeders' lines submitted by plant breeders from over 30 Prairie growing locations over eight growing seasons. The total database contains over 4,000 spectra.

Near infrared analysis

All samples were scanned at, or close to, the time of receipt on a Foss/NIRSystems Model 6500 visible/NIR scanning spectrophotometer. All scans were carried out using whole grains and the Foss/NIRSystems Natural Products sample cell. Check samples of the respective classes of wheat were used and of samples of CWRS wheat with well-defined weather damage (sprouting and frost). Software included Foss/NIRSystems NSAS and WINISI. The very rapid NSAS software was used to optimise mathematical treatments of the log $1/R$ signals. Data-sets processed by the optimum mathematical treatment were transposed to WINISI for evaluation of the influence of scatter correction (SNV/Detrend and Multiplicative Scatter Correction). Details of mathematical treatments are available.

Reference methods

Reference methods used in calibration development for composition, kernel texture and physical dough characteristics were Approved Methods of the American Association of Cereal Chemists.[3] Grade factors such as % Frost were determined by experienced inspectors of the CGC, using the Official Grain-Grading Guidelines. The Falling Number test was used in place of % sprouted kernels.

Results

The preliminary results are summarised in Table 2. Data are presented in Table 2A for parameters to be used in an electronic system, based on kernel colour and texture and gluten strength. Data are also included for calibrations developed for some barley and oat functionality parameters. Table 2B includes results for prediction by NIR spectroscopy of parameters used in the present visual grading system.

Discussion

All of the wheat parameters were predictable with accuracy and precision suitable for the development of an electronic grading system for classes of wheat other than CWRS and CWAD. The grades would be based mainly on kernel colour and texture. Wheat gluten strength parameters could be used to "fine-tune" the red wheat calibration. For example, the Farinograph stability of the CPSR usually lies between 5–8 minutes, while that of CWES wheat usually lies between 25–35 minutes. The *SEP* of 1.9 minutes would clearly serve to differentiate between these two classes. The Farinograph stability time and kernel texture of CWRW wheat are similar to those of CPSR wheat and these two existing classes could effectively be marketed as a single class.

Prediction of factors such as sprout, frost, % hard vitreous kernels and *Fusarium* damage would assist in establishing grades of all classes of wheat, including the CWRS and CWAD classes. Assessment of these grade factors is time-consuming. Inspectors have to divide a 1–2 kg sample down to about 30 g. The sample is weighed accurately and examined visually to identify the presence and degree of damaged kernels. These are hand-separated and the damaged kernels weighed and reported as "% sprouted", % severe sprouted", etc. This exercise takes about 10–15 minutes for all similar grade factors. The procedure is also subjective, because the results depend on the opinions of individual inspectors. An NIR calibration could reduce the time to the usual 40–50 seconds and provide a higher degree of precision. Calibrations for both % Frost and Falling Number have proved to be significantly superior in precision to the visual method and also to the precision of the Falling Number test.

Turning to wheat gluten "Strength" calibrations developed to date have been developed on whole wheat kernels and preliminary results, based on two growing seasons, have been reported elsewhere.[4] The Foss/NIRSystems Natural Product sample cell can be used at only ¼-full, which is equivalent to about 30 g of wheat. This would enable plant breeders to screen for gluten strength at as early as F3 and still retain selected lines for planting.

Advertise proposed electronic grading systems among potential users and their clients

All concerned agencies should be made fully aware of all aspects of the proposed electronic system well in advance (at least one and, preferably, two years) of its projected implementation. This includes the reasons, the methods to be employed, accountability, economics, potential benefits and potential pitfalls. This can be carried out via the media and by workshops and presentations at Town Halls and

Table 2. Results of prediction of grading factors using NIR spectroscopy.

A. Electronic system	r^2	*SEP*	Potential for use
Kernel colour1 (Minolta b*)[1]	0.96	0.30	Excellent
Kernel texture (PSI %)[1]	0.94	2.38	Excellent
Farinograph water abs. %[1]	0.91	1.43	Excellent
Gluten Strength[1]			
a. Farinograph dev. time	0.71	1.2	Good
b. Farinograph stab. time	0.80	1.9	Good
c. Farinograph mix. tol.	0.92	17.7	Very good
d. Extensigraph max. ht.	0.72	85	Good
e. CSP mixing energy	0.76	2.5	Good
HVK % (Durum wheat)	0.83	4.22	Very Good
Protein content (All)	—	—	Excellent
Moisture content (All)	—	—	Excellent
Malt fine HWE3 (barley)	0.52	1.00	Fair
True met. energy (barley)	0.61	0.15	Fair
Groat % (oats)	0.82	0.95	Good

[1]wheat only

B. Present (visual) system	r^2	*SEP*	Potential for use
Falling number (seconds)	0.85	42.5	Fair
Fusarium "Scab" (DON) ppm	0.76	1.10	Not recommended
Frost-damaged kernels %	0.82	4.62	Good
Plump kernel % (barley)	0.83	11.5	Good
Chlorophyll ppm (canola)	0.96	2.06	Excellent
Oil content (canola seed)	0.95	0.70	Excellent
Seed size (lentil)	0.92	9.58	Excellent
"Greenness: (lentil)	0.99	0.014	Excellent

other meetings. Feed-back from concerned people can be employed to introduce improvements before actual day-to-day use.

Implement systems for electronic grading

The new system would be best introduced via a small-volume crop. This would enable operational pitfalls to be resolved with minimal economic consequences. In Canada, a crop such as western white wheat would be suitable. The crop will exist in at least two distinct classes (Hard and Soft White Spring wheats), neither of which will be grown in high volume but high enough to enable all aspcts of the electronic system to become apparent. Another suitable crop would be canola seed. This is quite a high volume crop, but grading on the basis of electronically-determined chlorophyll content and segregation/marketing by oil content would rapidly realise the potential economic benefits of such a system.

Conclusions

(1) The visible/NIR spectra of grains and seeds contain enough information to enable these commodities to be classified on the basis of functionality
(2) A classification system could be developed to establish price in relation to the functionality of the commodities, rather than to their appearance
(3) Such a system, using NIR spectroscopy, would be faster and less subjective in operation than grading/pricing systems based on visual evaluation
(4) An electronic system that incorporated reduction in the number of classes of wheat would improve the efficiency of bin-space utilisation at elevators in Canada
(5) An electronic grading system would be attractive to wheat breeders (particularly in Canada), since the requirement for the kernel shapes of wheat classes to be visually distinguishable from one another would be eliminated and material could be selected on the basis of gluten strength in early generations.

References

1. J.E. Dexter, P.C. Williams, N.M. Edwards and D.G. Martin, *J. Cereal Sci.* **7,** 169 (1988).
2. J.E. Dexter, D.G. Martin, K.R Preston, K.H. Tipples and A.W. McGregor, *Cereal Chem.* **62,** 75 (1985).
3. American Association of Cereal Chemists, *Approved Methods of the Association*, 10th Edition. American Association of Cereal Chemists, St. Paul, Minnesota, USA (2000).
4. T. Pawlinsky and P.C. Williams, *J. Near Infrared Spectrosc.* **6,** 121 (1998).

Application of a multi-wavelength near infrared diode laser array for non-destructive food analysis

P. Butz, C. Merkel, R. Lindauer and B. Tauscher

Institute for Chemistry and Biology, Federal Research Centre for Nutrition, Haid-und-Neustr. 9, D-76131 Karlsruhe, Germany

Introduction

Near infrared (NIR) spectroscopy has become a widely used method in food and beverage analysis because of its speed, accuracy and the simplicity of sample preparation. If application of NIR spectroscopy is in the transmission technique the high-energy part of the near infrared between 1100 nm and 800 nm is preferred. The advantage is the larger penetration depth into the measuring medium. In comparison with mid infrared (MIR) one registers reduced NIR absorption around the factor 100, but due to the higher energy a 1000-fold larger penetration depth in the measuring medium is achieved.[1] For the wavelength area of 700–1100 nm, layer thickness from 1.0–20.0 cm is recommended to be still within the range of acceptable extinction.[2] Besides, it has to be taken into account that, due to the high energy, the sample under investigation might undergo structural changes. Clearly, biological material is particularly concerned if broad-band thermal radiation produced by an incandescent filament is employed. However, there are non-thermal types of emitter, which emit radiation from a much narrower range of wavelengths like laser diodes. The principal advantage of a non-thermal source is its efficiency: power consumption is much reduced lowering thermal dissipation. Diode lasers function the same way as lasers but linewidths are not as narrow as typical lasers. Since NIR absorption bands are relatively broad and only poorly dissolved due to overlays of overtones and combination bands, even the use of only relatively few laser diodes might already supply sufficient spectral information in the frequency range from 700 to 1100 nm. The broader linewidths of diode lasers are to be seen as advantageous for this application. Currently, standard diode lasers of adequate power output are not available for every wavelength in the range of 750–1100 nm and emitters produced for non-standard wavelengths are expensive. Thus, in this work, the suitability of an array of six (seven) standard diode lasers as light source for NIR transmission measurements in fruit and vegetables was investigated. Another requirement of NIR instruments is that detection must be efficient to enable measurements to be made in a reasonably short time, as for some applications (for example, sorting of fruits on a conveyor band) short response times are essential. For this purpose a fast diode array spectrometer (integration time from 3 ms) was used as the detector. The objective of this work was to investigate the ability of this experimental set-up to satisfactorily predict Brix values, firmness and internal defects in fruits and vegetables by non-destructive NIR transmission measurements.

Materials and methods

Materials

Apples (Rubinette variety), were from Obstgut Müller, Neuwied-Oberbieber, Germany, harvest season, 2000. Peaches (Concord variety) were from Staatliche Lehr- u. Forschungsanstalt, Neustadt / Weinstraße, Germany. Tomatoes (three different varieties) were purchased from a local market. For measurement of internal defects some of the Rubinette apples were artificially damaged by storing in different high carbon dioxide atmospheres which caused internal browning of different severity. Sucrose and deuterium oxide (D_2O) were from Merck, Germany.

Reference analysis of Brix value, firmness and internal defects

Cylindrical segments of about 2 cm diameter were taken from the fruits at the positions where the NIR measurements had been performed. After squeezing, the Brix value of the juice was measured using an AR 2008 digital refractometer (Krüss Optronic, Germany). Fruit firmness was measured by a TA-XT2i Texture Analyser (Stable Micro Systems, United Kingdom), firmness given as (kg cm^{-2}), force sensor area 3.14 mm.[2] For estimation of internal defects in apples the fruits were cut into halves and the degree of defective tissue was classified visually on a scale between 1 and 9, where the latter corresponded to the severest defects.

NIR diode laser array

A schematic view of the array is given in Figure 1. Temperature-controlled laser diode modules LGT (LG-Laser Technologies GmbH, Kleinostheim, Germany) were used. The modules fulfil the same specifications (round beam profile and high coherence length) as Helium–Neon lasers, but are available for many more wavelengths. Seven LGT for single mode laser diodes were employed with the following wavelengths and net output powers behind the focussing optics: 785 nm/40 mW, 808 nm/50 mW, 830 nm/30 mW, 850 nm/50 mW, 920 nm/50 mW, 980 nm/50 mW and 1060 nm/25 mW. The 920 nm diode laser was not yet in operation during the acquisition of the presented first results. The power stability (peak to peak, 4 h) was < 0.5% deviation, noise (peak to peak, 0 Hz–20 MHz) was < 0.5%. Operating voltage: 10–12 VDC, operating current: 0–3 A. The modules

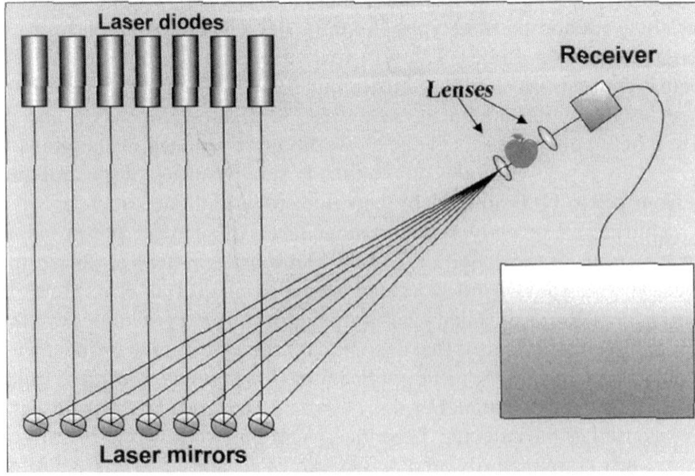

Figure 1. Schematic view of the NIR diode laser array.

were capable of modulation up to 20 MHz. A Basic-programmable interface (CIL Microsystems, United Kingdom) was used to control the lasers by TTL-pulses (+5 V: ON, 0 V: OFF) with a PC.

NIR transmission measurements

Near infrared 'spectra' were obtained using a transputer-integrated diode array spectrometer Tidas (J&M, Analytische Mess- und Regeltechnik GmbH, Aalen, Germany) containing a Zeiss MMS spectrometer module with 256 pixels. The spectral range was from 340 to 1150 nm, the minimal integration time was 1.4 ms. Measuring software was Spectralys (LabControl, Cologne, Germany). For measurements of sucrose concentrations a standard sample holder for 10 mm cuvettes was used connected to the detector via fibre-optics. For transmission measurements in fruits, using the experimental set-up shown in Figure 1, two spectra per fruit (in 90° angle) were recorded. Integration times were fixed for all measurements at three milliseconds for the reference (no sample in light path) and ten seconds for measurements of the samples. A typical spectrum of the reference is shown in Figure 2. *Quasi* absorption spectra (shown in Figure 3) were calculated by the measuring software and delivered in the format ".UVD".

Data analysis

Analysis of NIR data was performed by partial least squares (PLS) regression, using either Spectralys (LabControl, Cologne, Germany) or Opus NT (Bruker, Ettlingen, Germany). The latter automatically uses a variety of data subsets, spectral data point averaging, derivatives (1st and 2nd) and other data pre-treatments (for example, mean centring, multiplicative scatter correction and baseline correction) to determine the best data pre-treatment for each assay. The number of PLS factors used in the calibration was determined by the prediction residual error sum of squares F-statistic from the one-out cross-validation procedure. The accuracy of the developed calibration models was based on the high multiple correlation coefficient R^2 and low standard error of prediction (*SEP*) or the root mean square errors of prediction (*RMSEP*). In view of the fact that the NIR laser transmission measurements were done with varying path-lengths (diameters of the fruits) the data received from the reference method analytics were corrected for the manually determined path-lengths before entering them into the calibration process leading to unusual units like 'Brix per mm'.

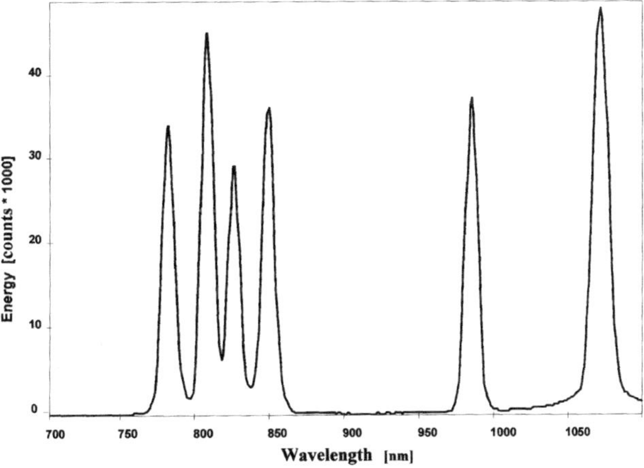

Figure 2. Typical reference spectrum (no material in the light path) of the laser diode array.

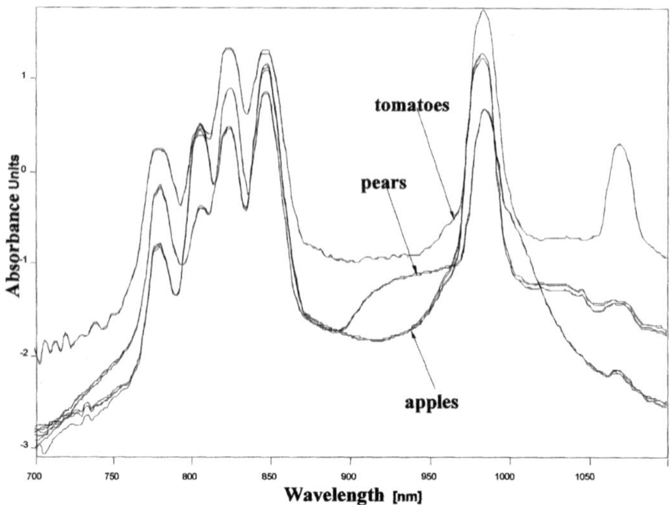

Figure 3. Typical 'absorbance spectra' of different fruits, recorded with the experimental set-up.

First results and discussion

Figure 2 shows a typical reference spectrum of the laser diode array. Note that the 920 nm diode laser was not yet in operation during the acquisition of the presented first results. The shown spectrum was recorded using a spectrometer module with 256 pixels over the spectral range from 340 to 1150 nm which means, dependent on the large spectral bandwidth of that detector (> 3.2 nm per pixel) the peaks appear broader than they actually are. Using a detector with a spectral resolution of < 0.8 nm

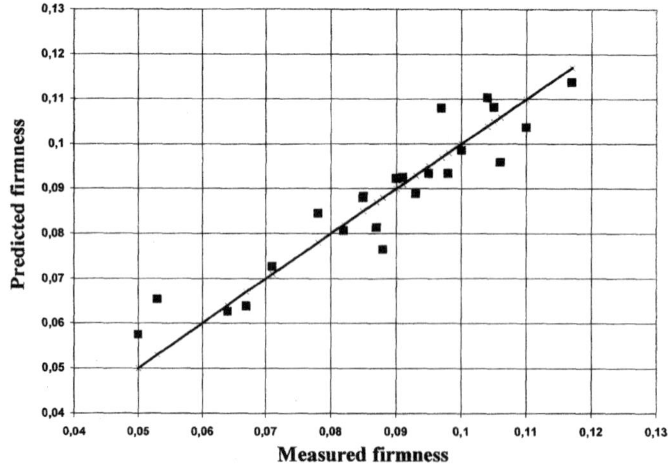

Figure 4. Correlation between fruit firmness obtained by the destructive mechanical method and that predicted by NIR.

Table 1. Comparison of calibration results of different assays.

			R^2 for single laser					
Assay	R^2	*RMSEP*	785 nm	808 nm	830 nm	850 nm	980 nm	1060 nm
Apples Internal defects (scale 0–9)	94	0.41	–02	59	66	64	91	77
Apples Firmness (range 3–9 kg cm^2)	88	0.43	47	58	74	35	52	61
Apples Brix (range 13–19%)	90	0.53	02	13	25	06	03	50
Pears Brix (range 10–18%)	93	0.49	03	10	12	21	31	33
Tomatoes Brix (range 4–7%)	97	0.24	36	36	65	66	29	15
Sucrose in D_2O (range 0–15%)	99	0.42	92	88	75	42	95	86

the peaks appear more narrow (data not shown): They cover about 5 nm at half height and about 10 nm at the basis. That means that fruits are illuminated by the laser diodes with light that usefully covers about 2–5 nm.

Typical 'absorbance spectra' of different fruits, recorded with the experimental set-up are shown in Figure 3. Note that they are only *quasi*-absorption spectra as they are calculated from spectra recorded with different—but fixed—integration times (see materials and methods). Differences between the 'absorption' spectra of tomatoes, pears and apples in the invisible range from 700 to 1100 nm can to be recognised recognised.

The spectra of apples seem to contain enough spectral information to make possible calibrations for non-destructive measurements of fruit firmness. Figure 4 shows the correlation between fruit firmness obtained by the destructive mechanical method and that predicted by NIR. The *RMSEP* was 0.43 (range 3–9 kg cm^{-2}). Good agreement between the conventional method and NIR was observed with $R^2 = 88$.

Table 1 lists the calibration results of further experiments carried out with apples, pears, tomatoes and sucrose solutions that were assayed for Brix values or sucrose content. Apples were also assayed for internal defects and firmness (see also Figure 4). In all cases good or acceptable values for *RMSEP* and R^2 were achieved. In rows 4 to 9 of the table the correlation coefficients for calibrations using only the spectral range of a single laser are listed. For the determination of Brix values the wavelengths 830, 850 and 1060 show the highest correlation. The measurement of sucrose in heavy water relies mostly on the absorption at 980 and 785 nm which has to be expected from conventional absorption spectra. Wavelengths of 980 and 1060 nm give the best information on internal defects in the investigated apples. There might be a correlation with the water content (peak at 980 nm). Apple firmness correlates best with absorption at 830 nm. The most valuable laser wavelengths of this investigation seem to be 830 and 1060 nm. Unfortunately, these laser diodes were commercially available only with about half of the output power of the residual used laser diodes. Some further improvement is to be expected from the 920 nm laser that was not yet in operation during the acquisition of the presented first results. This laser should be very suitable for transmission measurements in fruits containing much water since the high and broad water absorption around 980 nm is avoided.

Conclusions

Non-destructive food analysis (Brix value in apples, pears and tomatoes; Brix, firmness, internal defects in apples) using an array of NIR laser diodes in transmission mode, in combination with a fast diode array spectrometer as the detector, seems to be feasible. First calibration results are promising with respect to correlation coefficients and standard errors of prediction.

References

1. B.G. Osborne, T. Fearn, and P.H. Hindle, *Practical NIR spectroscopy: with applications in food and beverage analysis.* Longman Group UK Limited, Harlow, Essex, UK, 2nd Edn (1993).
2. L. Liekmeier, *Schwingungsspektroskopie im nahen Infrarotbereich.* Firmenschrift Perkin/Elmer Bodenseewerk, Überlingen, Germany (1984).

Mature instrument, immature technology: is near infrared spectroscopic analysis of high-moisture materials a serious proposition?

Nils Berding

BSES, PO Box 122, Gordonvale 4865, Australia

Introduction

Increasingly, near infrared (NIR) spectroscopic analyses are used for high-moisture materials. Zero or minimal sample preparation and immediacy of analytical results are real incentives for their development. Their availability facilitates at-line and on-line applications. However, development of calibrations, whether global or local, represents a considerable investment. Calibration viability largely relies upon instrument stability so that contemporarily collected spectra fall within the domain of the calibration population. Instrument changes that jeopardise this sustainability negate the investment inherent in calibration data. Stable reference cells that spectrally mimic the high-moisture material being analysed only can provide insurance against this. Such reference cells are rare. This paper discusses an example where two maintenance events radically altered an NIR instrument and rendered a large, global calibration database for fresh sugarcane useless.

Materials and methods

This paper focuses on an NIRSystems (Silver Spring, MD, USA) Model 6500 scanning monochromator (#3422-9405) acquired in August 1995. Applications were developed for Brix, commercial cane sugar (CCS), fibre, moisture and polarisation reading of disintegrated sugarcane stalk tissue.[1] Samples, drawn from clonal evaluation trials conducted on the northeast coast of Queensland (16° 15′ to 18° 15′ S Lat.), were scanned in a large cassette module (LCM).[1,2] An NIRSystems remote reflectance probe, with an 1800 mm fibre-optic bundle, provided the sample to instrument interface. Earlier research established the feasibility of analysing sugarcane stalk tissue, a high-moisture material, and demonstrated the impracticality of using NR7080 cells.[3,4] The instrument has been used primarily in this configuration but also has been used extensively when configured with either sample-transport or indexing-cup modules for analysis of dried materials.

A second NIRSystems Model 6500 scanning monochromator (# 6189-9812), fitted with auto-gain detectors and a "fast" motherboard, was acquired in September, 1999. The instrument was interfaced to a second LCM with an NIRSystems remote reflectance probe fitted with a 600 mm fibre-optic bundle.

Standardisation to "slave" the 1999 instrument to the 1995 instrument was undertaken in September 1999. A black-anodised cell, 100 mm (H) × 160 mm (L) × 125 mm (W), fitted with one quartz face, was used. Clean mature stalks from ten random clones were passed through a Dedini (Piracicaba, SP) disintegrator and mixed for 90 s. The cell was packed full with a sub-sample of this material so no voids were present on the window. The cell's contents were scanned via the remote reflectance probe to yield a mean spectrum. Each sample was scanned three times on each instrument, with sequential scans taken on alternate instruments. The first scan for sequential samples commenced on alternate instruments. A mean spectrum was generated for each sample's spectra from each instrument. A "SCORE" (WinISI, ISI, PA, USA) was generated for each instrument's spectra, and the sample (clone) with mean spectra nearest the population mean in each population identified.

A "10% check" population captured each year (1996–2000) allowed retrospective monitoring of the 1995 calibrations. Each population contained a random 10% of scanned samples plus significant H and t outliers. Supplementary populations were captured in 1996 and 1997.[5] Population structure and the relationship to the calibration domain were monitored using the "SCORE" function.

Seven spectral populations (1995–1998) were subjected to "evolutionary" calibration development.[5] A random 50% of each population was spectrally selected using the SCORE function and combined for calibration. The residual of each was combined for a "prediction" population.

In March 1999, the detectors in the remote reflectance probe were "adjusted" during a routine service. In June 2000, the PbS detectors in the remote reflectance probe were replaced.

Results and discussion

Internally, the pre-maintenance populations (1996–98), showed variation for median H value (0.32–0.93) and the number of components required for the SCORE (13–32; Table 1). The proportion of samples with a significant global H value (> 3.0) varied from 1 / 80 (L6, 1996) to 40 / 711 (10%–CHK, 1997). Relative to the 1995 base population, only the 10%–CHK (1998) population had an average $H > 3.0$. A relatively large number of samples (156/770 ≈ 20%) had an H value > 3.0. Unusual crop conditions cannot explain this.

The SCORE details for the post-maintenance populations (1999 and 2000) themselves are benign, with median H values less than the 1995 base population, showing each populations was quite cohesive. The number of components for the SCORE file equalled that for the 1995 base population (Table 2). Relative to the 1995 base population, the populations were in marked contrast to the pre-maintenance situation (Table 1). The two populations fell well outside the bounds of the 1995 base population, as indicated by H values of 12.4 and 20.4, respectively. All samples fell outside the bound set by the global $H = 3.0$. Cyclonic disturbances experienced in the pre-harvest period influenced crop conditions in 1999 and 2000, but these are not considered to have impacted the relationships of the check

Table 1. Details of pre-maintenance spectral populations used for calibration development and monitoring of NIR applications for analysis of quality components of mature sugarcane stalk tissue, with results of internal global distance (*H*) analyses for each population and results for *H* analyses for each population relative to the base calibration population.

Population			Internal			Relative to base population (1995)		
Year	Detail	N	Median H	# H > 3.0	# components for H	Average H	Median H	# H > 3.0
1995	Base	1,764	0.686	42	32	1.000	0.709	43
1996	D1X	252	0.687	10	28	1.055	0.890	3
	L5	471	0.655	18	22	0.603	0.530	0
	L6	80	0.927	1	17	1.023	0.803	2
	10%–CHK	332	0.518	17	19	1.434	0.861	22
1997	10%–CHK	711	0.366	40	32	1.573	0.881	58
	NIR	372	0.773	9	13	0.916	0.828	2
1998	10%–CHK	770	0.316	44	32	3.362	1.536	156

Table 2. Details of post-maintenance spectral populations used for monitoring of NIR applications for analysis of quality components of mature sugarcane stalk tissue, with results of internal global distance (*H*) analyses for each population, and results for *H* analyses for each population relative to the base calibration population.

Population			Internal			Relative to base population (1995)		
Year	Detail	N	Median H	# $H > 3.0$	# components for H	Average H	Median H	# $H > 3.0$
1995	Base	1,764	0.686	42	32	1.000	0.709	43
1999	10%–CHK	293	0.336	21	32	12.400	7.957	293
2000	10%–CHK	749	0.403	38	32	20.422	15.207	749

populations relative to the 1995 base population. Crop conditions in 1997 were influenced also by a cyclonic disturbance, yet spectral populations in that year fell within the domain of the 1995 base population (Table 1). The maintenance performed on the instrument in 1999 and 2000 is the most likely explanation for the marked shift in relationships relative to the 1995 base population.

Analysis of disintegrated, mature-stalk tissue of sugarcane using NIS has been successful, with marginal benefit being gained from evolutionary calibration development using supplementary data gathered from 1996 to 1998.[5] Population statistics for the amalgamated calibration population (Table 3) show that there was ample variation (for example, CV % $(= \sigma / \bar{x})$ values from 3.5 for moisture, to 12.6 for fibre) for all components. Calibration statistics generally were excellent, with R^2 values ranging from 0.94 to 0.99. Standard errors of calibration (*SEC*) were small, gauged either as a *CV* % value or as a ratio of the population standard deviation. Comparison of these calibration statistics with those for the base 1995 population[1] indicates only a marginal deterioration in key statistics. These results make two clear statements. The care taken in structuring the initial calibration population allowed de-

Table 3. Population statistics for five quality components of disintergrated, mature-stalk samples for a spectrally-selected portion of combined populations analysed from 1995–1998, together with a summary of calibration statistics.

Measure[a]	Brix	CCS	Fibre	Moisture	Pol. reading
	(g kg^{-1})	(g kg^{-1})	(g kg^{-1})	(g kg^{-1})	(°Z)
No. samples	2258	2266	2275	2134	2263
Mean	213.9	152.2	130.1	688.4	82.2
Minimum	103.9	63.4	85.7	617.7	38.6
Maximum	273.9	193.9	200.9	824.8	105.0
SD	19.45	17.31	16.34	24.24	9.39
No. terms	13	14	15	15	15
SEC	2.14	3.33	3.91	2.51	1.13
R^2	0.99	0.96	0.94	0.99	0.99
SECV	2.20	3.44	4.06	2.59	1.16

[a]*SD* = standard deviation; *SEC* = standard error of calibration; R^2 = multiple coefficient of determination; *SECV* = standard error of cross-validation

Table 4. Prediction statistics for five quality components of disintegrated mature-stalk sugarcane samples, resulting from application of calibrations developed on a spectrally-selected portion of the combined 1995–1998 populations, to the residual portion of the combined population.

Measure[a]	Brix	CCS	Fibre	Moisture	Pol. reading
	(g kg^{-1})	(g kg^{-1})	(g kg^{-1})	(g kg^{-1})	(°Z)
No. samples	2374	2368	2374	2210	2372
$\bar{x}$	213.76	151.90	129.16	689.79	82.06
Slope (b)	1.008	0.967	0.991	0.991	0.981
r^2	0.976	0.897	0.890	0.979	0.954
SEP	3.10	6.05	4.85	3.43	2.09
Bias	0.061	0.008	–0.170	0.299	–0.026
SEP(C)	3.10[b]	6.05[b]	4.85	3.42[b]	2.09[b]

[a]r^2 = simple coefficient of determination; *SEP* = standard error of prediction; *SEP(C)* = standard error of prediction, corrected for bias

[b]*SEP* and *SEP(C)* values fall above threshold of significance designated by WINISI

velopment of robust calibrations that really performed well in three seasons after 1995. There was little benefit in performing evolutionary calibration development because of this robustness. The predictive value of these calibrations on the residual 1995–1998 population shows their general acceptability (Table 4), with r^2 values of ≈ 0.89 for CCS and fibre and > 0.95 for the other components. Slope (b) values were near 1.0 for all components except CCS (0.967) and all bias values were small.

The standardisation exercise to slave the 1999 instrument to the 1995 instrument was unsuccessful. The only recourse was to incorporate the spectra used as a rep. file for use with the transferred calibrations, which would have been unsatisfactory. The reason for this failure is obvious. Comparable spectra from the master and slave populations suggest that the two instruments are radically different. The spectrum for the "master" is markedly compressed in the "*Y*" axis (Figure 1). This is after only the first of the two maintenance events. Obviously, this standardisation was attempted too late. A SCORE test revealed the standardisation populations fell well outside the bounds of the 1995 base population, as defined by a global $H = 3.0$. Alone, this could be considered a possible sampling artifact, but is unlikely as the calibrations displayed robustness prior to conduct of any instrument maintenance. The hypothesis that maintenance altered the instrument, so that post-maintenance spectral populations statistically fell outside the domain of the 1995 base population, appears increasingly real.

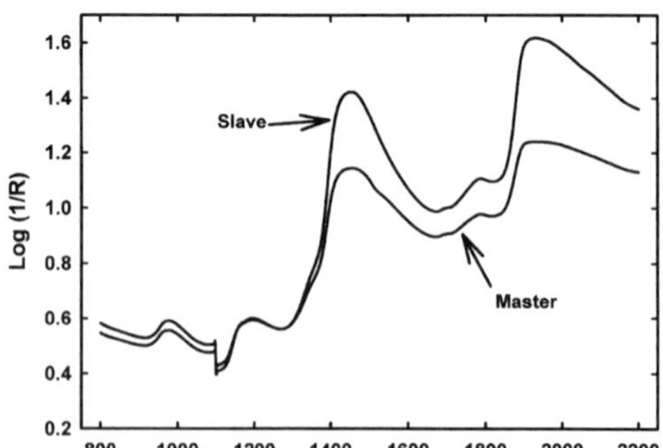

Figure 1. Comparative mean spectra for a sample of disintegrated sugarcane culm from the master (1995) and slave (1999) Model 6500 scanning monochromators.

Calibrations developed from the 1995–98 data (Table 3) performed poorly on the spectral data collected in 1999 (data not shown) and 2000 (Table 5). Few of the critical statistics were acceptable. This is not surprising given the relation-

Table 5. Prediction statistics for five quality components of disintegrated mature-stalk sugarcane samples, resulting from application of calibrations developed on a spectrally-selected portion of the combined 1995–1998 populations, to the 10% check population, 2000.

Measure[a]	Brix	CCS	Fibre	Moisture	Pol. reading
	(g kg^{-1})	(g kg^{-1})	(g kg^{-1})	(g kg^{-1})	(°Z)
No. samples	749	749	749	742	749
$\bar{x}$	211.80	154.13	127.34	692.68	82.31
Slope (b)	1.012	1.037	0.993	1.148	1.033
r^2	0.847	0.856	0.829	0.741	0.897
SEP	8.05	11.68	10.30	13.58	3.67
Bias	–5.11[b]	9.62[b]	7.51[b]	–5.71[b]	2.08[b]
SEP(C)	6.23[b]	6.63[b]	7.06[b]	12.33[b]	2.65[b]

[a] r^2 = simple coefficient of determination; *SEP* = standard error of prediction; *SEP(C)* = standard error of prediction, corrected for bias
[b] *SEP* and *SEP(C)* values fall above threshold of significance designated by WINISI

ship of these spectral populations to the 1995 base population (Table 2). Calibration redevelopment for the five components using the 1995–2000 data yielded statistics (Table 6) that were comparable to those resulting for calibration re-development using the 1995–1998 data (Table 3). A rigorous independent test of these calibrations was not possible. Use of the 1995–2000 calibrations on the residual

Table 6. Population statistics for five quality components of disintegrated, mature-stalk tissue for a spectrally-selected portion of combined populations analysed from 1995–1998 plus the 1999 and 2000 10% check populations, together with a summary of calibration statistics.

Measure[a]	Brix	CCS	Fibre	Moisture	Pol. reading
	(g kg^{-1})	(g kg^{-1})	(g kg^{-1})	(g kg^{-1})	(°Z)
No. samples	3235	3194	3243	2996	3199
Mean	212.70	152.85	129.11	689.56	82.15
Minimum	103.90	37.06	81.20	617.70	25.53
Maximum	273.90	193.90	200.90	824.80	104.98
SD	18.91	17.08	16.36	23.63	9.10
No. terms	13	14	15	13	14
SEC	2.36	3.40	4.11	2.89	1.18
R^2	0.984	0.960	0.937	0.985	0.983
SECV	2.41	3.62	4.34	3.20	1.26

[a] *SD* = standard deviation; *SEC* = standard error of calibration; R^2 = multiple coefficient of determination; *SECV* = standard error of cross-validation

1995–1998 population, not surprisingly, gave statistics comparable (data not shown) to those produced by use of the 1995–1998 calibrations on the same population. Application of the 1995–2000 calibrations to the 1999 or 2000 populations (data not shown) produced statistics that were not as bad as obtained from use of the 1995–1998 calibrations (Table 5). Obviously, they were less desirable than could be achieved for a population falling within the domain of the calibration population.

The above scenario demonstrates the vulnerability of calibrations for high-moisture materials, particularly when based on single instruments. There is a real need for the development of stable reference cells that spectrally mimic the high-moisture material of interest. Their existence would allow frequent checks of instrument functionality. Differences resulting from maintenance may be resolvable with the existence of such standards. In multi-instrument sites, or networks, the need for standard cells is not as critical, but still would be highly desirable.

In the absence of standard cells, instrument manufacturers, particularly for single-instrument or isolated installations, should exercise a greater duty of care. Such instruments should be slaved frequently (for example, six-monthly) to a protected "master" maintained by the manufacturer. This should provide insurance against an instrument being changed by maintenance. Within limits, standardisation can negate maintenance-related changes and protect investment in the development of high-moisture material calibrations. In the case reported here, "maintenance" altered an instrument. The most basic duty of care obligation should dictate that a manufacturer repeat maintenance until an acceptable instrument, i.e. one that can be standardised, and so remain true to existing calibrations, results.

References

1. N. Berding and G.A. Brotherton, in *Sugarcane: Research towards Efficient and Sustainable Production,* Ed by J.R. Wilson, D.M. Hogarth, J.A. Campbell and A.L. Garside. CSIRO Division of Tropical Crops and Pastures, Brisbane, Australia, p. 57 (1996).
2. N. Berding and G.A. Brotherton, *NIR news* **7(6),** 14 (1996).
3. N. Berding and G.A. Brotherton, in *Leaping Ahead with Near Infrared Spectroscopy,* Ed by G.D. Batten, P.C. Flinn, L.A. Welsh and A.B. Blakeney. Royal Australian Chemical Institute, Near Infrared Spectroscopy Group, Melbourne, Australia, p. 199 (1995).
4. N. Berding and G.A Brotherton, in *Near Infrared Spectroscopy: The Future Waves,* Proceedings 7th International Conference on Near Infrared Spectroscopy, Ed by A.M.C. Davies and P. Williams. NIR Publications, Chichester, UK, p. 648 (1996).
5. N. Berding and G.A. Brotherton, in Proceeding of 2nd International NIR Users Meeting for the Sugar and Alcohol Industries, 2–3 August 1999. Ribeirao Preto, Sao Paulo, Brazil, South America, 16 pp. (Proceeding un-paginated) (1999).

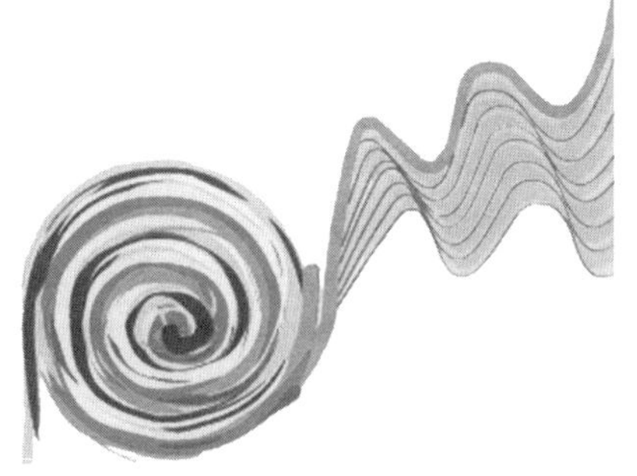

Delegates List

Delegates list

Lena Aberg, Foss Tecator AB, Box 70, SE-26321 Höganäs, Sweden, Tel: 46-42-361500, Fax: 46-42-340349, lena.aberg@foss.tecator.se

Daniel Alomar, Austral de Chile University, Institute Animal Production, Campus Isla Teja, PO Box 567, Valdivia, Chile, Tel: 56-63-221653, Fax: 56-63-221460, dalomar@uach.cl

Masahiro Amari, National Institute of Livestock and Grassland Science, 2 Ikenodai, Kukizakicho, Inashiki-gun, Ibaraki, Japan, Tel: 81-298-38-8667, Fax: 81-298-38-8606, amari@affrc.go.jp

Hiromichi Aoki, Fantec Research Institute, 317-67 Okubo, Yokoyama Kosai-city, Shizuoka, Japan, Tel: 81-53-573-2515, Fax: 81-53-573-2511

Woody Barton, USDA-ARS, Richard B. Russell Res. Center, PO Box 5677, Athens, GA 30604, USA, Tel: 1-706-546-3497, Fax: 1-706-546-3607, wbarton@qaru.ars.usda.gov

Graeme D. Batten, Charles Sturt University, School of Agriculture, LMB 588, Wagga Wagga, NSW 2650, Australia, Tel: 61-2-6933-4207, Fax: 61-2-6933-2812, gbatten@csu.edu.au

Nils Berding, Bureau of Sugar Experiment Station (BSES), PO Box 122, Gordonvale 4865, Australia, Tel: 61-7-4056-1255, Fax: 61-7-4056-2405, nberding@bses.org.au

Ronald J. Berry, Kwansei Gakuin Univ, 2-1 Gakuen, Sanda, Hyogo 669-1337, Japan, Tel: 81-798-54-6380, Fax: 81-798-51-0914, vsus9092@kwansei.ac.jp

Paolo Berzaghi, USDA Army Forage Research Center, 1925 Linden Dr, Madison, WI 53706, USA, Tel: 1-608-264-5232, Fax: 1-608-264-4157, pberzaghi@facstaff.wisc.edu

Cassandra Black, Victorian Institute for Dryland Agriculture, PB 260, Horsham, Victoria 3401, Australia, Tel: 61-3-5362-2111, Fax: 61-3-5362-2198, cassandra.black@nre.vic.gov.au

Jacqui Boschenok, Foss Pacific Pty Ltd., P.O. Box 1862, North Rayde 2113, Australia, Tel: 61-2-8877-8407, Fax: 61-2-8889-4989, jboschenok@foos.com.au

Paul Brimmer, Foss NIRSystems, 28A Mimosa Street, Oatley, NSW 2223, Australia, Tel: 61-2-9580-0536, Fax: 61-2-9580-0538, rstreamer@foss.com.au

Nils Bo Büchmann, Foss Tecator AB, Litteraturvagen 8, Box 70, SE-26321 Höganäs, Sweden, Tel: 46-42361507, Fax: 46-42331465, bo.buchmann@foss.com.se

Robert Burling-Claridge, AgResearch Limited, Ruakura MIRINZ Center, East Street, Private Bag 3123, Hamilton, New Zealand, Tel: 64-7-8385248, Fax: 64-7-8385625, robert.burlingclaridge@agresearch.co.nz

David Burns, McGill University, Otto Maass Building, Montreal, Quebec H3A 2K6, Canada, burns@chemistry.mcgill.ca

Michael Byström, Albihns Stockholm, Linegatan 2, SE-114 85 Stockholm, Sweden, Tel: 46 8 5988 7240, Fax: 46 8 5988 7300, michael.bystrom@home.se

Tiziana Maria Piera Cattaneo, Istituto Sperimentale Lattiero-Caseario, Via A. Lombardo, 11, I-26900 Lodi, Italy, Tel: 39-0371-45011, Fax: 39-0371-35579, cattaneo.ilc@pop.telware.it

Bogustawa Charnik-Matusewicz, Faculty of Chemistry, University of Wroclaw, F. Joliot-Curie 14, 50-383 Wroclaw, Poland, Tel: 71-3204203, Fax: 71-3282328, bc@wchuwr.chem.uni.wroc.pl

Jung-Hwan Cho, Sookmyung Women's University, College of Pharmacy, 608 Chungpa-dong, Yongsan-gu, Seoul 140-742, Korea, jhcho@sdic.sookmyung.ac.kr

Rae-Kwang Cho, Kyungpook National University, Department. of Agricultural Chemistry, 1370 Sankyukdong, Puk-gu, Taegu 702-701, Japan, Tel: 82-53-950-5719, Fax: 82-53-950-6280, rkcho@knu.ac.kr

Chang-Hyun Choi, Sungkyunkwan University, Department. of Bio-Mechatronic Engineering, Suwon 440-746, Korea, Tel: 82-31-290-7824, Fax: 82-31-290-7829, chchoi@yurim.skku.ac.kr

Hoell Chung, Asea Brown Boveri Ltd., 9 Floor Oksan Bldg, 157-33 Samsung-dong, Kangnam-ku, Kangnam, PO Box 1545, Seoul 135-090, Korea, Tel: 82-2-528-2395, Fax: 82-2-528-3590, hoeil@hotmail.com

Je-Hong Chung, LG-Caktex Oil, Tel: 82-61-680-2730, Fax: 82-61-680-2528, c12392@lgcaltex.co.kr

Phillip J. Clancy, NIR Technology Australia, 34 Clements Avenue, Bankstown, NSW 2200, Australia, Tel: 61-2-9790-6450, Fax: 61-2-9790-1552, lineart@zipworld.com.au

Robert P. Cogdill, Iowa State University, 1551 Food Science Bldg, Ames, IA 50014, USA, Tel: 1-515-294-6358, Fax: 1-515-294-6383, jed@iastate.edu

Ian Cowe, Foss Tecator AB, Box 70, SE-26321 Höganäs, Tel: 46-42-361500, Fax: 46-42-340349, ian.cowe@foss.tecator.se

Donald Dahm, Rowan University, 201 Mulica Hill Road, Glassboro, NJ 08028-1707, USA, Tel: 1-617-510-0156, Fax: , dburns@po-box.mcgill.ca

Pierre Dardenne, CRAGx, 24 Chaussee de Namur, B-5030-Gemloux, Belgium, Tel: 32-81-62-03-54, Fax: 32-81-62-03-88

Tony Davies, Norwich Near Infrared Consultancy, 75 Intwood Road, Cringleford, Norwich NR4 6AA, UK, Tel: 44-1603 451 073, Fax: 44-1603 250 328, tony_davies@nir-pub.demon.co.uk

Natasha de Bruyn, University of Stellenbosch, Department. of Food Science, Matieland 7602, South Africa, Tel: 27-21-808-3511, Fax: 27-21-808-3510, 12733989@narga.sun.ac.za

Begona De la Roza, SERIDA, Crta. De Oviedo, s/n Apdo 13, 33300 Villaviciosa Asturias, Spain, Tel: 34-985-89-0066, Fax: 34-985-89-1854, begonard@princast.es

Alessandro Del Bianco, CTR-Carinthian Tech Research, Badstubenweg 40, A-9500 Villach, Austria, Tel: 43 4242 56300 206 , Fax: 43 4242 56300 400 , Alessandro.DelBianco@CTR.at

Gab-Soo Do, The University of Tokyo, Yayoi 1-1-1, Bunkyo-ku, Tokyo 113-8657, Japan, Tel: 81-358-41-8973 , Fax: 81-358-41-5335, dogabsoo@iml.u-tokyo.ac.jp

Gerard Downey, Consumer Foods Department, TEAGASC, The National Food Centre, Dunsinea, Castleknock, Dublin 15, Ireland, Tel: 35-3-1-8059500, Fax: 35-3-1-8059550, g.downey@nfc.teagasc.ie

Thierry Droz-Georget, Solvias AG, Klybeckstrasse 191, PO Box WKL-127.6.06, CH-4002 Basel, Switzerland, Tel: 41-61-686-60-95, Fax: 41-61-686-60-96, thierry.droz-georget@solvias.com

Michael Esler, Australian Wine Research Institute, PO Box 197, Glen Osmond, SA 5064, Australia, Tel: 61-8-8303-6600, Fax: 61-8-8303-6601, mesler@awrl.com.au

Peter Flinn, Agriculture Victoria, DNRE Pastoral and Veterinary Institute, Private Bag 105, Hamilton, Victoria 3300, Tel: 61-3-5573-0915, Fax: 61-3-5571-1523, peter.flinn@nre.vic.gov.au

Stephen Flynn, Consumer Foods Department, TEAGASC, The National Food Centre, Dunsinea, Castleknock, Dublin 15, Ireland, Tel: 35-31-805-9500, Fax: 35-3-1-805-9550, sflynn@nfc.teagasc.ie

Daniel Fraser, Technology Development Group, Hortresearch TDG, Private Bag 3123, Hamilton, New Zealand, Tel: 64-7-8584742, Fax: 64-7-8584705, dgfraser@xtra.co.nz

Ana Garrido-Varo, University of Cordoba, ETSIAM, Avda. Menendez Pidal, s/n Apdo. 3049, 14080 Cordoba, Spain, Tel: 34-957-21-8531, Fax: 34-957-21-8563, pa1gavaa@uco.es

Paul L. Geladi, Umea University, Department of Chemistry, SE 901 87 Umea, Sweden, Tel: 46-90-786-6917, Fax: 46-90-138-8885, paul.geladi@chem.uu.se

Szilveszter Gergely, Department. of Biochemistry and Food Technology, Budapest University, Müegyetem Rkp. 3, H-1111 Budapest, Hungary, Tel: 36-1-463-1422, Fax: 36-1-463-3855, sziszi.bet@chem.bme.hu

Roberto Giangiacomo, Istituto Sperimentale Lattiero-Caseario, Via A. Lombardo, I-11 26900 Lodi, Italy, Tel: 39-0371-45011, Fax: 39-0371-35579, giangiacomo.ilc@telware.telware.it

Mirta Golic, Plant Sciences Group, Central Queensland University, Rockhampton 4702, Australia, Tel: 61-7-4930-9707, Fax: 61-7-4930-6536

Jon G. Goode, Bruker Optics, 19 Fortune Drive, Manning Park, Billerica, MA 01821, USA, Tel: 1-978-667-9580, Fax: 1-978-663-9177, jon.goode@bruker.com

Brian Gray, University of Stellenbosch, Department. of Food Scienc, Matieland 7602, South Africa, Tel: 27-21-808-3511, Fax: 27-21-808-3510, onefunkycow@yahoo.com

Colin Greensill, Plant Sciences Group, Central Queensland University, Rockhampton, Queensland 4702, Australia, Tel: 61-7-49-309707, Fax: 61-7-4930-6536

Wolfgang Guggenbichler, University of Innsbruck, Innrain 52a, A-6020 Innsbruck, Austria, Tel: 43-572-507-5779, Fax: 43-572-507-2965, wolfseng.guggenbichler@uibk.ac.at

John Guthrie, Plant Sciences Group, Central Queensland University, Rockhampton, Queensland 4702, Australia, Tel: 61-7-49-309707, Fax: 61-7-4930-6536

Takuhiro Hamada, Nagoya University, Furo-cho, Chikusa-ku, Nagoya 464-8601, Japan, Tel: 81-52-789-4157, Fax: 81-52-789-4155

Ikuko Hamagumi, Atago Bussan Co. Ltd, 1-7-8 Higashi-Kanda, Chiyoda-ky, Tokyo 101-0031, Japan, 81-3-5823-0140, 81-3-5823-0144, at-hamag@horiba.co.jp

Shuichi Hayano, Bruker Japan, 21-5 Ninomiya 3-chome, Tsukuba-city, Ibaraki, Japan, Tel: 81-298-52-1234, Fax: 81-298-58-0322
Jakob Heller, Büchi Labortechnik AG, Meierseggstr. 40, CH-9230 Flawil 1, Switzerland

Percha Hengtrakul, Sithiporn Associates Co., Ltd., 451 Sirinthorn Road, Bangplud, Bangkok, Thailand, Tel: 66-2-4338331, Fax: 61-2-4349510, sa-lab@sithiporn.co.th

Steve Holroyd, New Zealand Dairy Research Institute, Private Bag 11029, Palmerston North, New Zealand, Tel: 64-6-350-4649, Fax: 64-6-350-4607, steve.holrord@nzdri.ors.nz

David W. Hopkins, NIR Analysis, Analytical Chemistry, 472 South Moorland Drive, Battle Creek, MI 49015-3944, USA, Tel: 1-616-964-9920, Fax: 1-616-964-9900, dwh@pop.voyager.net

Christian Huck, University of Innsbruck, Innrain 52a, A-6020 Innsbruck, Austria, Tel: 43-542-507-5195, Fax: 43-512-507-2965, christian.w.huck@uibk.ac.at

Yun-Jeong Hur, Sookmyung Women's University, College of Pharmacy, 608 Chungpa-dong, Yongsan-gu, , Seoul 140-742, Korea, Tel: 82-2-710-9580, Fax: 82-2-3211-9580, k0131189@sdic.sookmyung.ac.kr

Harri Huttunen, University of Oulu, Technology Park 127, 87400 Kajaani, Finland, Tel: 35-8-6149638, Fax: 35-8-6149615, harri.huttunen@oulu.fi

Hiroyuki Ishizu, Maki Manufacfuring Co Ltd., 630 Sasakase-cho, Hamamatsu-shi, Shizuaka 435-8540, Japan, Tel: 81-534-21-8010, Fax: 81-534-22-3175, hiroyuki.ishizu@nifty.ne.jp

Hidekazu Ito, National Research Institute of Vegetables, Ornamental Plants and Tea, Kusawa, Ano, Age, Mie 514-2392, Japan, Tel: 81-592-68-4636, Fax: 81-592-68-1339, hito7@affrc.go.jp

Christoph Jansen, Büchi Labortechnik AG, Meierseggstr. 40, CH-9230 Flawil 1, Switzerland

Bruce B. Johnston, Elan Group Limited, 7/56 Simpson Street, Beerwah, Adelaide 4519, Australia, Tel: 61-7-5494-6689, Fax: 61-7-5494-0340, elanbio@elanbio.com

Hyun-Jin Joo, K-Mac, 104-11 Munji-dong, Yusung-Gu, Taejon 305-380, Korea, Tel: 82-42-868-6863, Fax: 82-42-868-6868, hjjoo@kmac.to

Bob Jordan, Technology Development Group, Hortresearch TDG, Private Bag 3123, Hamilton, New Zealand, Tel: 64-7-8584748, Fax: 64-7-8584705, jordanb@hort.cri.nz

Karoly J. Kaffka, Szent Istan University, Menési ut 45, H-1118 Budapest, Hungary, Tel: 36-1-372-6200, Fax: 36-1-372-6321, kkaffka@omega.kee.hu

Myoung-Gu Kang, LG-Caktex Oil, Tel: 82-61-680-2723, Fax: 82-61-680-2739, c14618@lgcaltex.co.kr

Yukiteru Katsumoto, Department. of Chemistry, School of Science, Kwansei Gakuin Univ, 2-1 Gakuen, Sanda, Hyogo 669-1337, Japan, Tel: 81-798-54-6380, Fax: 81-798-51-0914, katsumoto@kwansei.ac.jp

Sumio Kawano, National Food Research Institute, 2-1-12 Kannondai, Tsukuba 305-8642, Japan, Tel: 81-298-38-8088, Fax: 81-298-38-7996, kawano@nfri.affrc.go.jp

Miyuki Kawauchi, Saika Tecnological Institute, Kuroda, Wakayama City, Wakayama, Japan, Tel: 81-734-74-0860, Fax: 81-734-74-0862

Sandra E. Kays, United State Department. of Agriculture, 950 College Station Road, Athens, GA 30605, USA, Tel: 1-706-546-3338, Fax: 1-706-546-3607, sekays@garu.ars.usda.gov

Mark Kemper, Thermo Nicolet Corporation, 225 Verona Rd, Madison, Wisconsin 53711, USA, Tel: 1-608-273-5006, Fax: 1-608-273-5045, kemper@thermonicolet.com

Michael Kester, ABB Bomem Inc., 585 Charest Blvd East, Suite 3000, Quebec, Quebec G1K 9H4, Canada, Tel: 1-418-877-2944, Fax: 1-418-877-2834, garry.l.vail@ca.abb.com

Hwan Kim, Yokogawa Electric Korea Co, Tel: 82-3284-3058, Fax: 82-2-3284-3016, hwan_kim@korea.yokogawa.com

Hyo-Jin Kim, Dongduk Women's University, 23-1 Wolgok-dong, Sungbuk-gu, Seoul 136-714, Korea, Tel: 82-2-940-4525, Fax: 82-2-943-9578, hyojkim@www.dongduk.ac.kr

Yoen-Joo Kim, Samsung Advanced Institute of Tech, PO Box 111, Suwon 440-600, Korea, Tel: 82-31-280-6523, Fax: 82-31-280-9208, poons@samsung.co.kr

Gabriella Kisko, Szent Istvan University, Menési ut 45, H-1118 Budapest, Hungary, Tel: 36-1-372-6200, Fax: 36-1-372-6321, gkisko@hotmail.com

Jae-Yeon Koo, Agribrands Purina Korea Inc., Tel: 82-31-665-3240, Fax: 82-31-667-3642

Guergui Krivoshiev, Institute for Horticulture and Canned Foods, 154 Vassil Aprilov Blvd, Peovdiv 4000, Bulgaria, Tel: 35-9-32-952109, Fax: 35-9-32-952286, krivoshiev@hotmail.com

Akira Kudo, Fantec Research Institute, 317-67 Okubo, Yokoyama, Kosai-city, Shizuoka, Japan, Tel: 81-535-73-2515, Fax: 81-535-73-2511

Ken-ichi Kudo, The University of Tokyo, Yayoi 1-1-1, Bunkyo-ku, Tokyo 113-8657, Japan, Tel: 81-3-5841-6466 , Fax: 81-3-5800-6968, kkudoh@intellect.pe.u-tokyo.ac.jp

Rainer Kunnmemyer, University of Waikato, Private Bag 3105, Hamilton, New Zealand, Tel: 64-7-838-4630, Fax: 64-7-838-4219, rainer@waikato.ac.nz

Bogdan Kurtyka, Foss NIRSystems, 12101 Tech Road, Silver Spring, MD 20904, USA, Tel: 1-301-680-9600, Fax: 1-301-236-0134, bkurtyka@foss-nirsystems.com

Chul-Sung Kye, Kye Boram Corporation, Dong Suwon, PO Box 61, Suwon 442-600, Republic of Korea, Tel: 82-31-212-6430, Fax: 82-31-216-4839, cskye@chollian.net

Matthias Lau, Sentronic Ltd, Gostritzer Strasse 61-63, D-01217 Dresden, Germany , Tel: 49-173-567-4858, Fax: 49-351-871-8465, m.lau@sentronic.de

Donald Law, Halberd Pty Ltd, PO Box 7047, Toowcemba 4352, Australia, Tel: 61-7-4637-9960, Fax: 61-7-4637-9962, halbendassoc@blgpond.com

Sang-Hak Lee, Kyungpook National University, Department of Chemistry, 1370 Sankyukdong, Puk-gu, Taegu 702-701, Korea, Tel: 82-53-950-5338, Fax: 82-53-950-6330, shlee@kyungpook.ac.kr

Hyo-Won Lee, Korea Open University, Dongsungdong, Jongro-Ku, Seoul, Korea, Tel: 82-2-3668-4634, Fax: 82-2-3668-4187, hyowon@mail.knou.ac.kr

Ju-Hyeok Lee, Yullin Technology, 6th Floor Heungkuk BD. #6-7, Sungae-dong, Bundang-Ku, Sunanam, Kyeonggi-do, Korea, Tel: 82-31-713-8100, Fax: 81-31-713-8485, juhlee@yullin.com

Kang-Jin Lee, National Agricultural Mechanisation Research Institute, 249 Seodun-dong, Suwon, Korea, Tel: 82-31-290-1874, Fax: 82-31-290-1930, jini2002@rda.go.kr

Nam-Jin Lee, Agribrands Purina Korea Inc., Tel: 82-31-665-3240, Fax: 82-31-667-3642, njlee@agribrands.co.kr

Casper Leuewhagew, H. Lundbeck A/S, Ottiliavej 9, DK-2500 Valby, Denmark, Tel: 45-3630-1311, Fax: 45-3630-9951, linu@lundbeck.com

Jungi Li, SAIKA Tecnological Institute, Kuroda, Wakayama City, Wakayam , Japan, Tel: 81-734-74-0860, Fax: 81-734-74-0862, maps@mbp.sphere.ne.jp

Ling Liu, Beijing Vegetable Reseach Center, PO Box 2443, Beijing, China, lliu98@hotmail.com

Merccdcs G. Lopez , Cinveslav, Km. 9.6, Carretera Libramiento Norte Leon- Irepuato, Irepuato, Gto. , Mexico, Tel: 52-4-623-9600, Fax: 52-4-624-5996, mlopez@ira.cinvestav.mx

Karen Luchte, KES Analysis, 160 West End Ave, New York, NY 10023, USA, Tel: 1-212-595-7046, Fax: 1-212-787-3858

Line Lundsberg-Nielsen, H. Lundbeck A/S, Ottiliavej 9, DK-2500 Valby, Denmark, Tel: 45-3630-1311, Fax: 45-3630-9951, linu@lundbeck.com

Diane Malley, PDK Projects Inc, 365 Wildwood Park, Winnipeg, Manitoba R3T 0E7, Canada, Tel: 1-204-284-8047, Fax: 1-204-475-6090, dmalley@pdkprojects.com

Marena Manley , University of Stellenbosch, Department. of Food Science, Matieland 7602, South Africa, Tel: 27-21-808-3511, Fax: 27-21-808-3510, mman@maties.sun.ac.za

Teruyoshi Matoba, Nara Women's Univ, Kitauoya Nishimachi, Nara 630 8506, Japan, Tel: 81-742-20-3453, Fax: 81-742-20-3447, matoba@cc.nara-wu.ac.jp

W. Fred McClure, NC State University, PO Box7625, Raleigh, NC 27695-7625, USA, Tel: 1-919-515-6764, Fax: 1-919-515-7760, mcclure@eos.ncsu.edu

Ranjana Mehrotra, National Physical Laboratory, Kr. K.S. Krishnan Road, New Delhi 110012, India, Tel: 91-11-5787161(2577), Fax: 91-11-5852678, ranjana@csnpl.ren.nic.in

Ian Michael, NIR Pubrications, 6 Charlton Mill, Charlton, Chichester, West Sussex PO18 0HY, UK, Tel: 44-1243-811334, Fax: 44-1243-811711, ian.michael@nirpublications.com

Tai-Gi Min, Taegu University, Kyungsan, Kyungbuk 712-714, Korea, Tel: 82-53-850-6761, tgmin@biho.taegu.ac.kr

Meinhard Missbach, Labor chemie GmbH, Kanitzgasse 21, A-1230 Wien, Austria, Tel: 43-1-888-2601, Fax: 43-1-889-2355, mssb@ins.at

Mitsuru Mitsumoto, National Institute for Animal Industries, Tsukuba Norindanchi, PO Box 5, Ibaraki-ken 305-0901, Japan, Tel: 81-298-38-8690, Fax: 81-298-38-8609, mmitsuru@niai.affrc.go.jp

Yoshiko Mitsumoto, Natl. Inst. Anim. Ind., Tsukuba Norindanchi, PO Box 5, Ibaraki-ken 305-0901, Japan, Tel: 81-298-38-8690, Fax: 81-298-38-8609, mmitsuru@niai.affrc.go.jp

Tsuyoshi Miura, Bran+Luebbe, 15-17 Nishishinjuku 8-chome, Shinjukuku, Tokyo, Japan, Tel: 81-3-5330-1661, Fax: 81-3-5330-1665, miura@bran-luebbe.co.jp

Kumi Miyamoto, Wakayama Fruit Tree Exp. Station, 640-2, Do, Miyahara, Arida City, Wakayama 649-0436, Japan, Tel: 81-737-88-6112, Fax: 81-737-88-9710, kmiyamo@cypress.ne.jp

Masahiro Miyazawa , National Institute of Agrobiological Sciences, 1-2 Owashi, Tsukuba-shi, Ibaraki 305-8634, Japan, Tel: 81-298-38-6213, Fax: 81-298-38-6159, miyazawa@affrc.go.jp

Susumu Morimoto, Kubota Cooperation, 2-35 Jimmucho, Yao-shi, Osaka, Japan, Tel: 81-729-93-2096, Fax: 81-729-93-2453, ss-mormt@gkn.kubota.co.jp

Hidefumi Motohashi, Yokogawa Electric Corporation, 2-9-32- Naka-cho, Musashino-shi, Tokyo, Japan, Tel: 81-422-52-5694, Fax: 81-422-52-0622, hidefumi_motohashi@yokogawa.co.jp

Natsuga Motoyasu, Faculty of Agriculture, Yamagata University, 1-23 Wakaba-machi Tsuruoka 997-8555 , toko@tds1.tr.yamagata-u.ac.jp

Masahiro Muramatsu, Graduate School of Media and Govemance, Keio University, 5322 Endo, Fujisawa-shi, Kanagawa 252-0861, Japan, Tel: 81-466-88-3686, Fax: , muramatu@sfc.keio.ac.jp

Ian Murray, Scottish Agricultural College, Craibstone, Bucksburn, Aberdeen AB21 9YA, Scotland, UK, Tel: 44-1224-711201, Fax: 44-1224-276717, i.murray@ab.sac.ac.uk

Dae-Bok Na, Sookmyung Women's University, College of Pharmacy, 608 Chungpa-dong , Yongsan-gu, Seoul, 140-742, Korea, Tel: 82-2-710-9580, Fax: 82-2-3211-9580, k0031229@sdic.sookmyung.ac.kr

Yutaka Nakanishi, Saika Tecnological Institute, Kuroda, Wakayama City, Wakayama, Japan, Tel: 81-734-74-0860, Fax: 81-734-74-0862

Andreas Niemoller, Bruker Optik GmbH, Rudolf-Plank-Strasse 23, Abt. NIR, D-76275 Ettlingen, Germany, Tel: 49-7243-504-679, Fax: 49-7243-504-673, andreas.niemoller@bruker.de

Karl Norris, Consultant, 11204, Montgomery Rd, Beltsville, MD 20705, USA, Tel: 1-301937-7547, Fax: 1-301937-6536, knnirs@bellatlantic.net

Rukundo Nzabonimpa, Nestec Ltd, Nestle Research Center, Ver-Chez-Les-Blanc 100, Lausanne, Switzerland, Tel: 41-21-785-8902, Fax: 41-21-785-8555, rukundo.nzabonipa@rdls.neshe.com

Myoung-Jin Oh, LG-Caktex Oil, Tel: 82-61-680-2733, Fax: 82-61-680-2737, clone@lgcaltex.co.kr

Ulf Ohgren, Foss Tecator AB, Box 70, SE-26321 Höganäs, Sweden, Tel: 46-42-361682, Fax: 46-42-361780, ulf.ohgren@foss.tecator.se

Kazuo Osaki, Bran+Luebbe, 15-17 Nishishinjuku 8-chome, Shinjukuku, Tokyo, Japan, Tel: 81-3-5330-1661, Fax: 81-3-5330-1665, osaki@bran-luebbe.co.jp

Yukihiro Ozaki, Department of Chemistry, School of Sciences, Kwansei Gakuin University, 2-1 Gakuen, Sanda, Hyogo 669-1337, Japan, Tel: 81-798-54-6380, Fax: 81-798-51-0914, ozaki@kwansei.ac.jp

Hyesun Park, International Paper Company, 1422 Long Meadow Rd, Tuxedo, NY 10987, USA, Tel: 1-845-577-7345, Fax: 1-845-577-7307, hyesun.park@ipaper.com

Geung-Soon Park, Foss Korea, KVMA Bldg, 272-5 Seohyun-dong, Pundang-Ku, Sungnam-si, Kyunggi-do, Korea, Tel: 82-31-709-9591, Fax: 82-31-709-9594, gspark@foss.co.kr

Jun-Ho Park, Yullin Technology, 6th Floor Heungkuk BD #6-7, Sungae-dong, Bundang-Ku Sunanam, Kyeonggi-do, Korea, Tel: 82-31-713-8100, Fax: 81-31-713-8485, jhpark@yullin.com

Woo-Churl Park, Kyungpook National University, Department of Agricultural Chemistry, 1370 Sankyukdong, Puk-gu, Taegu 702-701, Korea, Tel: 82-53-950-5716, Fax: 82-53-953-7233, wcpark@knu.ac.kr

Young-Joo Park, Foss Korea, KVMA Bldg, 272-5 Seohyun-dong, Pundang-Ku, Sungnam-si, Kyunggi-do, Korea, Tel: 82-31-709-9591, Fax: 82-31-709-9594, yjpark@foss.co.kr

Celio Pasquini, Instituto de Quimica-UNICAMP, CEP:13083-970, CP 6154 Campinas SP, Brazil, 55-019-3788-3001, 55-019-3788-3033, pasquini@iqm.unicam.br

Sam Pavic, Foss Pacific Pty Ltd., PO Box 1862, North Ryde 2113, , Australia, Tel: 61-28-877-8407, Fax: 61-28-889-4989, spavic@foss.com.au

John C. Richmond, Bruker Optics Ltd, Banner Lane, Coventry CV4, 9GH, UK, Tel: 44-2476-855-200, Fax: 44-2476-46-5317, john.richmond@optics.bruker.co.uk

Craig Roberts, University of Missouri, 210 Waters Hall, Columia, MO 65211, Tel: 1-573-882-2001, Fax: 1-573-882-1467, robertscr@missouri.edu

Michael Rode, Carl Zeiss Jena Gmbh, 07740 Jena, Germany, Tel: 49-3641-64-3385, Fax: 49-3641-64-2485, rode@zeiss.de

Kwan-Shig Ryu, Taegu University, Department of Agricultural Chemistry, Jinryang Kyongsan, Kyongpuk 712-749, Korea, Tel: 82-53-850-6752, Fax: 82-53-850-6519, ryuks@taegu.ac.kr

András Salgó, Budapest University of Technology and Economics, Mügyetem rkp. 3, H-1111 Budapest, Hungary, Tel: 36-1-463-7422, Fax: 36-1-463-3855, salgo@chem.bme.hu

Sirinnapa Saranwong, Chiang Mai University, Chiangmai 50002, Thailand, Tel: 66-01-530-0688, Fax: 66-53-94-3351, ssaranwong@hotmail.com

Harumi Sato, Department. of Chemistry, School of Science, Kwansei-Gakuin University, 1-1-155, Uegahara, Nishinomiya 662-8501, Tel: 81-798-54-6380, Fax: 81-798-51-0914, 899418@kwansei.ac.jp

Laurence R. Schimleck, CSIRO forestry and Forest Products, Private Bag10, Clayton South MDC, Victoria 3169, Australia, Tel: 61-3-9545-2142, Fax: 61-3-9545-2248, l.schimleck@ffp.sciro.au

Angela Schmidt, Bruker Optik GmbH, Rudolf-Plank-Strasse 23, Abt. NIR, D-76275 Ettlingen, Germany, Tel: 49-7243-504-673, Fax: 49-7243-504-673, angela.schmich@bruker.de

Hartwig Schulz, Federal Centre of Breeding Research (BAZ), Neuer Weg 22-23, D-06484 Quedlinburg, Germany, Tel: 49-3946-47231, Fax: 49-3946-47234, H.Schulz@bafz.de

Chris Scott, GlaxoSmithKline, Tauwood House, 345, Rlislip Road, Southall, London, UK, Tel: 44-20-883-2502, Fax: 44-20-883-2438, rk17992@glaxowelcome.co.uk

Vegard H. Segtnan, Agricultural University of Norway, Department. of Food Science, PO Box 5036, N-1438 Ås, Norway, 47-64948558, 47-64943789, vegard.segtnan@inf.nlh.no

Kang-Il Seo, 2, Joongang-dong, Kwachon, Kyunggi-do 427-716, Korea, Tel: 82-2-509-7260

Zsolt Serégely, Metrika Research & Development Co., petzval J. u. 25, H-1119 Budapest, Hungary, Tel: 36-1-382-7314, Fax: 36-1-382-7311, zsolt.seregely@metrika.hu

Lorenzo Serva, Dipartimento Di Scienze Zootecniche, Universita'di Padova-Italy, Padova, Italy, Tel: 39-532-291171, Fax: 39-532-291168, pdlorenzo@libero.it

H.S.S. Sharma, Applied Plant Science, Agriculture & Food Science Centre, Newforge Lane, Belfast BT9 5PX, Northern Ireland, Tel: 44-28-9025-5248, Fax: 44-28-9025 5007, shekhar.sharma@dardni.gov.uk

John S. Shenk, Infrasoft International LLC, RD#1, 109 Sellers Lane, Port Matilda, PA 16870, USA, Tel: 1-814-237-9707, Fax: 1-814-237-0867, jws@vicon.net

Katsuaki Shibata, Quality AssuranceDivision, 2-1-1 Shimo-sueyoshi, Tsurumi-Ku, Yokohama-shi, Kanagawa 230-8504, Japan, Tel: 81-45-571-6144, Fax: 81-45-571-6166, k-shibata-ie@morinaga.co.jp

K. Shigefuji, Saika Tecnological Institute, Kuroda, Wakayama city, Wakayam Pref., Japan, Tel: 81-734-74-0860, Fax: 81-734-74-0862

Kimiyuki Shinohara, NIRECO, 2951-4 Hachioji-shi Ishikawa-machi Tokyo, Tel: 81-426-60-7344, Fax: 81-426-60-7326

Heinz W. Siesler, Department. of Physical Chemistry, University of Essen, Schutzenbahn 70, D-45117 Essen, Germany, Tel: 49-201-183-2927, Fax: 49-201-183-3967, hw.siesler@uni-essen.de

Erik Skibsted, Novo Nordisk & University of Amsterdam, Department. of Chemical Engineering, Process Analysis and Chemometrics, Nieuwe Achter gracht 166, NL-1018 WV Amsterdam, The Netherlands, Tel: 31-20-525-6991, Fax: 31-20-525-5604, skibsted@its.chem.uva.nl

Un-Tak Song, Foss Korea, KVMA Bldg, 272-5 Seohyun-dong, Pundang-Ku, Sungnam-si, Kyunggi-do, Korea, Tel: 82-31-709-9591, Fax: 82-31-709-9594, utsong@foss.co.kr

Lambert K. Sørensen, Steins Laboratorium, Ladelundvej 85, DK-6650 Brørup, Denmark, Tel: 45-75-381733, Fax: 45-75-383721, lks@steins.dk

Vincent Soudant, University of Oulu, Technology Park 127, 87400 Kajaani, Finland, Tel: 35-8-6149623, Fax: 35-8-6149615, vincent.soudant@oulu.fi

Ed W. Stark, KES Analysis, 160 West End Ave, New York, NY 10023, USA, Tel: 1-212-595-7046, Fax: 1-212-787-3853, starkdw@aol.com

Bo Stenberg, Swedish University of Agricultural Sciences, Department. Microbiology, Box 234, SE-532 23 Skara, Sweden, Tel: 46-511-67274, Fax: 46-511-67134, bo.stenberg@jvsk.slu.se

R. Streamer, Foss NIRSystems, 28A Mimosa Street, Oatley, NSW. 2223, Australia, Tel: 61-2-9580-0536, Fax: 61-2-9580-0538, rstreamer@foss.com.au

Ken-ichiro Suehara, Facalty of Information Sciences, Hiroshima City University, 3-4-1 Ozuka-Higashim, Asaminami-ku, Hiroshima 731-3194, Japan, Tel: 81-82-830-1681, Fax: 81-82-830-1681, suehara@im.hiroshima-cu.ac.jp

Jose-Ehilio Suerrero Ginel, University of Cordoba, Avda. Menendez Pidal, s/n Apdo 3049, 14080 Cordoba, Spain, Tel: 34-957-21-8555, Fax: 34-957-21-8563, pa1guji@uco.es

Junichi Sugiyama, National Food Research Institute, Postharvest Technology Divison, sugiyama@affrc.go.jp

Lloyd D. Sunders, Foss Tecator AB, 13 Gordon Ave, Castle Mill, Sydney 2154, Australia, Tel: 61-29-899-6348, lsaunders@foss.com.au

Nobutaka Suzuki, Shimonoseki University of Fisheries, Yoshimi, Shimonoseki 759-6595, Japan, Tel: 81-832-86-5111, Fax: 81-832-59-8826, suzukin@fish-u.ac.jp

Hitoshi Takamura, Nara Women's Univ, Kitauoya Nishimachi, Nara 630 8506, Japan, Tel: 81-742-20-3454, Fax: 81-742-20-3447, takamura@cc.nara-wu.ac.jp

Elena Tamburini, Chemistry Department, Univeristy of Ferrara, Via L. Borsari, 46, I-44100 Ferrara, Italy, Tel: 39-532-291171, Fax: 39-532-291168, vcg@unife.it

Tetsuya Tanabe, Fantec Research Institute, 317-67 Okubo, , Yokoyama Kosai-city, Shizuoka, Japan, Tel: 81-53-573-2515, Fax: 81-53-573-2511

Bernhard Tauscher, Federal Research Center for Nutrition, Haid-und-Neu-Strasse 9, D-76131 Karlsruhe, Germany, Tel: 49-721-6625-200, Fax: 49-721-6625-111, bernhard.tauscher@bfe.uni-karlsruhe.de

Yoko Terazawa, Tsukuba University, NFRI, 2-1-12 Kannondai, Tsukuba 305-8642, Japan, Tel: 81-298-38-8088, Fax: 81-298-38-7996, terazawa@nfri.affrc.go.jp

Apichart Thitikasemson, Sithiporn Associates Co., Ltd., 451 Sirinthorn Road, Bangplud, Bangkok, Thailand, Tel: 66-2-4338331, Fax: 66-2-4349510, sa-spd@sithiporn.co.th

Søren Thode, Department. of Analytical Chemistry, Research Centre Foulum, PO Box 50, DK-8830 Tjele, Denmark, Tel: 45-89991482, Fax: 45-89991460, soren.thode@agrsci.dk

Steven Thorwart, Infrasoft International LLC, R.D#1 109 Sellers Lane, Port Matilda, PA 16870, USA, Tel: 1-814-237-9707, Fax: 1-814-237-0867, isi@vicon.net

Peter Tillmann, VDLUFA, Am Veruschfeld 13, D-34128 Kassel, Germany, Tel: 49-567-9888, Fax: 49-561-8202935, peter.tillmann@t-conline.de

Masahiro Tomo, Yokogawa Electric Corporation, 2-9-32- Naka-cho, Musashino-shi, Tokyo, Japan, Tel: 81-422-52-5694, Fax: 81-422-52-0622

Hansen Torsten, Foss Tecator AB, Box 70, SE-26321 Höganäs, Sweden, Tel: 46-42361639, Fax: 46-42331465, torsten.hansen@foss.com.se

Simona Tosi, University of Ferrara, Chemistry Department.-Via L. Borsari 46, 44100 Ferrara, Italy, Tel: 39-532-291171, Fax: 39-532-291168, vcg@unife.it

Roumiana Tsenkova, Faculty of Agriculture, Kobe University, 1-1 Rokkodai, Kobe, Nada-ku 657-8501, Japan, Tel: 81-78-803-5911, Fax: 81-78-803-5911, rtsen@eng.ans.kobe-u.ac.jp

Satoru Tsuchikawa, Nagoya University, Furo-cho Chikusa-ku, Nagoya 464-8601, Japan, Tel: 81-52-789-4157, Fax: 81-52-789-4155, st3842@agr.nagoya-u.ac.jp, Mitsuhiro Miyazawa, National Institute of Agrobiological Sceinces, 1-2 Owashi, Tsukuba-shi, Ibaraki 305-8634, Japan, Tel: 81-298-38-6213, Fax: 81-298-38-6159, miyazawa@affrc.go.jp

Mizuki Tsuta, The University of Tokyo, Graduate School of Agricultural, and Life Sciences , Yayoi 1-1-1, Bunkyo-ku, Tokyo 113-8657, Japan, Tel: 81-3-5841-8973, Fax: 81-3-5841-5335, mizukit@affrc.go.jp

Garry Vail, ABB Bomem Inc., 585 Charest Blvd East, Suite 3000, Quebec, Quebec G1K 9H4, Canada, Tel: 1-418-877-2944, Fax: 1-418-877-2834, michael.kester@ca.abb.com

Maria Varadi, Central Food Research Institute, Herman 15, H-1022 Budapest, Hungary, Tel: 36-1-355-8982, Fax: 63-1-212-9853, m.varadi@cfri.hu

Kerry Walsh, Plant Sciences Group, Central Queensland University, Rockhampton, Queensland 4702, Australia, Tel: 61-7-4930-9707, Fax: 61-7-4930-6536, k.walsh@cqu.edu.au

Qian Wang, Bruker Optics, 19 Fortune Drive, Manning Park, Billerica, MA 01821, USA

Donald Webster, FOSS NIRSystems, 12101 Tech Road, Silver Spring, MD 20904, Tel: 1-301-680-9600, Fax: 1-301-236-0134, dwebster@foss-nirsystems.com

Hakan Wedelback, Foss Tecator AB, Box 70, SE-26321 Höganäs, Sweden, Tel: 46-42-361646, Fax: 46-42-361780, hakan.wedelsback@foss.tecator.se

Chris Weikert, Carl Zeiss Jena Gmbh, 07740 Jena, Germany, Tel: 49-3641-64-2882, Fax: 49-3641-64-2485, weikert@zeiss.de

Helmut Weiler, Bruker Optik GmbH, Rudolf-Plank-Strasse 23, Abt. NIR, D-76275 Ettlingen, Germany, Tel: 49-7243-504-673, Fax: 49-7243-504-673, helmut.weiler@bruker.de

Russell J. Wilkie, Elan Group Limited, 7/56 Simpson Street, Beerwah, Adelaide 4519, Australia, Tel: 61-7-5494-6689, Fax: 61-7-5494-0340, elanbio@elanbio.com

Phil Williams, Canadian Grain Commission, Grain Research Laboratory, 1404-303 Main Street, Winnipeg, Manitoba R3C 3G8, Canada, Tel: 204-983-3344, Fax: 204-983-0724, pwilliams@cgc.ca

William R. Windham, USDA-ARS, Richard B. Russell Res. Center, PO Box 5677, Athens, GA 30604, USA, Tel: 1-706-546-3513, Fax: 1-706-546-3633, rwindham@saa.ars.usda.gov

Takuo Yano, Faculty of information Sciences, Hiroshima City University, Ozukahigashi 3-4-1, Asaminami-ku, Hiroshima 731-3194, Tel: 81-82-830-1606, Fax: 81-82-830-1606, tyano@im.hiroshima-cu.ac.jp

Gil-Won Yoon, Samsung Advanced Institute of Technology, PO Box 111, Suwon 440-600, Korea, Tel: 82-31-280-6520, Fax: 82-31-280-9208, gyoon@sait.samsung.co.kr

Arne Zilian, Solvias AG, Klybeckstrasse 191, PO Box WKL-127.6.06, CH-4002 Basel, Switzerland, Tel: 41-61-686-60-95, Fax: 41-61-686-60-96, arne.zilian@solvias.com

Thomas Ziolko, Büchi Labortechnik AG, Meierseggstr. 40, Ch-9230 Flawil 1, Switzerland

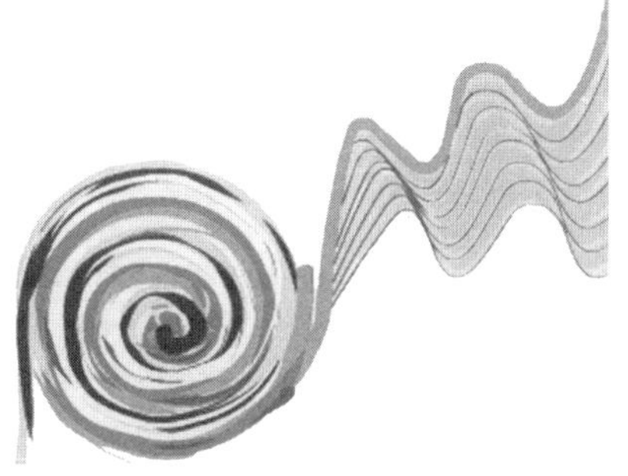

Author Index

Author index

G

H

I

J

K

L

M

N

O

P

R

S

T

V

W

Y

Z